*The Bering Land Bridge*

# The Bering Land Bridge

EDITED BY DAVID M. HOPKINS

Stanford University Press
Stanford, California

Based on a Symposium Held at the VII Congress of the
International Association for Quaternary Research,
Boulder, Colorado, August 30–September 5, 1965

Stanford University Press
Stanford, California
© 1967 by the Board of Trustees of the
Leland Stanford Junior University
Printed in the United States of America
ISBN 0-8047-0272-1
Original edition 1967
Last figure below indicates year of this printing:
90  89  88  87  86  85  84  83  82  81

*This book is dedicated to two pioneer Beringologists—*
ERIC HULTÉN, *whom I have admired from afar,*
*and* LOUIS GIDDINGS, *with whom I worked*

*Preface*

Eighteen years ago, stormbound at Wales village, I studied the mist smoking over a turbulent Bering Strait and wondered who, on this violent day, might be shouldering the wind on the Asian shore to share my search for traces of the past. Near me rose a peaty mound, the midden left by generation upon generation of Eskimos dwelling at the western tip of North America; behind me rose Cape Mountain, scarred by ancient glaciers, carved by ancient waves. Perhaps someone was at that moment sheltering his Cyrillic notes from the mist as he huddled on a terrace on East Cape, at the eastern tip of Siberia—or in an Eskimo burial ground at Uelen, Siberia's easternmost village.

That search—and the findings of my colleague on the other shore—have continued to fascinate me in the years since, and to draw substantially on my professional attention. For Bering Strait is the dramatic focal point of one of the world's great crossroads. The Strait itself is the one narrow avenue of sea communication between the North Pacific and the Arctic and North Atlantic Oceans, and, of course, it is the point where Asia lies only one day's umiak journey from North America. More significantly, it lies at the center of what has been, on several occasions in the past, the sole avenue for dry-land migrations between the Old and New Worlds—and thus at the center of what was, during these occasions, the barrier partitioning North Pacific and North Atlantic marine biotas. For these reasons, the history of Beringia (western Alaska, Northeastern Siberia, and the shallow parts of Bering and Chukchi Seas) has long excited the interest of geologists, biogeographers, anthropologists—and even medical geographers, for the first men to colonize North America brought their diseases and parasites over the Bering Land Bridge with them.

The investigation of Beringia and of its past landscapes is an international and multidisciplinary affair. While Western scientists have been investigating the history of the landscape of the North American segment of the Bering Land Bridge, their Soviet colleagues have, indeed, been looking into similar aspects of the Siberian segment, and scientists of both hemispheres have begun to learn something of the landscape of the central segment, which now lies hidden beneath the sea. But the published results of these recent studies are scattered through the specialized scientific journals of two nations, written in two languages. The partitions between disciplines are almost as serious a hindrance to communication as is the language barrier. Scientists in different disciplines turn up apparently conflicting data and develop mutually incompatible hypotheses, with little awareness of the paradox they have fostered. But paradox in science is itself a form of information—it tells us that data are incomplete or that hypotheses are in need of mutual adjustment.

The history of Beringia has been the unifying thread of my own research throughout the past 20 years, and during much of that period I have hoped to find occasion to persuade my colleagues in other nations, in other disciplines, or with other focal interests to think hard about our rapidly growing body of data, in order to identify and fill in the more glaring gaps in our knowledge of the Bering Land Bridge and in order to resolve numerous paradoxical interpretations. The long-sought occasion arose when I was invited to organize an all-day symposium entitled "Late Cenozoic History and Environments of the Bering Land Bridge" for the VII Congress of the International Association for Quaternary Research (INQUA), which was held at Boulder, Colorado, August 25 to September 5, 1965. Because INQUA is interdisciplinary as well as international in scope, the INQUA Land Bridge Symposium provided an ideal opportunity to bring together contributors whose disciplinary affiliations include geology, molluscan paleontology, vertebrate paleontology, micropaleontology, geophysics, oceanography, geochronology, botany, cytology, paleobotany, palynology, physical anthropology, and archaeology, and whose national affiliations include the Soviet Union, Germany, Great Britain, Iceland, Canada, and the United States.

This book is an outgrowth from, but not a record of, the INQUA symposium: eight of the essays included in the book were not given at the symposium, and two of the papers read at the symposium are not included in the book. Because I hoped that the INQUA symposium might represent a true dialogue among the contributors and that it might lead to a coherent and integrated series of essays prepared by persons fully aware of the ideas and findings of their colleagues, lengthy preliminary abstracts were exchanged *prior* to the meetings, an opportunity was provided for extensive revisions *after* the INQUA symposium, and my own editing offered a further opportunity to point out paradoxes and complementary findings to the individual contributors. The Soviet contributors were asked to submit their manuscripts in both Russian and English, and their papers were then extensively rewritten in order to achieve optimal clarity and readability in translation.

Many friends have contributed vital assistance in the preparation of this book. First, I must thank the 27 authors whose papers are included herein, both for their labors of creation and for the fortitude with which they bore my heavy-handed editing. My colleagues M. C. Blake and Michael Churkin, Jr., provided vital assistance by translating correspondence and much manuscript material from Russian to English. D. W. Scholl, E. D. Mitchell, C. A. Repenning, and J. A. Wolfe assisted me with the technical editing of certain specialized papers, and Elizabeth Spurr and W. W. Carver of Stanford University Press provided encouragement, advice, and assistance in the art of converting a symposium into a book. Helen C. Bailey and William Sanders of the United States Geological Survey Library in Menlo Park undertook the onerous task of tracking down and verifying or correcting a most diversified series of bibliographic citations. Charlene D. Preedy typed and retyped all manuscripts and kept this literary project going in a dozen large and small ways. Rebecca Cruickshank worked patiently and skillfully to prepare the Index. Helen Koelbl redrafted many of the illustrations with a skill I have long been acquainted with. I am of course indebted to the officers of the VII INQUA Congress, and especially to G. M. Richmond, H. E. Wright, and J. F. Lance, for their assistance in providing a forum for the Land Bridge Symposium at Boulder and thus for creating the stimulus that brought this book into being. I am most grateful to the Geological Survey for providing much of the time and facilities that I required to edit this book, and, finally, I must thank my family for their patience, for they saw very little of me during its preparation.

I feel well rewarded for the considerable time and effort that went into this book, because my own understanding of the history of Beringian landscapes and migrations has been considerably enhanced in the process. The fog over Bering Strait is, in fact, not quite so impenetrable as it was 18 years ago.

D.M.H.

*March 1, 1967*

# Contents

*The Bering Land Bridge*

# 1. Introduction

DAVID M. HOPKINS

On a foggy day in August 1728, Vitus Bering sailed north through the strait that now bears his name and thus established that the land areas of the Old and New Worlds are entirely disjunct. Yet had the day been fair, a continent would have been visible on either shore, and had Bering dropped a sounding line, he would have found that the two continents are separated by no great depths of water. Though the continents were proved disjunct, one might easily have imagined that they had once been interconnected.

The interhemispheric distribution pattern of many species of land animals and plants strongly suggests that the Old and New Worlds have indeed been joined somewhere at some time in the past. As Heilprin noted in 1887, the tropical biotas of the Old and New Worlds have relatively few elements in common; the temperate biotas differ less impressively; and the arctic biotas are nearly identical in Eurasia and North America. Thus, it has seemed likely that the interhemispheric land connection lay somewhere in the north.

Although plausible biogeographic arguments can be developed to support the hypothesis that eastern North America and Europe were joined early in Tertiary time by a land connection extending across the North Atlantic through Greenland, Iceland, the Faeroes, and the British Isles (Löve and Löve, 1963; Kurtén, 1966),[a] the shallow seas separating Alaska and Siberia in and near Bering Strait have long seemed the most likely site of more recent Eurasian–North American land connections (Wallace, 1876, I, p. 154). G. M. Dawson emphasized in 1894 that water depths are consistently

---

[a] The plausibility of an early Tertiary interhemispheric land connection across the North Atlantic is greatly strengthened by the recent application of paleomagnetic data to demonstrate that the ocean floor has apparently been actively spreading, widening the gap between the continents, for at least several million years (Vine, 1966).

less than 100 fathoms throughout the northeastern half of Bering Sea, as well as in Bering Strait and Chukchi Sea, and that these shallow seas "must be considered physiographically as belonging to the continental plateau region as distinct from that of the ocean basins proper." He saw every reason to believe that "in later geologic times more than once and perhaps during prolonged periods [there existed] a wide terrestrial plain connecting North America and Asia" (Dawson, 1894, p. 143–144). Finds of fossil mammoth remains on Unalaska and the Pribilof Islands, cited by Stanley-Brown (1892) and by Dall and Harris (1892), lent conviction to the notion that much of the Bering continental shelf had indeed once been dry land connecting these remote islands with the Alaskan and Siberian mainlands. Thus, by 1915, the importance for biogeography of the Bering Land Bridge was an article of faith to so stout a believer in the permanency of the continents and the ocean basins as W. D. Matthew.

Until recently, estimates of the time when the Bering Land Bridge might have existed have been based largely upon comparisons between synchronous Eurasian and North American fossil mammal faunas found in areas remote from Bering Strait (for example, Osborn, 1909; Willis, 1909; Simpson, 1947). But as early as 1910, Adolph Knopf was able to draw a remarkably accurate sketch of land-bridge history from the scanty geologic, paleontologic, and geomorphic evidence then available from the shores of Bering and Chukchi Seas. Until 1934, when Daly popularized the concept that sea level had fluctuated drastically during the Ice Ages in response to fluctuations in the amount of water stored on land as continental glaciers, geologists felt obliged to invoke tectonic upheavals to explain a past conjunction and a present separation of the continents at Bering Strait (for example, Knopf, 1910; P. S. Smith, 1927, 1934). But in a companion paper to Smith's 1933 discussion, the archaeologist Diamond Jenness (1934) suggested that Bering Strait might have been dry land during the last glacial interval, and in 1937 both Smith and Eric Hultén noted that a broad land connection must have joined Alaska to Siberia as a consequence of the eustatic reduction in sea level during the last glaciation.

John Muir (1884) believed, after participating in the 1881 voyage of the Revenue Cutter *Corwin*, that the entire continental shelf beneath Bering Sea, Bering Strait, and Chukchi Sea had been glaciated, but Dawson disputed this in 1894, stressing a complete absence of any traces of general glaciation there.[b] In fact, Dawson noted, large portions of the Alaskan mainland, too, had lain beyond the limits of the Cordilleran and Laurentide ice

[b] The dispute was revived in 1909 when O. H. Hershey reported that the Baldwin Peninsula in eastern Kotzebue Sound appears to represent the end moraine of a large glacier that originated in the Kobuk River valley. P. S. Smith, in 1912, expressed his

.masses of North America. A. G. Maddren observed in 1905 that the un-glaciated areas of central and western Alaska had at times been isolated from central North America by a wall of glacial ice that spanned north-western Canada from the Pacific to the Arctic Ocean, and he wondered how the time of this isolation might have been related to the time of existence of the Bering Land Bridge. Eric Hultén undertook to answer this question, 32 years later, in his now-classic "Outline of the History of Arctic and Boreal Biota during the Quaternary Period" (1937). Analysis of the dis-tribution of living plant species indicated that "Beringia," as he christened the vast arctic lowland that must have been exposed during the worldwide glacial epochs, had been a refugium in which most arctic and many boreal plant species were isolated while much of northern North America and parts of Siberia were covered with glacial ice. In a pioneer attempt to apply the knowledge of Pleistocene glacial chronology that had begun to emerge in the late 1930's, Hultén concluded that Beringia had lain exposed as a land bridge during both the Riss (or Illinoian) and Würm (or Wisconsin) Glacia-tions; he suggested that "isostatic pressure" might also have maintained the land connection "for some time after the glacial period" (p. 34). In this matter, we still have not progressed very far beyond Hultén's analysis; data are insufficient, even in 1967, to establish the precise time relationships between openings and closings of the land bridge and openings and closings of a passage from Alaska to North America between the Cordilleran and Laurentide ice sheets.

It has always seemed likely that the first human inhabitants of the Americas entered the continent from the northwest. E. N. Wilmsen notes in his excellent review "An Outline of Early Man Studies in the United States" (1965) that Fray José de Acosta postulated in 1590 a land bridge or narrow strait in high northern latitudes, over which small bands of hunters first entered North America; the hypothesis remained a favored one for the ensuing three and a half centuries. Some anthropologists have pos-tulated multiple waves of migration through Alaska and northwestern Canada in order to explain the distribution patterns of the rather diversified aborigi-nal populations found by the first explorers of the Western Hemisphere (for example, Hooton, 1931; Gladwin, 1947). From 1890 until 1925, man was thought to be such a recent addition to the North American fauna that his arrival would have taken place at a time when northern landscapes differed

---

belief that the land areas adjoining Kotzebue Sound have never been glaciated and that the till observed by Hershey must have been deposited by a large ice cap that Smith inferred to have once occupied the floor of Chukchi Sea. Time has proved Hershey right and Smith wrong; the Kobuk River valley was intensely glaciated, and the floor of Chukchi Sea was not (McCulloch, this volume).

little from those of the present time. However, general acceptance during the late 1920's and early 1930's of proof that man had arrived in central North America by late Pleistocene time, and that he had been a contemporary of the late Pleistocene fauna, led to a renewal of interest among anthropologists and archaeologists in the history and former landscape of the Bering Land Bridge (Jenness, 1934; Giddings, 1954; Sauer, 1957; Chard, 1958, 1959, 1960). Unfortunately, sites of human occupation dating unambiguously to land-bridge times have yet to be found within the limits of Beringia.

As we shall see later in this book, changes in the distribution of land and sea in Beringia have affected the biogeography of marine organisms quite as profoundly as they have affected the biogeography of land plants and animals. When open, Bering Strait has been a gateway for interoceanic migrations of marine organisms, whereas the Bering Land Bridge has at other times been an isolating barrier causing Pacific organisms on the one hand and Atlantic and Arctic organisms on the other to go their separate and divergent evolutionary ways. Soviet marine zoologists, having focused more intently than their Western colleagues upon the biota of the northern seas, have been well aware, for at least two decades, of the important biogeographical role of Bering Strait (Zenkevitch, 1947, 1963), but one is hard put to find more than the most desultory discussion (for example, Davies, 1934, p. 208–209; Ekman, 1953, p. 121–122) of the possible biogeographical significance of Bering Strait and Beringia in European and American treatises on marine biogeography and paleontology. A single important early exception is Soot-Ryen's (1932) discussion of the migrational history of several boreal and arctic mollusks. Much more recently, Elliott (1956) has discussed transarctic migrations of brachiopods, and J. L. Davies (1958a, 1958b) and I. A. McLaren (1960) have considered the roles of land barriers and seaways in Beringia in establishing past and present distribution patterns of sea mammals, while F. S. MacNeil has drawn attention to the partitioning during middle Tertiary time of Arctic and North Pacific mollusk faunas (1957) and to the subsequent dispersals through Bering Strait of several molluscan species of Pacific ancestry (1965).

MacNeil's imaginative use of the data of molluscan paleontology in interpreting the history of Beringia led directly to my own first attempt, in 1959, to develop the history of the Bering Land Bridge by synthesizing the available direct geological and geophysical data with the indirect clues that could be gleaned from the migrational histories of marine and terrestrial biota. But much has been learned in eight years, and it seems useful to examine once again the many lines of evidence available to us, and to attempt once more a chronology of the seaways and land bridges in Beringia and a reconstruction of the character of the seascapes and landscapes that have prevailed there in the past. The essays that follow summarize and analyze the data available in 1967—data assembled from findings on two continents and an island in the North Atlantic by scientists of six nations and a dozen disciplines—bearing on the history of the Bering Land Bridge.

## *REFERENCES*

Chard, C. S. 1958. New World migration routes: Anthro. Papers, Univ. Alaska, v. 7, p. 23–26.

—— 1959. New World origins: a reappraisal: Antiquity, v. 33, p. 44–48.

—— 1960. Routes to Bering Strait: Am. Antiquity, v. 26, p. 283–284.

Dall, W. H., and G. D. Harris. 1892. Correlation papers, Neocene: U.S. Geol. Survey Bull. 84, 349 p.

Daly, R. A. 1934. The changing world of the Ice Age: Yale Univ. Press, 271 p.

Davies, A. M. 1934. Tertiary Faunas: v. II, The sequence of Tertiary faunas: Murby and Co., London, 252 p.

Davies, J. L. 1958a. The Pinnipedia: an essay in zoogeography: Geogr. Review, v. 48, p. 474–493.

—— 1958b. Pleistocene geography and the distribution of northern pinnipeds: Ecology, v. 39, p. 97–113.

Dawson, G. M. 1894. Geologic notes on some of the coasts and islands of Bering Sea and vicinity: Geol. Soc. America Bull., v. 5, p. 117–146.

Ekman, Sven. 1953. Zoogeography of the sea: Sidgwick and Jackson, Ltd., London, 417 p.

Elliott, G. F. 1956. On Tertiary transarctic Brachiopod migrations: Ann. & Mag. Nat. Hist., Ser. 12, v. 9, no. 100, p. 280–286.

Giddings, J. L. 1954. Early man in the Arctic: Scientific American, v. 190, p. 82–88.

Gladwin, H. S. 1947. Men out of Asia: Whittlesey House, 390 p.

Heilprin, A. 1887. The geographical and geological distribution of animals: Internat. Science Series (Appleton, New York), 435 p.

Hershey, O. H. 1909. The ancient Kobuk Glacier of Alaska: Jour. Geology, v. 17, p. 83–91.

Hooton, E. A. 1931. Indians of Pecos pueblo: a study of their skeletal remains: Dept. Archaeology, Phillips Acad., Andover, Mass., Papers of the Southwestern Expedition, no. 4, Yale Univ. Press, 391 p.

Hopkins, D. M. 1959. Cenozoic history of the Bering Land Bridge: Science, v. 129, p. 1519–1528.

Hultén, Eric. 1937. Outline of the history of arctic and boreal biota during the Quaternary Period: Bokförlags Aktiebolaget Thule, Stockholm, 168 p.

Jenness, Diamond. 1934. Origin and antiquity of the American Aborigines: 5th Pacific Science Congress, 1933, Vancouver, Canada, Proc. v. 1, p. 753–758.

Knopf, Adolph. 1910. The probable Tertiary land connection between Asia and North America: Univ. Calif. Pub. Geol. Bull., v. 5, p. 413–420.

Kurtén, Björn. 1966. Holarctic land connections in the early Tertiary: Soc. Sci. Fennica, Comm. Biol., v. 29, no. 5, 5 p.

Löve, Doris, and Áskell Löve. 1963. North Atlantic biota and their history: Pergamon, Oxford, 430 p.

McLaren, I. A. 1960. Are the Pinnipedia biphyletic?: Syst. Zool., v. 9, p. 18–28.

MacNeil, F. S. 1957. Cenozoic megafossils of northern Alaska: U.S. Geol. Survey Prof. Paper 294-C, p. 99–127.

———— 1965. Evolution and distribution of the Genus *Mya*, and Tertiary migrations of Mollusca: U.S. Geol. Survey Prof. Paper 483-G, 51 p.

Maddren, A. G. 1905. Smithsonian Exploration in Alaska in 1904, in search of mammoth and other fossil remains: Smithsonian Misc. Coll., v. 44 (Publ. 1584), 117 p.

Matthew, W. D. 1915. Climate and evolution: Ann. N.Y. Acad. Sci., v. 24, p. 171–318 (republished, 1939, as p. 1–148 *in* New York Acad. Sci. Spec. Publ., v. 1, 223 p.).

Muir, John. 1884. On the glaciation of the Arctic and subarctic regions visited by the United States Steamer *Corwin* in the year 1881: p. 135–147 *in* Capt. C. L. Hooper, Report of the cruise of the U.S. Revenue Steamer *Thomas Corwin* in the Arctic Ocean, 1881, Washington: Govt. Printing Office, 147 p. (Also issued as U.S. 48th Cong., 1st sess., Senate Doc. 204, and U.S. Treasury Dept. Doc. 601. Reprinted p. 235–258 *in* John Muir. 1917. The Cruise of the *Corwin*: Houghton-Mifflin, Boston and New York, 279 p.)

Osborn, H. F. 1909. Cenozoic mammal horizons of western North America: U.S. Geol. Survey Bull. 361, 138 p.

Sauer, C. O. 1957. The end of the Ice Age and its witnesses: Geogr. Rev., v. 47, p. 29–43.

Simpson, G. G. 1947. Holarctic mammalian faunas and continental relationships during the Cenozoic: Geol. Soc. America Bull., v. 58, p. 613–687.

Smith, P. S. 1912. Glaciation in Northwestern Alaska: Geol. Soc. America Bull., v. 23, p. 563–570.

———— 1927. Some post-Tertiary changes in Alaska of climatic significance: Nat. Res. Council Bull. 61, p. 35–39.

———— 1934. Geographic and geologic evidence relating to the connection of Siberia and Northwestern Alaska: 5th Pacific Sci. Cong., 1933, Vancouver, Canada, Proc., v. 1, p. 753–758.

———— 1937. Certain relations between Northwestern America and Northeastern Asia, p. 85–92 *in* G. G. MacCurdy, ed., Early Man: J. P. Lippincott, Philadelphia and New York.

Soot-Ryen, Tron. 1932. Pelecypoda, with a discussion of possible migrations of Arctic pelecypods in Tertiary times: Norweg. N. Polar Exp. "Maud," 1918–1925, Sci. Results (pub. by Geofysk Inst., Bergen), v. 5, no. 12, 35 p.

Stanley-Brown, Joseph. 1892. Geology of the Pribilof: Geol. Soc. America Bull., v. 3, p. 496–500.

Vine, F. J. 1966. Spreading of the Ocean Floor: New evidence: Science, v. 154, p. 1405–1415.

Wallace, H. R. 1876. The geographical distribution of animals: Harper, New York, 2 vols.

Willis, Bailey. 1909. Paleogeographic maps of North America: *13*, Eocene-Oligocene North America: Jour. Geology, v. 17, p. 503–505; *14*, Miocene North America: Jour. Geology, v. 17, p. 506–508; Quaternary North America: Jour. Geology, v. 17, p. 600–602.

Wilmsen, E. N. 1965. An outline of early man studies in United States: Am. Antiquity, v. 31, p. 172–192.

Zenkevitch, L. A. 1947. Moriia SSSR: v. 2; Fauna i biologicheskaia produktivnost' Moryiia (Seas of the USSR: v. 2, Fauna and biological productivity of the seas): Sovietskaia Nauka, Leningrad, 587 p.

———— 1963. Biology of the seas of the U.S.S.R.: Interscience, 955 p.

## 2. Geology of the Floor of Bering and Chukchi Seas— American Studies

JOE S. CREAGER and DEAN A. MCMANUS
*Department of Oceanography, University of Washington*

Reconnaissance surveys of both the bottom sediments and the bathymetry of the continental shelves of Chukchi and Bering Seas have been made by American scientists (Figs. 1 and 2), but the samples collected in many parts of the area have not as yet been studied. In other parts the samples are so widely scattered that they are not germane to a general summary of the marine geology of Bering and Chukchi Seas as applied to the problem of the Late Cenozoic history of the Bering Land Bridge. Only the eastern portion of Chukchi Sea can be described with any degree of precision. However, by drawing on incomplete and unpublished analyses of samples north and east of St. Lawrence Island and in western Chukchi Sea, it is possible, in a preliminary fashion, to consider the post-Wisconsin history of the Bering "Sea" Land Bridge.

## Previous Information and Surveys

Until the late 1940's, the bottom sediments and bathymetry of Bering and Chukchi Seas were known only from observations of expeditions that were sampling for other information or were sampling over a much greater area. Samples from these expeditions were widely spaced, and the information obtained, such as that by the *Albatross* (Tanner, 1893; Rathbun, 1894), was meager and largely restricted to Bering Sea.

Then, from 1947 to 1949, the U.S. Navy Electronics Laboratory and the

Contribution No. 370, University of Washington, Department of Oceanography, Seattle. This research was supported by Contract AT-45-1-540 with the U.S. Atomic Energy Commission; Contract Nonr-477(37), Project NR 083 012, with the Office of Naval Research; and Grants GP-2457 and GA-808 with the National Science Foundation. Discussions with B. J. Enbysk, Department of Geology, Northwestern University, have been most helpful.

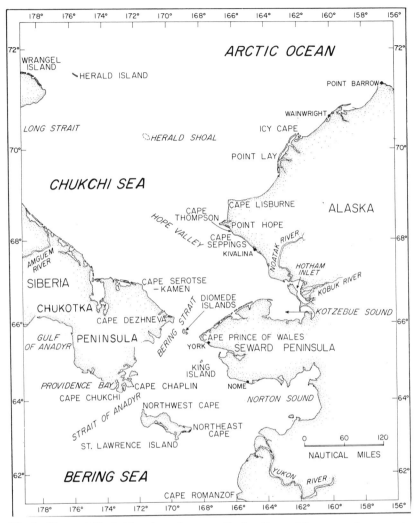

Fig. 1. Chart of Chukchi and northeastern Bering Seas with names of principal features along the coast.

Pacific Oceanography Group (Canada) carried on oceanographic surveys of Bering and Chukchi Seas that produced a description of the bottom sediments and bathymetry of eastern Bering Sea and eastern Chukchi Sea (Carsola, 1954a,b; and Dietz *et al.*, 1964).

Size analyses of sediments collected by a number of icebreakers have been reported in Technical Reports of the U.S. Hydrographic Office (1958a, 1959), the U.S. Naval Oceanographic Office (1960), and Informal Manuscript Reports of the U.S. Naval Oceanographic Office (1963a,b; 1964).

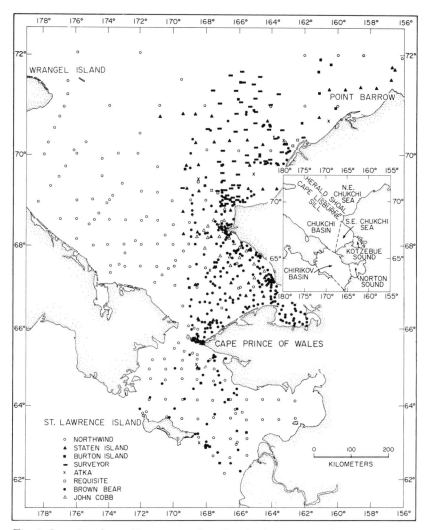

Fig. 2. Location chart of bottom samples taken by American workers in the Chukchi and northeastern Bering Seas. Inset chart shows the location of the seven depositional environments discussed in the text.

Additional samples have been collected from both Bering Sea and Chukchi Sea during cruises of the USS *Burton Island* (1960), USS *Staten Island* (1961), USCGC *Northwind* (1961, 1962, 1963), and USC&GS *Surveyor* (1962); these more recent samples and the soundings collected during the same cruises are presently being studied at the Department of Oceanography, University of Washington.

The Foraminifera from parts of Bering Sea have been studied by Cushman (1910–17) and by Anderson (1961), and those from Chukchi Sea have been studied by Cushman (1920), by Carsola (1955), and by Cooper (1964).

During 1959 and 1960, personnel of the University of Washington aboard the MV *Brown Bear* and USFWS *John N. Cobb* collected 286 grab samples, 60 gravity cores, one piston core, and soundings along approximately 7,000 kilometers of track in eastern Chukchi Sea (Figs. 2 and 3). Results of the

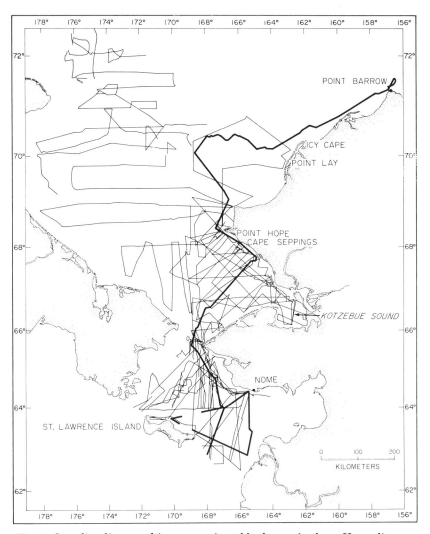

Fig. 3. Sounding lines used in preparation of bathymetric chart. Heavy line represents acoustic-reflection profile obtained by D. Moore (1964).

study of these samples and soundings have been reported by Cooper (1964), Creager (1963), Creager and McManus (1961, 1965, 1966), and McManus and Creager (1963).

The beaches and nearshore shelf between Kotzebue Sound and Point Hope have been studied by personnel of the U.S. Geological Survey (Kachadoorian *et al.*, 1960, 1961; G. Moore, 1960a,b; G. Moore and Cole, 1960; G. Moore and Scholl, 1961; and Scholl and Sainsbury, 1961).

An acoustic-reflection reconnaissance survey through the eastern Bering and Chukchi Seas (Fig. 3) was made in 1960 by D. Moore (1964).

## Oceanographic and Geologic Setting

Chukchi Sea and Northern Bering Sea are ice-covered for seven to eight months of the year as far south as 60° N. (U.S. Coast and Geodetic Survey, 1955, p. 638; U.S. Hydrographic Office, 1958b). By August and September, the ice has generally retreated northward to about 72° N., leaving the area north of St. Lawrence Island ice-free for only two to three months a year. During the ice-free periods, the waves in Chukchi Sea result from local winds, generally from the north. However, the shallowness of the water restricts the growth of large waves. Waves in the area north of St. Lawrence Island are also restricted in size by the shallow water and the short fetch lengths. The region east of St. Lawrence Island is open to swell from Bering Sea.

The tide range throughout the area is small, the mean range being 0.5 meter or less, and consequently the tidal currents are weak. The only short-term records from Chukchi Sea, taken at Kivalina, show a maximum change in mean sea level of about 1.8 meters, which is probably due to meteorologic causes (Fleming *et al.*, 1959).

The permanent currents near Alaska set toward the north along the contours of the bottom topography (Fig. 4). In east-central Chukchi Sea these currents are weak, their speeds measured near the surface and bottom being less than 5 to 25 cm/sec. In Bering Strait and between Cape Thompson and Point Hope, the speeds are greater and range from 15 to 34 cm/sec near the bottom and 15 to 72 cm/sec near the surface. Weaker currents set toward the south along the Chukotka Peninsula coast (Zenkevich, 1963, p. 261–269, 818–841).

A more complete description of the physical oceanography is given by Aagaard (1964), Creager and McManus (1966), and Fleming and Heggarty (1966).

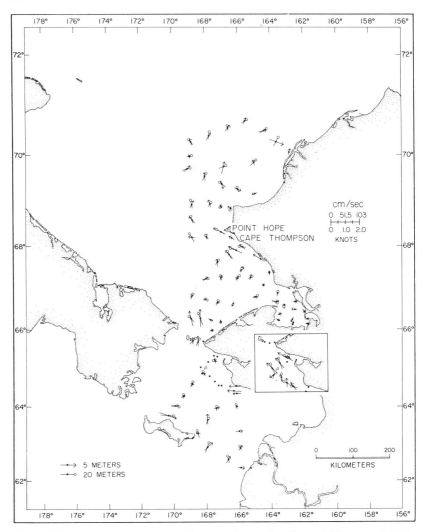

Fig. 4. Current measurements at 5-meter (*arrowhead*) and 20-meter (*circle-head*) depth obtained on cruises of MV *Brown Bear* (after Fleming and Heggarty, 1966). Currents near the bottom parallel the 20-meter currents but are weaker.

The land bordering the shores of the Chukchi and northern Bering Seas consists of generally flat, low, coastal-plain areas, with numerous lakes, alternating with areas of rolling hills rising a few tens to a few hundred meters near the coast. Much of the Chukchi shore along the Alaskan side and along the Siberian side northwest of Cape Serotse–Kamen consists of chains of barrier islands backed by large shallow lagoons, and interrupted by large shallow bays. In marked contrast, the Siberian coast between Cape Serotse–Kamen and Cape Chukchi is dissected by deep fjords penetrating far inland.

The Yukon, which empties into the northeastern Bering Sea, is the only large river in the region. Smaller rivers are the Amguem, which enters western Chukchi Sea from the Chukotka Peninsula, and the Kobuk and Noatak, which enter the eastern Chukchi Sea through Hotham Inlet. The sediment introduced into the seas by these rivers and other streams has been produced by polar weathering processes, and the resulting products of congelifraction (Bryan, 1946, p. 627) differ significantly from weathered products supplied to marine environments in temperate regions. Depending upon the pedogenic factors of time, topography, drainage, and composition, congelifraction produces, according to Hopkins and Sigafoos (1951, p. 59), "a well-sorted silty soil containing few large rocks and very few particles in the clay-size range."

The continental shelf of Chukchi Sea is a flat, almost featureless plain (Fig. 5) having average depths of 45 to 55 meters and regional gradients ranging from 2 minutes to unmeasurably gentle slopes (Creager and Mc-Manus, 1966). Local maximum gradients range up to 1°55′. The few soundings available suggest a similarly flat bottom in northeastern Bering Sea, although minor relief features may be more numerous. Excluding the slope between the land and the sea floor, the major relief features in Chukchi Sea are Herald Shoal, Hope Seavalley (Creager and McManus, 1965), and the Cape Prince of Wales Shoal (McManus and Creager, 1963); and in Bering Sea they are the fjords of the Chukotka Peninsula and the discontinuous trough paralleling the Chukotka Peninsula shore north of Northwest Cape, St. Lawrence Island (Udintsev *et al.*, 1959).

The major portion of Chukchi Sea is a shallow, northwest-southeast-oriented basin connected to the Arctic basin by a submarine canyon between Herald Island and Herald Shoal. The Chukchi Basin is separated from the Arctic basin by a sill between Herald Shoal and Cape Lisburne about 44 meters deep; it is delimited from the East Siberian Sea by a sill at least as shallow as 47 meters, and from Bering Sea by a sill about 48 meters deep north of the western end of Bering Strait. The Chirikov Basin or northeastern Bering Sea Basin (Udintsev *et al.*, 1959) is defined by a sill about 46 meters deep north of the Strait of Anadyr between St. Lawrence Island and Cape Chaplin, and by a sill about 32 meters deep to the east of St. Lawrence Island.

## Acoustic-Reflection Survey of the Subbottom

The only acoustic-reflection studies in the northern Bering and Chukchi Seas were made by D. Moore (1964). Location of the track is shown in Fig. 3. Both sonoprobe and arcer equipment were used, so that a preliminary picture of the structure of the bedrock and sediment overburden is available. The following discussion is based upon Moore's paper (1964).

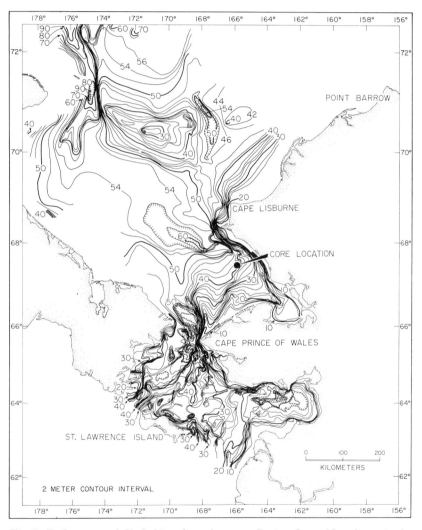

Fig. 5. Bathymetry of Chukchi and northeastern Bering Seas. Also shown is the location of the 7.4-meter core.

The acoustic-reflection records indicate, according to Moore (p. 333), that the bedrock beneath northeastern Bering Sea is probably similar to the volcanic and intrusive rocks of St. Lawrence Island,[a] whereas bedrock of the eastern Chukchi Sea is composed of stratified sedimentary rocks that "appear to be extensions of the huge Brooks Range Structure" (p. 347). The relief of the bedrock surface generally is low, seldom exceeding 10 meters and commonly amounting to only 1–3 meters. An exception to this is a broad valley just north of St. Lawrence Island, which has a relief of approximately 30 meters. The bedrock topography generally consists of local or regional broad depressions (possible valleys and interfluves off the mouth of the Yukon River and Kotzebue Sound) and gentle, low, rolling hills or swells and swales. Rare steplike changes in level of about 3 meters occur in the reflection traces and suggest faulting.

Over large portions of northeastern Bering Sea, bedrock is believed to be at or near the surface. Two-thirds of the subbottom reflection trace between St. Lawrence Island and the Yukon delta, and the greater portion of the track running from Nome to a point north of Bering Strait suggest that bedrock is exposed on the bottom. In marked contrast, the bedrock of southeastern Chukchi Sea is generally buried under approximately 10 meters of sediment. Along the shore between Cape Seppings and Cape Lisburne and between Icy Cape and Point Barrow, reflections suggest that bedrock is generally exposed, a suggestion confirmed by the nearshore marine geologic investigations of Scholl and Sainsbury (1961) in the Cape Thompson area. A traverse from Cape Lisburne to Herald Shoal indicates bedrock at or near the surface, but the track from Herald Shoal to Icy Cape suggests that bedrock varies from being exposed to being buried by as much as 12 meters of sediment.

In detail, much of the area traversed in northern Bering Sea and eastern Chukchi Sea has a "local, rough, subbottom topography" (p. 346). Subbottom reflections suggest that "the extreme flatness of these shelves on a local scale is directly the result of blanketing by sediment" (p. 346). However, "The very flat Bering-Chukchi platform, thousands of square miles in extent, can only be ascribed to erosional leveling" (p. 346).

The sediment blanket is generally less than 5 meters thick. "A relatively uniform, thin layer of sediment covers most of the shelf and is locally absent

[a] However, D. M. Hopkins points out (written communication, 1965) that the acoustical records illustrated by Moore merely record bedrock lacking planar reflectors. Tightly folded metamorphic rocks such as those exposed throughout most of southwestern Seward Peninsula might be represented by the records Moore illustrates for much of northeastern Bering Sea.

or thicker as modified by local topography" (p. 346). The sediment is locally thickest (about 30 meters) north of North Cape, St. Lawrence Island, and is regionally thickest (about 12 meters) throughout the embayment of southeastern Chukchi Sea.

## Post-Wisconsin Sediments[b]

### SUBSURFACE NONMARINE OR BRACKISH-WATER SEDIMENTS

D. Moore (1964, p. 346) indicates that the thin sediment-cover over much of the bottom of the Chukchi-Bering Sea area thickens over lows in the bedrock topography. Creager and McManus (1965) describe one of these lows as a Pleistocene valley system smoothed over in part by a greater thickness of sediment: the Pleistocene Hope Seavalley south of Cape Thompson, Alaska, is buried beneath approximately 6–10 meters of sediment. Creager and McManus (1965) and Colinvaux (1962) describe a 7.4-meter core taken at a depth of 39.4 meters that must have penetrated almost the entire valley fill (Fig. 5; Table 1). The sediments of the lower 5.5 meters of the core are described as brackish-deltaic in origin, probably deposited near the head of an estuary formed by the Kobuk and Noatak Rivers during a period of lower sea level. Radiocarbon dates from the bottom of the core and from 1.5 meters down from the top suggest a rate of deposition of approximately 1,200 mg/cm²/yr between 14,240 ± 600 years B.P. (before present) and 13,600 ± 650 years B.P. The upper 1.5 meters of the core consist of sediments deposited at the much slower rate of about 20/mg/cm²/yr and probably represent a marine environment. As reported by Creager and McManus (1965), the dates may contain an error, making the sediments in fact 1,500–2,000 years too old. The true ages of the deltaic sediments in the core probably range from about 11,500 to 12,500 years B.P.

Except for the one long core described above, all of the more than 150 cores collected by American workers have failed to penetrate more than 120–150 cm into the bottom sediments. Therefore, information on the sub-bottom lithology is scanty. The thinness of the sediment overburden, as interpreted from acoustic-reflection surveys, and the dates of 14,000 years B.P. or younger for the relatively thick sediment section south of Cape Thompson lead to the speculation that most, if not all, of the sediment over-

[b] In this paper, the end of the Wisconsin interval is considered to correspond to the last maximum lowering of sea level. Sediments deposited during the fluctuating general rise of sea level since the last major sea-level minimum are considered to be of post-Wisconsin age.

TABLE 1. *Radiocarbon Ages (Total Organic Carbon) and Determined Rates of Deposition of Sediments in Core Located in Fig. 5*

| Depth in Core (cm) | Radiocarbon Age Years B.P. | Estimated Rate of Deposition[1] mg/cm²/yr | Laboratory |
|---|---|---|---|
| 19–39 | 4,390 ± 210 | ⎫ 20 | Univ. of Miami |
| 162–194 | 13,600 ± 650 | ⎬ | Isotopes, Inc. |
| 288–340 | 15,500 ± 800 | ⎭ | Isotopes, Inc. |
| 344–359 | 14,050 ± 210 | ⎫ | Univ. of Miami |
| 536–569 | 13,600 ± 450 | ⎬ 1,200 | Isotopes, Inc. |
| 725–736 | 14,240 ± 600 | ⎭ | Isotopes, Inc. |

Source: Creager and McManus, 1965.

[1] Estimated rates of deposition are based on elapsed time between the radiocarbon age determination at the beginning of an indicated interval and the end of the interval, taking into account the statistical uncertainty in the older determinations.

burden in the Chukchi and northeastern Bering Seas is Holocene in age and represents deposits laid down during the post-Wisconsin transgressive sea-level rise.

## SURFACE SEDIMENTS

The total surface sediment distribution can be sketchily described by using the detailed studies of bottom sediments in the southeastern Chukchi Sea (Creager, 1963; Creager and McManus, 1966; and McManus and Creager, 1963) and the tentative findings from work in progress at the University of Washington on samples from the western and northeastern Chukchi Sea and the northeastern Bering Sea. The only information available at this time from the samples currently under study is the sediment texture based on the Shepard (1954) three-end-member classification (Fig. 6).

In general, grain size of the sediments decreases away from shore or downstream from the presumed sources. The coarsest gravel (not differentiated on Fig. 6 but noted for the southeastern Chukchi Sea by Creager, 1963) with two exceptions is located near cliffs and headlands or appears to be associated with bedrock outcrops on the sea floor. Coastal gravel is found near the cliffs of Wrangel Island, Cape Lisburne, the Point Hope–Cape Thompson coast, southern Kotzebue Sound, the Diomede Islands and Cape Prince of Wales, the coast southeast of York, near Nome, and along the northern coast of St. Lawrence Island. Gravel associated with presumed bedrock outcrops is on Herald Shoal, in eastern Bering Strait, and between King Island and Seward Peninsula. Unexplained occurrences of gravel are found in deeper water north of Northwest Cape and in shallow water offshore from the Point Lay–Icy Cape–Wainwright coast. The finer sedi-

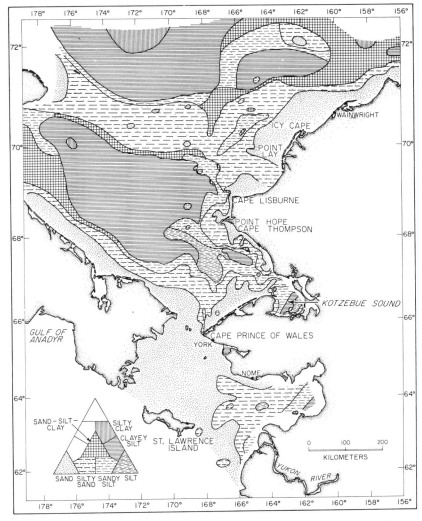

Fig. 6. Texture of bottom sediments expressed in terms of three-end-member relationship of Shepard (1954).

ment throughout the remainder of the area does not as a rule contain appreciable quantities of erratic gravel-size particles. Although ice-rafted gravel is reported from the region by Dietz *et al.* (1964) and from the adjacent Anadyr Gulf by Lisitsyn (1959), the transportation and deposition of gravel by floating ice do not appear important except in western Bering Sea, along the Siberian coast, and in northeastern Chukchi Sea.

In general, the sediment south of Bering Strait is sand, whereas north of the straits the bottom is covered mostly by silt or clayey silt. Within these principal divisions, the surface sediment appears to fall into seven depositional environments: (1) Kotzebue Sound, (2) southeastern Chukchi Sea, (3) Chukchi Basin, (4) Herald Shoal–Cape Lisburne sill, (5) northeastern Chukchi Sea, (6) Chirikov Basin, and (7) Norton Sound. Location of these regions may be found in the inset of Fig. 2.

*Kotzebue Sound.* The shallow bottom of Kotzebue Sound consists mainly of silt with various admixtures of clay- and sand-size particles, suggesting relatively quiet-water deposition. Local sources in the small rivers and cliffs located along the shores are suggested by the presence of sandier sediment to the north near the deltas of the Kobuk and Noatak Rivers, and by the sand and gravel to the south and southeast.

*Southeastern Chukchi Sea.* The large embayment forming southeastern Chukchi Sea has been studied in considerable detail. The following description is taken from Cooper (1964), Creager and McManus (1965, 1966), and McManus and Creager (1963). The sediment is primarily silt and sand with moderate-to-poor sorting. The distributional pattern is controlled by the bathymetry, proximity of source areas, currents, and waves. Sediment is introduced from the south and distributed through the central portion of the embayment by the Bering Strait current (Fig. 4). Secondary sources are the cliffs near Cape Thompson, which provide nearshore gravel, and the Kobuk and Noatak Rivers to the east, which presumably furnish silty sand to the embayment. The Bering Strait current flows in from the south, travels north and east around Cape Prince of Wales Shoal, and then north and west across the embayment. After passing along the north shore of the embayment, it is again deflected around a shoal, off Point Hope, before continuing northward toward the Arctic basin. Where the current is strongest and where the sediment source is nearby, as in Bering Strait and off Cape Thompson, the sediment is coarsest and most poorly sorted. Where the current is weakest and where the sediment source is distant, as in the west-central portion of the embayment, the sediment is finer and better sorted.

Cape Prince of Wales Shoal, which may have been constructed by deposition of sediment by the Bering Strait current, effectively shields the nearshore area to the east of the shoal from the current and sediment supply. This shallow-water region along Seward Peninsula is characterized by fine and very fine sands that are better rounded and better sorted than sediment elsewhere in the embayment and that contain more abundant heavy minerals. In addition, organic carbon constitutes a smaller percentage of the samples here than in any other area in the embayment. These sediments are considered to be reworked by wave action, for this area is exposed at present to the longest fetch directions and is also most isolated from sediment sources.

The Foraminifera consist of an Arctic fauna differentiated into ecologic units by differences in the percentages of *Eggerella advena, Buccella frigida,* and *Elphidium clavatum* (Cooper, 1964). In general, the distribution of ecologic units is most closely correlated with sediment type. A diverse calcareous assemblage was found with the coarse sediment, whereas an arenaceous assemblage of few species occurs with the fine sediment.

*Chukchi Basin.* The sediment of this broad, flat basin (depth 54–58 meters) is primarily clayey silt, which suggests that the depositional environment is one of quiet water. Erratics are rare, indicating lack of ice rafting. The prolonged yearly ice cover and relatively great depths of water must limit wave effects on the bottom and leave only the weak currents as an agent for the dispersal and deposition of clastic sediments. Northward toward Herald Shoal and southward toward Siberia the silt grades into sand with small amounts of gravel. Westward into the Long Strait, the silt grades into a sand-silt-clay mixture. Along the southeastern margin the silt passes into the varied sediments of the southeastern Chukchi Sea environment. The tongue of sandy silt and silt extending into the Chukchi Basin from the southeast is associated with Hope Seavalley.

*Herald Shoal–Cape Lisburne sill.* The stratigraphic and structural continuity of Wrangel Island, Herald Island, Herald Shoal, and Cape Lisburne is discussed by Hopkins (1959) and D. Moore (1964). Continuity in the form of a low ridge is seen in the bathymetry (Fig. 5). Acoustic-reflection soundings indicate that bedrock is exposed on the eastern portion of Herald Shoal (Moore, 1964, p. 360), where coarse sand and gravel occur. The sill between Herald Shoal and Cape Lisburne is covered by sandy silt that separates coarser sediments to the east from the finer sediments of the Chukchi Basin.

*Northeastern Chukchi Sea.* Sediment-size distribution throughout this region is diversified, ranging from gravel to clayey silt.

Incomplete studies of the samples and the meager information available on the physical environment preclude a discussion of the observed sediment-distribution patterns. It is enough to note, then, that gravel is associated with the cliffs and headland near Cape Lisburne, and that gravel off Point Lay and Icy Cape either may be a wave-reworked remnant from a previous environment or may represent ice transport from the beach.

*Chirikov Basin.* The sediment throughout the Chirikov Basin is moderately-to-well-sorted sand, but gravel occurs locally in Bering Strait, in patches between King Island and Seward Peninsula, off Nome, along the northern coastal headlands of St. Lawrence Island, and in the Strait of Anadyr. Coarse erratics are generally rare. A definite gradation is observed from iron-stained, chlorite-rich sands near the Yukon delta and offshore along the Seward Peninsula coast to chlorite-poor sands in the central portion of the basin to the west between St. Lawrence Island and Bering Strait. Although definite statements about the present sedimentary environment are not possible without detailed studies, the well-sorted sand of the central basin appears to be reworked sand having a nearby provenance, possibly the underlying bedrock. Currents and wave action are probably not active agents of present-day sediment transportation and deposition, because the currents in the central basin are relatively weak and the effects of waves are restricted by the depth of water (>30 meters) and the protection given by St. Lawrence Island. Silty sand from the Yukon River appears to be encroaching along the eastern margin of the basin, and reworking of this sand may also be important, for the eastern part of the basin has depths of 30 meters or less and is open to waves and swell from the long fetch to the south.

*Norton Sound.* Norton Sound, which is generally less than 30 meters deep, contains the Yukon delta in its southern portion. The sediment ranges from silty sand along the western margin of the sound, through silt, to clayey silt near the head of the sound. A decrease in sediment size and worsening of the sorting is also noted as the actively building delta is approached from the west or north.

## Post-Wisconsin History

### WISCONSIN GLACIATION

Valley glaciers were present during the Wisconsin Glaciation in some highland areas on Seward Peninsula and the Chukotka Peninsula, but glaciers approached the present coast only on northwestern Seward Peninsula and in some of the present fjords near Cape Chukchi and Cape Chaplin (Coulter *et al.*, 1965; Baranova and Biske, 1964; Sainsbury, this volume). Coarse bottom sediments on the sill just north of the Strait of Anadyr have no obvious present-day source in nearby shoreline cliffs; they may represent a submerged moraine marking the former terminus of a glacier that extended eastward onto the shelf from one of the glaciated valleys of the nearby coast of Chukotka Peninsula. Which glaciation this may represent is unknown, but probably it is pre-Wisconsin (see Hopkins, this volume, Fig. 2, p. 462). No other glacial drift was encountered at the depths reached by the available cores from Chukchi Sea and northern Bering Sea.

### SEA-LEVEL CHANGES

Although there is evidence of crustal instability during the early and middle Pleistocene along both the Siberian and Alaskan sides of the region

(Hopkins *et al.*, 1965; Sainsbury, this volume), the consensus is that there
have been only small and localized vertical movements during and since the
Wisconsin (Budanov *et al.*, 1957; Hopkins, 1959; Ostenso, 1962). There
are some suggestions of very minor tectonic activity at present, such as
low-magnitude earthquakes along the Seward and Chukotka Peninsulas and
just north of the Chukotka Peninsula (Gutenberg and Richter, 1949) and
the minor faulting inferred from the previously mentioned acoustic-reflection profiles. The three 3-meter ledges in acoustic-reflection horizons reported by D. Moore (1964) are not expressed in the bathymetry, and are
considered as predating post-Wisconsin sedimentation.

Inferences concerning post-Wisconsin sea-level changes have been made
by Creager and McManus (1965), based on the one long core from the
southeastern Chukchi Sea and the bathymetry. A summary of our discussion
is presented below.

The broad, flat floor of the Chukchi Basin between −54 and −58 meters
lies within the limits of a sea-level stillstand of approximately 55–60 meters
recognized by a number of investigators (Parker and Curray, 1956; Ludwick and Walton, 1957; Emery, 1958; and Fairbridge, 1961). Creager and
McManus (1965) tentatively suggest that this broad, flat floor resulted from
deltaic and marine depositional processes during a lengthy stillstand at these
depths between 12,000 and 17,000 years B.P. During a later minor fluctuation, about 10,000–11,000 years B.P., sea level may have again fallen to this
level for a short period of time.

Only two other stillstands can be suggested on the basis of Chukchi Sea
data. One is approximately coincident with present sea level, and the other
is approximately 38–44 meters below present sea level. This −38-meter stillstand coincides generally with the 32-to-40-meter platform of Fairbridge
(1961) and the −40 meter sea-level stand of Curray (1960, 1961) (Fig. 7).
As can be seen in Fig. 7, the dates for the stillstand do not coincide with
dates suggested by Curray and Fairbridge. Creager and McManus (1965)
suggest that the sediment samples used for radiocarbon dating of the Chukchi
stillstand may be contaminated and the dates too old. After applying a correction based upon the anomalously old surface date, a corrected date of
11,500–12,500 years B.P. was proposed for the −38-meter stillstand. This
correction would bring the date in line with the suggested stillstand depth
for the Two Creeks Glacial Interstade. The sea-level curve (Fig. 7) presented
by Curray (1960, 1961), modified possibly by a higher stand (−55 to −60
meters) during the period 14,000–18,000 years B.P. as suggested by Fairbridge's work (1961), thus seems fairly representative of sea-level history
in the Chukchi–northern Bering Seas region during the last 20,000 years.

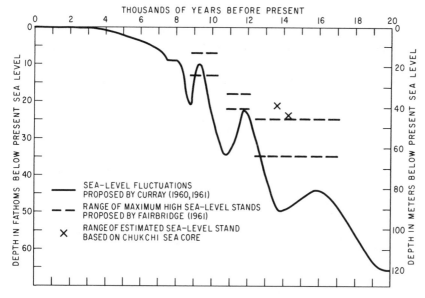

Fig. 7. Fluctuations in sea level (Creager and McManus, 1965).

## CHARACTER OF THE CHUKCHI AND NORTHEASTERN BERING SEAS DURING THE LAST REGRESSION OF SEA LEVEL

The most recent Bering Land Bridge was, in general, an exposed crystalline rock surface south of Bering Strait and an exposed sedimentary rock surface north of Bering Strait. When Wisconsin sea-level recession was at its maximum, more than 100 meters below present sea level, the entire floor of Chukchi Sea and northeastern Bering Sea must have been an exposed coastal plain that was being dissected and denuded by streams. That this denudation to bedrock apparently was complete is suggested by the position of this erosion surface only a few meters beneath the present sea floor, as shown by composition of the long Chukchi Sea core and the nature of the acoustic-reflection profiles. Indications that the erosion locally was not complete are the thick (up to 30 meters) sediment overburden north of St. Lawrence Island and the rugged bathymetry along the Siberian coast, which may indicate large moraines. Preservation of the low local relief of less than 9 meters that had been produced over most of the region was due in part to the tectonic quiescence of the region during late Quaternary time.

### TRANSGRESSIVE SEA SEDIMENTARY ENVIRONMENTS

For at least the last 15,000 years, sea level has been generally rising as the continental glaciers have melted. The rise has not been smoothly continuous, but has been marked by numerous minor fluctuations producing stillstands and periods of accelerated rise and fall of sea level. When sea level rose, the number and size of estuaries must have increased. When sea level stood at −54 meters (Fig. 8), there was a large northwest-southeast estuarine embayment in the Chukchi Basin, and large lakes lay in the vicinity of Bering Strait. Moreover, floodplains must have increased in area because of the increased runoff from melting glaciers; the deposition probably was localized in marine and lacustrine deltas. Not only runoff, but sediment sup-

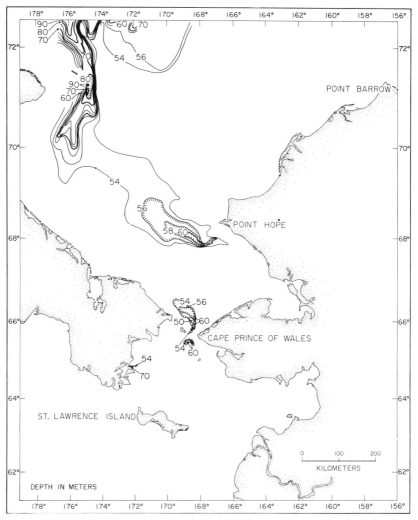

Fig. 8. Bathymetric chart of Chukchi and northeastern Bering Seas when sea level was at −54 meters.

ply, also, would have increased during intervals of rising sea level, thereby increasing the rate of deltaic, estuarine, and paludal deposition. These deposits, however, would be localized along the established drainage channels, producing the relatively thick sedimentary overburden in the bedrock depressions, which consist largely of alluvial fill or deltaic deposits. During stillstands, larger masses of deltaic sediments accumulated at the mouths of major streams. An example in Chukchi Sea was described by Creager and McManus (1965) from acoustic-reflection profiles and the lone long core from southeastern Chukchi Sea. Such concentrated valley filling is the first step in the smoothing of the bedrock erosional relief.

By the time sea level rose to the −38-meter stillstand level 13,600–14,050 years B.P. (or 11,500–12,500 years B.P., if our proposed correction is applied), the Bering Land Bridge had been broken at both the Strait of Anadyr and Bering Strait (Fig. 9). If, then, sea level fell to at least −48 meters, the land bridge would have been reconnected across the Strait of Anadyr until sea level again rose, finally flooding the land bridge. A fall in sea level after the Two Creeks Interstade is suggested by both Curray (1960, 1961) and Fairbridge (1961). After another minor fluctuation of between 20 and 40 meters during the interval 8,500–11,000 years B.P., sea level rose steadily, reaching its present position 3,000–4,000 years ago.

## MODERN MARINE SEDIMENTARY ENVIRONMENT

As the transgression has continued, more and more land has fallen permanently below sea level to mark the beginning of a marine depositional phase, characterized at least locally by a decrease in the rate of deposition. The rate of marine deposition may well have been greater in northeastern Bering Sea than in Chukchi Sea, because of the larger source areas, although the thickness of modern marine deposits in both seas is only about 1–3 meters. The marine deposits, together with the preceding and contemporaneous nonmarine and brackish-water deposits, have produced the present monotonously flat floors.

Once the waters of the Pacific and Arctic basins were reconnected through Bering Strait, the sedimentologic environment changed markedly. The present northerly-setting currents, which are strongest along the eastern margins of the seas, were established by the reconnection (Fleming and Heggarty, 1966). The modern sedimentary environment in the southeastern Chukchi Sea is discussed at length by Creager (1963) and Creager and McManus (1966). As sea level gradually covered the flat central and eastern parts of the Chirikov Basin, waves were able to rework the sediment and produce the relatively well-sorted sand.

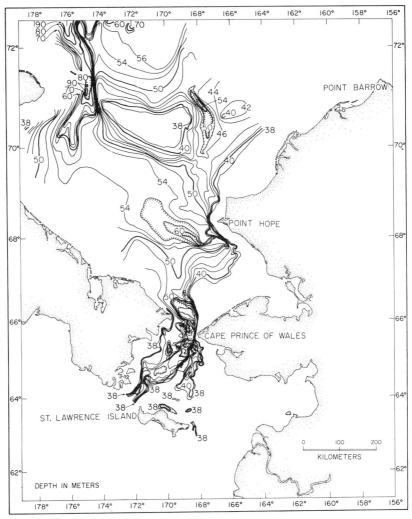

Fig. 9. Bathymetric chart of Chukchi and northeastern Bering Seas when sea level was at −38 meters.

The relatively swift Bering Strait current began depositing sediment on the northern margin of the strait, producing the spitlike shoal north of Cape Prince of Wales, which has continued to grow to this time (McManus and Creager, 1963). Similar spitlike shoals are found at Point Hope and probably at Northwest Cape on St. Lawrence Island. As the Cape Prince of Wales Shoal grew, the area to the east became more protected from the current, and waves there were able to rework the sediment, producing a relatively clean sand. As sea level passed over Herald Shoal, waves acted along the shoal crest to produce the coarse material that has remained exposed as a thin veneer because of the remoteness of the shoal from other sediment sources and because of continued wave and current action.

As sea level approached its present height, Kotzebue and Norton Sounds were formed and, together with the Chukchi Basin, became areas of quiet-water deposition either too protected or too deep to be affected by wave action, and all were sufficiently removed from the dominant current system. The Yukon River, which empties into Norton Sound, supplies much of the coarser-grained sediment there. Wherever there is resistant rock at the shore, waves are actively eroding cliffs and producing coarse-grained deposits. Little can be said at present about the northeastern Chukchi. It appears not to be similar to the remainder of Chukchi Sea in terms of the modern marine sedimentary environment. Possibly there is a greater amount of ice-rafted debris here than elsewhere.

## Implications

As often suggested before (e.g., Hellbom, 1957; Chard, 1958), and here supported by further information, the last link in the Bering Land Bridge was by way of St. Lawrence Island. The sill north of the Strait of Anadyr appears to lie between 46 and 48 meters, whereas the sill north of Bering Strait appears to lie between 48 and 50 meters. The sill east of St. Lawrence Island is at least 10 meters shallower. These sill depths, however, and hence interpretations of land bridge connections, are based on present-day bathymetry. Possibly all of the sills have significant thickness of Holocene marine sediments; if so, they would have been lower and would have been drowned earlier during the late Quaternary rise in sea level. D. Moore (1964) suggests that a good part of the region east of St. Lawrence Island is exposed bedrock. Therefore, unless there is an unknown, deep, filled channel present, we may disregard the area east of St. Lawrence Island in discussing the time and place of the final severing of the land bridge. Almost no work has been done in the vicinity of the two important sills, north of Bering Strait and north of the Strait of Anadyr. These sill depths are minimal and it is possible that they were deeper before post-Wisconsin deposition. However, because the sills are on the Siberian side of the seas, where glaciation is considered to have been both more intense and of longer duration (but pre-Wisconsin), these sills, particularly the one north of the Strait of Anadyr, may be morainal and not related to post-Wisconsin deposition. If this is the case, then the sills were probably established before the last transgression and can be used in considering the time of last severance of the land bridge.

Neither in Bering Strait nor in the Strait of Anadyr are the sill depths related to the stillstand depths proposed for the Chukchi and northern Bering Seas, and therefore the sills probably do not represent depositional features produced by rapid deltaic deposition. Bering Strait sill has vague bathymetric and sedimentary connections with the Cape Prince of Wales Shoal. The definite continuity of sediment sizes (Fig. 6) from the shoal across the sill suggests that the sill may have been formed at the same time. Therefore, the sill depth would have been greater during Wisconsin time and the land bridge severed much sooner in this area.

These questions cannot be solved with the data presently available. Nevertheless, using the minimal depths, the land bridge must have been severed before 11,500–12,500 years ago and perhaps as early as 13,600–14,050 years ago, although it may later have been reconnected for a short time by a minor regression. Present information suggests that the sill north of the Strait of Anadyr was most likely the last migration route to be breached, but probably this breaching took place soon after the drowning of the sill north of Bering Strait.

## Summary

Subbottom reflection profiles collected by D. Moore north of the latitude of St. Lawrence Island suggest that bedrock south of Bering Strait is structureless crystalline rock, whereas bedrock north of Bering Strait is folded sedimentary rock. Although the bedrock has been deeply dissected in some localities, such as the area north of St. Lawrence Island where local bedrock relief is as much as 30 meters, it generally spans no more than 3 to 9 meters in relief. Sediment cover is generally thin throughout the Bering and Chukchi Seas, occurring mainly as fill in the depressions in the buried bedrock topography.

Few long cores of the sediment overburden have been taken. A 7-meter piston core in southeastern Chukchi Sea, representing essentially the entire sediment fill of a buried valley, records a sea-level stillstand approximately 12,000 years ago at about −38 meters. The stillstand is recorded by the lower 5.5 meters of core, which was deposited within an interval of only about 500 years, and which records deltaic deposition in an estuarine environment. The upper 1.5 meters of sediment are marine. This information, together with the subbottom reflection profiles, suggests that the sediment overburden of both coastal plain and marine deposits in Bering and Chukchi Seas is largely of late Wisconsin and post-Wisconsin age.

The final break in the land link between the continents appears to have occurred when the post-Wisconsin transgression covered the sill west of

St. Lawrence Island. This sill 46 meters deep is shallower than the sills near Bering Strait. Its configuration associates it with a possible glacial origin, suggesting that the sill is not the result of post-Wisconsin deposition. The land bridge no longer existed when sea level stood at $-38$ meters 12,000 years ago.

## REFERENCES

Aagaard, K. 1964. Features of the physical oceanography of the Chukchi Sea: Unpublished master's thesis, Univ. Washington, Seattle, 28 p.

Anderson, G. J. 1961. Distribution patterns of Recent Foraminifera of the Arctic and Bering Seas: Unpublished master's thesis, Univ. Southern California, Los Angeles, 90 p.

Baranova, Iu. P., and S. F. Biske. 1964. Istoria Razvitii Rel'efa Sibiri i Dal'nogo Vostoka: Severo-vostok SSSR (History of the development of the relief of Siberia and the Far East: Northeastern USSR): Akad. Nauk SSSR, Sibirskoe Otdelenie, Inst. Geol. Geofiz., 289 p.

Bryan, K. 1946. Cryopedology—the study of frozen ground and intensive frost-action with suggestions on nomenclature: Am. Jour. Sci., v. 244, p. 622–642.

Budanov, V. I., A. T. Vladimirov, A. S. Ionin, P. A. Kaplin, and V. S. Medvedev. 1957. Modern vertical movements of the sea coast in the Far East: Dokl. Akad. Nauk SSSR., Geol. Sci. Sec., v. 116, p. 829–832 in Consultants Bureau running translation.

Carsola, A. J. 1954a. Recent marine sediments from Alaskan and Northwest Canadian Arctic: Am. Assoc. Petroleum Geologists Bull., v. 38, no. 7, p. 1552–1586.

———— 1954b. Microrelief on Arctic Sea floor: Am. Assoc. Petroleum Geologists Bull., v. 38, no. 7, p. 1587–1601.

———— 1955. Foraminifera from the Beaufort and Chukchi Seas: Jour. Paleontology, v. 29, no. 4, p. 738.

Chard, C. S. 1958. New world migration routes: Anthropological Papers of the University of Alaska, v. 7, p. 23–26.

Colinvaux, P. A. 1962. Environment of the Bering Land Bridge: Unpublished Ph.D. thesis, Duke Univ., Univ. Microfilms 63-3888, Ann Arbor, Mich., 104 p.

Cooper, S. C. 1964. Benthonic Foraminifera of the Chukchi Sea: Contr. from the Cushman Foundation for Foram. Research, v. 15, pt. 3, p. 79–104.

Coulter, H. W., *et al.* 1965. Map showing the extent of glaciations in Alaska: U.S. Geol. Survey Misc. Geol. Inv. Map I-415.

Creager, J. S. 1963. Sedimentation in a high energy, embayed, continental shelf environment: Jour. Sed. Petrology, v. 33, no. 4, p. 815–830.

Creager, J. S., and D. A. McManus. 1961. Preliminary investigations of the marine geology of the southeastern Chukchi Sea: Univ. Washington Dept. Oceanography Tech. Rept., v. 68, 46 p.

———— 1965. Pleistocene drainage patterns on the floor of the Chukchi Sea: Marine Geology, v. 3, p. 279–290.

———— 1966. Geology of the southeastern Chukchi Sea, p. 755–786 in N. J. Wilimovsky, ed., Environment of the Cape Thompson region, Alaska: U.S. Atomic Energy Commission.

Curray, J. R. 1960. Sediments and history of Holocene transgressions, continental shelf, northwest Gulf of Mexico, p. 221–266 in F. P. Shepard, F. B. Phleger, and T. H. van Andel, eds., Recent sediments, northwest Gulf of Mexico: Am. Assoc. Petroleum Geologists, Tulsa, Okla.

——— 1961. Late Quaternary sea level: a discussion: Geol. Soc. America Bull., v. 72, p. 1707–1712.

Cushman, J. A. 1910–1917. A monograph of Foraminifera of the North Pacific Ocean. Pt. I, Astrorhizidae and Lituolidae; II, Textulariidae; III, Labenidae; IV, Chilostomellidae, Globigerinidae, Nummulitidae; V, Rotaliidae; VI, Miliolidae: U.S. Nat. Mus. Bull. 71, p. 134, 108, 125, 46, 87, 108.

——— 1920. Foraminifera: Canadian Arctic Exped., 1913–1918, Report, v. 9, pt. M, King's Printer, Ottawa, 13 p.

Dietz, R. S., A. J. Carsola, E. C. Buffington, and C. J. Shipek. 1964. Sediments and topography of the Alaskan shelves, p. 241–256 in R. L. Miller, ed., Papers in marine geology: Shepard commemorative volume: Macmillan, New York.

Emery, K. O. 1958. Shallow submerged marine terraces of southern California: Geol. Soc. America Bull., v. 69, p. 39–60.

Fairbridge, R. W. 1961. Eustatic changes in sea level: physics and chemistry of the Earth, v. 4, Pergamon Press, New York, p. 99–185.

Fleming, R. H., and D. E. Heggarty. 1966. Physical and chemical oceanography of the southeastern Chukchi Sea, p. 697–754 in N. J. Wilimovsky, ed., Environment of the Cape Thompson region, Alaska: U.S. Atomic Energy Commission.

Fleming, R. H., et al. 1959. Oceanographic survey of the eastern Chukchi Sea, 1 August to 2 September 1959: Seattle, Univ. Washington, Department of Oceanography Preliminary Report of Brown Bear Cruise No. 236, 36 p.

Gutenberg, B., and C. F. Richter. 1949. Seismicity of the Earth and associated phenomena: Princeton Univ. Press, Princeton, N.J., 273 p.

Hellbom, A. B. 1957. Indians, Eskimos, and Whites: 32d Internat. Cong. Americanists (Copenhagen), August 7–14, 1956, Jour. Ethnos, nos. 1–2.

Hopkins, D. M. 1959. Cenozoic history of the Bering Land Bridge: Science, v. 129, p. 1519–1528.

Hopkins, D. M., F. S. MacNeil, R. L. Merklin, O. M. Petrov. 1965. Quaternary correlations across Bering Strait: Science, v. 147, p. 1107–1114.

Hopkins, D. M., and R. S. Sigafoos. 1951. Frost action and vegetation patterns on Seward Peninsula, Alaska: U.S. Geol. Survey Bull., no. 974-C, p. 51–101.

Kachadoorian, R., et al. 1960. Geological investigations in support of Project Chariot in the vicinity of Cape Thompson, northwestern Alaska—Preliminary Report: USAEC Report TEI-753, U.S. Geol. Survey, 94 p.

——— 1961. Geological investigations in support of Project Chariot, phase III, in the vicinity of Cape Thompson, northwestern Alaska—Preliminary Report: USAEC Report TEI-779, 104 p.

Lisitsyn, A. P. 1959. Bottom sediments of the Bering Sea, p. 65–179 in P. L. Bezrukov, ed., Geographical description of the Bering Sea—bottom relief and sediments: Akad. Nauk SSSR, Trudy Instituta Okeanologii, v. 29, Moscow: transl. pub. by Israel Program for Scientific Translations, Jerusalem, 1964.

Ludwick, J. C., and W. R. Walton. 1957. Shelf-edge calcareous prominences in northeastern Gulf of Mexico: Am. Assoc. Petroleum Geologists Bull., v. 41, p. 2054–2101.

McManus, D. A., and J. S. Creager. 1963. Physical and sedimentary environments on a large spit-like shoal: Jour. Geology, v. 71, p. 498–512.

Moore, D. G. 1964. Acoustic-reflection reconnaissance of continental shelves: eastern Bering and Chukchi Seas, p. 319–362 in R. L. Miller, ed., Papers in marine geology: Shepard commemorative volume: Macmillan, New York.

Moore, G. W., and J. Y. Cole. 1960. Coastal processes in the vicinity of Cape Thompson, Alaska, from May 3 to May 9, 1960: USAEC Report TEI-764, U.S. Geol. Survey, p. 24–28.

———— 1960b. Recent eustatic sea-level fluctuations recorded by Arctic beach ridges: U.S. Geol. Survey Prof. Paper 400-B, p. 335–337.

Moore, G. W., and J. Y. Cole. 1960. Coastal processes in the vicinity of Cape Thompson, Alaska: USAEC Report TEI-753, U.S. Geol. Survey, p. 41–55.

Moore, G. W., and D. W. Scholl. 1961. Coastal sedimentation in northwestern Alaska: USAEC Report TEI-779, U.S. Geol. Survey, p. 43–65.

Ostenso, N. A. 1962. Geophysical investigations of the Arctic Ocean Basin: Univ. Wisconsin, Geophys. and Polar Res. Center Rept. 62–4, 124 p.

Parker, R. H., and J. R. Curray. 1956. Fauna and bathymetry of banks on continental shelf, northwest Gulf of Mexico: Am. Assoc. Petroleum Geologists Bull., v. 40, p. 2428–2439.

Rathbun, R. 1894. Summary of the fishery investigations conducted in the North Pacific Ocean and Bering Sea from July 1, 1888, to July 1892, by the U.S. Fish Commission steamer *Albatross*: U.S. Bur. Fish. Bull., v. 12, p. 127–201.

Scholl, D. W., and C. L. Sainsbury. 1961. Marine geology and bathymetry of the Chukchi Shelf off the Ogotoruk Creek area, northwest Alaska, *in* G. O. Raasch, ed., Proc. First Internat. Symp. on Arctic Geology: Univ. Toronto Press, Toronto, v. 1, p. 718–732.

Shepard, F. P. 1954. Nomenclature based on sand-silt-clay ratios: Jour. Sed. Petrology, v. 24, no. 3, p. 151–158.

Tanner, Z. L. 1893. Report upon the investigations of the U.S. Fish Commission steamer *Albatross* from July 1, 1889, to June 30, 1891: U.S. Fish and Wildlife Service, Report of the Commissioner of Fish and Fisheries, 1889–1891, p. 207–342.

Udintsev, G. B., I. G. Boichenko, and V. F. Kanaev. 1959. Bottom relief of the Bering Sea, p. 14–64 *in* P. L. Bezrukov, ed., Geographical description of the Bering Sea—bottom relief and sediments: Akad. Nauk SSSR, Trudy Instituta Okeanologii, v. 29, Moscow: trans. pub. by Israel Program for Scientific Translations, Jerusalem, 1964.

U.S. Coast and Geodetic Survey. 1955. United States Coast Pilot 9, Alaska 1954. Cape Spencer to Arctic Ocean; 6th ed.: U.S. Govt. Prtg. Off., Washington, D.C., 673 p.

U.S. Hydrographic Office. 1958a. Oceanographic survey results, Bering Sea area, Winter and Spring 1955: TR-46, 95 p.

———— 1958b. Oceanographic atlas of the Polar Seas. Part II, Arctic: U.S. Hydrographic Office Pub. 705, 149 p.

———— 1959. Oceanographic observations, Arctic waters task force five and six, Summer-Autumn 1957: TR-59, 145 p.

U.S. Naval Oceanographic Office. 1960. Oceanographic observations, Arctic waters task force five and six, Summer-Autumn 1956; TR-58, 89 p.

———— 1963a. Oceanographic Data Report, Arctic 1958: Informal Manuscript Report NO. 0–26–63, Unpublished Manuscript, 200 p. U.S. Naval Oceanographic Office, 1963b, Oceanographic Data Report, Arctic 1959: Informal Manuscript Report NO. 0–43–63, Unpublished Manuscript, 245 p. U.S. Naval Oceanographic Office, 1964, Oceanographic Data Report, Arctic 1960: Informal Manuscript Report NO. 0–62–63, Unpublished Manuscript, 154 p.

Zenkevich, L. 1963. Biology of the seas of the U.S.S.R.: Interscience, Wiley, New York, 955 p.

# 3. Late Quaternary Sediments of Bering Sea and the Gulf of Alaska

D. E. GERSHANOVICH
*Research Institute of Marine Fisheries and*
*Oceanography (VNIRO), Moscow*

An attempt to elucidate the characteristic geologic features of the Quaternary Period must include careful investigation of the upper layers of the sedimentary mantle of the sea floor. The most fruitful approach to studying the stratigraphy of Quaternary marine sediments is based on taking long cores of bottom deposits, using the recently designed coring tubes now widely used in oceanographic expeditions. Lithologic and stratigraphic investigations of such cores show that sedimentation processes of a great variety were active during the Quaternary Period. The causative factors involved both in the processes and in their frequent variation are of great interest, because we wish not only to understand changes in the sedimentation regime at sea but also to define more accurately the physical and geographical conditions that existed throughout large regions—conditions that determined the whole process of sedimentation. Bottom sediments generally reflect with great accuracy the natural conditions that have determined their peculiarities of composition and their mode of occurrence. Thus, knowledge of sea-floor deposits is of great importance to paleogeographic reconstruction of the epochs when they accumulated.

In recent years, rich material has been collected from the Quaternary deposits of a number of regions in the extreme northern and northeastern parts of the Pacific Ocean (Menard, 1953; Bezrukov and Lisitsin, 1961; Saidova, 1961; Skorniakova, 1961; Jousé, 1962). Investigations conducted in Bering Sea and the Gulf of Alaska from 1958 to 1962 have provided new data on the main features of Late Quaternary sedimentation there (Gershanovich, 1962, 1963b). Earlier investigations of the central and eastern areas of Bering Sea and the Gulf of Alaska were very limited, but these areas are now known in great detail; about 1,000 samples have been obtained, including 300 cores from 1.5 to 6.0 meters long. Most of the cores were taken in deep-water areas in depths as great as 4,500 meters, but some were taken in shallow-water areas where the sediments are thin and can be completely penetrated by the coring tubes (Fig. 1). Thus, relatively detailed geologic data are available concerning the Late Quaternary sediments of almost all the major physiographic provinces in the extreme northern part of the Pacific Ocean and in the neighboring seas (Gershanovich, 1963a), including the epicontinental shelf of Bering Sea, the shelf of the geosynclinal areas of Bering Sea and the Gulf of Alaska,[a] the continental slope of Bering Sea and of the Gulf of Alaska, the Commander-Aleutian Ridge, the Kamchatka and Aleutian Basins of the Bering Sea, and the oceanic floor within the Gulf of Alaska. Still poorly studied are the deposits of the submarine ridges of Bering Sea (Bowers Bank and the Shirshov or Olyutorskiy Ridge), the submarine mountains and hills that are characteristic features of the Gulf of Alaska, and the eastern extremity of the Aleutian Trench where depths are less than 4,500 meters.

## Structure and Stratigraphy of the Late Quaternary Sediments

The Quaternary deposits of Bering Sea and the Gulf of Alaska are generally stratified. Stratification is normally expressed by changes in the lithologic and granulometric composition of the sediments, color changes, and large variations in the amount and nature of organic remains. Two prevailing types of stratification are observed. The type most frequently encountered reflects lithologic differences associated with general changes in the conditions of sedimentation. The second type, local in nature and determined by local processes such as landslides or turbidity currents, affects sediments of the continental slope and adjacent areas of the abyssal plain. In some areas, the stratification is determined by pulsations in the amount of clastic rock material of diverse granulometry that has derived from the adjacent continents. Similar stratification is produced near the Aleutian Islands and near the submarine volcanoes of the Gulf of Alaska by volcanic rock deposited mostly near the eruptive centers. This second, local type of stratification is observed everywhere, superimposed on the regional stratification of the first type, so that the stratigraphy of the surface sedimentary strata penetrated by the corers is very complicated.

Microstratification (Fig. 2) can also be seen in some areas, mainly in those remote from sources of clastic sediments. Apparently, microstratifica-

[a] The term "shelf of the geosynclinal areas," as used by Dr. Gershanovich, refers to those areas that are still tectonically active and in which the continental shelf is carved across recently deformed sediments or volcanic rocks. Ed.

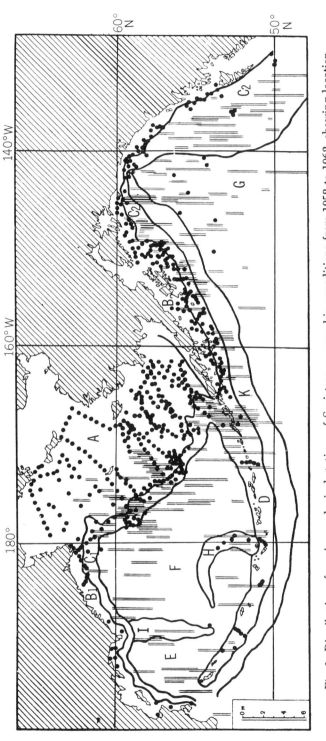

Fig. 1. Distribution of marine geological stations of Soviet oceanographic expeditions from 1958 to 1962, showing location of dredge samples (*circles*) and deep-sea cores (*upper ends of double lines*; *length of double line indicates relative length of core*). The scale of core lengths is shown at lower left.

*A.* Epicontinental shelf of the Bering Sea. *B.* Shelf of the geosynclinal areas: ($B_1$) of the Bering Sea, and ($B_2$) of the Gulf of Alaska. *C.* Continental slope: ($C_1$) of the Bering Sea, and ($C_2$) of the Gulf of Alaska. *D.* Commander-Aleutian Ridge. *E.* Kamchatka Basin. *F.* Aleutian Basin. *G.* Oceanic floor in the Gulf of Alaska. *H.* Bowers Ridge. *I.* Shirshov Ridge. *K.* Aleutian Trench.

Fig. 2. Microstructure of deep-water sediments in Bering Sea from a depth of .1,264 meters: (*a*) 270–275 cm, and (*b*) 299–304 cm, station 183. Light-colored bands are diatomaceous sediment. Scale is indicated by 1-mm grid. Location of station is given in Gershanovich (1962).

tion results primarily from changes in relative amount and character of organic material (mostly diatoms) and clastic particles deposited during comparatively short periods. In all of the physiographic provinces studied, microstratification appears only in the late glacial[b] sediments, according to the data collected thus far. Microstratification is most distinct in the sediments of deep-water areas near the Commander-Aleutian Arc; the most frequent and most diversified changes in sedimentation conditions appear to have occurred there.

The Late Quaternary deposits are characterized by the presence of a rather strongly marked boundary between late glacial sediments and overlying sediments formed in postglacial time (Fig. 3). The postglacial deposits vary widely in granulometric composition from one area to another; they consist of greenish-gray sediments of diverse granulometry. The surface zone is commonly of a brownish color, owing to oxidizing conditions at the sediment-water interface. In deep-water areas remote from shore, the thickness of this oxidized layer locally reached 10 to 20 cm or even more.

A characteristic feature of the postglacial sediments in many areas is the presence of appreciable quantities of diatom tests. Thus, bottom sedi-

[b] "Late glacial" as used in this paper describes sediments deposited during the last cold cycle. Their precise age is unknown; they may include deposits of both Late ("classical") Wisconsin and Early Wisconsin age.

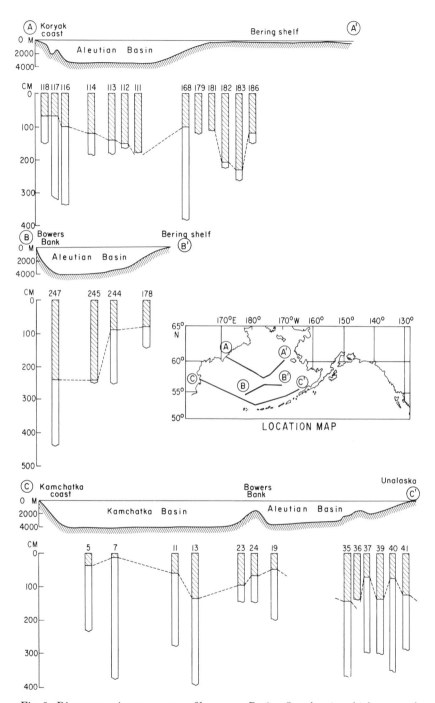

Fig. 3. Diagrammatic east-west profiles across Bering Sea showing thicknesses of the postglacial sediments (*shaded*). Station numbers are given at top of profile.

ments in deep water near the margin of the Bering shelf generally contain 10 to 15 per cent or more of authigenic silica, and in the center of the deep part of Bering Sea the authigenic silica content reaches 30 to 40 per cent, with a maximum of 48.9 per cent in the samples analyzed. The thin shelf sediments of Bering Sea contain about 5 to 10 per cent authigenic silica. Bottom sediments in the Gulf of Alaska contain less authigenic silica; the average concentration is only about 2 per cent in coastal areas, but as much as 5 to 7 per cent can be found in some deep-sea sediments. A maximum content of 15.45 per cent was observed in the eastern part of the gulf.

Recent investigations have shown that the deposits of turbidity currents are relatively abundant on the continental rise in the eastern part of the Gulf of Alaska. Several turbidite cycles can be recognized in some of the cores; they are easily recognized by their textural characteristics. Each cycle is represented by a series of sedimentary bands of differing granulometric characteristics. The lowermost band is composed of coarser material (sand or silty sand) resting with sharp contact upon the underlying layer. The coarse material grades upward into finer sediments, and the cycle generally ends with clayey silt. In some cores, brown-clay sediments are overlain by coarser deposits typical of shallow water. The thickness of individual turbidite units ranges from a few centimeters to a few tens of centimeters. Turbidites were observed in late glacial strata as well as in the postglacial sediments. They are less extensively developed on the continental slope and near the submarine ridges of Bering Sea.

The late glacial sediments are characterized by a lack of lateral variation in composition, especially on the shelf and on the upper part of the continental slope. Throughout the region, these sediments contain only a negligible quantity of diatom tests; concentrations of authigenic silica seldom exceed 3 to 5 per cent, even in Bering Sea, and are generally lower than 2 per cent in the Gulf of Alaska.

## Distribution of the Principal Types of Postglacial Sediments

The postglacial deposits of Bering Sea and the Gulf of Alaska have now been studied well enough to define the distribution of the principal types of sediment (Fig. 4). Vast areas of the epicontinental shelf of northern and eastern Bering Sea are covered by clastic sediments. Coarse clastic deposits prevail in coastal areas, and finer deposits predominate on the central part of the shelf. Along the outer edge of the shelf, the post-Pleistocene deposits are again commonly represented by coarser sediments—sand and silty sand. The clastic sediments grade into slightly diatomaceous deposits on the continental slope and may contain 10 to 30 per cent authigenic silica at depths

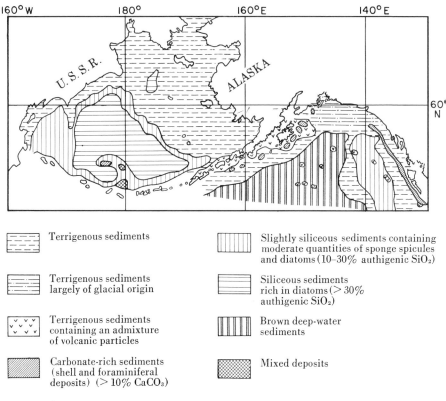

Fig. 4. Schematic distribution of the principal types of postglacial deposits in Bering Sea and the Gulf of Alaska.

below 1,000 meters. Slightly diatomaceous and spongoid deposits prevail on the crest and slopes of the Shirshov (Olyutorskiy) Ridge. Postglacial deposits on Bowers Bank are more diverse: spongoid, foraminiferal, and slightly diatomaceous deposits are abundant; less common are sediments with a rather high content of volcanic particles. Fine-grained, slightly diatomaceous deposits compose the uppermost layers of the Quaternary sedimentary mantle in the Kamchatka Basin of Bering Sea; similar sediments are encountered in the marginal regions of the Aleutian Basin as well as on the continental rise at depths below 2,500 meters. Diatomaceous sediments with an authigenic silica content exceeding 30 per cent occur in the center of the Aleutian Basin, and are much more abundant there than in the neighboring areas of the Pacific Ocean.

Terrigenous sediments prevail in the area of the Commander-Aleutian Arc, but terrigenous-volcanic deposits are abundant in the eastern zone on both sides of the Aleutian Islands and in the Gulf of Alaska. The shelf of the Gulf of Alaska is characterized by an irregular distribution of sediments; coarse clastic deposits, locally rich in shell detritus, prevail on the plateaus, whereas submarine valleys are occupied by terrigenous silt. Deposits consisting mainly of loam particles of glacial origin are observed almost everywhere in the northern and northeastern parts of the gulf, not only on the shelf but also on adjacent areas of the continental slope. These deposits were introduced mainly by rivers draining from the glaciated areas of Alaska, and represent one of the most distinctive features of sedimentation in the Gulf of Alaska. They are grayish in color and structurally and texturally homogeneous; the authigenic silica content is extremely low, about 2.7 per cent on the average and nowhere exceeding 3.6 per cent. Foraminiferal, slightly diatomaceous, volcanic, and terrigenous sediments are found on the continental slope of the Gulf of Alaska. Postglacial deposits at oceanic depths are mainly terrigenous. The surface layer of these deposits is commonly oxidized to a brown color and can be regarded as a transitional stage to "oceanic red clay." Volcanic and foraminiferal sediments are also encountered at oceanic depths; an admixture of volcanic material is observed more commonly in these deposits than in the sediments of any other physiographic provinces in the gulf. Solid volcanic rocks are exposed on the slopes and flat summits of submarine mountains; postglacial sediments are rare and thin on these mountains, and generally consist of volcanic debris and planktonic foraminifera.

## Thickness of the Postglacial Sediments

The thickness of the postglacial deposits varies rather widely, from 0.3 to 3 meters. Figure 5 is a schematic map showing the thickness of postglacial deposits in the areas of Bering Sea where the most detailed data are available. No investigations have been conducted in the coastal areas of the Bering Shelf, and data from the Gulf of Alaska are too scanty to permit us to do more than offer a general description.

On the epicontinental shelf of Bering Sea, the maximum thickness of sedimentary material was found in the central part, south of the Gulf of Anadyr and west of Bristol Bay. The peculiarities of the bottom contours and of water circulation in this area have resulted in an accumulation mainly of fine particles. The thickness of the postglacial sequence reaches 2 meters. Postglacial sediments are thinner, generally about a meter thick, on the outer edge of the shelf in depths of from 120 to 150 meters and in many southeastern areas of the shelf. One can infer from the available data that the narrow shelf zones in the southwestern and western parts of Bering Sea and along the Commander-Aleutian Arc have a distinctively thin cover of post-

D. E. GERSHANOVICH

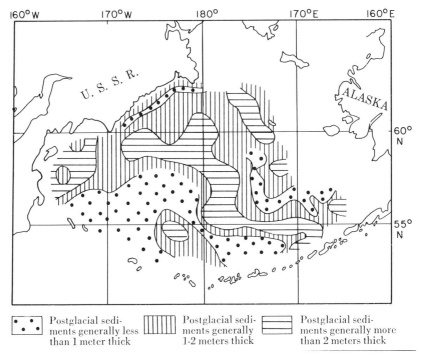

Fig. 5. Thickness of postglacial sediments in Bering Sea.

glacial sediments. These areas are characterized by extensive bottom erosion and subsequent elutriation of sedimentary material, which has exposed outcrops of solid rock; consequently, a very thin layer of coarse clastic detritus and shell debris mantles the solid rock in these areas.

The picture is quite different in the submarine valleys that cut deeply into the Gulf of Alaska shelf; there, fine clastic material has accumulated in great abundance, especially near the glaciated areas in the mountains of southern Alaska. The process is still going on; cores 2 to 5 meters long in this area failed to penetrate below the post-Pleistocene stratum. Clay particles are also accumulating everywhere on those positive relief elements of the shelf that are in immediate proximity to existing glaciers; this process renders the shelf zone of southern Alaska markedly different from adjacent areas.

Notable differences are observed in the thicknesses of postglacial sediments on the continental slope of Bering Sea, which ranges in depth from 150 to 2,500 meters, and on the continental rise in depths from 2,500 to 3,500 meters. The continental slope, a rather steep incline indented by submarine valleys and canyons, is covered by a layer of postglacial deposits ranging roughly from 1 to 2 meters in thickness. Postglacial deposits are absent only on the flanks of canyons, on some ridges, and on the surface of outcrops of hard pre-Quaternary rocks. Outcrops are most common near the Pribilof Islands, along the Koryak coast of Bering Sea, and on the slopes and summits of the Commander-Aleutian Arc. The thickness of postglacial sediments increases on the gently sloping sections of the continental slope, especially within the large tectonic depression west of Bristol Bay and in the depression south of the Gulf of Anadyr. The continental rise is character- ized nearly everywhere by a considerable thickness of postglacial deposits— 2 to 3 meters or even more—and thus constitutes the principal site of ter- rigenous and biogenic sedimentary accumulation. Large thicknesses of post- glacial deposits are also observed in the peripheral zones of the abyssal basins near the continental rise and alongside some of the submarine ridges. The smooth insular slopes on the Bering Sea side of the easternmost Aleu- tian Islands constitute a sort of borderland characterized by especially rapid sedimentation, by the presence of the greatest thicknesses of postglacial sedi- ments observed in Bering Sea (up to 3 meters), and by the presence of much volcanic debris mixed with the terrigenous matter in the bottom sediments.

Thickness variations in the area of the continental slope in the Gulf of Alaska present a rather complicated picture. Postglacial sediments reach their maximum thickness (2 meters or more) in the northern and north- eastern parts of the gulf, in the areas where glacial debris is supplied most abundantly. Sedimentation is not so intensive in other areas of the conti- nental slope, but the thickness of the postglacial deposits increases again to 1.5 or 2.0 meters or even more on the continental rise, which occupies a large area in the eastern part of the gulf; turbidite beds are especially com- mon there.

A further decrease in the thickness of postglacial deposits is observed on the abyssal plains. The minimum thicknesses registered in Bering Sea were found in the central and southern part of the Aleutian Basin, where the sediments are less than 1 meter thick. This basin is one of the largest abyssal plains in the northern Pacific region; it has resulted from extreme leveling by accumulated sediment. The pattern is similar in the southern and south- western parts of the Gulf of Alaska. Limited investigations in the eastern end of the Aleutian Trench, in depths up to 4,500 meters, indicate that the postglacial sediments there are more than 1.5 meters thick. The postglacial deposits on the slopes of the Shirshov (Olyutorskiy) Ridge and Bowers Bank in Bering Sea range from 1 to 2 meters in thickness.

Thus the major physiographic provinces differ in the average thickness of their postglacial sediments, although wide variations in local thicknesses are found in each province. The data now available prove that sedimentation was quite heavy in most areas of the Bering Sea and the Gulf of Alaska during postglacial time.

## Rates of Sedimentation

Heusser (1960) believes that the duration of the post-Pleistocene period in Alaska was 9,000 years. Yu. P. Degtiarenko (1961) assumes that the post-Pleistocene period in the vicinity of the Koryak coast along the northwest shore of Bering Sea has lasted about 8,000 years. From these data and from the established values for the thickness of the postglacial deposits, we can estimate roughly the rates of sedimentation during the postglacial period.

The rate of sedimentation on the shelf ranges from 2 to 30 cm/1,000 years. The sedimentation rate varies even more widely on the continental slope and rise, the highest rate being as much as 40 cm/1,000 years. The average rate on the slope is 15 to 20 cm/1,000 years, and on the rise, about 30 cm/1,000 years. These rates of sedimentation would change only slightly if the duration of the post-Pleistocene period were assumed to be 10,000 to 12,000 years rather than 8,000 to 9,000 years. Thus the rates of sedimentation during the postglacial period have been considerably higher almost everywhere in Bering Sea and in the Gulf of Alaska than, for example, in the central Polar Basin (Belov and Lapina, 1961), or than in most other areas of the ocean, for that matter (Sachs, 1950).

Great attention should be focused on the intensity of sedimentation in the vicinity of the continental slope, where terrigenous and biogenic (mainly diatomaceous) sediments accumulate in great quantities. This aspect of the zone of the continental slope seems to be of great importance, and its study can produce information that will help to solve many problems concerning the geology and morphology of the sea floor. Several sections of the continental slope and rise in Bering Sea and the Gulf of Alaska are characterized by the highest rates of sediment accumulation known in the Gulf of Alaska or Bering Sea region. This high rate of sedimentation may be not merely a local phenomenon; it may prove to be a more or less common feature of the pattern of marine sedimentation in open ocean basins.

The data available for shelf areas are inadequate for a comprehensive estimate of the rate of postglacial sedimentation there. It is evident, however, that much of the shelf is characterized by low sedimentation rates, due to the configuration of the bottom and to the action of bottom currents that carry a considerable part of the clastic particles off the shelf. Sediments have accumulated on a large scale only in depressions where tectonic subsidence has been in progress. One must also consider that the shallow-water areas of the shelf were emergent until the postglacial transgression, and that the duration of sedimentation has therefore been shorter in these areas than in areas of deeper water.

## Transition from Late Glacial to Postglacial Sediments

The transition from late glacial to postglacial sediments is recorded within a layer ranging in thickness from a few centimeters to 10 cm. In many cores, the transitional layer is characterized by a coarser granulometric composition, expressed mainly in a high content of silt, sand, or even gravel-sized particles. Evidently, weathered surficial deposits were reworked by waves and currents during the transgression produced by the postglacial eustatic rise in sea level. A concentration of stony material in postglacial shelf sediments in the northwestern part of Bering Sea and in many areas of the Commander-Aleutian Arc is evidently the result of these processes. Transportation of clastic material into various areas in the sea was the cause of the increase in the content of coarse-grained matter in the bottom deposits of the continental slope and the abyssal plains.

## Deposits of Late Glacial Age

Deposits of late glacial age are lithologically more homogeneous than those of the postglacial period. They were probably formed by processes basically similar to those that formed the Holocene glaciomarine sediments in the northern and northeastern parts of the Gulf of Alaska. Glaciers were much more extensive during late glacial time, and extended over parts of the present shelf area, augmenting the total inflow of clastic matter considerably. Thus, terrigenous sediments were accumulating in regions of Bering Sea and the Gulf of Alaska where biogenic (mainly diatomaceous) sediments were also being deposited. Consequently, differences between the late glacial and the postglacial sediments in the various physiographic provinces are expressed principally by differences in their granulometric composition, rather than by differences in the qualitative nature of their material. The effect of volcanic processes continued, of course, and may even have intensified during late glacial times.

We did not observe a high content of pebble-sized particles in the late glacial sediments. Apparently, gravel was not dispersed by ice any more intensively during the last glacial period than during later times within the area of investigation. Gravel accumulation was always localized and associated primarily with strong currents; as at the present time, the composition of the sediments in areas remote from the shore was not appreciably affected by the dispersal of gravel-sized material.

We may assume that the lithology of the late glacial sediments differs from that of the postglacial sediments primarily because of differences in the weathering processes in the continental drainage basins from which the sediments were derived. For example, during cold periods, mechanical weathering increased and chemical weathering declined in importance. Sediment particles reaching the sea were relatively unaltered, and were apparently more resistant to the effects of the marine environment during cold periods than during the warm postglacial epoch and the interglacial epochs. As mentioned above, the amount of terrigenous material introduced into the sea during glacial periods was considerable; moreover, it was distributed over a smaller water area (Hopkins, 1959), the area of Bering Sea, for example, diminishing approximately 30 per cent during the last glacial period. This resulted in an accelerated sedimentation rate and in a slackening in diagenetic changes in the composition of the particles.

Investigations in the Barents Sea and the White Sea have shown large differences in the organic content between late glacial and postglacial deposits. The late glacial deposits are commonly poor in organic matter. A similar pattern has been observed in the Baltic Sea; although the postglacial silty sediments of the Baltic are rich in organic carbon (2 to 4 per cent), glacial drift and varved clay deposits of Late Pleistocene age contain on the average only 0.23 and 0.20 per cent organic carbon, respectively (Gorshkova, 1963). The concentration of organic carbon in late glacial gray clay in the Barents Sea is 0.11 to 0.20 per cent, several times lower than in greenish-gray postglacial deposits of the same granulometry (Klenova, 1948).

There is no difference in organic content between sediments of Bering Sea and sediments in the Gulf of Alaska. Late glacial sediments are generally characterized by a lower concentration of organic carbon than are postglacial sediments, but the differences, as a rule, are expressed in tenths of a per cent. This decrease corresponds in a general way to the decrease in organic carbon content that distinguishes the contemporary glaciomarine deposits in the northern and northeastern parts of the Gulf of Alaska from other granulometrically similar deposits. For example, clayey Holocene sediments in Bering Sea have an average organic carbon content of about 1.25 per cent, and those in the Gulf of Alaska about 0.90 per cent, whereas the glaciomarine deposits contain about 0.83 per cent organic carbon. According to our data, the concentration of organic carbon in late glacial clay-silt in Bering Sea ranges from about 0.50 to 0.70 per cent. Similar values have been established for the Gulf of Alaska.

The observed differences in organic content between Bering Sea and the Gulf of Alaska, on the one hand, and the Barents, White, and Baltic Seas, on the other, can be explained in two ways. First, continental glaciation was not as extensive in northeastern Asia and Alaska as in northern Europe (Flint, 1957; Sachs *et al.*, 1955), and this could not fail to influence sedimentation conditions and the accumulation of organic matter in bottom sediments. Second, biological processes and the hydrological and hydrochemical factors that caused them developed in such a way that organic matter was produced in relatively great quantities in certain regions; because this organic matter was then buried in the sediments, comparatively high contents of organic carbon have resulted. It may be suggested that upwelling, which enhances biological productivity, was intensive enough in Bering Sea and the Gulf of Alaska during the late glacial interval to overcome the inhospitable effects of general cooling. General changes in climatic conditions have of course influenced the abundance and species composition of the plankton population, the main source of the organic matter contained in bottom sediments. Analyses of the diatom content of the sea and ocean sediments by A. P. Jousé (1962) and the establishment of stratigraphic divisions based on these analyses have contributed greatly to the elucidation of this question.

## REFERENCES

Belov, N. A., and N. N. Lapina. 1961. Donnye otlozheniia Arkticheskogo basseina (Bottom deposits of the Arctic Basin) : Leningrad, Izd-vo. Morsk. Transport, 152 p.

Bezrukov, P. L., and A. P. Lisitsin. 1961. Osnovnye cherty osadkoobrasovaniia v dal'nevostochnykh moriakh v chetvertichnoe vremia (Main features of the process of the sedimentation in the Far Eastern Seas during Quaternary time) : Akad. Nauk SSSR, Materialy Vsesoyusnogo soveshchaniia po isucheniiu chetvertichnogo perioda, t. 1, Moscow.

Degtiarenko, Yu. P. 1961. Drevnee oledenenie Koriakskoi gornoi sistemy (Ancient glaciation of the Koriak Mountain system) : Vsesoiuzn. Nauchno-Issled. Geol. Inst. (VSEGEI), Tr. 64, Leningrad, p. 135–140.

Flint, R. F. 1957. Glacial and Pleistocene Geology: Wiley, New York, 553 p.

Gershanovich, D. E. 1962. Novye dannye o sovremennykh otlozheniiakh Beringova Moria (New data on recent deposits in the Bering Sea) : Vsesoiuzn. Nauchno-Issled. Inst. Morskogo Rybnogo Khosiaistva i Okeanografii (VNIRO), Tr. 46, p. 128–164.

Gershanovich, D. E. 1963a. Rel'ef osnovnykh rybopromyslovykh raionov (Shel'f, materikovyi sklon) i nekotorye cherty geomorbologii Beringova Moria (Topography of the major fishing areas [the shelf, the continental slope] and some geomorphological features of the Bering Sea): Vsesoiuzn. Nauchno-Issled. Inst. Morskogo Rybnogo Khosiaistva i Okeanografii (VNIRO), Tr. 48, p. 13–76. In Russian; English summary translation available from U.S. Dept. Comm. Clearinghouse for Fed. Sci. Tech. Info.

Gershanovich, D. E. 1963b. Shel'fovye otlozheniia Zaliva Aliaska i usloviia ikh obrazovaniia (Shelf deposits in the Gulf of Alaska and the conditions of their formation): p. 32–38 *in* Del'tovye i melkovodnomorskie otlozheniia (Delta and shallow-water sea deposits): Akad. Nauk SSSR, Komm. po Osadochnym Porodam, Reports to VI Intern. Sedimentological Congress, Moscow.

Gorshkova, T. I. 1963. Donnye osadki Baltiiskogo moria i ego zalivov (Bottom sediments in the Baltic Sea and its bays): p. 14–21 *in* Del'tovye i melkovod-nomorskie otlozheniia (Delta and shallow-water sea deposits): Akad. Nauk SSSR, Komm. po Osadochnym Porodam, Reports to VI Intern. Sedimento-logical Congress, Moscow.

Heusser, C. J. 1960. Late-Pleistocene environments of North Pacific North America: Am. Geogr. Soc. Spec. Pub. 35, 308 p.

Hopkins, D. M. 1959. Cenozoic history of the Bering Land Bridge: Science, v. 129, p. 1519–1528.

Jousé, A. P. 1962. Stratigraficheskie i paleogeograficheskie issledovaniia v severo-zapadnoi chasti Tikhogo Okeana (Stratigraphic and paleogeographic investigations in the northwestern part of the Pacific Ocean): Akad. Nauk SSSR, Inst. Okeanologii, Moscow, 260 p.

Klenova, M. V. 1948. Geologiia moria (Marine geology): Uchpedgiz, Moscow, 493 p.

Menard, H. W. 1953. Pleistocene and Recent sediment from the floor of the northeastern Pacific Ocean: Geol. Soc. America Bull., v. 64, p. 1279–1294.

Sachs, V. N. 1950. O skorosti nakopleniia sovremennykh morskikh osadkov (The rate of accumulation of Recent marine sediments): Priroda, no. 6, p. 24–33.

Sachs, V. N., N. A. Belov, and N. N. Lapina. 1955. Sovremennoe predstavlenie o geologii tsentralnoi Arktiki (Our present concepts of the geology of the Central Arctic): Priroda, no. 7, p. 13–22.

Saidova, H. M. 1961. Ekologiia foraminifer i paleogeografiia dal'nevostochnykh morei SSSR i severo-zapadnoi chasti Tikhogo Okeana (Ecology of foramin-ifera and paleogeography of the Far East Seas of the USSR and the north-western part of the Pacific Ocean): Akad. Nauk SSSR, Inst. Okeanologii, 232 p.

Skorniakova, N. S. 1961. Donnye otlozheniia severo-vostochnoi chasti Tikhogo Okeana (Bottom deposits in the northeastern part of the Pacific Ocean): Akad. Nauk SSSR, Inst. Okeanologii, Tr. 45, p. 22–64.

# 4. Quaternary Marine Transgressions in Alaska

DAVID M. HOPKINS
*U.S. Geological Survey, Menlo Park, California*

The Quaternary Period has been a time of oscillating sea level due to the intermittent storage of large volumes of water on the continents in the form of ice caps and large glaciers. Sea level has fallen as much as 100–150 meters below its present level during glacial intervals, and has risen at least 20 meters, and perhaps as much as 100 meters, above its present position when the continental glaciers have melted almost completely away. The history of these fluctuations holds special interest for scientists concerned with the history of Beringia, because a reduction in sea level of less than 40 meters would result in a continuous land connection between Alaska and Siberia, and a reduction of 100 meters would create a land connection nearly as wide as the present-day north-to-south extent of Alaska.

These oscillations in sea level have left their marks upon the coasts of Alaska and Siberia, and probably upon the submerged continental shelf as well, in the form of flights of terraces, abandoned wave-cut cliffs, old beach ridges, and sheets of fossiliferous marine sediments or submarine pillow lavas separated from one another by unconformities, weathering profiles, and subaerial deposits. Because these eustatic sea-level transgressions and regressions affected wide areas simultaneously, they provide a convenient basis for subdividing Quaternary marine deposits.

Eventually it will be possible, no doubt, to relate these sea-level events and their deposits to the classical glacial sequences of western Europe and eastern North America. At present, however, the ages of the Alaskan marine transgressions are known more precisely than are the ages of any except the latest of the classical Pleistocene glaciations. Furthermore, there are more high-sea-level events recorded in Alaska than can be fitted into any classical glacial-interglacial sequence; the number of high-sea-level episodes recorded in Alaska accords more nearly with the large number of glacial

Publication authorized by the Director, U.S. Geological Survey.

and interglacial events recorded in Iceland (Einarsson *et al.*, this volume). Consequently, it has seemed advisable to christen with provincial names the marine transgressions recognized in Alaska, so that they can be discussed without prejudice to their possible relationship to the Quaternary climatic events recognized thus far in other parts of the world.

In a paper written in 1963 and published in Russian in 1965, I defined and described the Beringian, Anvilian, Kotzebuan, Pelukian, Woronzofian, and Krusensternian marine transgressions (Table 1) and presented the evidence then available concerning their radiometric ages and correlation. Continuing studies in 1964 and 1965 have refined the geochronometric dating and have shown that the deposits ascribed to the Beringian trans-gression accumulated over a long time span and probably during at least two distinct episodes of high sea level. In addition, a much later transgres-sion, herein named the Einahnuhtan transgression, has been recognized. Some of these more recent interpretations are summarized in Hopkins *et al.* (1965) and Péwé *et al.* (1965), but the basic evidence for, and definitions of, the Alaskan marine transgression sequence have remained unpublished in English.

This paper defines and describes, for English-speaking readers, the Alas-kan marine transgression sequence and summarizes the data available in June 1966 that bear on the age and character of the individual transgres-sions. Knowledge of the Quaternary history of Alaska and its neighboring seas is advancing rapidly, and the scheme put forth here, like the earlier schemes, must therefore be regarded as a working hypothesis subject to con-tinuing refinement and change. Knowledge of the timing of the Anvilian transgression and of its interrelationships with the preceding Beringian and succeeding Einahnuhtan transgressions remains incomplete; and, as Karl-strom has noted in a paper now in press, correlations continue to be un-certain between the early/middle Pleistocene marine deposits of western Alaska and corresponding deposits along the Pacific Coast of Alaska.

The events described in this report as "transgressions" represent epi-sodes when relative sea level rose high enough to flood the Bering Strait; they were separated by episodes when sea level fell enough to restore the land connection between Siberia and Alaska. Although most of the Alaskan transgressions represent the local expressions of worldwide eustatic rises in sea level in response to the melting of continental glaciers, the Beringian transgression in western Alaska probably represents a local flooding due to the subsidence of a peneplain on the present site of the continental shelves of Bering and Chukchi Seas (Hopkins, 1959). The Beringian transgression evidently began during Pliocene time, but it was not terminated by a eustatic reduction in sea level until well after the Quaternary Period had begun.

The deposits of the Quaternary marine transgressions are separated in most emergent areas in Alaska by deposits indicating clearly the existence of a subaerial environment during intervals of lowered sea level, but it is obvious that deep-water areas must have continued to accumulate marine sediments during the regressive intervals. Thus, the late Tertiary and early Quaternary geosynclinal sequence on Middleton Island in southern Alaska contains a continuous marine sedimentary record that probably spans several of the marine transgressions described here, as well as the intervening regressive intervals. Similar records of continuous marine sedimentation are likely to be found elsewhere in southern Alaska.

Radiometric dating has played an important role in establishing the ages and correlations of the Alaskan marine transgressions. Of the several types of radiometric data discussed, I place greatest confidence in the radiocarbon age determinations and in the potassium-argon age determinations for the older rocks. The potassium-argon age determinations for rocks younger than 500,000 years suffer from relatively large uncertainties, but nevertheless provide some useful data. Less confidence can be accorded the determinations based on the uranium-decay products in fossil shells. Most of the determinations cited here were pioneer efforts; some did not include determinations of $Ra_{226}/Th_{230}$ ratios, and only one included determinations of the concentrations of $Th_{232}$ and $U_{234}$. These additional determinations are necessary to indicate whether or not the basic assumptions of the $Th_{230}/U_{238}$ method of age determination are fulfilled (Broecker, 1965).

The discussion of Quaternary molluscan faunas is based upon identifications by F. S. MacNeil (1957, 1965; MacNeil *et al.*, 1943; and unpublished data), and that of the foraminifer and ostracode faunas upon determinations by J. A. Cushman (1941), Ruth Todd (Cushman and Todd, 1947; and unpublished data), Patsy B. Smith (1963; and unpublished data), Ruth A. M. Schmidt (1963; and unpublished data), and R. W. Faas (1962a, 1962b, 1964). Most of the pollen floras were identified by Estella B. Leopold (Wolfe and Leopold, this volume; and unpublished data).

Of the localities discussed in this paper (Fig. 1), I have studied those at Kivalina, around the shores of Kotzebue Sound and Seward Peninsula, on the Pribilof Islands, in Cook Inlet, and on Middleton Island. Stratigraphic and paleontologic data for some of the other localities have been taken from unpublished studies by J. B. O'Sullivan (Arctic coastal plain); P. V. Sellman and Jerry Brown (Point Barrow); W. W. Patton (Unalak-

TABLE 1. *Quaternary Marine Transgressions Recorded on Alaskan Coasts*

| Transgression | Type Locality | Altitude of Shoreline | Climate as Compared with the Present | Archaeological or Radiometric Dating | Correlation North America | Correlation Europe |
|---|---|---|---|---|---|---|
| Krusensternian | Recent beach ridges at Cape Krusenstern | Within 2 meters of present sea level for deposits <4,000 yrs. old | Same | <5,000 yrs. at Cape Krusenstern; up to 10,000 yrs. for terraces along Gulf of Alaska coast | Late Wisconsin and Recent | Late Würm and Recent |
| Woronzofian | Bootlegger Cove Clay near Point Woronzof, Anchorage area (Miller and Dobrovolny, 1959) | Probably a few meters below present sea level | Water colder Air colder | <48,000 yrs.; >25,000 yrs. | Middle Wisconsin interstade | Middle Würm interstade |
| Pelukian | Second Beach at Nome (Hopkins et al., 1960) | Two distinct high-sea-level stands at +7–10 meters | Water warmer Air slightly warmer | Ca. 100,000 yrs. | Sangamon Interglaciation | Broerup Interstade (?) and Riss-Würm Interglaciation |
| Kotzebuan | Marine beds below Illinoian drift along eastern shore of Kotzebue Sound (McCulloch et al., 1965) | Probably ca. +20 meters | Water same Air unknown | 170,000 yrs.; 175,000 yrs. | Pre-Illinoian interglaciation | Mindel-Riss Interglaciation |

TABLE 1 (*Continued*)

| Transgression | Type Locality | Altitude of Shoreline | Climate as Compared with the Present | Archaeological or Radiometric Dating | Correlation North America | Correlation Europe |
|---|---|---|---|---|---|---|
| Einahnuhtan | Einahnuhto Bluffs, St. Paul Is. (Cox et al., 1966) | Probably *ca.* +20 meters | Water same Air unknown | <300,000 yrs.; >100,000 yrs. | | Pre-Mindel interglaciation |
| Anvilian | Third Beach–Intermediate Beach at Nome (Hopkins et al., 1960) | Probably much higher than Kotzebuan and Einahnuhtan: <+100 meters; >+20 meters | Water warmer Air warmer | Probably <1,900,000; >700,000 | Middle Pleistocene interglaciation | |
| Beringian | Submarine Beach at Nome (Hopkins, et al., 1960) | Two distinct episodes during which sea level was higher than at present but probably lower than Anvilian sea level | Water much warmer Air much warmer | Last episode *ca.* 2,200,000 yrs. on St. George Is. | Late Pliocene and early Pleistocene | |

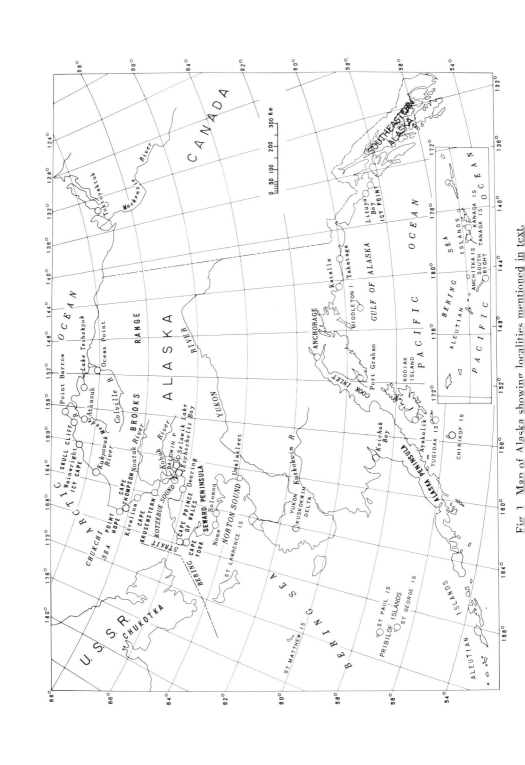

Fig. 1. Map of Alaska showing localities mentioned in text.

leet) ; W. H. Condon and J. M. Hoare (Yukon Delta) ; P. A. Colinvaux (St. Lawrence Island) ; G. W. Moore (Tugidak Island and western Kodiak Island) ; and G. Plafker (Middleton Island and coast of the Gulf of Alaska). I thank all of these colleagues for making their data available to me.

## The Beringian Transgression

The so-called "Beringian transgression" (Hopkins, 1965) was the episode of submergence that brought Bering and Chukchi Seas into existence in approximately their present forms. Thus, deposits of the Beringian transgression include the oldest marine sediments recognized along Alaskan coasts that adjoin the continental shelves of the Bering and Chukchi Seas. The Beringian transgression was complex and included several high-sea-level episodes. The earliest of these evidently took place late in the Pliocene, but the latest was in progress about 2.1 million years ago, well after the beginning of the Pleistocene.

*Type locality at Nome.* The type locality for the Beringian transgression is designated as the area between the Snake River and Bering Sea that is included in the Nome B-1 topographic quadrangle (1:63,360, U.S. Geol. Survey, 1950). Here the Beringian transgression is represented by gold-bearing and richly fossiliferous marine sediments known locally as "Inner Submarine Beach" and "Outer Submarine Beach." The transgression is named after Bering Sea.

The deposits of the Beringian transgression at the type locality near Nome were described as "marine sand and clay of Submarine Beach" by Hopkins *et al.* (1960, p. 46–47). They consist of a few meters of sand and shell-rich clay resting on an undulating bedrock surface and terminating inland at the edge of a buried canyon beneath the Snake River (Fig. 2). Bedrock is appreciably higher on the north side of the Snake River channel; evidently the shore cliff that marked the innermost extent of the Beringian transgression lay beneath the present course of the Snake River at an altitude about equal to present sea level. The Beringian deposits at Nome are covered by till of the Iron Creek Glaciation of early Pleistocene age.

*Beringian beds on St. George Island.* The Beringian transgression is represented on the east coast of St. George Island by a complex sequence of tuffaceous and detrital marine beds, pillow breccias, and basaltic lava flows (Fig. 3). This sequence is discussed in detail below, because it provides, better than any other sequence studied thus far, evidence concerning the age and duration of the Beringian transgression. That it was a protracted and complex event is indicated by the presence of a geomagnetic polarity

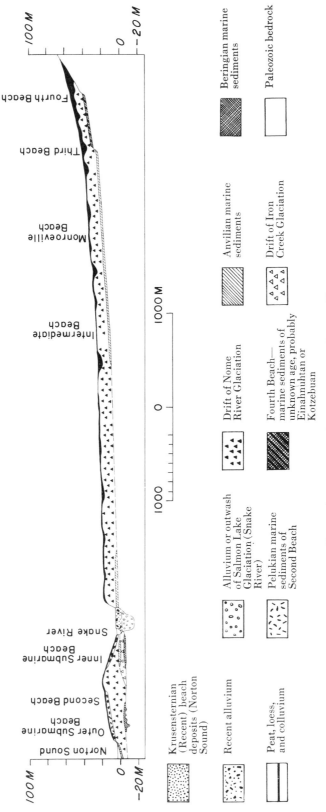

Fig. 2. Cross section through coastal plain at Nome.

Krusensternian
(Recent) beach
deposits (Norton
Sound)

Recent alluvium

Peat, loess,
and colluvium

Alluvium or outwash
of Salmon Lake
Glaciation (Snake
River)

Pelukian marine
sediments of
Second Beach

Drift of Nome
River Glaciation

Fourth Beach—
marine sediments of
unknown age, probably
Einahnuhtan or
Kotzebuan

Anvilian marine
sediments

Drift of Iron
Creek Glaciation

Beringian marine
sediments

Paleozoic bedrock

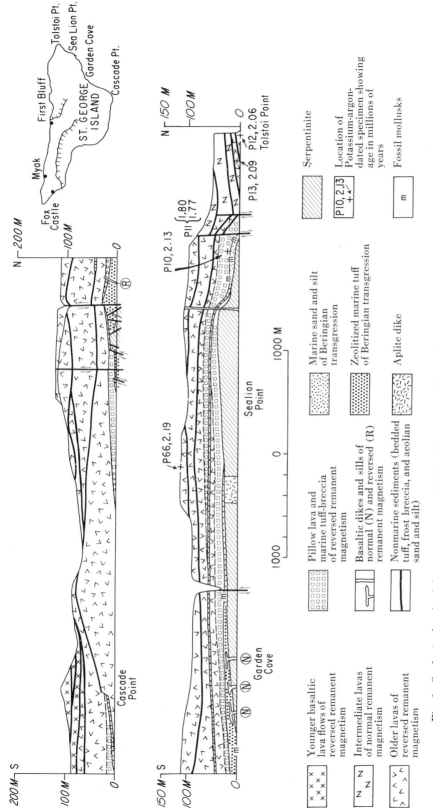

Fig. 3. Geologic sketch of the eastern coastal cliffs of St. George Island. Late Quaternary drift and Pelukian and Kotzebuan marine sediments at Garden Cove and Sea Lion Point are omitted. Hachured line on location map shows 90-meter beach of possible Anvilian age.

change[a] recorded by dikes and intercalated lava flows, by the presence of two unconformities, and by an interbedded series of nonmarine basaltic lava flows.

The Beringian beds on St. George Island rest on a wave-planed basement of serpentinite intruded by an aplite dike (Barth, 1956). In most places, the basement is mantled by a layer of zeolitized tuff that commonly shows foreset bedding and that locally reaches a thickness in excess of 30 meters. The tuff grades laterally into marine sand, and in one place is separated from the serpentinite basement by a few meters of hard, shell-bearing mudstone containing well-rounded cobbles and boulders. The marine tuff contains volcanic bombs and small basaltic dikes and sills having normal remanent magnetism. I have not found identifiable fossils in these beds, but they may be the source of a small collection made by Barth (1956, p. 119) at Garden Cove.

The marine tuff is overlaid unconformably by an eruptive sequence consisting of a basal foreset-bedded marine pillow breccia, locally more than 20 meters thick, and an overlying pile of rather thin nonmarine basaltic lava flows that reaches a total thickness exceeding 80 meters near the site of the former volcanic vent. The pillow lava and the overlying lava flows are of reversed remanent magnetism.

Another layer of fossiliferous littoral sand and gravel, locally as much as 18 meters thick, rests upon a wave-cut surface beveled across part of the pile of thin-bedded lavas and locally across the underlying serpentinite basement. This layer contains a Beringian fauna in exposures near Tolstoi Point. The fossiliferous beds are covered by a thick lava flow of reversed magnetism, the lower part of which consists of marine pillow lava. Higher in the cliffs and in other parts of St. George Island, the lava flows of reversed magnetism associated with the Beringian beds are overlain by younger lavas of normal magnetism, and these by still younger lavas of reversed magnetism (Cox et al., 1966).

The basal marine tuff of the Beringian sequence is cut by faults that are truncated by an erosion surface at the base of the lower pillow breccia, indicating that an episode of deformation followed the accumulation of the tuff. The boundary between pillow lava and nonmarine basaltic lava marks the position of the water-air interface at the time the lava complex was erupted; this boundary in the lower lava complex arches from below sea level at Cascade Point to +60 meters near Sea Lion Point, indicating renewed crustal warping after the pillow lava accumulated. A new rise in sea level,

---

[a] The sequence of Quaternary geomagnetic polarity epochs and events is described and illustrated elsewhere in this volume by Einarsson et al.

followed again by uplift and faulting, is indicated by the fact that the higher marine layer in the Beringian sequence rests on a tilted wave-planed surface beveled across nonmarine basaltic lavas covering the lower marine beds.

The pillow lava covering the youngest part of the Beringian sediments near Tolstoi Point has a potassium-argon age of 2.13 ± 0.06 million years, and a flow slightly higher in the sequence has a potassium-argon age of 2.19 ± 0.10 million years. (P-10 and P-66, Fig. 3) (Cox *et al.*, 1966). These flows were erupted during the early part of the Matuyama reversed-geomagnetic-polarity epoch. The marine tuff comprising the lower part of the Beringian sequence and the dikes and sills that intrude the tuff evidently date from an earlier interval of normal geomagnetic polarity. These lower beds can be no younger than the Gauss normal-geomagnetic-polarity epoch, which lasted from 3.35 to 2.4 million years ago. They may be much older.

*Beringian beds on Middleton Island.* Correlations are difficult between Pliocene and early and middle Pleistocene marine beds along the coast of the Gulf of Alaska and those of Bering Sea, because there are and evidently always have been pronounced differences between contemporary molluscan faunas in the two areas. Because glaciomarine sediments are a major component, one is tempted to ascribe a relatively young age to the upper part of the folded late Cenozoic sequence that is exposed in many places along the coast of the Gulf of Alaska from Icy Point to Tugidak Island. However, some of the glaciomarine beds are associated with molluscan faunas that unambiguously indicate ages ranging down into the early Pliocene and possibly into the late Miocene (Miller, 1957; MacNeil *et al.*, 1961).

The youngest part of this folded late Cenozoic sequence seems to be represented by the 1,500 meters of beds exposed on Middleton Island. The Middleton Island section was partly described by D. J. Miller in 1953, but additional beds became exposed after an uplift of the adjoining reefs during the great Alaskan earthquake of 1964. The lower part of the Middleton Island section contains molluscan faunas that MacNeil (in press) considers to be of late Pliocene age; he suggests that the Pliocene–Pleistocene boundary lies within stratigraphic unit 17 of Miller (1953) at the level of the highest occurrence of *Chlamys (Leochlamys) tugidakensis* and the lowest occurrence of *C. (Chlamys) plafkeri* (MacNeil, in press). These Pliocene beds and the overlying early Pleistocene beds are probably correlative with the Beringian transgression. Higher in the section are beds, discussed later, that contain molluscan faunas suggestive of the Anvilian transgression; the uppermost beds may possibly be as young as the Einahnuhtan transgression.

*Other localities.* Other bodies of marine sediments assigned to the Beringian transgression on the basis of their marine faunas and stratigraphic relationships are found near Ocean Point on the lower Colville River, possibly at Skull Cliff near Point Barrow, along the lower course of the Kivalina River, and near the town of Solomon on the Seward Peninsula. Faunas associated with volcanic rocks on Tanaga Island and Kanaga Island in the Aleutian chain suggest that these rocks were also emplaced during the Beringian transgression, and Beringian beds are evidently represented in the folded marine clastic rocks on Tugidak Island and in the upper part of the Yakataga Formation near Lituya Bay and in the Chaix Hills near Yakataga.

The Beringian beds at Ocean Point constitute the basal part of the Gubik Formation (Schrader, 1904) in that area; they consist of richly fossiliferous cross-bedded sand and sandy gravel that in places is more than 15 meters thick (O'Sullivan, 1961). MacNeil (1957) describes and figures the molluscan fauna of these beds. The Beringian beds rest on bedrock of Cretaceous age and are covered by about 10 meters of gravel and sand that probably was deposited during the Anvilian transgression. The Beringian beds at Ocean Point wedge out 40 km inland from the present coast at an altitude of about 30 meters; farther inland, the deposits of the Anvilian transgression rest directly on bedrock.

The fossiliferous marine beds exposed at Skull Cliff have been visited repeatedly by fossil collectors, but the stratigraphy has never been studied in detail. The faunas (Meek, 1923; MacNeil, 1957) indicate that at least part of the sequence was deposited during the Anvilian transgression. However, two stratigraphic units are present, separated by an unconformity (O'Sullivan, 1961), suggesting the possibility that Beringian beds may also be present.

The Beringian beds at Kivalina have been described by Hopkins and MacNeil (1960) and by McCulloch (this volume). The deposits there stand about 10 meters above present sea level.

At Solomon the Beringian beds are not exposed at the surface, but they were penetrated by mine shafts in the delta of the Solomon River about 1.8 km east of Solomon village. The reported depths of the shafts at Solomon indicate that the Beringian beds lie slightly below present sea level.

The beds tentatively assigned to the Beringian transgression on Kanaga and Tanaga Islands consist of the youngest part of a gently folded sequence of marine volcanic rocks that ranges in age from Miocene to early Pleistocene (Fraser and Barnett, 1959, p. 217–224). Beds of Beringian age are probably included in the highest part of the thick Pliocene sequence on

Tugidak Island (MacNeil *et al.*, 1961); an early Pleistocene molluscan fauna in beach boulders found by G. W. Moore on nearby Chirikof Island (F. S. MacNeil, written communication, January 21, 1963) suggests that Beringian beds may be present there, as well. Faunas indicative of a late Pliocene and early Pleistocene age are found high in the Yakataga Formation in the Chaix Hills near Yakataga and near LaPerouse Glacier between Lituya Bay and Icy Point (George Plafker, unpublished data). These beds are tentatively correlated with the Beringian transgression.

*Fauna and flora.* The most significant marine fossils found thus far in deposits of the Beringian transgression are listed in column 1 of Table 2. The Beringian faunas "have a decidedly modern aspect, and most of the species are comparable to modern species; nevertheless, in more than half of them there are subtle differences that will have to be recognized when the faunas are formally described. The pectinid assemblages, especially, are unlike any modern ones, and most of the *Neptuneas* and *Astartes* are unlike any recent variants" (F. S. MacNeil, written communication, 1962).

Among the extinct forms, some are, as far as is known at present, unique to deposits of Beringian age and to the Alaskan areas. Several others occur in older Tertiary beds in Alaska and elsewhere, but not in younger Pleistocene beds. The Beringian (?) beds on Tugidak Island contain pectinid assemblages that most nearly resemble those of the late Pliocene beds in Japan, and pectinids in the middle part of the Middleton Island sequence are most closely related to pectinids known from beds of Pliocene or early Pleistocene age in California. *Neptuneas* in these Middleton Island beds are similar to and evidently closely related to those found in the "Cardium" zone (latest Pliocene or earliest Pleistocene) at Tjörnes, Iceland (Einarsson *et al.*, this volume). *Fortipecten hallae* (Dall), found at Solomon and Kivalina, belongs to a genus known outside of Alaska only in beds considered to be of Pliocene age in Kamchatka, Sakhalin, and Japan. *Yoldia koluntunensis* (Slodkewitsch), found at Nome, is known only from the Kavran Series of Kamchatka, also considered to be of Pliocene age.

Several living forms that are common in younger beds make their first recorded appearances on the shores of Alaska in beds of Beringian age (Table 2, column 1).

At least nine molluscan species found in Beringian beds at Nome are now confined to areas south of the winter limit of sea ice, and four species found at Ocean Point no longer range north of Bering Strait. These range changes suggest that Bering and Chukchi Seas were warmer than at present and that winter sea ice did not extend as far south as Bering Sea during at least the earlier parts of Beringian time. However, the fauna of late Ber-

TABLE 2. *Stratigraphic Distribution of the More Significant Fossils Found in Marine Beds of Pleistocene Age in Alaska*

*Column headings:* (1) Beringian, (2) Anvilian, (3) Einahnuhtan, (4) Kotzebuan, (5) Pelukian, (6) Woronzofian, (7) living in Alaskan waters at present time. *Symbols:* Tapered ends of line indicate first or last recorded appearance in Alaska; solid lines indicate presence of animal firmly established; queried lines indicate questionable identifications or questionable age assignment for enclosing beds; dashed lines indicate animal probably present but has not been collected. *Notes:* E—animal extinct; EC—animal extinct but closely related to a living form; A—nearest relative now confined to North Atlantic; J—nearest relative now confined to Japanese or Eastern Siberian waters.

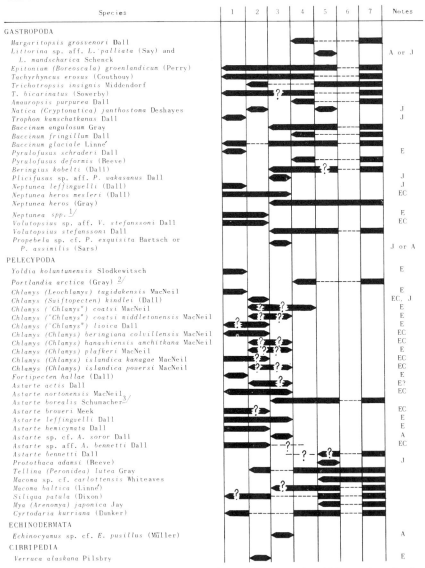

| Species | 1 | 2 | 3 | 4 | 5 | 6 | 7 | Notes |
|---|---|---|---|---|---|---|---|---|
| **GASTROPODA** | | | | | | | | |
| *Margaritopsis grosvenori* Dall | | | | | | | | |
| *Littorina* sp. aff. *L. palliata* (Say) and *L. mandschurica* Schenck | | | | | | | | A or J |
| *Epitonium (Boreoscala) groenlandicum* (Perry) | | | | | | | | |
| *Tachyrhyncus erosus* (Couthouy) | | | | | | | | |
| *Trichotropsis insignis* Middendorf | | | | | | | | |
| *T. bicarinatus* (Sowerby) | | | | | | | | |
| *Amauropsis purpurea* Dall | | | | | | | | |
| *Natica (Cryptonatica) janthostoma* Deshayes | | | | | | | | J |
| *Trophon kamschatkanus* Dall | | | | | | | | J |
| *Buccinum angulosum* Gray | | | | | | | | |
| *Buccinum fringillum* Dall | | | | | | | | |
| *Buccinum glaciale* Linné | | | | | | | | |
| *Pyrulofusus schraderi* Dall | | | | | | | | E |
| *Pyrulofusus deformis* (Reeve) | | | | | | | | |
| *Beringius kobelti* (Dall) | | | | | | | | |
| *Plicifusus* sp. aff. *P. wakasanus* Dall | | | | | | | | J |
| *Neptunea leffingwelli* (Dall) | | | | | | | | J |
| *Neptunea heros mesleri* (Dall) | | | | | | | | EC |
| *Neptunea heros* (Gray) | | | | | | | | |
| *Neptunea* spp. [1] | | | | | | | | E |
| *Volutopsius* sp. aff. *V. stefanssoni* Dall | | | | | | | | EC |
| *Volutopsius stefanssoni* Dall | | | | | | | | |
| *Propebela* sp. cf. *P. exquisita* Bartsch or *P. assimilis* (Sars) | | | | | | | | J or A |
| **PELECYPODA** | | | | | | | | |
| *Yoldia koluntunensis* Slodkewitsch | | | | | | | | E |
| *Portlandia arctica* (Gray) [2] | | | | | | | | E |
| *Chlamys (Leochlamys) tugidakensis* MacNeil | | | | | | | | EC, J |
| *Chlamys (Swiftopecten) kindlei* (Dall) | | | | | | | | E |
| *Chlamys ("Chlamys") coatsi* MacNeil | | | | | | | | E |
| *Chlamys ("Chlamys") coatsi middletonensis* MacNeil | | | | | | | | E |
| *Chlamys ("Chlamys") lioica* Dall | | | | | | | | EC |
| *Chlamys (Chlamys) beringiana colvillensis* MacNeil | | | | | | | | EC |
| *Chlamys (Chlamys) hanashiensis amchitkana* MacNeil | | | | | | | | E |
| *Chlamys (Chlamys) plafkeri* MacNeil | | | | | | | | EC |
| *Chlamys (Chlamys) islandica kanagae* MacNeil | | | | | | | | EC |
| *Chlamys (Chlamys) islandica powersi* MacNeil | | | | | | | | E |
| *Fortipecten hallae* (Dall) | | | | | | | | E? |
| *Astarte actis* Dall | | | | | | | | EC |
| *Astarte nortonensis* MacNeil [3] | | | | | | | | |
| *Astarte borealis* Schumacher | | | | | | | | |
| *Astarte broweri* Meek | | | | | | | | EC |
| *Astarte leffingwelli* Dall | | | | | | | | E |
| *Astarte hemicymata* Dall | | | | | | | | E |
| *Astarte* sp. cf. *A. soror* Dall | | | | | | | | A |
| *Astarte* sp. aff. *A. bennetti* Dall | | | | | | | | EC |
| *Astarte bennetti* Dall | | | | | | | | |
| *Protothaca adamsi* (Reeve) | | | | | | | | J |
| *Tellina (Peronidea) lutea* Gray | | | | | | | | |
| *Macoma* sp. cf. *carlottensis* Whiteaves | | | | | | | | |
| *Macoma baltica* (Linné) | | | | | | | | |
| *Siliqua patula* (Dixon) | | | | | | | | |
| *Mya (Arenomya) japonica* Jay | | | | | | | | |
| *Cyrtodaria kurriana* (Dunker) | | | | | | | | |
| **ECHINODERMATA** | | | | | | | | |
| *Echinocyamus* sp. cf. *E. pusillus* (Müller) | | | | | | | | A |
| **CIRRIPEDIA** | | | | | | | | |
| *Verruca alaskana* Pilsbry | | | | | | | | E |

[1] Beringian and Anvilian beds contain complexes of *Neptuneas*, in addition to those listed above that are ancestral to but distinct from several modern species.

[2] *Portlandia arctica* has not been found in Alaskan beds older than the Kotzebuan transgression, but it is present in Chukotka in the Pinakul' beds of Einahnuhtan age.

[3] *Astarte borealis* is found throughout the Pleistocene marine sequence of Tjörnes, Iceland (Einarsson *et al.*, this volume, but this stock is represented in the older part of the Alaskan Pleistocene sequence by the closely related *A. nortonensis*.

ingian age on St. George Island consists largely of species that still live in nearby waters, and includes one species, *Trichotropis bicarinatus* (Sowerby), that may no longer range as far south as the Pribilof Islands. Evidently, water temperatures in late Beringian time differed little from present-day temperatures.

Foraminiferal faunas from Beringian beds are richer and more varied than those found in the deposits of later transgressions, but most, if not all, of the species are within their modern geographic ranges.

Fossil pollen and wood floras indicate that during Beringian time *Picea* and *Larix* reached the Arctic coast at Ocean Point (O'Sullivan, 1961) and *Pinus* and *Picea* were present at Kivalina. Pollen grains of *Pinus, Picea, Abies, Tsuga, Larix,* and *Carya* have been identified in Beringian sediments at Nome, but pollen grains diagnostic of early Tertiary beds are present as well, suggesting that the sediments may be contaminated by pollen redeposited from older beds (Wolfe and Leopold, this volume). However, the fossil plants do suggest that the climate of western and northern Alaska was considerably warmer during the Beringian transgression than at present.

*Position of sea level.* The absolute position of sea level during the Beringian transgression cannot be determined, because the Beringian deposits have been deformed by crustal movements in most and probably all of the places where they are known. However, deposits of the Anvilian transgression lie at higher levels and extend farther inland on the Arctic coastal plain, at Nome, and probably at Solomon and on St. George Island. These observations suggest that sea level was not as high during the Beringian transgression as it was during the Anvilian transgression.

*Correlation.* The deposits of the Beringian transgressions are considered to have been laid down during late Pliocene and early Pleistocene time, because their molluscan faunas, especially the *Neptunea* and *Chlamys* complexes, suggest that they were deposited at the same time as beds considered to be of late Pliocene and early Pleistocene age in England, Iceland, California, Japan, Sakhalin, and Kamchatka. Deposition of the Beringian sequence on St. George Island was terminated about 2.1 million years ago, during the early Matuyama geomagnetic polarity epoch, well after the Pleistocene began (Einarsson *et al.*, this volume). The Pliocene–Pleistocene boundary in Alaska evidently lies within the Beringian sequence.

## The Anvilian Transgression

The name "Anvilian transgression" (from Hopkins, 1965) has been chosen to designate a marine transgression that took place after the first glacial advance recorded on the coastal plain in the area of Nome. Deposits of the Anvilian transgression are distinguished from those of the preceding Beringian and the following Einahnuhtan transgressions on the basis of their contained faunas and, in places where all three sets of deposits are present, on the basis of their stratigraphic relationships. Anvilian faunas generally contain appreciable numbers of extinct species, but they resemble modern faunas more closely than do most Beringian faunas. The age of the Anvilian transgression is uncertain, but the transgression probably took place during the later part of the Matuyama geomagnetic polarity epoch, i.e., between 1.9 and 0.7 million years ago.

*Type locality at Nome.* The type locality for the Anvilian transgression is designated as the segment of the Nome coastal plain shown on the Nome B-1 topographic quadrangle (1:63,360, U.S. Geol. Survey, 1950), where the Anvilian transgression is represented by fossiliferous and richly auriferous marine sediments known locally as "Third Beach," "Monroeville Beach," and "Intermediate Beach." The transgression is named after Anvil Creek, along whose middle course the gold placers of Third Beach of Anvilian age were discovered.

The deposits of the Anvilian transgression at Nome (Fig. 2) consist of a discontinuous sheet, 2–3 meters thick, of sandy and gravelly beach and littoral sediments that extends inland 5 km from the present coastline, wedging out against a buried sea cliff at an altitude of 22 meters (Hopkins *et al.*, 1960). The deposits of the Anvilian transgression rest on a smoothly sloping bedrock platform in most places, but they rest on drift of the Iron Creek Glaciation of early Pleistocene age in the area between the lower course of the Snake River and Norton Sound. The Anvilian deposits are covered in most places by drift of the Nome River Glaciation.

*Other localities.* Fossiliferous marine sediments tentatively assigned to the Anvilian transgression are also found at Skull Cliff, at Atkasuk on the Meade River on the Arctic coastal plain, at Black Bluffs and Tolstoi Point on St. Paul Island, near South Bight on Amchitka Island, and in the upper part of the Middleton Island sequence. Old beach lines at altitudes of 90 meters or more on the Arctic coastal plain, near Solomon, and on St. George Island were also probably formed during the Anvilian transgression.

Sediments of the Anvilian transgression probably constitute most of the Gubik Formation in areas of the Arctic coastal plain that lie inland from the highest beaches of the Einahnuhtan transgression (the middle Pleistocene transgression of McCulloch, this volume). The marine portion of the Gubik Formation extends inland about 120 km from the Arctic coast and to altitudes of 100 meters or more. Marine faunas that I consider to be diagnostic of the Anvilian transgression have been obtained much lower and nearer the coast at Atkasuk and Skull Cliff (MacNeil, 1957). Typical measured sections of portions of the Gubik Formation that I consider to be of Anvilian age consist of 1–2 meters of gravel, gravelly clay, or clay, overlain by 3–20 meters of well-bedded sand (Black, 1964; O'Sullivan, 1961). The Anvilian deposits generally rest on Cretaceous bedrock and in most places are covered by several meters of windblown sand, silt probably of aeolian origin, and silty and peaty deposits of thaw lakes. The Anvilian deposits seem to have been removed by marine erosion in most of the areas on the Arctic coastal plain that were submerged during later transgressions (McCulloch, this volume).

The inner edge of the coastal plain at Solomon, 55 km east of Nome, is delimited by a wave-cut cliff and a linear accumulation of wave-rounded quartz gravel lying at an altitude of about 90 meters. No fossils have been found in the beach gravel, but it lies far inland from the Beringian deposits near Solomon village, and the intervening coastal plain area is deeply dissected. It seems likely that the high beach line was carved during the Anvilian transgression rather than during a later interval of high sea level, but if this is true, there has been a differential uplift of some 65 meters in the area between the Anvilian(?) deposits at Solomon and those at Nome.

Although the faunas are not diagnostic, certain fossiliferous deposits on St. Paul Island probably were deposited during the Anvilian transgression.[b] These consist of beds of cross-bedded beach sand grading laterally into fossiliferous silt and unconformably overlain by coarse beach gravel of the Einahnuhtan transgression at Tolstoi Point, on St. Paul Island (Fig. 4). Boulders of calcareous mudstone containing a similar nondiagnostic fauna and also thought to be derived from Anvilian sediments are incorporated in marine volcanic agglomerate of unknown age (mistaken for glacial till by Barth, 1956, p. 106) at Black Bluffs, near the southeast corner of St. Paul Island. The fossiliferous sediments of probable Anvilian age at Tolstoi Point, St. Paul, extend from present sea level to a height of 3 or 4 meters. A pronounced soil profile is developed in the upper part, and narrow fossil frost cracks that extend down into the Anvilian(?) sediments are truncated by the overlying deposits of the Einahnuhtan transgression.

No fossiliferous sediments younger than the Beringian transgression are recognized on St. George Island, but an extensive wave-cut scarp that now

[b] The lower beds at Tolstoi Point, St. Paul, and the fossiliferous boulders at Black Bluffs were tentatively assigned to the Beringian transgression by Hopkins (1965) because the overlying Einahnuhtan beds had not yet been recognized as representing a separate, post-Anvilian transgression.

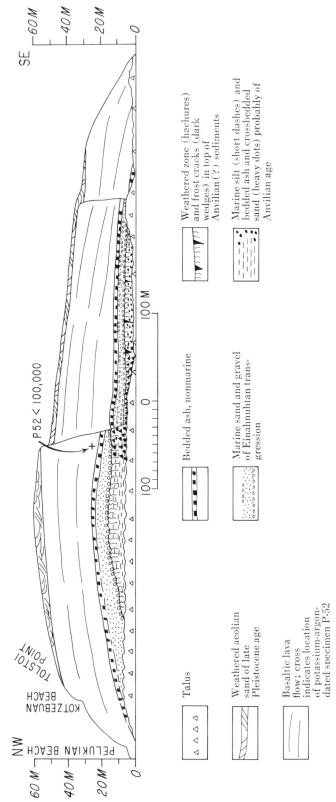

Fig. 4. Sketch of the coastal cliffs at Tolstoi Point, St. Paul Island.

lies at an altitude of 90 meters (inset, Fig. 3) is probably of Anvilian age, because its position in the paleomagnetic sequence indicates clearly that it is older than the Einahnuhtan transgression and because it lies stratigraphically far above the fossiliferous Beringian beds of St. George. The scarp is carved in most places in lava flows of reversed magnetism. At First Bluff, it can be traced into a volcanic complex of reversed remanent magnetism consisting of subaquatic pillow lava and tuff breccia below 90 meters and of subaerial cone agglomerate and basaltic crater fill above. Lava flows and pillow breccias from a volcanic fissure extending from Fox Castle to Myak (insert, Fig. 3) display similar relationships to the 90-meter beach. Evidently these volcanoes were active while the scarp was being cut. The reversely magnetized lava flows into which the 90-meter beach is carved, and the reversely magnetized volcanic rocks erupted from the vents at First Bluff, Fox Castle, and Myak all lie stratigraphically above lava flows of normal remanent magnetism erupted during the Olduvai event (Cox *et al.*, 1966), indicating that the 90-meter beach was carved during the later part of the Matuyama geomagnetic polarity epoch, more than 0.7 and less than 1.9 million years ago.

Anvilian beds are probably included in a terrace-gravel sequence at South Bight on Amchitka Island, described by Powers *et al.* (1960). This Aleutian Island sequence consists of about 22 meters of fossiliferous coarse sand and gravel that terminates inland as an ancient beach deposit at an altitude of 40 meters. The sequence rests unconformably upon older tilted beds of sand and gravel of unknown age and origin, and is covered by a thin mantle of glacial drift. The small molluscan fauna from these beds includes several extinct species, including *Chlamys coatsi* MacNeil, *C. coatsi middletonensis* MacNeil, *C. hanashiensis amchitkana* MacNeil, *Astarte actis* Dall, and *A. leffingwelli* Dall (Powers *et al.*, 1960; MacNeil, in press). *C. coatsi* has turned up in one collection from the Gubik Formation that had been assigned by McCulloch to the Kotzebuan transgression (his "pre(?)-Illinoian" transgression, this volume). *C. coatsi middletonensis* and *C. hanashiensis amchitkana* are found in the upper part of the Middleton Island section; *Astarte actis* is otherwise known only from the type Anvilian deposits at Nome, and *A. leffingwelli* is common in Beringian, Anvilian, and Einahnuhtan beds in western Alaska, but is unknown in younger beds. The South Bight, Amchitka, fauna can hardly be younger than Einahnuhtan; it seems most likely to be of Anvilian age.

Shell-bearing beds high in the Middleton Island section (in stratigraphic unit 20 of Miller, 1953) contain a *Chlamys* assemblage similar to that in the Anvilian(?) beds on Amchitka and sharply different from the *Chlamys* assemblage found in older beds on Middleton Island (MacNeil, in press). *Astarte leffingwelli* and a complex of extinct *Neptuneas* are also present. Efforts have been made to establish the geochronometric age of this part of the Middleton Island sequence by measuring the ratios of the various uranium daughter products contained in fossil shells. In an earlier attempt, two large pectinid shells from slightly different levels were determined to have $Th_{230}/U_{238}$ ratios indicative of ages of 214,000 ± 18,000 and 221,000 ± 17,000 years, respectively, and $Ra_{226}/U_{238}$ ratios indicative of ages of 187,000 ± 15,000 and 210,000 ± 13,000 years, respectively (Blanchard, 1963). A more recent analysis (L-1051G) of a pectinid shell from approximately the same level "indicates only that the age is greater than 200,000 years. Although the $U_{234}/U_{238}$ ratio suggests an age of 300,000 ± 50,000 years, we have found this ratio to be a highly unreliable age indicator in mollusks" (W. Broecker, Lamont Geol. Observatory, Columbia Univ., written communication, May 19, 1966). The character of the faunas indicates that the beds that yielded Blanchard's and Broecker's specimens can be no younger than the Einahnuhtan transgression; they seem more likely to be of Anvilian age.

*Fauna and flora.* The marine faunas found in fossiliferous Anvilian deposits consist largely of forms living in adjacent waters at the present time; several mollusks that are common in younger beds make their first recorded appearance near Alaskan shores in beds of Anvilian age (Table 2, column 2). An appreciable number of extinct forms, chiefly species of *Neptunea*, *Chlamys*, and *Astarte*, are present as well. Most of the extinct forms found in Anvilian beds are also found in deposits of the Beringian transgression, but *Astarte actis* Dall, *Chlamys (Swiftopecten) swifti kindlei* (Dall), and the extinct barnacle *Verruca alaskana* Pilsbry, have been found thus far only in Anvilian beds.

Most of the fossil faunas collected in Anvilian deposits contain a few species whose presence suggests that water temperatures were warmer than at present. *Natica janthostoma* Deshayes, found at Nome and Skull Cliff, is most closely related to species now limited to waters adjoining Japan, Kamchatka, and the Commander Islands. This snail is also common in Pelukian beds, but it has not been found in deposits of the Einahnuhtan and Kotzebuan transgressions. *Chlamys (Swiftopecten) swifti kindlei* (Dall), abundant at Nome, is closely related to a living Japanese species. *Macoma* sp. aff. *M. inquinata* (Deshayes), obtained at Skull Cliff, is most closely related to species now limited to the North Pacific Ocean; and *Pododesmus macroschisma* (Deshayes) and *Pholadidea penita* (Conrad), found in Anvilian beds at Nome, now reach the northern limits of their ranges in southern Bering Sea at the average southern limit of winter sea ice. Winter sea ice probably did not reach Nome during the warmer part of the Anvilian transgression. However, a discovery by J. B. O'Sullivan (1961) of a cluster of erratic boulders near Atkasuk suggests that "ice islands" (large icebergs from Canadian ice shelves) may have reached the Arctic coast.

The northern part of the Gulf of Alaska may have been colder than at present, perhaps because of local cooling by glacial meltwater. The beds believed to be of Anvilian age on Middleton Island contain one species, *Plicifusus arcticus* Philippi, now limited to the Aleutian Islands and waters to the northward, and two others, *P. verkruzeni* (Kobelt) and *Volutopsius* sp. cf. *stefanssoni* (Dall), that no longer range south of Bering Sea.

Sparse paleobotanical data suggest that *Tsuga* may have reached the Arctic coast and that *Picea* and *Pinus* were present in the Nome area during the Anvilian transgression. Evidently, air temperatures in Alaska were warmer than at present.

*Position of sea level.* The position of sea level during the Anvilian transgression cannot be determined with certainty, because the Anvilian shorelines may have been affected by crustal warping in all of the places where they are known. Sea level appears to have risen higher during the Anvilian than during the Beringian transgression. It probably stood at least 20 meters higher than at present and may have been as high as +100 meters.

*Correlation.* The Anvilian transgression took place during an interglacial interval early in middle Pleistocene time. Anvilian beds rest on drift of the Iron Creek Glaciation at Nome, and beds tentatively correlated with the Anvilian transgression on St. Paul Island were disrupted by ice wedges during a later cold interval that preceded the Einahnuhtan transgression. If the 90-meter shoreline on St. George Island is of Anvilian age, then the Anvilian transgression took place during the later part of the Matuyama reversed-geomagnetic-polarity epoch, between 0.7 and 1.9 million years ago.

## The Einahnuhtan Transgression

In formulating the Alaskan transgression sequence in 1963 (Hopkins, 1965) I tentatively correlated certain beds on St. Paul Island and on the Kukpowruk River in northwestern Alaska (Fig. 1) with the Anvilian transgression because, although their faunas had a distinctly modern aspect, they contained a few species that differed slightly from the most closely related modern species. Deposits of the Kotzebuan transgression, to which I was tempted at the time to assign these beds, were not known to contain any

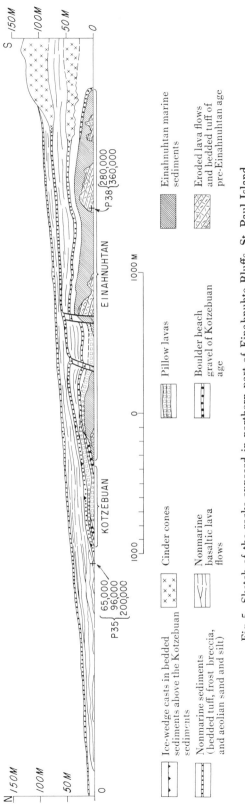

Fig. 5. Sketch of the rocks exposed in northern part of Einahnuhto Bluffs, St. Paul Island. Crosses indicate locations of potassium-argon-dated specimens.

extinct taxa. However, a comparison with the stratigraphic relationships and faunas of a series of Pleistocene marine beds in Chukotka soon indicated that the Alaskan beds in question represent a previously unrecognized interval of high sea level younger than the Anvilian and older than the Kotzebuan transgression (Hopkins *et al.*, 1965). Later field observations have supported this conclusion. The new high-sea-level event is here named the Einahnuhtan transgression.

*Type locality on St. Paul Island.* The name "Einahnuhtan transgression" is applied to a middle Pleistocene marine transgression younger than the Anvilian and older than the Kotzebuan transgressions. The type locality is in the Einahnuhto Bluffs on St. Paul Island, Pribilof Islands, where the Einahnuhtan transgression is represented by a sequence of fossiliferous beach and littoral sediments about 30 meters thick (Fig. 5). Deposits of this transgression bury an ancient sea cliff cut in older basaltic lava flows; a single specimen (P-38) from one of these flows has yielded potassium-argon age determinations of 280,000 and 360,000 years (Cox *et al.*, 1966).

The Einahnuhtan sediments in the Einahnuhto Bluffs are overlain by basaltic lava flows and bedded tuff more than 150 meters thick that record three separate volcanic eruptions widely spaced in time. The lowest volcanic complex includes pillow lavas that rest on the Einahnuhtan beds, thus recording a volcanic eruption near the end of the Einahnuhtan transgression. In the northern part of the bluffs, the Einahnuhtan sediments and the associated pillow lava are truncated by an erosion surface that is covered at low altitudes by shelly boulder gravel representing the Kotzebuan transgression; the marine boulder gravel can be traced laterally into a layer of bedded tuff and aeolian sand and silt that separates two groups of basaltic lava flows. The Kotzebuan sediments are covered, in turn, by a volcanic sequence of Kotzebuan age that includes pillow lava at low altitudes. Still higher in the section is another layer of bedded tuff and windblown sediment traceable from sea level to a point near the top of the bluffs; this layer contains the casts of several ice wedges, and evidently was formed at a time when permafrost was present and winters were extremely cold. A still later eruption is recorded by lava flows that cover the frost-disturbed layer; a single specimen (P-35) of the lowest of these flows has yielded potassium-argon age determinations of 65,000, 96,000, and 200,000 years (Cox *et al.*, 1966). Exposures on the north coast of St. Paul Island show that the highest of the flows in this upper complex is overlain by marine sediments deposited during the Pelukian transgression.

Einahnuhtan beds are also exposed on the south coast of St. Paul Island at Zapadni Point and Tolstoi Point. The Einahnuhtan beds at Zapadni Point lack molluscan fossils, but those at Tolstoi Point contain a molluscan fauna essentially identical to the fauna in the Einahnuhto Bluffs. The Einahnuhtan beds of both Zapadni and Tolstoi Points are covered by younger lava flows. Two wave-cut scarps, probably representing the Kotzebuan and Pelukian transgressions, are carved on the lava flows covering the Einahnuhtan beds on both Tolstoi Point (Fig. 4) and Zapadni Point, and mollusk-bearing sand forms a discontinuous mantle on the basaltic lava below the level of the highest scarps. A specimen (P-52) of the flow covering the Einahnuhtan beds on Tolstoi Point (Fig. 4) has yielded a potassium-argon age determination of less than 100,000 years (Cox *et al.*, 1966); however, the geologic evidence indicates that this age determination is probably younger than it should be, because the lava flow is clearly older than at least one and probably two later intervals of high sea level.

All of the exposures of Einahnuhtan beds on St. George Island have been tectonically warped or disturbed by faults; some displacements have been as much as 45 meters.

*Other localities.* The only deposits of the Einahnuhtan transgression that are recognized with certainty away from St. Paul Island are on the western part of the Arctic coastal plain and in Chukotka.

Ancient beach ridges at an altitude of 30 meters on the Kukpowruk and Epizetka Rivers, described by McCulloch (this volume) as "deposits of a middle Pleistocene transgression," contain molluscan faunas very similar to those of the Einahnuhtan beds on St. Paul Island. They lie inland and at higher altitudes than the raised beaches deposited during the Kotzebuan transgression (pre(?)-Illinoian transgression of McCulloch, this volume). The Pinakul' Suite of the Chukotka Peninsula, Siberia (Petrov, this volume), is correlative with the pre-Kotzebuan beds in the Einahnuhto Bluffs (Merklin *et al.*, 1964; Hopkins *et al.*, 1965), and thus is of Einahnuhtan age.

Marine terraces and bodies of sediment in several other places in Alaska may have been formed during the Einahnuhtan transgression. Fourth Beach at Nome, which lies at an altitude of 30 meters (Hopkins *et al.*, 1960) is probably either of Einahnuhtan or Kotzebuan age. The small molluscan fauna recovered from the surface of the deformed York Terrace of western Seward Peninsula (Sainsbury, this volume) indicates that the terrace cannot be as old as the Anvilian transgression; it must have been carved during either the Einahnuhtan or the Kotzebuan transgression. An equally well preserved but strongly warped terrace extending along the coast of Norton Sound south of Unalakleet is probably of the same age.

*The "Middletonian transgression" of Karlstrom.* The name "Middletonian transgression" has been proposed by Karlstrom (in press; and 1965, Table 1) for a high-sea-level episode supposed to have taken place approximately within the time of the Einahnuhtan transgression as defined here. The type section is designated as stratigraphic unit 20 in the Middleton Island section of Miller (1953), and the dating is based upon Blanchard's $Th_{230}/U_{238}$ age determinations discussed earlier (p. 66). I have not adopted the name "Middletonian transgression" because Broecker's later determination of the radiometric age of these beds failed to confirm Blanchard's determinations, and because the molluscan faunas suggest that these beds are likely to be of Anvilian age. However, if future paleontological and geochronometric studies confirm that Miller's stratigraphic unit 20 on Middleton Island does indeed record the same transgression as that represented by the pre-Kotzebuan beds in the Einahnuhto Bluffs, then the name "Middletonian" will have priority over the name Einahnuhtan and will be applicable to all of the deposits described here as representing the Einahnuhtan transgression.

*Fauna and flora.* The Einahnuhtan faunas are essentially modern in character (Table 2, column 3). With a few exceptions, the molluscan faunas consist of species living at the present time. Most of the species and varieties of *Neptunea* are identical with living forms, but the *N. heros* (Gray) lineage is represented by populations intermediate between the ancestral *N. heros mesleri* (Dall) and the living *N. heros*. The *Astarte* complexes, too, consist mostly of living species, but *A. leffingwelli* Dall is present in Einahnuhtan beds on the western part of the Arctic coastal plain, and the large *A. borealis* (Schumacher) populations in Einahnuhtan beds both there and on St. Paul Island include a few individuals that are more like the extinct *A. nortonensis* MacNeil than are any individuals found in later assemblages. Unfortunately, no *Chlamys* have been found in beds clearly identified as Einahnuhtan. The presence of *C. coatsi* in a collection of presumed Kotzebuan age on Kuk Lagoon near Wainwright suggests that varieties of *Chlamys* that are now extinct may have persisted through Einahnuhtan time.

Water temperatures during Einahnuhtan time evidently differed little from those of the present. Most of the fossil mollusks are of species that still live in adjoining waters, but *Liomesus nux* Dall, found in the Einahnuhto Bluffs, no longer ranges north of the Aleutian Islands, and *Astarte elliptica* (Brown), found on the northwestern part of the Arctic coastal plain, no longer ranges north of Bering Strait. *Plicifusus* sp. aff. *P. wakasanus* Dall, found in the Einahnuhto Bluffs, is closely related to a species now living on the coast of Japan.

A brief invasion of North Atlantic species into Bering Sea is indicated by the presence in the Einahnuhtan beds on St. Paul of the mollusks *Prope-bela* sp. cf. *P. assimilis* (Sars) or *P. exquisita* Bartsch and *Astarte soror* Dall and the echinoid *Echinocyamus* sp. cf. *pusillus* (Müller). The *Prope-bela* is related to species now living along the coasts of Norway and Japan; *Astarte soror* now lives in Greenland waters; and *Echinocyamus pusillus* in the vicinity of Iceland. None of these species have been found previously, either living or as fossils, in western North America.

Thus far, no plant remains have been found in Einahnuhtan beds.

*Position of sea level.* The position of sea level during the Einahnuhtan transgression is unknown. The Einahnuhtan beds on St. Paul Island have been faulted and warped, and the Arctic coastal plain, where Einahnuhtan beds are as high as +20 meters, may have undergone a regional uplift since Einahnuhtan time (McCulloch, this volume).

*Correlation.* The Einahnuhtan transgression of Alaska took place during an interglacial interval late in middle Pleistocene time. The stratigraphic relationships in the Einahnuhto Bluffs (Fig. 5) indicate that the Einahnuh-tan transgression took place between 320,000 years ago (the mean age based on the potassium-argon age determinations of the underlying lava flow) and 175,000 years ago (the probable age of the overlying Kotzebuan deposits). The presence of frost-disturbed beds under the Einahnuhtan beds on Tolstoi Point (Fig. 4) indicate that the Einahnuhtan transgression was preceded by an interval cold enough to cause permafrost to form at low altitudes on St. Paul Island. No information is available, thus far, concerning climatic events during the interval that separated the Einahnuhtan and Kotzebuan transgressions, but the results of recent potassium-argon dating of terraces of the Rhine River (Frechen and Lippolt, 1965) suggest the possibility that the Einahnuhtan beds correspond to the interglaciation that preceded the Mindel Glaciation of the Alps.

## The Kotzebuan Transgression

The name "Kotzebuan transgression" (Hopkins, 1965) has been used to designate a marine transgression that took place during the interglacial interval that immediately preceded the Illinoian Glaciation. Radiometric dating indicates that it probably took place about 175,000 years ago.

Deposits of the Kotzebuan transgression generally lie inland from, and higher than, the well-preserved wave-cut cliffs and ancient beach ridges formed during the following Pelukian transgression. The Kotzebuan deposits contain faunas nearly identical with modern faunas in nearby waters, where-as deposits of the Pelukian transgression commonly contain a few molluscan species that are displaced northward from their present limits. The deposits of the Kotzebuan transgression are difficult to separate from the deposits of the Einahnuhtan transgression in many areas because sea level seems to have reached about the same position during both transgressions, and Kotzebuan and Einahnuhtan shorelines may have nearly coincided. However, large Einahnuhtan faunas contain a few extinct molluscan taxa that are ancestral to the modern taxa found in Kotzebuan faunas.

*Type locality on Baldwin Peninsula.* The type locality for the Kotzebuan transgression is designated as the sea cliffs facing Kotzebue Sound between latitudes 65° 32′ N and 65° 35′ N, on the west shore of Baldwin Peninsula (Hopkins, 1965).[c] This area is shown on the Selawik C-6 topographic quadrangle (1:63,360, U.S. Geol. Survey, 1951). Stratigraphic relationships there are described and illustrated by Hopkins (1965, Fig. 6). McCulloch discusses them as "marine sediments of the Yarmouth Interglaciation" (McCulloch et al., 1965) and as "sediments of the pre(?)-Illinoian transgression" (McCulloch, this volume).

The deposits of the Kotzebuan transgression on the Baldwin Peninsula are at least 100 meters thick and consist chiefly of thick-bedded silty clay and thin-bedded peaty silt of deltaic origin, but also include smaller quantities of well-sorted sand and gravel that accumulated on or near a beach. The base of these beds is not exposed. Two shells from the Kotzebuan beds have yielded $Th_{230}/U_{238}$ age determinations of 170,000 ± 17,000 and 175,000 ± 16,000 years, respectively (Blanchard, 1963).

At the type locality and in many other places in the Kotzebue Sound area, the Kotzebuan deposits have been folded and faulted as a result of glacial overriding. They are covered by drift comparable in both topographic expression and stratigraphic setting to the drift of the Nome River (Illinoian) Glaciation at Nome. The drift is covered by a thick sequence of loess, colluvium, and pond sediments that locally include a buried forest layer and a buried soil profile recording the Sangamon Interglaciation. In the southern part of the type locality, the buried forest layer and its accompanying buried soil profile can be traced laterally into beach deposits of the Pelukian transgression (Hopkins, 1965, Fig. 6).

The original position of the Kotzebuan shoreline cannot be determined precisely in the Kotzebue Sound area, but it probably lay between +21 and +36 meters. Relatively undisturbed estuarine sediments of the Kotzebuan transgression, covered by glacial drift, extend to an altitude of 21 meters on

[c] The type locality for the Kotzebuan transgression is rather inaccessible, but an excellent reference section is well exposed and easily reached at Cape Blossom, 18 km south of Kotzebue.

the southwestern shore of Selawik Lake; alluvial sediments probably correlative with the Kotzebuan transgression are found mantled by drift at
altitudes as low as 36 meters in stream valleys to the southeast. Minor faulting has disturbed both the Kotzebuan and the Pelukian sediments in parts
of the Kotzebue Sound area.

*Other localities.* The Kotzebuan transgression is represented on the western part of the Arctic coastal plain by old beach ridges, fossiliferous marine
sediments, and an old shoreline at an altitude of about 33 meters (McCulloch, this volume). The York terrace of the western Seward Peninsula
(Sainsbury, this volume) is probably of Kotzebuan age, although the possibility cannot be excluded that it is as old as Einahnuhtan. The uplifted but
equally well-preserved terrace south of Unalakleet and Fourth Beach at
Nome are also probably of either Kotzebuan or Einahnuhtan age. Marine
beds apparently deformed by glacial overriding on the shores of Kvichak
Bay (Spurr, 1900, p. 173–174) contain a molluscan fauna suggestive of the
Kotzebuan transgression, but no firm age assignment can be made until the
locality has been revisited and studied further.

The Kotzebuan transgression is represented on the Pribilof Islands by
shelly boulder gravel that rests unconformably on Einahnuhtan sediments
in the Einahnuhto Bluffs (Fig. 5), and by the higher of two sets of wave-cut
scarps found in low-lying areas on both islands. I have noted earlier (p. 69)
that the Kotzebuan beds in the Einahnuhto Bluffs are overlain by bedded
tuff and aeolian sediments disturbed by "fossil" ice wedges and in turn by
a volcanic complex older than the Pelukian transgression. The mean of three
potassium-argon age determinations on a specimen (P-35) from the base
of the volcanic complex is 120,000 years (Cox *et al.*, 1966).

Low-lying areas on St. George and St. Paul Islands generally display
two wave-cut scarps, each adjoined by discontinuous sheets of fossiliferous
sand interspersed among wave-rounded boulders. The lower scarp generally
lies between +5 and +8 meters and probably represents the Pelukian transgression. The higher scarp generally lies between +14 and +20 meters, but
it is as high as +33 meters on Tolstoi Point, St. Paul, where it is cut on the
flank of a faulted anticline (Fig. 4); the higher scarp was probably carved
during the Kotzebuan transgression. A wave-cut terrace sloping up to an
old shoreline at +18 meters north of Garden Cove, St. George Island (omitted on Fig. 3), is covered by glacial drift believed to be of Illinoian age
(Hopkins and Einarsson, 1966).

The marine beds making up the lower member of the Kresta Suite (Petrov, this volume) of Chukotka are correlative with the deposits of the Kotzebuan transgression in Alaska (Merklin *et al.*, 1964; Hopkins *et al.*, 1965).

*Fauna and flora.* The molluscan faunas of the deposits of the Kotzebuan transgression consist almost entirely of species living in nearby waters at the present time (Table 2, column 4). The extinct pectinid *Chlamys coatsi* has been collected in beds presumably of Kotzebuan age on Kuk Lagoon, near Wainwright. No other extinct forms have been found, and other rapidly evolving genera, such as *Neptunea* and *Astarte,* are represented by modern forms. A few molluscan species collected on Baldwin Peninsula and on the western part of the Arctic coastal plain no longer range north of Bering Strait, but other Kotzebuan mollusks, such as *Portlandia arctica* (Gray), are found near the southern limits of their modern ranges. *P. arctica* and several other molluscan species make their first recorded appearance in western Alaska in Kotzebuan beds (Table 2), but *P. arctica* is present in the Pinakul' beds of Chukotka, which are equivalent to the Alaskan deposits of the Einahnuhtan transgression.

Plant remains are abundant in Kotzebuan beds in the Kotzebue Sound area, but as yet have received little study. *Pinus* and *Populus* have been reported as wood in the deltaic beds there (R. A. Scott, written communication, April 15, 1965) and *Pinus* as pollen in silt filling a bison skull that probably came from Kotzebuan beds (Péwé and Hopkins, this volume). Pine no longer ranges into central and western Alaska.

*Position of sea level.* Kotzebuan shorelines lie at about +33 meters on the Arctic coastal plain, between +21 and +36 meters in the Kotzebue Sound area, and between +15 and +20 meters on the Pribilof Islands. Because all of these areas may have undergone tectonic disturbance, the former position of sea level cannot be determined with certainty. The evidence of regional uplift of the Arctic coastal plain suggests that Kotzebuan shorelines may be unusually high there. Sea level probably lay near +20 meters during the peak of the Kotzebuan transgression.

*Correlation.* The Kotzebuan transgression of Alaska and the initial phase of the Kresta transgression in Chukotka (Petrov, this volume) took place during the interglacial interval that preceded the Maximum (or Riss) Glaciation of Siberia and the Nome River (or Illinoian) Glaciation of western Alaska. Radiometric dating on St. Paul Island and on the Baldwin Peninsula indicates that the Kotzebuan transgression took place more than 120,000 years ago, and probably about 175,000 years ago. The results of recent potassium-argon dating of terraces along the Rhine River suggest the possibility that the Kotzebuan transgression corresponds to the Holstein Warm Period, which took place 140,000–150,000 years ago and which intervened between the Mindel and the Riss Glaciations in the Alps (Frechen and Lippolt, 1965).

## The Pelukian Transgression

The name "Pelukian transgression" (Hopkins, 1965) has been chosen to designate a marine transgression that took place during the Sangamon (Riss–Würm) Interglaciation. Radiometric dating indicates that it took place about 100,000 years ago. The Pelukian transgression is represented by well-preserved marine terraces and barrier-bar complexes that lie immediately inland from the Holocene coastal deposits of the Krusensternian transgression. They are found in most coastal areas that remained free of glacial ice during the Wisconsin Glaciation. However, the Pelukian deposits are separated from the present coast by deposits of the Woronzofian transgression in areas that have undergone uplift during late Quaternary time.

*Type locality at Nome.* The type locality for the Pelukian transgression is designated (Hopkins, 1965) as the area between the Snake River and Bering Sea shown on the Nome B-1 topographic quadrangle (scale 1:63,360, U.S. Geol. Survey, 1950). The Pelukian transgression is represented there by fossiliferous gold-bearing sand and gravel known locally as "Second Beach." The transgression was named after Peluk Creek, 3 km east of Nome, along whose course Pelukian sediments have been mined for gold.

Deposits of the Pelukian transgression at the type locality (Fig. 2) consist of a sheet 3–7 meters thick of marine sand and gravel extending seaward from an ancient sea cliff carved at an altitude of 10–12 meters in drift of the Nome River Glaciation, and a layer of estuarine sandy silt and black organic clay 2–4 meters thick resting on drift of the Nome River Glaciation on the south side of the Snake River valley (Hopkins *et al.*, 1960, p. 52–53). The Nome River drift bears an ancient weathering profile about 3 meters thick buried beneath a thin mantle of aeolian silt and colluvium just above the limits of the Pelukian transgression; the drift is not notably weathered where it is covered by the Pelukian deposits. The Pelukian sediments are also covered by as much as 3 meters of aeolian silt and colluvium of late Wisconsin and Recent age. The highest Pelukian deposits at Nome lie at altitudes of 10–12 meters.

Shells from Pelukian deposits at Nome have a radiocarbon age of more than 38,000 years (Hopkins *et al.*, 1960, p. 53), a $Th_{230}/U_{238}$ age of 100,000 $\pm$ 8,000 years, and a $Ra_{226}/U_{238}$ age of 78,000 $\pm$ 5,000 years (Blanchard, 1963).

*Other localities.* Low-lying coastal areas throughout northern and western Alaska and on St. Lawrence Island contain evidence, recognizable in aerial photographs, of multiple transgressions of the sea. A typical sequence consists of Recent (Krusensternian) beaches and barrier-bar complexes near the present strand, followed farther inland by subdued plains consisting of low ridges that represent ancient barrier bars alternating with low lake-dotted areas on the sites of former lagoons, followed still farther inland by low areas whose present topography consists of thaw lakes and small shallow stream valleys. The surfaces of the former barrier bars commonly terminate inland at altitudes between 7 and 15 meters. These are thought to represent areas in which deposits of the Pelukian transgression lie near the surface. Pelukian (Sangamon) marine deposits from the western part of the Arctic coastal plain and the Kotzebue Sound area are described by McCulloch (this volume).

On steeper coasts, the Pelukian transgression is commonly represented by narrow wave-cut terraces such as the unnamed coastal terrace near Cape Thompson (Sainsbury *et al.*, 1965) and the Lost River terrace of western Seward Peninsula (Sainsbury, this volume). The Pelukian transgression is represented on the Pribilof Islands by fossiliferous sediments on low terraces that terminate in wave-cut scarps at altitudes of 6–8 meters. A Pelukian shoreline at Garden Cove on St. George Island is incised in glacial drift that is probably of Illinoian age (Hopkins and Einarsson, 1966).

The Pelukian transgression is probably represented by some of the marine and deltaic sediments older than the last glaciation that are exposed in several places along the northwestern coast of Canada between the Nicholson Peninsula, 300 km east of the mouth of the MacKenzie River, and Herschel Island, 450 km to the west (Mackay, 1956; 1959). A finite radiocarbon date for fossil driftwood in a pingo near Toktoyaktuk (Müller, 1962) suggests, however, that Woronzofian deposits are also present along this part of the Canadian coast.

Some of the dissected and faulted marine terraces reported by Powers (1961, p. 380) on the Aleutian Islands are probably of Pelukian age, and the Pelukian transgression may be represented by some of the low-lying drift-covered marine terraces found on the small islands west of Kodiak Island. Some of the highest dissected marine terraces along the coast of the Gulf of Alaska between Katalla and Icy Point were probably carved during the Pelukian transgression.

The Val'katlen marine beds of Chukotka occupy a stratigraphic position similar to the Pelukian marine beds in Alaska (Petrov, this volume) and evidently are also of Sangamon (Riss–Würm) age. The Val'katlen beds are found at altitudes as high as 30 meters, indicating that the Chukotka Peninsula has undergone considerable uplift since the Sangamon Interglaciation.

*Fauna and flora.* The molluscan faunas found in the deposits of the Pelukian transgression consist entirely of living forms (Table 2, column 5).

*Mya japonica* Jay and *Macoma* sp. cf. *M. carlottensis* Whiteaves make their first appearance in Alaska in Pelukian beds, and *Protothaca adamsi* (Reeve), a species now limited to the northwestern Pacific Ocean, has been found in Pelukian beds at Nome and on Kotzebue Sound but never in older or younger Pleistocene beds in Alaska. *Natica janthostoma* Deshayes, a species now living in the northwestern Pacific Ocean and southern Bering Sea, again extended its range to the Kotzebue Sound area, as it had done previously during the Anvilian transgression. These four mollusks are fairly common and, when present, assist in determining the age of Pelukian beds.

The presence of the two Pacific forms as far north as Kotzebue Sound and of *Pododesmus macroschisma* (Deshayes) and *Pholadidea penita* (Conrad) at Nome indicate that water temperatures were warmer than at present and suggest that during some intervals sea ice did not extend south of Bering Strait.

A single specimen found at Nome of *Littorina* sp. aff. *L. palliata* (Say) or *L. mandschurica* Schenck represents the sole Alaskan occurrence of a representative of two closely related gastropod species that now live in the North Atlantic and in Japan.

Walrus remains are fairly common in the Pelukian deposits of the Nome area, and remains of three species of fish—the flounders *Platichthys stellatus* (Pallas) and *Lepidopsetta bilineata* (Ayres) and a small cod tentatively identified as *Microgadus proximus* (Girard) (identified by W. I. Follett, California Academy of Science; C. L. Hubbs, Scripps Institution of Oceanography; and J. E. Fitch, California State Fisheries Laboratory)—were found in the Pelukian estuarine deposits. The two species of flounder still live in northern Bering Sea, but the cod is now confined to waters south of the Alaska Peninsula.

Plant remains associated with sediments of the Pelukian transgression indicate that, at times, forests similar to the present forests extended 50–80 km west of their present limits in the Nome and Kotzebue areas (Colinvaux, this volume). However, the presence of an extensive frost-disturbed layer within a sequence of Pelukian sediments exposed on the north shore of Eschscholtz Bay (McCulloch, this volume) indicates that the Pelukian warm interval was interrupted by a protracted interval during which sea level fell below its present position and the climate grew cold enough for permafrost to form and ice wedges to grow.

*Position of sea level.* Two sets of wave-cut scarps of probable Pelukian age are found in a few places on the Pribilof Islands, at Nome, and on the south coast of western Seward Peninsula. These and the presence of a frost-disturbed layer containing large ice-wedge casts in the middle of the beach sediments of a Pelukian barrier bar on Eschscholtz Bay (McCulloch, this volume) indicate that the Pelukian transgression was a complex event consisting of two distinct episodes of high sea level, separated by a slight regression.

The uppermost deposits of the Pelukian transgression lie at altitudes of 6–12 meters along the more stable parts of the Alaskan coast. Pelukian sea level probably lay 2–4 meters below the highest beach and barrier-bar sediments; consequently, sea level probably lay between +7 and +10 meters at the height of the Pelukian transgression.

*Correlation.* The stratigraphic position, the radiometric dating, the presence of fossil mollusks north of their present limits, and the observation that buried forest beds can be traced laterally into marine beds of the Pelukian transgression in areas lying beyond the present forest boundaries all indicate that the peak of the Pelukian transgression coincided with the Sangamon or Riss–Würm Interglaciation, on the order of 100,000 years ago. The early and late phases of the Pelukian transgression probably correspond to the Monastir I and Monastir II shorelines of southern Europe and North Africa. The cold episode that accompanied the slight regression during Pelukian time may correspond to a previously unrecognized cold interval within the Sangamon Interglaciation. Alternatively, this cold episode may correspond to the first cold episode of the Wisconsin (or Würm) Glaciation; and the younger Pelukian deposits may correspond to the Broerup Interstade of western Euorpe, as Müller-Beck has suggested elsewhere in this volume for the Monastir II shorelines.

## The Woronzofian Transgression

The name "Woronzofian transgression" (Hopkins, 1965) has been chosen to designate a eustatic rise in sea level that took place during an interstade within the Wisconsin Glaciation. Radiometric dating indicates that it took place about 25,000 to 30,000 years ago. Woronzofian deposits are known only in areas of the Alaskan coast believed to have undergone late Quaternary uplift; sea level probably rose high enough to flood Bering Strait, but it probably did not rise as high as present sea level. The small molluscan faunas obtained thus far from Woronzofian deposits do not differ from modern faunas in nearby waters, but the foraminiferal faunas are depauperate, contain species displaced southward from their present southern limits, and suggest water temperatures considerably colder than those of the present time.

*Type locality at Anchorage.* The type locality for the Woronzofian transgression is designated as the coastal bluffs of Bootlegger Cove, 1.7 km east of Point Woronzof (shown in the Anchorage A-8 and Tyonek A-1 topographic quadrangles, 1:63,360, U.S. Geol. Survey, 1953 and 1952), where the Woronzofian transgression is represented by the Bootlegger Cove Clay. The Bootlegger Cove Clay, originally named and described by Miller and Dobrovolny (1959), is probably the most thoroughly studied Quaternary deposit in Alaska because of its well-recognized stratigraphic and paleoclimatic significance (Karlstrom, 1960; 1964; 1965; Schmidt, 1963; Trainer and Waller, 1965), because of its effects on ground-water distribution in the Anchorage area (Cederstrom *et al.*, 1964), and because of the disastrous landslides that developed in it during the great Alaskan earthquake of 1964 (Hansen, 1965).

The Bootlegger Cove Clay consists of a sequence of silty clay, and clay beds containing a few thin lenses of sand. Angular and rounded pebbles, cobbles, and boulders, some bearing glacial striae, are fairly abundant in the upper and lower part but are absent, or nearly so, in the middle part. The sequence is generally 20–30 meters thick, but reaches thicknesses in excess of 70 meters in a few places. The highest deposits now lie at +36 meters. The Bootlegger Cove Clay rests upon drift of the Knik Glaciation, and interfingers near Point Woronzof with the upper part of a sequence of stratified sand and gravel representing a proglacial outwash delta of Knik (early Wisconsin) age. It extends northward beneath end moraines of the Naptowne (late Wisconsin) Glaciation.

Foraminifera and marine ostracodes are found throughout the formation, and mollusks are fairly abundant in the middle part. The foraminifera indicate normal salinities during the deposition of the lower part of the sequence and low salinities during the deposition of the upper part. The $Th_{230}/U_{238}$ ratio for a shell from the Bootlegger Cove Clay indicates an age between 48,000 and 33,000 years (Sackett, 1958).

The stratigraphic relationships, the vertical distribution of ice-rafted stones, the indicated variations in salinity, and variations in the thixotropic sensitivity of the clay at different stratigraphic levels suggest that deposition of the Bootlegger Cove Clay began while ice of Knik age still lay nearby. Deposition continued during the Knik–Naptowne interstade, and ended only after Naptowne ice had advanced into the region.

*Other localities.* Deposits and erosional features can presently be assigned to the Woronzofian transgression only on the basis of stratigraphic relationships with overlying and underlying drift sheets or on the basis of radiometric dating. Woronzofian sediments are recognized at Point Barrow and in several places in southern Alaska, and are probably present in southeastern Alaska, as well. Woronzofian deposits have not been recognized in western Alaska, and may not be present above sea level there.

The presence of Woronzofian deposits at Point Barrow was established by P. V. Sellman, J. Brown, and R. A. M. Schmidt. These deposits consist of a blanket of marine sand and gravel several meters thick mantling areas within several kilometers of the present coast. A prominent Woronzofian beach ridge about 2 km inland trends parallel to the present coastline and rises to a maximum of 7 meters above sea level. Water-laid organic fibers in the upper section of the beach ridge have a radiocarbon age of 25,000 $\pm$ 2,300 years (Brown, 1965). Wood and peat fragments from a depth of 5 or 6 meters in marine sediments about 3 km inland from the beach ridge are 32,000 $^{+3,600}_{-2,400}$ years old (I-1604, J. Brown and P. V. Sellman, written communication, September 20, 1965).

Erratic boulders foreign to northern Alaska, some of them striated, are scattered in low-lying areas along the Arctic coast of Alaska from the Canadian boundary to the vicinity of Wainwright (MacCarthy, 1958). Most of the boulders found thus far are in areas that lie below 7.5 meters, with the notable exception of a cluster found in an area of Anvilian sediments near Atkasuk (O'Sullivan, 1961). These boulders were probably deposited during the Woronzofian transgression by large grounded icebergs similar to the present-day "ice islands" that originate from ice shelves of Ellesmere Island in Canada.

Woronzofian sediments are exposed in southern Alaska on the west coast of Kodiak Island near Ayakulik (Maddren, 1919, p. 306–307). These sediments rest on drift correlated with the Knik Glaciation, and are covered by drift correlated with the Naptowne Glaciation (Karlstrom, 1964).

The Woronzofian transgression is evidently represented in Chukotka by the Amguem beds, which form a marine terrace that now stands 8–10 meters above sea level (Petrov, this volume).

Fossiliferous marine sediments on a marine terrace older than the last glaciation at Coal Cove near Port Graham on Cook Inlet (Hopkins, unpublished field work, 1962) probably also represent the Woronzofian transgression. Other features probably of Woronzofian age are the marine till containing mollusk shells found in several places in southeastern Alaska (Twenhofel, 1952; Sainsbury, 1961, p. 330–331) and some of the terraces that stand above the highest Krusensternian (Recent) terraces along the Gulf coast of Alaska from Katalla to Icy Point (Miller, 1958; Heusser, 1960, p. 19–21). Some low terraces older than the last local glaciation, found in many places along the south coast of the Alaska Peninsula, on the islands to the south and on the Aleutian Islands, also were very probably carved during the Woronzofian transgression.

*Fauna and flora.* The small molluscan faunas enclosed in the Woronzofian deposits (Table 1, column 5) are similar to the modern molluscan faunas living in nearby waters. However, foraminiferal faunas differ. Those in the Woronzofian deposits at Point Barrow are suggestive of very cold water, in contrast to the richer and less "arctic" foraminiferal assemblages in underlying beds that I consider to be probably of Pelukian age (R. A. M. Schmidt, unpublished data). The Bootlegger Cove Clay contains foraminifera and ostracodes that are now limited to areas north of Bering Strait. The foraminiferal faunas and the presence of ice-rafted boulders suggest that water temperatures were colder than at present during the Woronzofian transgression. The presence of arctic taxa in the Bootlegger Cove Clay suggests that sea level was high enough to flood Bering Strait during at least part of the duration of the Woronzofian transgression.

Plant remains have not yet been studied from Woronzofian marine deposits.

*Position of sea level.* The Woronzofian transgression evidently records a moderate rise in sea level during the series of relatively warm intervals that punctuated the last major glaciation roughly between 25,000 and 35,000 years ago (Kind, this volume; Müller-Beck, this volume). A rise in sea level separating regressions during early and late Wisconsin time was first recognized by Hopkins (1959) on the basis of published radiocarbon dates from many parts of the world. Better documentation was provided by Curray (1961), who suggested that this mid-Wisconsin transgression rose only to about $-15$ meters.

Karlstrom (1964) concludes that the Woronzofian deposits in southern Alaska record a sea level rise that closely approached and probably exceeded present sea level. However, the Cook Inlet and Kodiak areas are tectonically active and underwent appreciable changes in level during the 1964 Alaskan earthquake. Additional isostatic changes in level may have taken place in response to the advance and retreat of glaciers during the Knik and Naptowne Glaciations. Consequently, the Woronzofian deposits near Anchorage and on western Kodiak Island probably do not lie at their original altitudes, and no inferences can be drawn from them concerning the position of sea level during the Woronzofian transgression.

The Arctic coastal plain may also have undergone regional uplift during Quaternary time, and it is quite possible that Woronzofian deposits lie above sea level there because of tectonic movements in post-Woronzofian time. The absence of recognized Woronzofian deposits in unglaciated parts of western Alaska leads me to believe that sea level remained lower than at present during the Woronzofian transgression. However, the presence of arctic foraminifera and ostracodes in the Bootlegger Cove Clay suggests that sea level was high enough to connect the Pacific and Arctic Oceans by way of Bering Strait. This conclusion is consistent with Curray's evidence (1961) that sea level in the Gulf of Mexico stood at about −15 meters approximately 30,000 years ago.

*Correlation.* The stratigraphic relationships and radiometric dating of the Bootlegger Cove Clay and of Woronzofian deposits at Point Barrow demonstrate that the Woronzofian transgression took place during some part of the interval between 25,000 and 35,000 years ago during an interstade within the Wisconsin (or Würm) Glaciation. The Woronzofian evidently corresponds to the "Freeport transgression" of Müller-Beck (this volume).

## The Krusensternian Transgression[d]

The name "Krusensternian transgression" has been used (Hopkins, 1965) to designate the transgression that resulted from the melting of continental glaciers during late Wisconsin and Recent time. In most parts of Alaska the deposits of this transgression lie within a few meters of present sea level and are less than 5,000–6,000 years old; however, elevated marine terraces in southern and southeastern Alaska that are dated by pollen or radiocarbon analyses as being 6,000–11,000 years old also are considered to have been carved during an early phase of the Krusensternian transgression.

*Type locality at Cape Krusenstern.* The broad barrier bar that separates a large lagoon from Chukchi Sea at Cape Krusenstern was chosen as the type locality for the Krusensternian transgression because it promised to provide an exceptionally complete stratigraphic and geomorphic record of sea-level positions, climatic fluctuations, and human history during the 5,000–6,000 years that have elapsed since sea level approached its present position. The Krusensternian deposits at Cape Krusenstern consist of sand

---

[d] The Krusensternian transgression, which takes its name from Cape Krusenstern, was inadvertently misspelled as "Kruzensternian" in my earlier published discussions of the Quaternary marine sequence of Alaska (Hopkins, 1965; Hopkins *et al.*, 1965; Péwé *et al.*, 1965).

and gravel making up a sharply curving barrier bar about 18 km long and ranging in width from 0.1 to 3.0 km. The wider areas on the barrier bar have a complex microtopography, consisting of some 114 beach ridges separated from one another by low swales and linear lakes. The most recent beach ridges—those closest to the present shore—are sites for present-day Eskimo camps; some of the ridges most distant from the modern strand yield artifacts of the Denbigh Flint Culture, estimated to be 4,000–5,000 years old; and on the intervening beaches are a sequence of occupation sites and cultural materials of intermediate age (Giddings, 1960; Giddings and Bandi, 1962; Moore and Giddings, 1962).

*Other localities.* Deposits of the Krusensternian transgression are, by definition, present along the coasts throughout Alaska. In many places they consist of sediments in the active beaches, but older Krusensternian materials are present wherever the coast has been prograding during Krusensternian time and in many places where the coast has been undergoing active uplift during the last 10,000–11,000 years.

Beach-ridge sequences, less complete than the sequence at Cape Krusenstern but still of impressive dimensions, are widely distributed along the northern and western coasts of Alaska; archaeological studies and radiocarbon dating of driftwood suggest that progradation began, in different places, from 1,000 to 4,500 years ago (Moore, 1960; Giddings, 1962; 1964; Ackerman, 1964; Hume, 1965; Brown and Sellman, 1966). Deltaic complexes are widely distributed but have been little studied. Most of the Krusensternian deltas appear to have had rather simple histories, but the vast joint delta of the Yukon and Kuskokwim Rivers consists of a complex of sub-deltas that have developed following each of a series of radical diversions of major distributaries during Krusensternian time (Hoare and Condon, 1966). The Krusensternian transgression is represented along the coasts of southwestern, southern, and southeastern Alaska by elevated wave-cut terraces and elevated accumulations of littoral sediments, as well as by the deposits of the modern beaches (Twenhofel, 1952,[e] Miller, 1953; 1958; Plafker and Miller, 1958; Powers, 1961; Plafker, in press). The oldest of these terraces were formed as much as 11,000 years ago, and the highest are about 285 meters above present sea level (Heusser, 1960, p. 20–22, 93–97).

*History of sea level.* Recent studies of the features of continental shelves in other parts of the world show that sea level rose from a position near

---

[e] Twenhofel's summary of evidence recording Recent shoreline changes in southern Alaska must be treated with caution because some of the deposits that he cites are certainly of Woronzofian age and others are likely to be of Woronzofian or Pelukian age.

−115 meters about 20,000 years ago to a position only slightly below present sea level about 5,000 years ago (Curray, 1961; Shephard, 1964). This late Quaternary rise in sea level, which is represented in Alaska by the Woronzofian transgression, evidently was irregular and marked by still-stands and brief regressions. A temporary stillstand at −38 meters—high enough to reopen Bering Strait—is represented by a submerged delta in southern Chukchi Sea; the delta is at least 12,000 and possibly 14,000 years old (Creager and McManus, this volume). A submerged bed of peaty silt found at a depth of −18 meters near Nome by Wendell Gayman and Andrew Stancioff, Ocean Science and Engineering, Inc., is 9,700 ± 350 years old (W-1800, Meyer Rubin, oral communication, May 1966). The bed is deeper, and probably older, than two ancient barrier-bar complexes shown at −16 meters and −10 meters, respectively, south of Cape York (Fig. 1) on Coast and Geodetic Survey chart 9369 (1957 ed.), and the two submerged barrier bars are probably both more than 5,000 years old.

Sea level has remained within a few meters of its present position since about 3,000 B.C., but minor oscillations of 3 or 4 meters are recorded during the last 5,000 years at Barrow (Hume, 1965; Brown and Sellman, 1966), at Point Hope, and on the shores of Kotzebue Sound (Moore, 1960; J. L. Giddings, written communication, 1958). Sea level seems to have been several meters below its present level during the interval 2,000–3,000 B.C., at least one meter higher than at present during part of the interval from 700 B.C. until 250 A.D., below −2 meters at some time between 500 and 800 A.D., as high as +1 meter at some time between 800 and 1100 A.D., and below −1 meter around 1400–1500 A.D.

Stratigraphic studies of coastal peat bogs in the Cook Inlet area indicate that the sea stood high relative to local shorelines between 5,000 and 6,000 years ago and around 1,500 years ago (Karlstrom, 1964, p. 47–51, 61–62). Karlstrom (in press) has proposed the name Kasilofian transgression for the earlier interval and Girdwoodian transgression for the later one. Because the Cook Inlet area has been undergoing active tectonic deformation during Holocene time (Plafker, in press), these transgressive events may be of only local significance.

## Summary

At least nine intervals during which sea level stood high enough to flood Bering Strait are recorded in western Alaska by deposits ranging in age from late Pliocene to Recent. They can be divided into seven age classes—

named here "marine transgressions" (Table 1) — on the basis of stratigraphic relationships, distinctive molluscan faunas, position in the geomagnetic polarity-reversal sequence, and radiometric dating. They can be correlated with a similar but less complete Quaternary marine sequence in
Siberia.

The Beringian transgression spanned late Pliocene time and part of
Pleistocene time. Most Beringian deposits are probably older than the first
continental glaciation, but Beringian deposits on St. George Island may be
younger than the first widespread glaciation in Iceland (Einarsson *et al.*,
this volume). The Anvilian deposits are younger than a major glaciation
at Nome, but they cannot be related to the glacial-interglacial events recognized in the conterminous United States and Europe. The Einahnuhtan
transgression probably corresponds to the interglaciation immediately preceding the Mindel Glaciation in the Alps, and the Kotzebuan to the Mindel–
Riss Interglaciation in the Alps and the interglaciation immediately preceding the Illinoian Glaciation in the United States. The early part of the
Pelukian transgression corresponds to the Riss–Würm Interglaciation of
the Alps and the Sangamon Interglaciation of the United States; the later
part may possibly correspond to the Broerup Interstade of northern Europe.
The Woronzofian transgression took place during a middle Wisconsin interstade, and the Krusensternian transgression records the postglacial rise in
sea level.

The Pliocene and Pleistocene Beringian transgression and the late Pleistocene Pelukian transgression are complex; each includes at least two episodes of high sea level separated by marine regressions of substantial duration. Sea level probably reached its highest position in late Cenozoic time
during the middle Pleistocene Anvilian transgression. Sea level during the
late Pleistocene Woronzofian transgression was probably lower than present
sea level, but high enough to bring Bering Strait into existence during part
of the interval between 25,000 and 35,000 years ago.

Water temperatures were considerably warmer than at present in most
places during the Beringian transgression, and were slightly warmer than
at present during the Anvilian, Kotzebuan, and Pelukian transgressions. The
Woronzofian transgression was a time of cold seas; the foraminifera and
ostracodes of the Woronzofian beds indicate that water temperatures were
lower than those of the present time. Transarctic migrations are indicated
by first appearances and unique appearances of Atlantic mollusks in Alaska
during several transgressions, suggesting that conditions in the Arctic Ocean
were more favorable for molluscan life during Pleistocene interglaciations
than at present.

## REFERENCES

Ackerman, R. E. 1964. Prehistory in the Kuskokwim–Bristol Bay region, Southwestern Alaska: Washington State Univ., Lab. Anthropology, Rept. Inv. 26, 48 p.

Barth, T. F. W. 1956. Geology and petrology of the Pribilof Islands, Alaska: U.S. Geol. Survey Bull. 1028-F, p. 101–160.

Black, R. F. 1964. Gubik Formation of Quaternary age in northern Alaska: U.S. Geol. Survey Prof. Paper 302-C, p. 59–91.

Blanchard, R. L. 1963. Uranium decay series disequilibrium in age determination of marine calcium carbonates: Ph.D. thesis, Washington Univ., St. Louis, 164 p.

Broecker, W. S. 1965. Isotope geochemistry and the Pleistocene climatic record, p. 737–753 *in* H. E. Wright and D. G. Frey, eds., The Quaternary of the United States: Princeton Univ. Press, 922 p.

Brown, Jerry. 1965. Radiocarbon dating, Barrow, Alaska: Arctic, v. 18, p. 37–48.

Brown, Jerry, and P. V. Sellman. 1966. Radiocarbon dating of a buried coastal peat, Barrow, Alaska: Science, v. 153, p. 299–300.

Cederstrom, D. J., F. W. Trainer, and R. M. Waller. 1964. Geology and ground-water resources of the Anchorage area, Alaska: U.S. Geol. Survey Water-Supply Paper 1773, 108 p.

Cox, Allan, D. M. Hopkins, and G. B. Dalrymple. 1966. Geomagnetic polarity epochs: Pribilof Islands: Geol. Soc. America Bull., v. 77, p. 883–909.

Curray, J. R. 1961. Late Quaternary sea level—a discussion: Geol. Soc. America Bull., v. 72, p. 1707–1712.

Cushman, J. A. 1941. Some fossil foraminifera from Alaska: Cushman Lab. Foram. Research Contr., v. 17, pt. 2, p. 33–38.

Cushman, J. A., and Ruth Todd. 1947. A foraminiferal fauna from Amchitka Island, Alaska: Cushman Lab. Foram. Research Contr., v. 23, no. 297, p. 60–72.

Faas, R. W. 1962a. Foraminiferal paleoecology of the Gubik (Pleistocene) Formation of the Barrow area, northern Alaska: Iowa Acad. Sci., v. 69, p. 354–361.

———— 1962b. Micropaleontology of some Quaternary sediments from the Barrow area, northern Alaska: M.S. thesis, Iowa State Univ. Science and Technology, Ames, 38 p.

———— 1964. A study of some late Pleistocene estuarine sediments near Barrow, Alaska: Ph.D. thesis, Iowa State Univ. Science and Technology, Ames, 190 p.

Fraser, G. D., and H. F. Barnett, Jr. 1959. Geology of the Delarof and westernmost Andreanof Islands, Aleutian Islands, Alaska: U.S. Geol. Survey Bull. 1028-I, p. 211–248.

Frechen, J., and H. J. Lippolt. 1965. Kalium-Argon Daten zum Alter des Laacher Vulkanismus, der Rheinterrassen und der Eiszeiten: Eiszeitalter und Gegenwart, v. 16, p. 5–30.

Giddings, J. L. 1960. First traces of man in the Arctic: Nat. History, v. 69, no. 9, p. 10–19.

———— 1962. Eskimos and old shorelines: The American Scholar, v. 31, p. 585–594.

———— 1964. The archaeology of Cape Denbigh: Brown Univ. Press, 331 p.

Giddings, J. L., and H. G. Bandi. 1962. Eskimo-archäologische strandwallunter-suchungen auf Kap Kruzenstern, Nordwest-Alaska: Germania, v. 40, p. 1–21.

Hansen, W. R. 1965. The Alaska earthquake, March 27, 1964: Effects on com-munities: Anchorage: U.S. Geol. Survey Prof. Paper 542-A, 68 p.

Heusser, C. J. 1960. Late-Pleistocene environment of North Pacific North Amer-ica: Am. Geog. Soc. Spec. Pub. 35, 308 p.

Hoare, J. M., and W. H. Condon. 1966. Geologic map of the Kwiguk and Black Quadrangles, western Alaska: U.S. Geol. Survey Misc. Geol. Inv. Map I-469.

Hopkins, D. M. 1959. Cenozoic history of the Bering Land Bridge (Alaska): Science, v. 129, p. 1519–1528.

———— 1965. Chetvertichnye morskie transgressii na Alyaske (Quaternary ma-rine transgressions in Alaska), in Antropogenovye period v Arktike i sub-arktike (Anthropogene Period in the Arctic and Subarctic): Nauchno-Issled. Inst. Geol. Arktiki Trudy, v. 143, p. 131–154. In Russian; translation available from Am. Geol. Inst.

Hopkins, D. M., and Th. Einarsson. 1966. Pleistocene glaciation on St. George Island, Pribilof Islands: Science, v. 152, p. 343–345.

Hopkins, D. M., and F. S. MacNeil. 1960. A marine fauna probably of late Pliocene age near Kivalina, Alaska, in Short papers in the geological sciences: U.S. Geol. Survey Prof. Paper 400-B, p. B339–B342.

Hopkins, D. M., F. S. MacNeil, and Estella B. Leopold. 1960. The coastal plain at Nome, Alaska—a late Cenozoic type section for the Bering Strait region: Rept., 21st Internat. Geol. Cong. (Copenhagen), 1960, pt. 4, p. 46–57.

Hopkins, D. M., F. S. MacNeil, R. L. Merklin, and O. M. Petrov. 1965. Quater-nary correlations across Bering Strait: Science, v. 147, p. 1107–1114.

Hume, J. D. 1965. Sea-level changes during the last 2,000 years at Point Barrow, Alaska: Science, v. 150, p. 1165–1166.

Karlstrom, T. N. V. 1960. The Cook Inlet, Alaska, glacial record and Quaternary classification, in Short papers in the geological sciences: U.S. Geol. Survey Prof. Paper 400-B, p. B330–B332.

———— 1964. Quaternary geology of the Kenai Lowland, Alaska, and glacial history of the Cook Inlet region, Alaska: U.S. Geol. Survey Prof. Paper 443, 69 p.

———— 1965. Upper Cook Inlet area and Matanuska River Valley: Internat. Assoc. Quaternary Res. (INQUA), 7th Congress (Boulder), 1965, Guide-book for Field Conference F: Central and south-central Alaska, p. 114–141.

———— In press. The time scale of the ice age—a current Quaternary problem: in R. B. Morrison and H. E. Wright, Jr., eds., Means of correlation of Quater-nary successions: Univ. of Utah Press.

MacCarthy, G. R. 1958. Glacial boulders on the Arctic coast of Alaska: Arctic v. 11, p. 71–85.

McCulloch, D. S., D. W. Taylor, and M. Rubin. 1965. Stratigraphy, nonmarine mollusks, and radiometric dates from Quaternary deposits in the Kotzebue Sound area, western Alaska: Jour. Geology, v. 73, p. 442–453.

Mackay, J. R. 1956. Deformation by glacier-ice at Nicholson Peninsula, N.W.T., Canada: Arctic, v. 9, p. 219–228.

———— 1959. Glacier ice-thrust features of the Yukon coast: Geogr. Bull., no. 13, p. 5–21.

MacNeil, F. S. 1957. Cenozoic megafossils of northern Alaska: U.S. Geol. Survey Prof. Paper 294-C, p. 99–126.

———— 1965. Evolution and distribution of the Genus *Mya*, and Tertiary migra-tions of Mollusca: U.S. Geol. Survey Prof. Paper 483-G, 51 p.

———— In press. Cenozoic pectinids of Alaska and related regions: U.S. Geol. Survey Prof. Paper 553-A.

MacNeil, F. S., J. B. Mertie, Jr., and H. A. Pilsbry, 1943. Marine invertebrate faunas of the buried beaches near Nome, Alaska: Jour. Paleontology, v. 17, p. 69–96.

MacNeil, F. S., J. A. Wolfe, D. J. Miller, and D. M. Hopkins. 1961. Correlation of Tertiary formations in Alaska: Bull. Am. Assoc. Petroleum Geologists, v. 45, p. 1801–1809.

Maddren, A. G. 1919. The beach placers of the west coast of Kodiak Island, Alaska: U.S. Geol. Survey Bull. 692, p. 299–319.

Meek, C. E. 1923. Notes on stratigraphy and Pleistocene fauna from Peard Bay, Arctic Alaska: Univ. Calif. Dept. Geol. Sci. Bull., v. 14, no. 13, p. 409–422.

Merklin, R. L., O. M. Petrov, D. M. Hopkins, and F. S. MacNeil. 1964. Popytka korrelyatsii pozdnekaynozoyskikh morshikh osadkov Chukotki, Severovostochnoy Sibiri i zapadnoy Alyaski (An attempted correlation for the late Cenozoic marine deposits of Chukotka, northeastern Siberia, and western Alaska): Akad. Nauk SSSR, Izv., Ser. Geol., 1964 no. 10, p. 45–57.

Miller, D. J. 1953. Late Cenozoic marine glacial sediments and marine terraces of Middleton Island, Alaska: Jour. Geology, v. 61, p. 17–40.

———— 1957. Geology of the southeastern part of the Robinson Mountains, Yakataga district, Alaska: U.S. Geol. Survey Oil and Gas Inv. Map OM-187, scale 1:63,360.

———— 1958. Anomalous glacial history of the northeastern Gulf of Alaska region (abs.): Geol. Soc. America Bull., v. 69, p. 1613–1614.

Miller, R. D., and E. Dobrovolny. 1959. Surficial geology of Anchorage and vicinity, Alaska: U.S. Geol. Survey Bull. 1093, 128 p.

Moore, G. W. 1960. Recent eustatic sea-level fluctuations recorded by Arctic beach ridges, *in* Short papers in the geological sciences: U.S. Geol. Survey Prof. Paper 400-B, p. B335–B337.

Moore, G. W., and J. L. Giddings. 1962. Record of 5,000 years of Arctic wind direction recorded by Alaskan beach ridges, *in* Abstracts for 1961: Geol. Soc. America Spec. Paper 68, p. 232.

Müller, F. 1962. Analysis of some stratigraphic observations and radiocarbon dates from two pingos in the MacKenzie Delta area, N.W.T.: Arctic, v. 15, p. 279–288.

O'Sullivan, J. B. 1961. Quaternary geology of the Arctic coastal plain, northern Alaska: Ph.D. thesis, Iowa State Univ., 191 p.

Péwé, T. L., D. M. Hopkins, and J. L. Giddings. 1965. The Quaternary geology and archaeology of Alaska, *in* H. E. Wright and D. G. Frey, eds., The Quaternary of the United States: Princeton Univ. Press, p. 355–374.

Plafker, G. In press. Holocene (Recent) vertical displacements in coastal south-central Alaska: a possible approach to prediction: U.S.–Japan Conf. Res. Related to Earthquake Prediction (Lamont Geol. Obs., Columbia Univ.).

Plafker, G., and D. J. Miller. 1958. Glacial features and surficial deposits of the Malaspina district, Alaska: U.S. Geol. Survey Misc. Geol. Inv. Map I-271, scale 1:125,000.

Powers, H. A. 1961. The emerged shoreline at 2–3 meters in the Aleutian Islands, *in* Pacific island terraces: eustatic? (a symposium), Zeitschr. Geomorphol. Supp. 3, p. 36–38.

Powers, H. A., R. R. Coats, and W. H. Nelson. 1960. Geology and submarine physiography of Amchitka Island, Alaska: U.S. Geol. Survey Bull. 1028-P, p. 521–554.

Sackett, W. M. 1958. Ionium-Uranium ratios in marine-deposited calcium carbonates and related materials: Ph.D. thesis, Washington Univ., St. Louis.

Sainsbury, C. L. 1961. Geology of part of the Craig C-2 quadrangle and adjoining areas, Prince of Wales Island, southeastern Alaska: U.S. Geol. Survey Bull. 1058-H, p. 299–362.

Sainsbury, C. L., R. Kachadoorian, R. H. Campbell, and D. W. Scholl. 1965. Marine platform of probable Sagamon age, and associated terrace deposits, Cape Thompson, northwestern Alaska: Arctic, v. 18, p. 230–245.

Schmidt, Ruth A. M. 1963. Pleistocene marine microfauna in the Bootlegger Cove Clay, Anchorage, Alaska: Science, v. 141, p. 350–351.

Schrader, F. C. 1904. A reconnaissance in northern Alaska across the Rocky Mountains, along Koyukuk, John, Anaktuvuk, and Colville Rivers and the Arctic coast to Cape Lisburne, in 1901, with notes by W. J. Peters: U.S. Geol. Survey Prof. Paper 20, 139 p.

Shepard, F. P. 1964. Sea level changes in the past 6,000 years: possible archaeological significance: Science, v. 143, p. 574–576.

Sigafoos, F. S. 1958. Vegetation of northwestern North America, as an aid to the interpretation of geologic data: U.S. Geol. Survey Bull. 1061-E, p. 165–185.

Smith, P. B. 1963. Possible Pleistocene-Recent boundary in the Gulf of Alaska, based on benthonic foraminifera: U.S. Geol. Survey Prof. Paper 475-C, p. 73–77.

Spurr, J. E. 1900. A reconnaissance in southwestern Alaska in 1898: U.S. Geol. Survey 20th Ann. Rept., pt. 7, p. 31–264.

Trainer, F. W., and R. M. Waller. 1965. Subsurface stratigraphy of glacial drift at Anchorage, Alaska: U.S. Geol. Survey Prof. Paper 525-D, p. 167–174.

Twenhofel, W. S. 1952. Recent shoreline changes along the Pacific coast of Alaska: Am. Jour. Sci., new ser., v. 250, p. 523–548.

# 5. Quaternary Geology of the Alaskan Shore of Chukchi Sea

D. S. MC CULLOCH
*U.S. Geological Survey, Menlo Park, California*

Six Pleistocene marine transgressions, two major glacial advances, and two important postglacial warm intervals (*ca.* 10,000–8,300 and *ca.* 6,000–3,000 years B.P.) are recorded by unconsolidated marine sediments exposed along the Alaskan coast of Chukchi Sea. The history constructed from the study of these deposits and from our knowledge of the Chukchi Sea floor (Creager and McManus, this volume) remains a loose and incomplete framework of events for the early and middle Pleistocene, but for the later Pleistocene and Recent time the history is relatively complete. The latter part of this history will help to set the stage onto which man walked as he entered the New World.

The chronology of the Pleistocene and Recent events along the eastern shore of Chukchi Sea that I shall present is based upon changes in marine molluscan faunas and demonstrable stratigraphic succession. Through the work of F. S. MacNeil of the U.S. Geological Survey, and the acquisition of new collections from known and new sites along the Alaskan coast, it is now evident that, despite the short span of time involved, faunal changes represented by the extinction of old forms and the evolution or migration of new forms have taken place in the waters adjacent to Alaska during the Pleistocene (Hopkins, this volume). The stratigraphic succession that greatly assists in this historic reconstruction is known from the Kotzebue Sound area, where fossiliferous deposits of two marine transgressions are found interbedded with continental glacial deposits (McCulloch *et al.*, 1965).

The Pleistocene and Recent history presented in this paper (summarized in Table 1) is based in part upon the published work of others and in part upon my own field examinations of the sediments and collections of marine

Publication authorized by the Director, U.S. Geological Survey.

TABLE 1. *Summary of the Quaternary Geology of the Alaskan Shore of Chukchi Sea*

| Event | History and Detailed Evidence | Radiometric Dates or Estimated Age Range |
|---|---|---|
| Postglacial thermal maximum | Development of weak alder peak in pollen profiles on northern coastal plain (1). Maximum retreat of valley glaciers in north-central Brooks Range (2). | <5,890 ± 170 <6,000 to >3,000 |
| Brief climatic cooling | End of warm interval in Kotzebue Sound–Seward Peninsula area (3). Glacial readvance in north-central Brooks Range (2). | |
| Early Recent warming | Expansion of forest and beaver range, melting of ground ice, and development of weathering profile in Kotzebue Sound–Seward Peninsula area (3). | 10,200 ± 800 to 8,350 ± 200 |
| | Formation of organic layer now found below permafrost table on northern coastal plain (4,5). | 10,900 ± 280 to 8,200 ± 200 |
| | Period of rapid retreat of last major glaciation in north-central Brooks Range (2). | 11,000–10,000 to 8,300 |
| Wisconsin Glaciation and mid-Wisconsin transgression | Loess deposition in climate favorable for growth of ground ice in Kotzebue Sound area (6). | 34,000 ± 2,000 to >38,000 |
| | Development of valley glaciers less than in preceding glaciation; four substages recognized in north-central Brooks Range (2). | |
| | Marine sediments and ice-rafted boulders deposited by mid-Wisconsin transgression on edge of northern coastal plain (5). Deposits raised at least 8 meters above sea level by later uplift. | 25,300 ± 2,300 |
| | Chukchi Sea floor exposed to subaerial erosion by sea withdrawal (7, 11). | |
| Marine transgression of Sangamon age | Fossiliferous marine sediment associated with buried forest; evidence for melting of ground ice and weathering profile found between loess deposits of Illinoian and Wisconsin age in Kotzebue Sound area (6). | >42,000 |
| | Marine sediment resting on wave-cut terrace along most of Alaskan coast of Chukchi Sea (7,8,9). | >26,000 ± 400 and >38,000 |
| | Molluscan fauna of 39 species, all living; *Protothaca adamsi*, *Natica janthostoma* and *Trichotropis insignis* now live south of Bering Strait, suggesting warm Sangamon Sea. | Correlative beds at Nome are 100,000 ± 8,000–78,000 ± 5,000 |

| | | |
|---|---|---|
| Illinoian Glaciation | Period of maximum glaciation in northern Alaska; sea level low and deposits of pre(?)-Illinoian marine transgression overridden and deformed by glaciers (6). | |
| Marine transgression of pre(?)-Illinoian age | Marine delta in Kotzebue Sound area containing 65 species of marine mollusks (6). Transgressive beach deposits on northern coastal plain with fauna of 57 species. First appearance of extant species *Amauropsis purpurea, Pyrulofusus deformis, Neptunea borealis*, and an unidentified species of *Buccinum*. Present ranges of mollusks and foraminifera found as fossils indicate water temperatures about as at present. | 170,000 ± 17,000 and 175,000 ± 16,000 |
| Period of unknown length for which there is no record. | | |
| Marine transgression of middle Pleistocene age | Most extensive transgressive deposits on northern coastal plain; possibly two transgressions separated by subaerial erosion and nonmarine deposition on marine platform cut by first transgression. Aggregate fauna of 48 species; extinct *Neptunea heros mesleri, Astarte leffingwelli* not found in younger deposits. | |
| Period of unknown length for which there is no record. | | |
| Marine transgression of late Pliocene(?)—early Pleistocene(?) age | Marine sediment on wave-cut bedrock platform at Kivalina (10). Fauna of 22 species; extinct species *Neptunea* n. sp. aff. *N. despecta* and *Astarte hemicymata* not found in younger deposits, and *Fortipecten hallae* of early-to-late Pliocene age. | |

References:

1 Livingstone, 1955, 1957.
2 Porter, 1964.
3 McCulloch and Hopkins, 1966.
4 Douglas and Tedrow, 1960.
5 Brown, 1965.
6 McCulloch *et al.*, 1965.
7 Scholl and Sainsbury, 1961a,b.
8 Moore and Scholl, 1961.
9 Sainsbury *et al.*, 1965.
10 Hopkins and MacNeil, 1960.
11 Creager and McManus, 1965.

mollusks in three widely separated areas along the coast. The most detailed investigations for this study have been done along the shores of Kotzebue Sound and the shore of Selawik Lake. Work in the other two areas, Kivalina and along the northwestern edge of the northern coastal plain, has been of a reconnaissance nature.

## Late Pliocene or Early Pleistocene Transgression

The oldest known marine sediments of possible Pleistocene age along the Chukchi Sea coast are found in the gullies of several small tributaries of the Kivalina River 11 km north-northwest of Kivalina (Fig. 1). Hopkins and MacNeil (1960) described the exposed sediments and the marine fossils collected by Hopkins in 1959. I revisited the site in 1963.

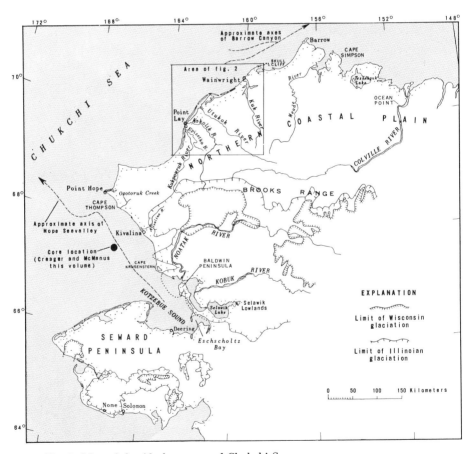

Fig. 1. Map of the Alaskan coast of Chukchi Sea.

Although poorly exposed in the slumped gully walls, the deposit is divisible into three units. The lowest is about 3 meters thick and grades upward from a dark pebbly clay, containing some pholad-bored limestone boulders, to a sandy silt. Fossils occur throughout this unit. Resting on the sandy silt are about 5½ meters of pebble gravel of limestone and green and black chert. Fossils are less abundant in this gravel than in the underlying sediments, but occur throughout the unit. The gravel is covered by 3 meters of olive-gray massive silt, which also blankets the surrounding hillsides.

The lowermost of the three depositional units is clearly marine, and, lacking evidence that the fossils were recycled into it, we can infer that the overlying pebble gravel is probably also marine. The silt forming the uppermost unit is identified as loess by its grain size, by its massive character and color, and by its disposition as a mantle over the rolling topography.

The base of the marine sediments lies at an altitude of about 10 meters on what may be a marine wave-cut platform eroded across Paleozoic limestone. The platform can be seen sloping gently seaward in exposures along the Kivalina River. About 4 km northwest of the exposed marine sediment there is a seaward-facing scarp about 3 km long that has been breached by modern streams. The crest of the scarp is composed of a buff siliceous limestone. Hopkins and MacNeil (1960) suggested that this scarp may be an ancient wave-cut cliff formed at the inland edge of the wave-cut marine platform upon which the marine sediments were deposited. If this is the case, it marks the position of the transgressive shoreline.

The age of the marine deposits is not well defined by the available geologic evidence. On their map of the surficial deposits of Alaska, Karlstrom *et al.* (1964) show a lobe of glacial moraine about 35 km up the Kivalina River from the fossil site. They correlate the moraine with the Nome River drift, which is presumably of Illinoian age. Hopkins and MacNeil (1960) have suggested that the loess blanketing the marine deposits was blown from the floodplain of the Kivalina River as it carried meltwater from this ancient glacier. Accepting this, we would conclude that the marine deposits are of pre-Illinoian age.

The age of the marine deposits, although not exactly known, is more precisely established by the fossil assemblage. Hopkins' original collection contained two extinct mollusks, *Astarte hemicymata* (Dall) and *Fortipecten hallae* (Dall). *A. hemicymata* (Dall) does not occur in other faunas from the Alaskan Chukchi Sea coast, but it has been found to the south, in deposits of the first two marine transgressions recognized at Nome—the Submarine Beach and Intermediate Beach deposits (type localities, respectively, for the Beringian transgression of late Pliocene and/or early Pleistocene age and the Anvilian transgression of early Pleistocene age; Hopkins, this volume).

*Fortipecten hallae* (Dall) is known from only one other place in Alaska, at its type locality 20 km east of Nome, a mine shaft near Solomon on the south side of Seward Peninsula. No other diagnostic fossils occur with *F. hallae* at Solomon. The genus is otherwise known only in beds of Pliocene age in East Asia (Durham and MacNeil, this volume)—and there the last known species occurs in beds currently assigned to the late Pliocene.

Additional collecting by the present author increased the Kivalina molluscan fauna from 13 to 22 species, and turned up an undescribed species of *Neptunea* closely related to *N. despecta*. This undescribed species is believed by MacNeil to be at least as old as early Pleistocene.

The *Astarte* and the undescribed *Neptunea* species in the Kivalina beds suggest that the transgression might have been of late Pliocene or early Pleistocene age, but the *Fortipecten* is not known in deposits younger than late Pliocene. Thus the transgression cannot as yet be more closely dated than late Pliocene or early Pleistocene.

## Younger Pleistocene Transgressions

Evidence for the next younger and for some of the succeeding marine transgressions comes from the vast seaward-sloping tundra-covered northern coastal plain of Alaska. The southern edge of the plain is bounded in many places by an abrupt escarpment resembling a wave-cut cliff where it abuts the northern foothills of the Brooks Range. Much of the coastal plain is mantled by unconsolidated deposits of late Pliocene(?) and Pleistocene age, to which the name Gubik Formation (Schrader, 1904, p. 91–93) is often applied. Because of the enormous area involved, the difficulty of travel, and the generally scarce and poor exposures of sediments that are in any case usually unfossiliferous and not greatly dissimilar, the history of the coastal plain is not well known. For example, the two most comprehensive studies of the Gubik Formation (Black, 1964; O'Sullivan, 1961) have different interpretations.

Black divides the Gubik Formation into three lithologic units, each of which has a marine transgressive facies. Relying primarily upon geomorphic and lithologic arguments, he assigns the transgressions Illinoian, Sangamon, and Wisconsin ages (p. 88–89). O'Sullivan divides the Gubik Formation into a series of surfaces and lithologic units. He suggests that the oldest surface at the toe of the southern escarpment may represent one or more transgressions, possibly of late Pliocene and Aftonian age. He also proposed that there were transgressions of Yarmouth and Sangamon age. The Yarmouth age assignment is based upon the occurrence of the mammoth, *Mammuthus primigenius*, although Hibbard (1958, p. 20) cites *M. primigenius* as first appearing in the New World during the Wisconsin Glaciation.

In the following discussion, evidence is presented for a revision of the historical reconstruction of the western end of the northern coastal plain. This interpretation is based on (1) ages assigned to new marine molluscan collections, as inferred from the sequence of Alaskan Pleistocene marine molluscan faunas, (2) gross geomorphic expression of the marine deposits, and (3) the assumption, partly demonstrable in the Kotzebue Sound area, that two of the marine transgressions occurred during nonglacial intervals. This historical reconstruction recognizes Pleistocene transgressions of middle Pleistocene, pre(?)-Illinoian, Sangamon, and mid-Wisconsin age. The transgressive deposits veneer an eroded bedrock surface. The highest altitude at which the deposits occur decreases as their ages decrease, and beach ridges, clearly distinguishable on the youngest deposits, are less distinct on older deposits.

## Middle Pleistocene Transgression

Deposits of a transgression believed to be middle Pleistocene in age are exposed along the Kukpowruk and Epizetka Rivers (Fig. 2). They have been mapped (as the Gubik Formation) along the Kukpowruk River by Chapman and Sable (1960), who drew their inshore limit at an altitude of about 60–70 meters, close to the base of the escarpment that bounds the coastal plain. According to my field observations, the marine sediments are composed of well-sorted sand and gravel lying directly on bedrock and have a total thickness of less than 5 meters. Typically, they are covered by the thin layer of brown silt that supports the modern tundra. There has been considerable solution and redeposition of iron and lime in the sediments. Some pebbles are weathered to ghosts, and the sand and gravel are often deeply colored by iron oxides. The small shell fragments that occur throughout the sand and gravel are soft and corroded, and many pebbles have limy coatings. The low percentage or absence of fine-grained material in the well-sorted sand and gravel makes them relatively immune to the frost stirring that is usually restricted to contact with the overlying silt.

On the Kukpowruk River, two somewhat irregular linear hills stand above the surface of the marine deposits. The ridges may be beaches formed along the retreating shore of the middle Pleistocene sea (Fig. 2). The ridge

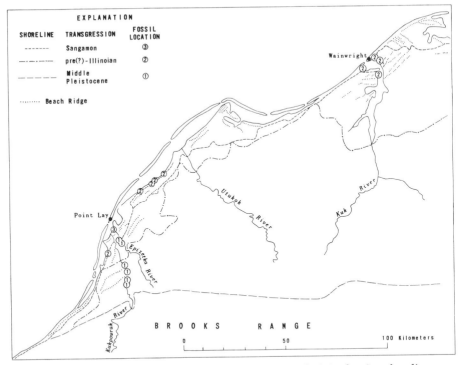

Fig. 2. Map of the western end of the northern coastal plain showing shorelines of marine transgressions, fossil localities, and probable beach ridges.

tops are at an altitude of about 30 meters; they lie 30–40 meters below, and about 19 km seaward of, the edge of the highest marine deposits of the Gubik Formation exposed on the Kukpowruk River. Similar ridges found on the Epizetka River lie still further seaward, and may also be beach ridges. Where the ridges are cut through by the Kukpowruk and Epizetka Rivers, they are composed of 10 to 12 meters of unconsolidated sediments resting upon a nearly planar erosion surface cut on the underlying bedrock. The section exposed on the east bank of the Kukpowruk River as it cuts through the westernmost ridge is composed of an upper nonmarine unit, a middle marine unit, and a basal nonmarine unit. The upper part of the section is one-half to 3 meters of silt. A few isolated pebbles near the base of the silt may have been stirred in by frost action from the underlying marine gravel. The silt has the same massive character and color as the silt that covers the adjacent lower land; both may be a single loess sheet.

The marine sediments beneath the silt can be divided into (1) an upper member about 4 meters thick of pebble gravel that becomes coarser on the inland side, (2) a middle member 1½ to 2 meters thick of iron-oxide-stained sand, interbedded silt and sand, and thin beds of detrital coal fragments, and (3) a basal 3 to 10 meters of finely and evenly bedded olive-black silt containing small, well-rounded pieces of driftwood. Fossil shells are either entirely destroyed or extremely weathered in the upper part of the marine section. The highest identifiable fossils occur in the interbedded silt and sand of the middle unit, and are found from there to the base of the marine section.

A thin nonmarine section directly underlies the olive-black marine silt at the northwest edge of the ridge. This section consists of about 20 cm of black autochthonous peat containing roots, stems, and leaves. The peat layer undulates with a relief of about 30 cm. Directly beneath the peat lies at least a meter of highly frost-contorted iron-oxide-stained sand and olive-gray silt lacking marine fossils but containing pockets of small twigs. The twig ends are angular, suggesting that they were not worked by waves or currents. This and the lack of marine fossils suggest that the unit is nonmarine. Because of extensive slumping at the base of the sediments, it was not determined whether the nonmarine unit rests directly on the bedrock that crops out several meters below, or whether it is underlain by marine sediments.

On the Epizetka River the exposures are very poor, but the section through the ridges is similar to the Kukpowruk section. Here there is unconsolidated sediment of about 12 meters, consisting in descending order of an upper unit of tan silt containing a few pebbles, a middle unit of gravel, sand, and silt, and finally, a basal light-gray clay. The clay overlies an erosion surface cut across deformed sandstone and coal. No basal nonmarine sediment was found. Fossil marine mollusks were collected from the scars and debris of small landslides that slid on the clay. The fossils were collected at the most upstream part of the river visited.

The locations of the fossil sites indicate the general distribution of the middle Pleistocene transgressive deposits (Fig. 2), but they do not define the position of the transgressive shoreline. Traverses were made for approximately 15 km up the Kokolik River and 60 km up the Kuk River, but no fossiliferous sites were found inland from those shown on Fig. 2. As previously mentioned, sediments probably belonging to this transgression are found along the Kukpowruk River almost to the escarpment that bounds the coastal plain. It has been suggested that this escarpment is a wave-cut cliff (Hopkins, 1959a), but its age is not well defined.

The nonmarine deposits lying beneath the marine sediments on the Kukpowruk River and the existence of a probable stream valley eroded into the bedrock beneath the marine sediments of the Gubik Formation north of the escarpment and west of the Colville River (Brosgé and Whitting-ton, 1967) suggest that the wave-cut platform and cliff may have been formed during an earlier transgression and subjected to subaerial erosion before they were covered by the middle Pleistocene transgressive deposits. The only marine deposits on the coastal plain known to predate the middle Pleistocene transgression are those forming the lowest part of the Gubik Formation at Ocean Point on the Colville River. The invertebrate marine fauna from this locality (MacNeil, 1957) is of late Pliocene or early Pleis-tocene age. However, as O'Sullivan has shown, these sediments pinch out several tens of kilometers north and perhaps 100 meters below the altitude of the base of the escarpment. Thus the shoreline of this early transgression probably lies many kilometers seaward of the escarpment. In the absence of evidence for deposits intermediate in age between the beds exposed at Ocean Point and those of the middle Pleistocene transgression, it is proposed that deposits assigned to the middle Pleistocene transgression may in fact have been deposited during two transgressive events that were separated by a regression of unknown duration during which the coastal plain was exposed to subaerial erosion. If this interpretation is correct, the shoreline of the earliest of the middle Pleistocene transgressions can be drawn along the base of the escarpment.

There is also evidence from the marine molluscan faunas for a trans-gression intermediate in age between the late Pliocene(?)–early Pleisto-cene(?) transgression that deposited the sediments at Ocean Point and the middle Pleistocene transgression represented by the faunas from the Kuk-powruk and Epizetka Rivers. This intervening transgression (the Anvilian, Hopkins, this volume) may be represented in mollusk collections from the Meade River 90 km south of Barrow and from Skull Cliff, southwest of Barrow (Fig. 1).

Both Black and O'Sullivan recognized that the central portion of the es-carpment bounding the southern edge of the northern coastal plain has been differentially upwarped. Unconsolidated sediments containing ostracodes and thought to be marine sediments of the Gubik Formation have been re-covered from an altitude of about 165 meters from a drill hole just north of the escarpment near the east edge of the coastal plain shown on Fig. 1 (Brosgé and Whittington, 1967). If these beds are contemporaneous with the marine sediments of the Gubik Formation that occur at an altitude of about 60–70 meters near the escarpment on the Kukpowruk River, the up-warping may be as much as 100 meters. If the escarpment is a marine cliff cut by the middle Pleistocene transgressive sea, the deformation must have occurred at some more recent time.

Neither the transgression that deposited the oldest known sediments on the western end of the northern coastal plain nor the possible earlier transgression that cut the scarp bounding the southern edge of the coastal plain can be closely dated by the evidence now available. This possibly double transgression is assigned a middle Pleistocene age because (1) deposits probably belonging to this transgression overlie older transgressive deposits with a late Pliocene or early Pleistocene fauna at Ocean Point, and (2) the deposits of this transgression contain extinct mollusks not found in younger pre(?)-Illinoian marine beds in the Kotzebue Sound area and their probable correlatives on the northern coastal plain.

The largest mollusk collections (36 species) from these deposits are from the ancient beach ridges transected by the Kukpowruk and Epizetka Rivers. Two forms, *Neptunea heros mesleri* and *Astarte leffingwelli,* are extinct and are not found in beds presumed to be younger. The fauna also lacks species that make their first appearance in younger transgressive deposits. The sites of other collections in which one or both of the extinct forms occur are shown on Fig. 2, and their associated mollusks bring the aggregate fauna to 48 species.

## Pre(?)-Illinoian Transgression

*Kotzebue Sound area.* Deposits of a still younger Pleistocene transgression underlie glacial drift of probable Illinoian age in the Kotzebue Sound area. These deposits form the type locality of the Kotzebuan trangression of Hopkins (this volume). The stratigraphy and age of these deposits, shown diagrammatically in Fig. 3, have been discussed by McCulloch *et al.*

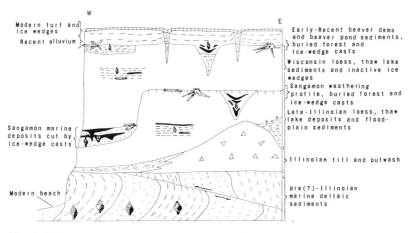

Fig. 3. Diagrammatic cross section of the Baldwin Peninsula.

(1965), and are only briefly reviewed here. The lowermost deposit exposed in the area is a highly fossiliferous marine deltaic sediment that was overridden by glacial ice from the valleys of the Kobuk and Noatak Rivers and from an icefield in the Selawik lowlands. The advancing ice deformed the marine sediment into a push moraine and deposited the overlying till and outwash. The push moraine is thought to have been formed during the Illinoian Glaciation because the weathering of the till and the modification of the morainal topography are comparable to those of the glacial deposits of the Nome River Glaciation (Hopkins et al., 1960) of Illinoian age. Moreover, the push moraine lies many kilometers downvalley from less modified moraines to which Fernald (1964) assigns a Wisconsin age.

The molluscan fauna from the marine sediments contains 65 species (identified by F. S. MacNeil) and is the oldest known Alaskan fauna composed entirely of living species. The fauna also differs from older faunas by the first appearance of *Macoma balthica*. The sediments have yielded a foraminifer fauna of 13 shallow brackish-water species (identified by Patsy B. Smith). All but *Elphidiella groenlandica*, which may not have survived the end of the Pleistocene, live in the area today. Similarly, all the mollusks are found living in the adjacent waters.

The present ranges of the mollusks found as fossils indicate that the temperature of the transgressive sea was like that of the present Kotzebue Sound. Because, as indicated by the foraminifera, the water was brackish and yet as warm as it is at present, it seems probable that the transgression occurred during a nonglacial period; if the transgression had been glacial, the brackish water in the bay would have been cooled considerably by meltwater from the extremely large ice masses that occupied the Selawik lowlands and the Noatak and Kobuk River valleys.

The fact that the marine sediment is found to an altitude of about 20 meters also suggests that the transgression did not occur during a period of glaciation, for there is no evidence for the substantial uplift that would have been necessary to raise glacially contemporaneous sediments, presumably deposited during a low stand of sea level, to the height at which these sediments are found.

Because of their position beneath till thought to be of Illinoian age, and the evidence that they were deposited in a high warm sea, these sediments were assigned to the Yarmouth Interglaciation (McCulloch et al., 1965). Shells from the sediments have yielded radiometric ages (Blanchard, 1963) of $170,000 \pm 17,000$ and $175,000 \pm 16,000$ years by $Th_{230}/U_{238}$ ratios, ages in accord with Emiliani's (1961) estimate of Yarmouth time.

These dates are, however, considerably younger than the 1.36-million-year date cited by Evernden *et al.* (1964) for a basalt associated with sediments containing a *Mammuthus*-bearing fauna of Irvingtonian age. Hibbard *et al.* (1965) equate the Irvingtonian land mammal age with the latter part of Kansan and the whole of Yarmouth time. Barring errors in the radiometric age determinations, the great difference between these dates suggests either that there is a large range in the ages of events assigned to the Yarmouth or that the marine sediments postdate the Yarmouth Interglaciation. Further work may show that the Illinoian Glaciation of Alaska was interrupted by one or more interstadials, and that these marine sediments were deposited during the Illinoian; however, until such evidence is available, and because these marine deposits are beneath till of probable Illinoian age, the transgression is assigned a pre(?)-Illinoian age.

*Northern coastal plain.* There are no transgressive deposits on the northern coastal plain that can be correlated with complete confidence with the pre(?)-Illinoian sediments of the Kotzebue Sound area; however, there are transgressive deposits for which this age seems most likely. These deposits generally occur as horizontally bedded sand, silt, and fine gravel, often lying on bedrock, and are found to altitudes of about 33 meters— considerably higher than marine deposits thought to be of Sangamon age, from which they are occasionally separated by an ancient wave-cut cliff. Beach ridges are clearly visible on the Sangamon and Recent deposits, but are less distinct on the deposits of probable pre(?)-Illinoian age.

Pre(?)-Illinoian transgressive deposits are well exposed on the west bank of the mouth of the Kuk River (Fig. 2). In descending order, the section consists of (1) 1 meter of brown sand containing marine shells, with a deflationary lag pebble gravel at its top; (2) 2 meters of pebble- and cobble-gravel containing few shell fragments; (3) 8 meters of interbedded sand and silty sand with a few pebble beds and abundant marine shells; and (4) bedrock to a height of 11 meters above the river.

Locally, these pre(?)-Illinoian deposits lie seaward of a break in slope that occurs at an altitude of about 33 meters. As suggested by O'Sullivan, this break in slope to the east of Kuk River is probably an ancient shoreline. On recently published topographic maps a break in slope at about the same altitude can be found to the southwest, and with less assurance to the northeast. The 33-meter altitude is about 10 meters higher than the highest pre(?)-Illinoian marine sediments found thus far in the Kotzebue Sound area; however, the marine sediments near Kotzebue were deposited at some distance from shore and were overridden and presumably eroded by Illinoian

glaciers. Where these deposits are found seaward of the shoreline of the succeeding Sangamon transgression, they are distinguished from the younger
sediments by minor differences in their molluscan faunas.

These deposits have yielded an aggregate fauna of 57 species, collected
from the sites indicated on Fig. 2. This fauna is distinctly younger than the
fauna of the middle Pleistocene marine transgression, because it contains
no extinct species but does contain the now-living species *Amauropsis pur-
purea, Pyrulofusus deformis, Neptunea borealis,* and a species of *Buccinum*
not found in the middle Pleistocene deposits. F. S. MacNeil (written communication, December 2, 1963) says that these species, regardless of their
ultimate origins, make their first appearance in Alaska at this time.

## Illinoian Glaciation

The Illinoian Glaciation is locally represented by the period of maximum
glaciation along the Chukchi Sea coast. Large valley glaciers formed in the
Brooks Range. At the eastern end of the north side of the range, where the
foothill belt is narrow, ice flowed for a short distance out onto the northern
coastal plain. To the west, glaciers on the north side of the range terminated
in the wider foothill belt, and the northern coastal plain remained ice-free
(Fig. 1 and Karlstrom *et al.,* 1964). Glaciers flowing southward from the
range were much more extensive. Piedmont glaciers came to within 15 km
of the coast near Kivalina, and what were possibly the largest valley glaciers
ever to exist in northern Alaska filled the valleys of the Kobuk and Noatak
Rivers. The Selawik lowland, lying east of Kotzebue Sound, was covered
by an ice sheet more than 200 km long and 75 km wide. Altitudes of marginal meltwater channels and of sediments ponded in the valleys of mountain
streams dammed by this ice sheet indicate that the margin of the ice sheet
was at least 65 meters thick. It was ice from this icefield and from the Noatak
and Kobuk valley glaciers that overran the pre(?)-Illinoian sediments in
Kotzebue Sound (Fig. 3). As these valley glaciers retreated, loess blown
from the valley trains of their meltwater streams was deposited as a blanket
over the moraines. Buried ice-wedge casts and the deposits of thaw lakes
within the loess indicate that the climate was cold during deglaciation.

Creager and McManus (1961, 1965, and this volume) have shown that
a submarine valley (Hope Seavalley), cut in bedrock of the Chukchi Sea
floor, can be traced discontinuously from the mouth of Kotzebue Sound
northwestward to the Arctic Ocean basin (Fig. 1). They suggest that this
valley was carved by subaerial erosion during one or more intervals of low

sea level. Some of the erosion may have been accomplished during the low
sea level that must have accompanied this period of maximum glaciation.

## Sangamon Interglacial Transgression

*Kotzebue Sound area.* The clearest stratigraphic evidence for a marine
transgression of Sangamon age, and the best clues to the character of the
Sangamon climate along the Chukchi Sea coast are found in the Kotzebue
Sound area (Fig. 3). The harsh permafrost climate that had existed during
the accumulation of Illinoian loess gave way to a more genial climate, and
loess deposition ceased. Ice wedges that had been growing in the loess were
melted, and the permafrost table was lowered sufficiently to allow a weathering profile to form on the Illinoian loess and till. The forest expanded into
this now-treeless tundra area, as is recorded by the stumps of small trees
that are found rooted in Illinoian loess and covered by Wisconsin loess.

Sea level was perhaps 10 to 12 meters higher than at present, for on the
southwest side of the Baldwin Peninsula, beach sands containing marine
mollusks and driftwood are found to an altitude of 12 meters at the toe of
a low marine bluff cut into Illinoian loess. On the southeast coast of the
Baldwin Peninsula, facing Eschscholtz Bay, fossiliferous marine beach sands
found to an altitude of about 6 meters occur as two beds separated by a peat
layer and peat-filled ice-wedge casts containing twigs and fresh-water mollusks. Wood from one of the ice-wedge casts has a radiocarbon age of
>42,000 years (W-1251). The ice-wedge casts enclosed between marine
beach sediments suggest that the rise of sea level was irregular to its peak
in Sangamon time, and that the climate during the interruption of the rise
in sea level was severe enough to promote the growth of ice wedges.

As is shown diagrammatically in Fig. 3, the contemporaneity of the
marine deposits, the interglacial weathering profile, the buried forest, and
the ice-wedge casts are established both by their position at the top of the
Illinoian loess and by the fact that they are covered by a younger loess
blanket thought to be of Wisconsin age.

Fossiliferous marine sediments of probable Sangamon age are also exposed in the low bluffs on the south shore of Kotzebue Sound, a few kilometers west of Deering. When visited, these exposures were poor, and the
deposits have not been studied in detail; however, the limited observations
made suggest that the marine sediments transgress unconsolidated sediments
as old as middle Pleistocene in age, or older. The fauna from the marine
sediments consists of only eight forms, all of which live today in the Arctic

and in Bering Sea. The marine sediments occur to an altitude of about 5 meters, and the shoreline angle may be preserved where the sandy marine sediments terminate laterally against a woody peat in which there are rooted stumps up to 15 cm in diameter.

*Coastal areas between Kotzebue Sound and the northern coastal plain.* Marine deposits also of probable Sangamon age are found elsewhere along the shore of Kotzebue Sound and at many places northward along the shore of Chukchi Sea. These deposits occur to an altitude of about 13–14 meters and often lie seaward of a marine cliff cut in bedrock or a break in slope in unconsolidated sediments. Many of the deposits are fossiliferous.

Moore and Scholl (1961) collected a small molluscan fauna (seven species) from marine sediments exposed at the north end of the lagoon behind Cape Krusenstern (Fig. 1). My collecting from the same and nearby sites in 1962 increased the fauna to 22 species. The exposed marine sediments are composed of 11 meters of interbedded fine-to-medium-grained, well-sorted sand and well-sorted pea-gravel. Most of the fossils were found in the lower 1.3 meters. *Protothaca adamsi*, obtained in the original and subsequent collections, is a form that now ranges from central to northern Japan and is not found along the coast of North America. Its only other fossil occurrence in Alaska is in beds ascribed to the Sangamon transgression (Second Beach of Hopkins *et al.*, 1960; Pelukian transgression, Hopkins, this volume) at Nome. On the basis of this fossil, a radiocarbon age of 26,000 ± 400 years for peat overlying the marine sediment (which they cite to show that the marine deposit has some antiquity), and the fact that sediment was deposited during a high stand of the sea, Moore and Scholl (1961) assign these deposits to the Sangamon Interglaciation. They also assign two beach ridges at Point Hope to the Sangamon high sea level, primarily on the basis of their altitudes of 3 and 12 meters. South of Point Hope, near Ogotoruk Creek (Fig. 1), marine sediments called the Chariot Gravels occur to an altitude of about 14 meters on an elevated wave-cut bedrock marine platform fronting an ancient marine cliff that approximately parallels the modern shoreline (Campbell, 1967). Similar deposits are found on an ancient bay-mouth bar across the valley floor at the mouth of Ogotoruk Creek (Sainsbury *et al.*, 1965). The marine sediment has yielded only a few fossils—all of them forms that live in the area today. West of Ogotoruk Creek, a driftwood log with a radiocarbon age of 38,000 years was found near the top of 6 meters of marine beach sand and gravel that lie on a wave-cut bedrock platform 7.5 meters above sea level.

*Northern coastal plain.* Still farther to the north, along the west end of the northern coastal plain, marine sediments probably dating from the

Sangamon transgression are exposed in the low coastal bluffs and occasionally lie seaward of a low escarpment thought to be a wave-cut cliff. The suggested shoreline for the Sangamon transgression in this area (Fig. 2) is drawn along this topographic break at an altitude of about 13 meters.

Molluscan faunas from the northern coastal plain deposits are small and contain only living species. All but *Trichotropis insignis,* found in a collection at the mouth of the Epizetka River, live in the area at present. *T. insignis* now ranges only as far north as Bering Strait (Burch, 1944–46).

The marine deposits on the west edge of the northern coastal plain are assigned to a Sangamon transgression, because (1) the modern aspect of their faunas indicates that they postdate older transgressions; (2) the deposits commonly lie seaward of a wave-cut escarpment, and sometimes lie upon a wave-cut bedrock surface, and thus are transgressive; and (3) they are found at about the same altitude as deposits at Kotzebue Sound and Nome, for which there is additional stratigraphic evidence for a Sangamon age assignment.

A sheet of marine deposits associated with beach ridges at altitudes of about 10 meters near Barrow and the south shore of Teshekpuk Lake (Fig. 1) contains a mollusk fauna similar to the modern Arctic coast fauna (Mac-Neil, 1957, and written communication, undated, 1961). These deposits are probably of Sangamon age; however, they have not been studied in detail, and this age assignment is speculative.

Possible further evidence for transgressive deposits of Sangamon age is the occurrence of the fossil *Bathyarca glacialis* Gray (identified by F. S. MacNeil) found in a drill hole on Cape Simpson, 150 km east-southeast of Barrow (Robinson, 1964, Simpson core test 13). The fossil was recovered at a depth of 10–10.5 meters (altitude approximately sea level) from within 25 meters of unconsolidated fossiliferous marine sediments of the Gubik Formation that rests upon the more highly indurated Seabee Formation of Late Cretaceous age. This is the only known Alaskan occurrence of *B. glacialis.* This same species is reported by Merklin *et al.* (1962) to occur in glaciomarine deposits of Middle Quaternary age (middle horizon of Kresta Suite) and in the deposits of the succeeding transgression (Val'katlen beds), to which they assign an "upper" Quaternary age. Assuming that transgressions resulted from high interglacial sea levels, and that *B. glacialis* Gray lived along the Alaskan coast of the Arctic Ocean during the time represented by the deposits on the Chukotka Peninsula, we may conclude that part of the Gubik Formation can be correlated with the Val'katlen beds assigned by Hopkins *et al.* (1965) to the Sangamon Interglaciation.

*Summary of Sangamon deposits.* The high Sangamon sea transgressed the eastern Chukchi Sea coast, depositing often-fossiliferous marine sediments to about 14 meters above modern sea level. The rise in sea level seems to have been interrupted by at least one lowering. A weathering zone, the expansion of the forest, and the melting of ice wedges in the Kotzebue Sound area show that the climate was less favorable for the growth of permafrost than at present.

The aggregate marine molluscan fauna from the Sangamon deposits contains only 39 forms. All are living forms, and all but three—*Trichotropis insignis*, *Natica janthostoma*, and *Protothaca adamsi*—live in the area today. The northern limit of these three forms now lies south of Chukchi Sea. *T. insignis* is found living south of Bering Strait. *P. adamsi* and *N. janthostoma* no longer inhabit the Alaskan coast; both are found off central and northern Japan, the latter ranging as far north as the Commander Islands. *P. adamsi* and *N. janthostoma* are both found as far north as Cape Krusenstern in the Sangamon deposits, and at present range only as far north as the winter limit of sea ice. The presence of these three forms and similarly the presence of *Pododesmus macroschisma* and *Pholadidea penita* in Sangamon deposits at Nome (Hopkins, this volume) suggest that the Bering and Chukchi Seas of Sangamon time were warmer than their modern counterparts.

Correlation of the Sangamon deposits north of Bering Strait with Sangamon deposits at Nome, often possible by one or more of the criteria of stratigraphic position, present altitude, or water temperature indicated by the molluscan fauna, makes it possible to assign a finite date to these deposits. Shells from the Sangamon deposits at Nome have yielded age determinations of $100,000 \pm 8,000$ years by $Th_{230}/U_{238}$ and $78,000 \pm 5,000$ years by $Ra_{226}/U_{238}$ ratios (Blanchard, 1963). These two dates lie within the range of radiometric dates determined for material of supposed Sangamon age from deep-sea cores (Broecker, 1965).

## Wisconsin Glaciation

The history of Wisconsin Glaciation in northern Alaska is complex. Four stades, all of "classical Wisconsin" age, have been recognized on the north side of the Brooks Range (Porter, 1964), and two Wisconsin stades are recognized on the south side (Fernald, 1964). However, at their maximum extent the Wisconsin glaciers were considerably smaller than their Illinoian predecessors (Fig. 1). For example, Illinoian glaciers filled the valleys of the Kobuk and Noatak Rivers, but the moraines of the Wisconsin glaciers lie close to, or within, the mountains at the valley heads. The northern coastal plain again remained free of glacial ice, and the moraines of glaciers that grew on the north side of the northern ranges lie upvalley from pre-Wisconsin moraines. To the south, on the western end of Seward Peninsula, Wisconsin glaciers extended far enough from the mountains to cover the Sangamon beach line in several places (Sainsbury, this volume).

In the Kotzebue Sound area, the Wisconsin Glaciation was a time of lower sea level, loess deposition, and thaw lake development (Fig. 3). On the Baldwin Peninsula, a blanket of Wisconsin loess as much as 10 meters thick was deposited on the weathered surface of the Illinoian loess and till, upon the Sangamon marine sediment, which was by that time exposed by the retreating sea, and upon the forest thought to be of Sangamon age. We may presume that permafrost conditions returned to this area during loess deposition, for thaw-lake deposits and ice-wedge casts are found throughout the loess. Two radiocarbon ages of >38,000 years (W-1256, W-1257) and an age of 34,000 ± 2,000 years (W-1262) determined from twigs from within the loess are compatible with a Wisconsin age assignment.

Far to the south, in the Anchorage area, there are fossiliferous marine sediments thought to have been deposited during a mid-Wisconsin high stand of sea level (Woronzofian transgression of Hopkins, this volume). On the northern coastal plain there is also evidence for a high mid-Wisconsin sea level. Near Barrow, organic fibres from a thin, discontinuous organic layer in the sandy gravel of a beach ridge standing at an altitude of about 7 meters have yielded a radiocarbon age of 25,300 ± 2,300 years (I-1384; Brown, 1965). These beds contain an impoverished molluscan fauna and a small foraminiferal fauna indicating cold, shallow water (Faas, 1962, 1964). The deposits are underlain by marine beds that have a much richer and more diversified fauna; I consider the lower beds to represent the Sangamon transgression, but Faas suggested that they are of Tertiary age.

The mid-Wisconsin transgression probably accounts for the ice-rafted boulders that have been found along the northern edge of the coastal plain (MacCarthy, 1958). These boulders are composed of exotic lithologies, and some are faceted and striated. They occur to an altitude of 8 meters, about the same altitude as the dated elevated beach ridge, and they have not been covered by later marine sediments.

If, as is suggested by Curray (1965), the highest position of the sea during the mid-Wisconsin transgression was several meters below its present level, the Barrow area has undergone more than 8 meters of uplift to raise the marine sediments and ice-rafted boulders to their present altitudes.

The lowering of sea level during the Wisconsin Glaciation drained the Chukchi Sea and exposed its floor to subaerial erosion. If we use Shepard's (1964) and Curray's (1961, 1965) estimates of the position of sea level, the entire Chukchi Sea floor would have been exposed until perhaps 12,000 to 14,000 years ago. Then, for the succeeding 10,000 years, sea level rose irregularly to near its present position. The available data give only a limited picture of what was happening on the exposed sea floor during the Wisconsin Glaciation and the succeeding time. Probably most of the westward-draining rivers were tributary to the Hope Seavalley (Creager and McManus, 1965, and this volume) and to a submarine valley that extends southwest along the coast from Barrow for about 150 km (Carsola, 1954; Lepley, 1962). Ogotoruk Creek (Fig. 1) was probably tributary to Hope Seavalley; Scholl and Sainsbury (1961a,b) have shown that during a low stand of sea level Ogotoruk Creek extended and downcut its valley floor across a now sub-merged bedrock marine platform. They suggested that some or possibly all of this downcutting may have occurred during the Wisconsin Glaciation.

Some of the details in the irregular rise in postglacial sea level are sug-gested by Creager and McManus (this volume), who attribute interruptions in the continuity of Hope Seavalley to deltaic deposition that took place during periods of stillstands or slower sea-level rise. The earliest such period is shown by an extensive flat area at a depth of 54 to 58 meters that Creager and McManus suggest resulted from marine and deltaic deposition during a lengthy stillstand. They assign an age of between 12,000 and 17,000 years to this surface by correlating it with low stands of sea level at about this depth described by other investigators in other areas that have been postulated to have occurred between 12,500 to 13,300 and 16,000 to 17,000 years ago. The next interruption in rising sea level is recorded by brackish-water deltaic and marine sediments filling the Hope Seavalley at a depth of 38 meters. Creager and McManus recovered a 7.4-meter core from these deposits about 100 km northwest of Cape Krusenstern. The core consists of soft gray mud, and at all levels examined it contains benthic and pelagic foraminifera, as well as spores of the freshwater alga *Pediastrum*; pollen of a freshwater aquatic herb, *Potomogeton*, and a water lily, *Nuphar*; and the eggs of a nonmarine tardigrade arthropod (Colinvaux, 1964, and this volume). The presence of both freshwater and saltwater biota suggests to Creager and McManus that the sediments in the lower 5.5 meters of the core were deposited in an estuarine delta. They conclude from radiocarbon dates that the sediments in the upper 2.2 meters of the core were deposited more slowly, and are probably marine. Colinvaux examined the pollen in seven samples spaced about a meter apart throughout the length of the core; in all but the sample at the very top of the core, the pollen suggests a grass and tussock tundra with dwarf birch, and only minor changes in the percentage of the constituents throughout the core. Minor floral elements in the lower

6 meters of the core suggest to Colinvaux that the flora lived near the banks of a watercourse. The flora in the surface sample shows an increase in alder and spruce. Six radiocarbon ages have been determined for organic carbon throughout the core. These ages, which range from 4,390 ± 210 to 15,500 ± 800 years B.P., are so disarranged that no consecutive dates are juxtaposed, and the oldest age determination is from the 2.88-to-3.40 meter interval (Creager and McManus, this volume, Table 1). This disarrangement of radiocarbon ages suggests that the delta sediments are to some degree composed of recycled sediment. By disregarding the 15,500 ± 800–year age as being inconsistent with the other radiocarbon ages, and by assuming that contamination has made the remaining dates too old by about the same amount that the radiocarbon age (4,390 ± 210 years B.P.) near the top of the core exceeds usual surface sediment dates, Creager and McManus conclude that the delta was formed about 12,000 years ago.

## Recent Time

The beginning of Recent time nearly coincides with the start of a warm period that began about 10,000 years ago and lasted until at least 8,300 years ago. Evidence for this warming comes from the shores of Kotzebue Sound and Seward Peninsula, from the Barrow area, and from the north-central Brooks Range. The most complete evidence for this warm period has been found in the Kotzebue Sound–Seward Peninsula area (McCulloch and Hopkins, 1966), which during the Wisconsin Glaciation was largely treeless tundra, much as it is today; the permafrost table lay near the surface of the ground, ice wedges were actively growing, and thaw lakes existed in areas underlain by fine-grained sediments (Fig. 3). As the climate ameliorated, birch and spruce began to grow beyond their present limit, spreading onto the Baldwin Peninsula and into some stream valleys on Seward Peninsula. Poplar, which today grows on well-drained ice-free sites in these areas, spread to sites underlain by fine-grained sediments at present poorly drained because of a shallow permafrost table that reduces the permeability of the ground.

The expansion of the forest in this coastal area suggests that there was an increase in the number of warm summer days, for, as shown by Hopkins (1959b), the edge of the spruce forest in Alaska is fixed by the total length of time during which the temperature equals or exceeds 50° F. (10° C.), and is probably independent of the severity of winter temperatures.

With the expanding forest came the beaver. Logs from buried beaver dams in the Kotzebue Sound area have been dated at 8,550 ± 400 (W-1249) and 9,480 ± 160 years (Y-1351) (J. Ostrum, written communication, September 29, 1964). On Seward Peninsula, wood found with beaver-gnawed twigs in a valley fill has a radiocarbon age of 8,350 ± 200 years (L-117C) and wood from a buried beaver dam of large logs has an age of 9,400 ± 750 years (L-137N). The melting of ground ice that accompanied the expansion of the forest resulted in the development of ice-wedge casts and allowed the formation of a weathering profile that now lies below the permafrost table. The radiocarbon age of a small piece of wood from an ice-wedge cast, upon which the weathering profile had developed, suggests that the weathering took place some time after 11,340 ± 400 years ago (W-1254). Wood from an ice-wedge cast associated with a beaver pond, and thought to have been formed during the development of the weathering profile, has a radiocarbon age of 9,020 ± 400 years (W-1255). Three additional radiocarbon dates of woody material thought to represent a warmer climate (10,200 ± 800, L-137G; 9,690 ± 400, W-48S; 8,800 ± 1,000, L-117E) make a total of eight relatively closely clustered dates for this warmer interval. These dates suggest that it began about 10,000 years ago and lasted until at least 8,300 years ago.

A few of the many ice-wedge casts formed during this warming have inactive wedge ice in their lower portions, which may be relict and may predate the melting. These old ice wedges were nearly twice the size of modern active ice wedges. Assuming a rate of ice-wedge growth similar to the present rate, the greater size of the older ice wedges suggests that the warm interval was preceded by a considerably longer period of low ground temperature than the present period of active ice-wedge growth.

The nonmarine molluscan fauna in the Kotzebue Sound area seems to have expanded during this period. Eleven collections from deposits of late Illinoian through Wisconsin age have yielded a total of only nine species, whereas eight collections from deposits associated with this early Recent warming have an aggregate fauna of 19 species (McCulloch et al., 1965).

Sometime after about 8,300 years ago, the climate in the Kotzebue Sound–Seward Peninsula area again changed. The number of warm summer days decreased, forcing the edge of the forest inland. With the retreating forest went the beaver. The depth of summer ground thawing decreased, the permafrost table was raised to within a meter or less of the present surface, and a new generation of ice wedges began to grow.

Because the amelioration of the climate occurred about 10,000 years ago, it is attributed to the worldwide climatic warming, for which there is abundant, diversified, and widespread evidence (Deevey and Flint, 1957; Broecker et al., 1960). Corroborative evidence that the change in worldwide

climate affected northern Alaska at about this time comes from the pedo-
logic studies of Douglas and Tedrow (1960) and Brown (1965) on the
northern coastal plain. These workers report a buried, somewhat discontinuous, organic horizon within the permafrost. Six radiocarbon ages determined for this material range from 10,900 to 8,200 years B.P. Douglas
and Tedrow suggest that this organic material may have been formed in
a period of warmer climate during which there was a decrease in ground ice.

Additional evidence for warming of the northern Alaskan climate at
about this time is found in the dating of the events in the glacial history of
the north-central Brooks Range. Porter (1964) has shown that the last
major glaciation (Antler Valley readvance) reached its maximum extent
perhaps 10,000 to 11,000 years ago, and was followed by a period during
which there was a considerable reduction in the size of the valley glaciers.
This period of ice recession was terminated by a minor readvance (Anivik
Lake) about 8,300 years ago.

In other areas of northern and southern Alaska (Heusser, 1960) there
is evidence that the postglacial thermal maximum was later than this brief,
early Recent warming. For example, studies by Livingstone (1955, 1957) of
fossil pollen from radiocarbon-dated cores from lakes and a valley-fill deposit on the north side of the central Brooks Range suggest that an alder
peak that appears shortly after 5,890 ± 170 years B.P. represents the period
of highest postglacial temperatures. Similarly, Porter's (1964) reconstruction
of the glacial history of the central Brooks Range suggests that maximum
glacial retreat occurred sometime between 6,000 and 3,000 years ago. Evidence for a well-developed warming of this age has not been found in the
coastal Kotzebue Sound–Seward Peninsula area, and it has been suggested
(McCulloch and Hopkins, 1966) that as sea level approached its present
level at the end of the period of rapid postglacial sea-level rise (Shepard,
1964; Curray, 1961), the climate in this coastal area became distinctly
maritime. The maritime influence, then as now, so reduced summer temperatures that there was no marked increase in the depth of annual ground
thawing, and the forest and beaver could not expand as they had done in
the earlier brief warm period. In short, the maritime climate prevented the
development of the kinds of evidence by which the earlier warm period can
be recognized. Two of the radiocarbon-dated occurrences of trees in this
now treeless coastal area do postdate the end of the early Recent warming
(7,270 ± 350, W-1250; 3,600 ± 500, L-117E). The older date lies between
the cluster of dates for the early Recent warming and those for the later
postglacial thermal maximum; it may reflect a local change in edaphic conditions. The younger date, however, seems to be compatible with the thermal

maximum, suggesting that there was a limited response in this coastal area to the last period of postglacial warming.

Since the thermal maximum, the climate has had minor warm and cold periods. These climatic fluctuations are reflected in the glacial history of the north-central Brooks Range, established by Detterman *et al.* (1958) and revised with respect to time and sequence by Porter (1964). The thermal maximum ended with the onset of a cool period, during which small valley glaciers of the Alapah Mountain Glaciation formed in small tributaries of larger valleys previously occupied by ice of the early Recent Anivik Lake Glaciation. Porter recovered organic material with a radiocarbon age of 2,830 ± 120 years (Y-771) from outwash he believes to have been deposited at the maximum of the·Alapah Mountain Glaciation. This glaciation was followed by a brief warm interval, which in turn was terminated by the cooling that accompanied the growth of Fan Mountain I glaciers. These glaciers were considerably smaller than the Alapah Mountain glaciers, and their moraines lie well up the tributary valleys. Porter estimates that their age may fall within the range of 1,000–1,500 years B.P. Once again there was a brief warm interval, starting with the recession of Fan Mountain I glaciers and ending with the regrowth of still smaller glaciers belonging to Fan Mountain II Glaciation. These glaciers were essentially restricted to the cirque floors, and Porter correlates them with the glacial advance that reached its maximum in North America in the early eighteenth century.

## Summary

The Quaternary history of the Alaskan shores of Chukchi Sea is primarily a history of successive marine transgressions, for it is not until the latter part of Pleistocene time that there is any record of intertransgressive intervals. Part of the interpretation of the stratigraphic record is based on the evidence that two transgressions occurred during nonglacial high-sea-level stands. This is demonstrable for the last two transgressions in the Kotzebue Sound area, where marine sediments, deposited during climates as mild as the present climate or milder, are found interbedded with glacial deposits. Although transgressive deposits can be correlated to some degree on the basis of their altitude and the extent to which their surfaces have been modified, the principal tool used to separate these deposits in any given area, and to correlate between areas, is the small but presumably significant differences in their marine molluscan faunas. Evidence for the melting of permafrost and expansion of the forest found in deposits of Sangamon and early Recent age in the Kotzebue Sound area suggests an amelioration of the climate.

The main events in the history are shown in Table 1, and are briefly summarized below.

The first event of possible Pleistocene age for which there is a stratigraphic record is the marine transgression that deposited fossiliferous sediments upon a wave-cut bedrock platform near Kivalina. Although this is the only evidence for this transgression along the Alaskan shores of Chukchi Sea, marine deposits exposed along the Colville River on the eastern part of the northern coastal plain may also have been deposited by this transgression. The molluscan fauna of the Kivalina deposits indicates that the transgression was of late Pliocene or early Pleistocene age.

Between the withdrawal of this transgression and the advance of the succeeding middle Pleistocene transgression there was a period of unknown length for which there is no recognized stratigraphic record. The middle Pleistocene transgression covered the northern coastal plain, and the escarpment at the southern edge of the plain may be a wave-cut cliff formed by this transgressive sea. The transgression may have had two periods of high sea level separated by a regression during which there was subaerial erosion of the upper portion of the exposed wave-cut bedrock platform and deposition of nonmarine sediments. Frost-stirring of the nonmarine sediments suggests that the climate during some part of the regression was conducive to the growth of seasonal frost. Mollusks from the marine deposits of this middle Pleistocene transgression can be separated into two slightly differing faunas, which also suggests that the sediments were deposited by two transgressions. Upwarping has since raised the marine sediments differentially at least 100 meters in the south-central part of the northern coastal plain.

Following the retreat of this transgression, there was again a period for which there is no stratigraphic record; however, sea level may have been low enough during this time to drain Chukchi Sea and to expose its floor to subaerial erosion, initiating the erosion of the now-submerged Hope Seavalley.

Sea level rose again in pre(?)-Illinoian time, depositing marine sediments both on the northern coastal plain and in the Kotzebue Sound area. The present ranges of the marine mollusks found as fossils in the deposits of this transgression in Kotzebue Sound indicate that the pre(?)-Illinoian Chukchi Sea was at least as warm as it is today.

The Illinoian Glaciation, the period of maximum glaciation in northern Alaska, followed the withdrawal of the high pre(?)-Illinoian transgressive sea. Large glaciers from the valleys of the Kobuk and Noatak Rivers and ice from an icefield in the Selawik lowlands overrode the newly exposed marine deposits and deformed them into a large push moraine, the present-day Baldwin Peninsula. As these glaciers receded, loess blown from their valley trains was deposited over the till and outwash. Permafrost conditions existed during loess deposition, as shown by buried deposits of successive thaw lakes and ice-wedge casts throughout the thickness of the loess.

Loess deposition ceased, and the climate ameliorated sufficiently to melt some of the ground ice. The forest spread onto the Baldwin Peninsula, and a weathering profile developed on the Illinoian deposits. With the change in climate came the transgression of Sangamon age. Deposits of this transgression are found discontinuously along the entire Alaskan coast of Chukchi Sea and south of Bering Strait. Ice-wedge casts interbedded between beach deposits of this transgression on the Baldwin Peninsula suggest that as the Sangamon transgression rose irregularly to its high position it was interrupted by a regression during which the climate was favorable for the growth of ground ice.

The Sangamon marine deposits contain a fauna composed entirely of living taxa, all but three of which live in the area today. The three extra-limital forms live today south of Bering Strait, and their presence in this fauna suggests that the transgressive sea was warmer than the present Chukchi Sea.

During the Wisconsin Glaciation, small valley glaciers grew in the Brooks Range. Although these glaciers did not reach the coast, loess blown from their outwash valleys mantled the marine sediment, forest, and weathering profile of Sangamon age on the Baldwin Peninsula. Ground-ice features in the loess are evidence that the glacial climate was more severe than it had been during the preceding interglaciation.

As suggested by a radiocarbon-dated elevated beach ridge near Barrow, part of the northern coastal plain was covered by a mid-Wisconsin transgression. Glacially faceted and striated boulders of exotic lithology that are found along the coast of the northern coastal plain, at about the same altitude as the dated beach ridge, were probably ice-rafted in this high mid-Wisconsin sea. Uplift since the mid-Wisconsin transgression has raised these deposits about 8 meters above present sea level. The sea, which had been sufficiently low to expose the Chukchi Sea floor probably both before and after the mid-Wisconsin transgression, rose during late Wisconsin time. Interruptions in the continuity of the submerged Hope Seavalley, attributed to periods of estuarine-deltaic and marine deposition, suggest that there were slight pauses or minor regressions during the rise of sea level.

At about the beginning of Recent time (10,000 years B.P.), summer temperatures increased and conditions became less favorable for the growth of ground ice in northwestern and northern Alaska. During this period the forest and beaver spread into the now treeless tundra of the Kotzebue Sound–Seward Peninsula area. Eight radiocarbon dates on buried beaver dams and organic material from ice-wedge casts in this area suggest that this warm period ended about 8,300 years ago. Six radiocarbon dates, ranging from 10,000 to 8,200 years B.P. for a buried organic horizon at Barrow thought to have been formed during a warmer climate, and an interval of marked glacial recession in the north-central Brooks Range that began about 10,000 to 11,000 years ago and ended with a minor readvance about 8,300 years ago, suggest that this brief warming affected other parts of northern Alaska.

Studies of radiocarbon-dated fossil pollen and evidence of a period of glacial recession in the north-central Brooks Range suggest that the postglacial thermal maximum occurred in northern Alaska some time between 6,000 and 3,000 years ago, well after the end of the early Recent warm interval. Because of the lack of evidence for this younger warming in the coastal Kotzebue Sound–Seward Peninsula area, it is suggested that the climate in this area was made distinctly maritime as the seashore approached its present position at the end of the rapid rise of sea level 7,000 years ago. The maritime climate may have reduced summer temperatures along the coast during the time when the rest of Alaska was experiencing the postglacial thermal maximum.

Brief fluctuations in the climate since the postglacial thermal maximum are suggested by events in the glacial history of the north-central Brooks Range. Cooler periods produced glacial advances at about 2,800 and 1,000–1,500 years ago and during the first half of the eighteenth century; intervening periods of glacial recession and the time since the last advance have been warmer.

## REFERENCES

Black, R. F. 1964. Gubik formation of Quaternary age in northern Alaska: U.S. Geol. Survey Prof. Paper 302-C, p. 59–91.

Blanchard, R. A. 1963. Uranium decay series disequilibrium in age determination of marine calcium carbonates: Ph.D. thesis, Washington Univ., St. Louis, 164 p.

Broecker, W. S. 1965. Isotope geochemistry and the Pleistocene climatic record, p. 737–753 *in* H. E. Wright Jr., and D. C. Frey, eds., The Quaternary of the United States—a review volume for the VII Congress of the International Association for Quaternary Research: Princeton Univ. Press, Princeton, N.J.

Broecker, W. S., W. M. Ewing, and B. C. Heezen. 1960. Evidence for an abrupt change in climate close to 11,000 years ago: Am. Jour. Sci., v. 258, no. 6, p. 429–448.

Brosgé, W. P., and C. L. Whittington. 1967. Geology of the Umiat-Maybe Creek region, northern Alaska: U.S. Geol. Survey Prof. Paper 303-H, p. 501–638.

Brown, J. 1965. Radiocarbon dating, Barrow, Alaska: Arctic, v. 18, no. 1, p. 36–48.

Burch, J. Q., ed. 1944–46. Distributional list of the west American marine mollusks from San Diego, California, to the Polar Sea: Conchological Club of Southern California Minutes, no. 54, p. 36.

Campbell, R. H. 1967. Areal geology in the vicinity of the Chariot site, Lisburne Peninsula, northwestern Alaska: U.S. Geol. Survey Prof. Paper 395, 71 p.

Carsola, A. J. 1954. Submarine canyons on the Arctic slope: Jour. Geology, v. 62, no. 6, p. 605–610.

Chapman, R. M., and E. G. Sable. 1960. Geology of the Utukok-Corwin region, northwestern Alaska: U.S. Geol. Survey Prof. Paper 303-C, p. 47–167.

Colinvaux, P. A. 1964. The environment of the Bering land bridge: Ecol. Monogr., v. 34, p. 297–329.

Creager, J. S., and D. A. McManus. 1961. Preliminary investigations of the marine geology of the southeastern Chukchi Sea: Univ. Washington Dept. Oceanography, Tech. Rept. 68, 46 p.

———— 1965. Pleistocene drainage patterns on the floor of the Chukchi Sea: Marine Geol., v. 3, p. 279–290.

Curray, J. R. 1961. Late Quaternary sea level—a discussion: Geol. Soc. America Bull., v. 72, no. 11, p. 1707–1712.

———— 1965. Late Quaternary history, continental shelves of the United States, p. 723–735 in H. E. Wright, Jr., and D. C. Frey, eds., The Quaternary of the United States—a review volume for the VII Congress of the International Association for Quaternary Research: Princeton Univ. Press, Princeton, N.J.

Deevey, E. S., and R. F. Flint. 1957. Postglacial hypsithermal interval: Science, v. 125, no. 3240, p. 182–184.

Detterman, R. L., A. L. Bowsher, and J. T. Dutro, Jr. 1958. Glaciation on the Arctic slope of the Brooks Range, northern Alaska: Artic, v. 11, no. 1, p. 43–61.

Douglas, L. A., and J. C. F. Tedrow. 1960. Tundra soils of arctic Alaska: Trans. 7th Internat. Cong. Soil Sci. (Madison, Wis.), 1960, v. 4, p. 291–304.

Emiliani, C. 1961. Cenozoic climatic changes as indicated by the stratigraphy and chronology of deep-sea cores of *Globigerina*-ooze facies: Ann. N.Y. Acad. Sci., v. 95, art. 1, p. 521–536.

Evernden, J. F., D. E. Savage, G. H. Curtis, and G. T. James. 1964. Potassium-argon dates and the Cenozoic mammalian chronology of North America: Am. Jour. Sci., v. 262, no. 2, p. 145–198.

Faas, R. W. 1962. Foraminiferal paleoecology of the Gubik (Pleistocene) formation of the Barrow area, northern Alaska: Iowa Acad. Sci., v. 69, p. 354–361.

———— 1964. A study of some late Pleistocene estuarine sediments near Barrow, Alaska: Ph.D. thesis, Iowa State Univ. Science and Technology, Ames, 190 p.

Fernald, A. T. 1964. Surficial geology of the central Kobuk River valley, northwestern Alaska: U.S. Geol. Survey Bull. 1181-K, 31 p.

Heusser, C. J. 1960. Late-Pleistocene environments of North Pacific North America—an elaboration of late-glacial and postglacial climatic, physiographic, and biotic changes: Am. Geog. Soc. Spec. Pub. 35, 308 p.

Hibbard, C. W. 1958. Summary of North American Pleistocene mammalian local faunas: Michigan Acad. Sci. Papers, 1957, v. 43, p. 3–32.

Hibbard, C. W., D. E. Ray, D. E. Savage, D. W. Taylor, and J. E. Guilday. 1965. Quaternary mammals of North America, p. 509–525 *in* H. E. Wright, Jr., and D. G. Frey, eds., The Quaternary of the United States—a review volume for the VII Congress of the International Association for Quaternary Research: Princeton Univ. Press, Princeton, N.J.

Hopkins, D. M. 1959a. Cenozoic history of the Bering land bridge (Alaska): Science, v. 129, no. 3362, p. 1519–1528.

———— 1959b. Some characteristics of the climate in forest and tundra regions in Alaska: Arctic, v. 12, no. 4, p. 215–220.

Hopkins, D. M., and F. S. MacNeil. 1960. A marine fauna probably of late Pliocene age near Kivalina, Alaska: U.S. Geol. Survey Prof. Paper 400-B, art. 157, p. B339–B342.

Hopkins, D. M., F. S. MacNeil, and E. B. Leopold. 1960. The coastal plain at Nome, Alaska—a late Cenozoic type section for the Bering Strait region *in* Chronology and climatology of the Quaternary: Proc. 21st Internat. Geol. Cong. (Copenhagen), 1960, pt. 4, p. 46–57.

Hopkins, D. M., F. S. MacNeil, R. L. Merklin, and O. M. Petrov. 1965. Quaternary correlations across Bering Strait: Science, v. 147, no. 3662, p. 1107–1114.

Karlstrom, T. N. V., H. W. Coulter, A. T. Fernald, J. R. Williams, D. M. Hopkins, T. L. Péwé, H. Drewes, E. H. Muller, and W. H. Condon. 1964. Surficial geology of Alaska: U.S. Geol. Survey Misc. Geol. Inv. Map I-357, scale 1:1,584,000.

Lepley, L. K. 1962. Submarine topography of the head of Barrow Canyon: Deep-Sea Research (Great Britain), v. 9, p. 214–217.

Livingstone, D. A. 1955. Some pollen profiles from Arctic Alaska: Ecology, v. 36, no. 4, p. 587–600.

———— 1957. Pollen analysis of a valley fill near Umiat, Alaska: Am. Jour. Sci., v. 255, no. 4, p. 245–260.

MacCarthy, G. R. 1958. Glacial boulders on the Arctic coast of Alaska: Arctic, v. 11, no. 2, p. 79–85.

McCulloch, D. S., and D. M. Hopkins. 1966. Evidence for an early Recent warm interval in northwestern Alaska: Geol. Soc. America Bull., v. 77, p. 1089–1108.

McCulloch, D. S., D. W. Taylor, and M. Rubin. 1965. Stratigraphy, non-marine mollusks and radiometric dates from Quaternary deposits in the Kotzebue Sound area, western Alaska: Jour. Geology, v. 73, no. 3, p. 442–453.

MacNeil, F. S. 1957. Cenozoic megafossils of northern Alaska: U.S. Geol. Survey Prof. Paper 294-C, p. 99–126.

Merklin, R. L., O. M. Petrov, and O. V. Amitrov. 1962. Atlas-guide of mollusks of the Quaternary deposits of the Chukotsk Peninsula: USSR Acad. Sci., Comm. Study of the Quaternary Period, Moscow, 56 p. In Russian; translation available from Office of Technical Services, U.S. Dept. of Commerce, and from American Geological Institute.

Moore, G. W., and D. W. Scholl. 1961. Coastal sedimentation in northwestern Alaska: U.S. Geol. Survey TEI-779, p. 43–65.

O'Sullivan, J. B. 1961. Quaternary geology of the Arctic coastal plain, northern Alaska: Ph.D. thesis, Iowa State Univ. Science and Technology, Ames, 191 p.

Porter, S. C. 1964. Late Pleistocene glacial chronology in north-central Brooks Range, Alaska: Am. Jour. Sci., v. 262, no. 4, p. 446–460.

Robinson, F. M. 1964. Core tests, Simpson area, Alaska: U.S. Geol. Survey Prof. Paper 305-L, 730 p.

Sainsbury, C. L., R. Kachadoorian, R. H. Campbell, and D. W. Scholl. 1965. Marine platform of probable Sangamon age, and associated terrace deposits, Cape Thompson area, northwestern Alaska: Arctic, v. 18, no. 4, p. 230–245.

Scholl, D. W., and C. L. Sainsbury. 1961a. Marine geology and bathymetry of the Chukchi shelf off Ogotoruk Creek area, northwest Alaska, p. 718–732, in G. O. Raasch, ed., Geology of the Arctic: Proc. 1st Internat. Symposium on Arctic Geology (Calgary, Alberta), 1960, v. 1.

———— 1961b. Subaerially carved Arctic seavalley under a modern epicontinental sea: Geol. Soc. America Bull., v. 72, no. 9, p. 1433–1435.

Schrader, F. C. 1904. A reconnaissance in northern Alaska across the Rocky Mountains, along Koyukuk, John, Anaktuvuk, and Colville Rivers and the Arctic coast to Cape Lisburne, in 1901: U.S. Geol. Survey Prof. Paper 20, 139 p.

Shepard, F. P. 1964. Sea level changes in the past 6,000 years—possible archaeological significance: Science, v. 143, no. 3606, p. 574–576.

# 6. Quaternary Geology of Western Seward Peninsula, Alaska

C. L. SAINSBURY
*U.S. Geological Survey, Denver*

During 1961–64, I did detailed geologic mapping of Quaternary deposits of the western Seward Peninsula (Fig. 1). The work throws some light on problems discussed in the literature since the works of Brooks *et al.* (1901) and Collier (1902). Recent literature (in particular, Hopkins, 1959a) has reflected a continuing interest in the geomorphic history of the western Seward Peninsula, especially as that history bears upon the existence of the Bering Land Bridge. According to Hopkins (1959a, p. 1527), development of the chronology of the Pleistocene history of the western Seward Peninsula has been hindered by the lack of detailed information on the deposits on the marine terraces. This paper contributes needed information on these deposits and on the glacial history of the York Mountains.

## Previous Studies

Brooks *et al.* (1901) visited the western Seward Peninsula and described some of the erosion levels discussed herein. They considered the York terrace (their York bench) a probable elevated marine platform correlative with their York Plateau (p. 52), an erosion surface about 200 meters in altitude west of the York Mountains, which they also considered to be of marine origin.

Collier (1902) concluded that the York terrace is a warped marine platform. However, because knobs of resistant rock rise well above the general level of the York Plateau, he questioned whether this plateau is a marine erosion surface.

Publication authorized by the Director, U.S. Geological Survey.

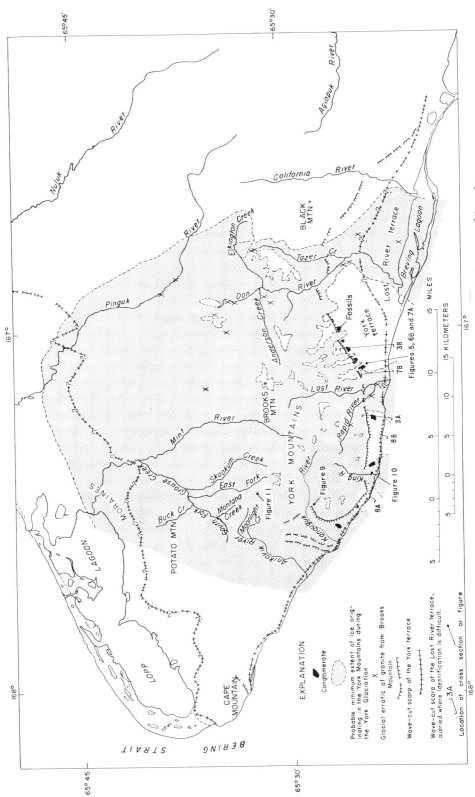

Fig. 1. Map of western Seward Peninsula, Alaska, showing principal geomorphic units.

Knopf (1910) was the first to discuss specifically the Bering Land Bridge in relation to the geomorphic history of the western Seward Peninsula. He pointed out that the area seems to have been deformed differentially and that generalizations embracing large areas are likely to be in error. Knopf concluded that the shoreline of the Seward Peninsula was approximately the same at the beginning of the Pliocene as at present, that the York terrace was older than the Nome beach deposits, which he considered to be of pre-late Miocene age,[a] and that Asia and North America were in land communication at times during the Cenozoic. Knopf found the topographic youth of the York terrace difficult to reconcile with its assumed age, an observation with which I heartily agree. Steidtmann and Cathcart (1922) discussed the geomorphology of the area, but made no original contributions to the geomorphic history. However, they did describe extrusive rocks that now bear on the question of the age of the York terrace. R. R. Coats apparently was the first to notice conglomerate on the York terrace (oral communication to D. M. Hopkins, 1961).

Hopkins (1959a) summarized his ideas on the history of the Bering Land Bridge, and correlated shoreline features over much of northwestern Alaska. He also worked out the stratigraphy and Pleistocene chronology of the Nome area, some 145 km southeast of my area of study (Hopkins *et al.*, 1960), and briefly visited my area in 1961 to study the marine terraces. During this visit, we discussed the conglomerates on the York terrace. In 1965, Hopkins and I traced glacial moraines as far north as the shoreline of Lopp Lagoon. I now believe that these moraines are the terminal moraines of the York Glaciation.

The cited workers contributed important information on the geomorphic history of the area, but their interpretations are open to question, inasmuch as none of them described the deposits on the terraces. The discussion that follows will strengthen some of their cited observations, question others, and present new observations based on the Pleistocene stratigraphy as outlined herein.

In preparing this paper, I have benefited greatly from discussions with D. M. Hopkins and D. S. McCulloch, U.S. Geological Survey geologists.

## Geomorphology

The main geomorphic units of the western Seward Peninsula are shown in Figs. 1 and 2. The oldest geomorphic unit comprises the main mass of

---

[a] For recent information on the Nome beach deposits, see Hopkins *et al.* (1960).

| 0 | | 1 | | 2 | | 3 | | 4 MILES |
| 0 | | 1 | | 2 | | 3 | | 4 KILOMETERS |

Fig. 2. Aerial photograph showing the three main geomorphic units of the York Mountains. If relief appears to be inverted, rotate page 180°. Areas of conglomerate on York terrace are labeled cg. Note bare rock slopes of York Mountains, bedding traces on bare rock platform of the York terrace, and the gravel fans being built on the Lost River terrace. Glacial moraine (York drift) on terrace east of Lost River retains hummocky topography. Compare moraine with moraine east of King River (Figure 10). Line X-X′ is line of section of Figure 7a.

Fig. 3. Profiles across the York terrace based on Teller B-3 and B-4 quadrangles (1:63,360, U.S. Geol. Survey, 1950). Locations are shown on Figure 1. Bedding planes in limestone are generalized. Dashed lines indicate areas mantled by Quaternary deposits. Lines with altitudes show exact location of topographic contour.

the York Mountains above the base of a line of ancient sea cliffs marking the landward margin of the York terrace (Figs. 1 and 2). The rocks above this level have been exposed to subaerial erosion continuously since an unknown time predating the Yarmouth(?) Interglaciation. The bedrock consists mostly of carbonate rocks, but the highest mountains are granite or thermally metamorphosed pelitic rocks. Except for small areas of talus or glacial deposit, the York Mountains are an area of denudation characterized by bare rock slopes covered with a veneer of frost-riven fragments. Earlier workers assumed that the York Mountains were carved from an older marine erosion surface, but there is no clear-cut evidence for such a surface within the area shown in Figs. 1 and 2.

The York terrace is a most striking topographic feature along the south front of the York Mountains, bounded landward by a line of ancient sea cliffs and seaward by the sea cliffs of a lower marine platform, the Lost River terrace (Fig. 4). Beveled across limestone beds tilted landward, the York terrace varies in width from a few hundred meters to almost 8 km. It has been exposed to subaerial erosion since a tectonic uplift that took place after the Yarmouth(?) Interglaciation and before the Wisconsin Glaciation. Most of the covering deposits have been stripped off, and streams have incised their valleys to or below the level of the lower marine platform, the Lost River terrace, which was cut during the Sangamon Interglaciation.

The York terrace is best preserved in limestone areas, especially those south of the York Mountains (Fig. 3); there it can be traced laterally for some 30 km as a nearly continuous surface. Erosion has largely destroyed the terrace in the drainage areas of the main rivers, in glaciated valleys, and in areas where bedrock is predominantly shale or slate. Thus, the York terrace is unrecognizable north of the York Mountains, where bedrock is mostly shale or argillaceous limestone. Glaciation has largely obliterated

Fig. 4. Air view across the York terrace east of Lost River (location omitted on Figure 1). Mountain front in background is the abandoned sea cliff of the York terrace. In the foreground is the sea cliff of the Lost River terrace, adjoined by alluvial fans that cover marine terrace gravel.

the terrace north of the mountains and also between the Don and California Rivers.

The terrace is deformed into a broad arch centering west of Lost River, where it reaches altitudes ranging from 175 to 225 meters (Fig. 3). Differential uplift amounts to about 5 meters per km on the flanks of this arch. The altitude of the terrace is uncertain east of Lost River because of the effects of glacial erosion. I consider a terrace cut on granite at an altitude of about 150 meters on Cape Mountain, at the western tip of Seward Peninsula, to be correlative with the York terrace.

The York terrace is characterized on limestone terranes by frost-riven bare rock covered locally with glacial moraine, talus, and slope wash, and on shale or slate terranes by tundra and thin soil. Along the southern margin of the York Mountains, conglomerate consisting of ancient beach and deltaic and alluvial gravel forms disconnected patches as large as 2.5 km² on the York terrace. Elsewhere, the terrace surface bears numerous pebbles and cobbles of exotic rocks foreign to the York Mountains, of which polished vein quartz, quartz-mica schist, and scoriaceous black lava are most striking.

Evidence presented later shows that marine deposits on the terrace are of Yarmouth(?) age, and thus that the cutting of the York marine terrace was completed during the Yarmouth(?) Interglaciation.

The Lost River terrace, a second and lower terrace almost as striking as the York terrace, can be traced continuously for more than 65 km along the coast (Figs. 1 and 2), along the south front of the York Mountains. Previous writers have called this feature a coastal plain (Brooks *et al.*, 1902, p. 16; Knopf, 1910, p. 418; Steidtmann and Cathcart, 1922, p. 36); my study has shown it to be an undissected marine platform completely covered by unlithified surficial deposits.

The Lost River terrace ranges in width from less than 30 meters to nearly 5,000 meters. It terminates landward against ancient sea cliffs as high as 150 meters (Fig. 3), and it extends seaward to the modern beach. The surface of the unconsolidated sediments on the terrace ranges in altitude from sea level to about 30 meters, but the planed bedrock surface that can be seen beneath the sediments in streams and along the modern beach cliff generally lies about 5–6 meters above sea level. The bedrock terrace is covered by alluvial fans, talus cones, colluvium, and deltaic deposits supporting an extensive cover of vegetation. Thaw lakes and polygonal ground are developed in the unconsolidated cover.

I have traced remnants of sea cliffs marking the landward border of the Lost River terrace for hundreds of kilometers around western Seward Peninsula, but the sea cliffs have been either eroded or overlaid by drift in areas covered by the ice of the York Glaciation. Stratigraphic relationships described later show that the Lost River terrace was carved during the Sangamon Interglaciation.

## Quaternary Deposits

The deposits of Quaternary age comprise cemented conglomerate, glacial deposits, outwash gravel, and terrace deposits of Pleistocene age, and stream gravel, talus, beach deposits, and local glacial deposits of Recent age. In this paper, only those deposits pertinent to the major Pleistocene events are discussed in detail. Table 1 summarizes the deposits and the geologic events correlative with them.

### CONGLOMERATE ON THE YORK TERRACE

Conglomerate, the oldest Pleistocene stratigraphic unit, is restricted to the inner margin of the York terrace, immediately adjacent to the ancient sea cliffs (Figs. 1 and 2). Locally, the conglomerate extends up valleys into the mountains, high above present stream level. The conglomerate rests on beveled limestone bedrock (Fig. 5). It is thickest near the old sea cliffs, and thins gradually seaward to isolated blocks and then to a veneer of cobbles. The conglomerate is about 35 meters thick east of Lost River; elsewhere,

TABLE 1. *Stratigraphic Record of the Main Quaternary Events
in the York Mountains, Alaska*

| Age | Deposits | Geologic Events |
|-----|----------|-----------------|
| Recent | Alluvium, talus, and small moraines | Minor glaciation; eustatic rise of sea level with marine planation |
| Wisconsin | Moraine, talus, and terrace gravels | Mint River Glaciation |
| | Moraine, talus, and outwash gravels | York Glaciation |
| Sangamon | Deltaic gravels | Eustatic lowering of sea level, deposition on Lost River terrace |
| | Beach gravels | Eustatic rise of sea level with marine plantation |
| Yarmouth(?) | Continental conglomerate | Eustatic lowering of sea level, followed by vertical uplift and erosion of the York terrace |
| | Deltaic conglomerate Marine conglomerate | High sea level, marine plantation, and delta formation below sea level |

especially away from former mouths of stream valleys, it is commonly only about 3 meters thick.

Detailed study of the conglomerate between Lost River and the Don River has shown a consistent vertical sequence consisting of a lower marine part and an upper part that is probably a subaerial deltaic or alluvial fan deposit (Figs. 6, 7). The lower, marine part of the conglomerate is firmly lithified, moderately well bedded, and jointed. Foreset and topset cross-bedding are seen locally (Fig. 5). The pebbles, cobbles, and boulders are well rounded, commonly polished, and smaller than 0.6 meters in diameter. Pebbles and cobbles in the conglomerate consist mostly of the local limestone, but there are pebbles of iron oxides ("limonite") with a density of 3.70 and a few rounded cobbles of porphyritic biotite granite similar to that at Brooks Mountain (Fig. 1). In addition, a small number of pebbles and cobbles of rocks not found in the York Mountains, including gneiss, schist, gabbro, black scoriaceous basalt, and highly polished vein quartz, are found in all exposures of the lower, marine part of the conglomerate (Table 2). These exotic clasts could only have been transported to their present positions by longshore drift along a beach or by ice-rafting. The matrix of the marine conglomerate consists of polished quartz grains, cassiterite and other heavy minerals, carbonate, and abundant mica. Marine fossils were obtained in weathered marine conglomerate immediately above limestone bedrock 7.5 km north-northeast of the mouth of Lost River.

The lower, marine part of the conglomerate grades upward into conglomerate of deltaic or alluvial-fan origin. The transition is marked by a

Fig. 5. Angular unconformity between limestone dipping north and conglomerate dipping gently south, York terrace east of Lost River valley. Outcrop is at head of snowbank shown on Figure 6.

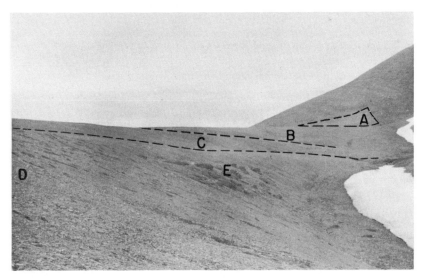

Fig. 6. Topographic expression of the Pleistocene and Recent deposits on the York terrace; Figure 7a is a diagrammatic section of the deposits shown here. *A* is a lobe of slope wash; *B* is partly disintegrated conglomerate formed of deltaic and fan gravels; *C* is stratified marine deltaic conglomerate; *D* is frost-riven bedrock; *E* is slump blocks of marine deltaic conglomerate. Glacial moraine is out of sight over the skyline.

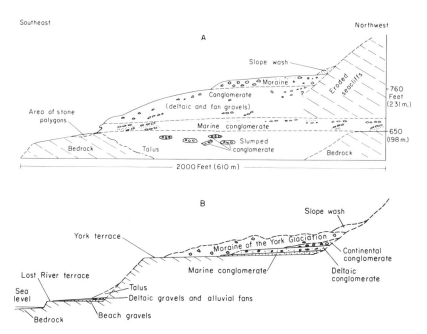

Fig. 7. Stratigraphic units on the York terrace east of Lost River. (*a*) Schematic
diagram showing units visible in Figure 6; line of section shown on Figure 2.
(*b*) Schematic cross section showing the similarity of deposits on the Lost River
and York terraces; line of section shown on Figure 1.

decrease in the number of foreign clasts and in the degree of stratification.
The upper part of the conglomerate is moderately well bedded, especially
near the boundary with the marine conglomerate, and contains sand lenses
capped by subrounded but tabular blocks of limestone. The upper conglomerate consists of rock types that crop out in nearby valleys; exotic pebbles
and cobbles are entirely lacking (Table 2). Boulders of limestone more than
60 cm in diameter are common, and locally cobbles of porphyritic biotite
granite as much as 25 cm in diameter are present. Some of the boulders and
cobbles are fresh; others are rotten. Pebbles of iron oxide are present but
are less dense (density 2.60) than those in the lower part of the conglomerate. The sand matrix consists principally of dolomite and calcite, and
includes relatively little quartz and mica. The conglomerate is cemented by
secondary calcium carbonate, which coats the coarse fragments as well as
the sand matrix. The vertical variation in lithology, particularly of the
matrix, is similar to that described by other workers in deltas (Shepard,
1960; Scruton, 1960).

The age assigned to the conglomerate is based upon a small fauna of

TABLE 2. *Pebble Counts Showing Lithologic Variations Between*
*Lower and Upper Parts of Conglomerate*[1]

| Rock Type | Upper Part | Lower Part | Bedding Plane on Lower Part |
|---|---|---|---|
| Vein quartz . . . . . . . . . . . | 0 0 0 0 | 3 0 1 3 | 2 |
| Metamorphic rocks . . . . . . . | 0 0 0 0 | 1 1 2 1 | 4 |
| Granite . . . . . . . . . . . . . | 0 0 0 0 | 1 1 0 0 | 3 |
| Olivine gabbro and basalt . . . . . | 0 0 0 0 | 0 0 0 0 | 5 |
| Diabase, rhyolite porphyry, and serpentinized gabbro . . . . | 0 0 0 0 | 1 1 1 1 | 2 |

[1] Pebble counts made in an exposure 7.5 km north-northeast of mouth of Lost River (see Fig. 7) by marking out areas of one square yard (0.84 square meter) within which were counted all rocks of lithology foreign to the York Mountains and larger than one-half inch (1.27 cm) long.

marine fossils that I collected from disintegrated conglomerate. The fossils were identified by F. Stearns MacNeil, U.S. Geological Survey, as follows (written communication, 1963):

Gastropoda: *Neptunea heros* (Gray); *Liomesus ooides canaliculatus* (Dall); *Boreoscala groenlandica* (Perry); *Crytonatica clausa* (Broderip and Sowerby); *Lunatia* cf. *L. groenlandica* (Möller); and *Colus* cf. *C. spitsbergensis* (Reeve).

Pelecypoda: Fragments of *Macoma*(?) and *Astarte*(?).

Age: This small collection does not contain a species that is taken to characterize any of the beaches at Nome. It is most like the upper bed at St. Paul in the Pribilofs [the Einahnuhtan of Hopkins, this volume]. The *Neptunea*, however, I would say is typical *N. heros*, a Recent species, whereas the one at St. Paul is *N. heros mesleri*, a subspecies described from the Intermediate ("Third") Beach at Nome.

At any rate, this fauna is not indicative of the earliest Pleistocene of the Bering Sea region. I would not date it as older than middle Pleistocene; it may be younger.

Three other items of evidence also suggest that the conglomerate is relatively young:

1. Fragments of black, scoriaceous olivine basalt lava are everywhere present in the lower, marine part of the conglomerate. No rocks of this type crop out in the York Mountains, but Steidtmann and Cathcart (1922, p. 31) described flows of black scoriaceous olivine basalt at Black Mountain and in the California River valley, some 19–24 km east of Lost River, stating that "the flows at Black Mountain were formed when the topography was essentially the same as present." If these lavas were extruded when the topography was essentially the same as now, they must be fairly young, and the presence of pebbles probably derived from them in the marine gravel of the York terrace suggests that the terrace, too, is relatively young.

2. As pointed out by Knopf (1910, p. 420), "the splendid state of pres-

ervation" of the York terrace is difficult to reconcile with great age. Smaller streams cross the terrace in sharp canyons graded to the Lost River terrace (Figs. 2 and 5). The interfluves are but slightly modified except along fault zones where shallow gullies have developed. If the terrace were as old as early Pleistocene, all the streams should have incised their canyons below the sea level of the Lost River terrace, which they did not do. During the Sangamon Interglaciation they would then have alluviated these deeper valleys. At present, however, most of the streams flow on bedrock until they debouch onto the gravel fans of the Lost River terrace, and no buried channels have been found.

3. Only minor erosional retreat of the old sea cliffs has occurred, as is indicated by marine conglomerate that extends up stream valleys inland beyond the abandoned sea cliffs, and by marine conglomerate lying on the marine platform within a few feet of the present mountain front.

The lower, marine part of the conglomerate evidently is of middle or late Pleistocene age, but the York marine platform was uplifted prior to the cutting of the Lost River terrace during the Sangamon Interglaciation. The lack of a discernible marine platform higher than the York terrace, and its geographic position near the present shoreline, indicate that terracing occurred during an interglaciation, and the age of the York terrace is therefore believed to be Yarmouth(?). The reasoning for this conclusion is as follows: Two marine platforms separated by a time span of about 100,000 years are well preserved. If the York terrace was cut during a low or intermediate stand of sea level, then at some time not exceeding the time span between a high interglacial sea-level stand and a low glacial stand, a still higher platform would have been cut, and remnants of it should have remained recognizable.

UNCONSOLIDATED DEPOSITS ON THE LOST RIVER TERRACE

The deposits on the Lost River terrace, which truncates the seaward margin of the York terrace, also reflect a change from marine to subaerial deposition, followed by glaciation. These deposits were studied between Lost River and the Kanauguk River, where the marine platform is exposed along the modern sea cliffs, and in lesser detail east of Lost River, where the platform is completely covered by tundra, alluvial fans, and talus cones (Figs. 7b, 8). West of Lost River, the planed bedrock surface of the Lost River terrace stands about 6 meters above sea level. Beach gravel containing shell fragments covers the planed bedrock to a depth of as much as 9 meters, and this in turn is covered by colluvium, which, against the old sea cliff, gives

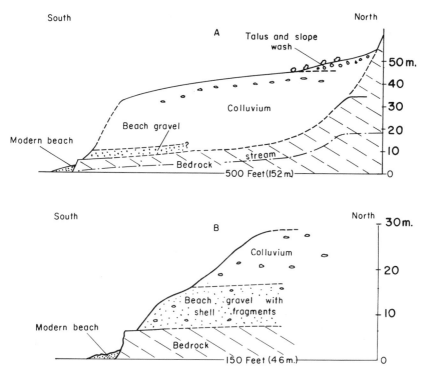

Fig. 8. Stratigraphic units on Lost River terrace (*a*) west of King River, (*b*) east of King River. Locations are shown on Figure 1.

way to talus cones containing large blocks of limestone fallen from the abandoned sea cliff (Fig. 8). No diagnostic fossils have been recovered from the beach gravel.

East of Lost River (see Fig. 2), large alluvial fans cover the Lost River terrace. The fans are composed of rocks derived from the drainage basins of the stream valleys and consist for the most part of carbonate rocks in a matrix of fine-grained carbonate. An occasional pebble or cobble of granite, vein quartz, and scoriaceous lava can be found in fans whose parent streams contain conglomerate within their drainage basins. Modern streams locally are trenching the fans, but none are entrenched deeply enough to expose underlying marine gravel, which must be present. Gravel in the exposed parts of the fans is almost identical to the upper part of the conglomerate on the York terrace, which is believed to represent older, cemented, subaerial deposits.

At the mouths of Lost and King Rivers, till of the York Glaciation is exposed below sea level. The York till at Lost River lies in a steep-walled stream bed that trenches the marine platform of the Lost River terrace. These exposures prove that these streams deepened their valleys below the Lost River terrace prior to or during the York Glaciation.

The Lost River terrace is considered to be of Sangamon age, on the strength of the following three points:

1. The altitude of the terrace is relatively constant throughout the western Seward Peninsula, which indicates that its position reflects a stand of sea level 9–12 meters above the present. If it were a platform tectonically uplifted over such a wide area, it would have been warped, as was the York terrace.

2. The Lost River terrace is younger than the York terrace, of Yarmouth(?) age, and older than a major glaciation of the York Mountains.

3. It can be traced with reasonable certainty into the Nome area, where it has been dated as Sangamon in age (Hopkins, *et al.*, 1960).

### DRIFT OF THE YORK GLACIATION

Drift of a widespread glaciation is preserved at widely scattered localities in the York Mountains. This glaciation is here named the York Glaciation, and the moraine on the York terrace east of Lost River is designated as the type locality.

In some places the York drift completely covers bedrock, as in the type locality on the York terrace, as well as in areas east of King River (Fig. 10), near the mouth of the Mint River, and in the divide between Elkington and Tozer Creeks (Fig. 1). More commonly, the York Glaciation is represented by erosional topography (parts of Fig. 10) and scattered erratics.

Glacial erratics of granite and tactite from Brooks Mountain assist in identifying the drift in many places. Granite erratics ranging from 0.3 to 5 meters across are found locally on hilltops north and east of Brooks Mountain, in the bed of Rapid River several miles upstream from its confluence with Lost River, and in drift exposed in the divide between Tozer and Elkington Creeks. Locations of a few of the larger erratics are shown in Fig. 1. Glacial drift is more difficult to recognize west of the Mint River–Lost River divide, because there it consists entirely of limestone cobbles that quickly lose facets and striae upon exposure at the surface.

End moraines of the York Glaciation are present at the type locality on the York terrace east of Lost River and at the south shore of Lopp Lagoon, 40 km north of the York Mountains. The moraine on the York terrace is about 10 meters thick near its eastern margin, and contains angular blocks up to 1.8 meters across of granite and metamorphic rock from Brooks

Mountain as well as fragments of the mineralized rocks associated with the tin deposits in Lost River valley. The moraine shows kame and kettle topography; two kettles about 18 meters across display bedrock in their floors. A few fragments of the underlying conglomerate occur in the moraine. Stone polygons are well developed and are illustrated by Steidtmann and Cathcart (1922, Fig. 6b), who described the moraine as "slope wash."

The relations between the deposits of the York Glaciation and those of the Lost River terrace are well displayed at the mouth of Lost and King Rivers. The mouth of the preglacial valley of Lost River is choked by drift of the York Glaciation, which displays kame-kettle topography and probably represents a recessional moraine. Lost River bypasses this moraine in a short but deep canyon that probably originated as a marginal drainage channel when ice of the York Glaciation occupied the center of the valley. Drift choking the preglacial valley is characterized by cobbles of biotite granite, dike rock, and contact rock, some of them faceted and striated. Nearly all of the granite cobbles are fresh, but a few are rotten throughout. About 0.6 km inland, the hummocky surface of the moraine grades into a smooth, seaward-sloping, gravel sheet, which buries the old shoreline of the Lost River terrace. Because it grades into a moraine, the gravel sheet is interpreted to be outwash. This and a sheet of gravel at the mouth of Lost River canyon mantle the bedrock platform of the Lost River terrace and thus are younger than the Lost River terrace.

Indurated till with a tan matrix enclosing striated cobbles is exposed in the bed of Lost River at its confluence with Rapid River, and similar till is exposed during minus tides beneath modern beach gravel at the mouth of Lost River. Because the degree of induration and oxidation of this till is noticeably greater than that of the drift of the York Glaciation discussed above, the till may represent drift of a separate and older ice advance. Its position in a stream bed trenching the marine platform of the Lost River terrace dates it, however, as younger than the Lost River terrace, or post-Sangamon; hence, it is tentatively assigned to the York Glaciation. Further study may show that the York drift records more than one glacial advance.

Wave erosion has exposed glacial drift beneath stream terrace gravel west of the mouth of King River. The glacial drift is composed of both angular and polished, striated limestone cobbles in a clayey matrix. The drift underlies stream terrace gravel, and, as at Lost River, it buries the sea cliff of the Lost River terrace. These two facts establish definitely that the York Glaciation is younger than the Lost River terrace of Sangamon age. I consider the York Glaciation to be early Wisconsin in age.

During the York Glaciation, ice spread northward from the Brooks

Mountain area to form a broad piedmont glacier as the ice coalesced with other glaciers in valleys to the west. Glacial erratics and drift at altitudes of 400 meters on the northernmost hills show that ice then no longer confined within valleys was at least 300 meters thick along the north front of the mountains. The ice surface probably sloped north, northwest, and northeast from the Brooks Mountain area, and thick ice spreading west along the lowlands north of the mountains forced north-flowing valley glaciers to turn west along the mountain front. A well-defined train of erratics carried by a glacier that originated in the cirque at the head of the East Fork of Grouse Creek can be traced along the headwaters of the Anikovik River, and south around the mountain front. The western limit of this piedmont lobe is not known; its northern limit may be marked by the moraines and outwash gravels along the present coastline between the Mint and Pinguk Rivers.

Ice from the Brooks Mountain area spread east and southeast into the Don River, and probably reached at least to the California River and south to the present shoreline of Bering Sea. East of California River, the Lost River terrace and an old offshore bar of the terrace are well preserved, and the east margin of the York Glaciation therefore lies near the California River.

Throughout the western Seward Peninsula, the destruction of the Lost River terrace can be accounted for, wherever the terrace is missing, by glacial erosion or deposition. On the basis of the known extent of drift and the destruction of the Lost River terrace, a minimum estimate of the extent of ice of the York Glaciation has been prepared (see Fig. 1). It is assumed that the highest peaks of the York Mountains remained as nunataks above the ice, but this is by no means certain.

### ALLUVIAL FANS AND STREAM TERRACE GRAVEL

Following the disappearance of ice from the main valleys of the York Mountains, large alluvial fans were deposited at the mouths of side streams. Although these fans are widespread in the York Mountains, they are especially well developed in the valleys of the Kanauguk, King, and Rapid Rivers. In valleys that contained ice during the following Mint River Glaciation, the fans are notched by several terraces; in other valleys they are notched by the present streams. Similar terraces are present but poorly developed in Lost River, Mint River, and Anderson Creek, and in the valleys of several other streams flowing northward from the York Mountains.

Three sets of paired terraces standing 5.5, 3.5, and 2.7 meters above present stream level are present on the upper Kanauguk River. Some of these are shown in Fig. 10. In many exposures, these terraces can be seen to be underlaid by stratified shingle gravel, lightly coated with iron oxide, resting on till of the York Glaciation consisting of an unsorted, unstratified mixture of polished pebbles and cobbles of limestone in a clay-rich matrix. Many segments of the terraces lack vegetation or fine-grained cover, but some of the lower terraces bear up to 0.5 meters of dark, fine-grained material that has been washed down from higher surfaces. Large alluvial fans at the mouths of small streams are notched by the upper terrace.

At one point in the upper Kanauguk valley, a glacier from a southern tributary seems to have forced the Kanauguk River against the north valley wall during the time that the terraces were formed. The former position of the glacier is marked by a terminal moraine of the Mint River Glaciation of late Wisconsin age, discussed later (Fig. 9). In most places the terraces are constructional, but opposite the moraine, they are erosional features cut on the bedrock of the north side of the valley and only thinly veneered with gravel. This suggests that the glacier blocked the former valley bottom to the south.

Terraces in the valley of King River also consist of washed gravel overlying till of the York Glaciation. The stream terraces there continue unbroken across the Lost River terrace, burying the old sea cliff (Fig. 10).

The stratigraphic and geomorphic relationships indicate that the terraces

Fig. 9. View of small hummocky end moraine of Mint River glaciation in the valley of the Kanauguk River. Location shown in Figure 1.

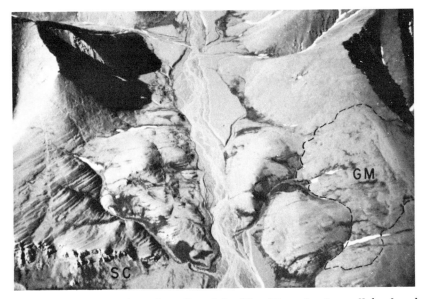

Fig. 10. Air view north up the valley of the King River showing well-developed terraces, moraine of the York glaciation (*GM*) on bench east of river, and sea cliffs (*SC*) of the Lost River terrace in the lower left corner.

are younger than the York Glaciation, but are not younger than the Mint River Glaciation. The fact that the highest terrace on the Kanauguk River notches large alluvial fans at the mouths of tributary valleys shows that terracing followed a period of alluviation younger that the York Glaciation. The relationships near the Mint River end moraine of the Mint River Glaciation in the upper Kanauguk valley (Fig. 9) indicate that at least some of the terraces are contemporaneous with the Mint River Glaciation. A lack of buried soils in any of the terraces suggests that they were formed in a relatively short time.

### DRIFT OF THE MINT RIVER GLACIATION

The York Mountains contain drift recording a glaciation later and less extensive than the York Glaciation. This later glaciation is here named the Mint River Glaciation after moraines on the Mint River, but the type locality is designated as the area in the western headwaters of the East Fork of Grouse Creek, shown in Fig. 11.

Drift of the Mint River Glaciation can be distinguished from drift of the York Glaciation by the position of its moraines nearer to the cirques in which ice originated, by its better-preserved topography (Fig. 9), and by the presence in two places of polished and striated limestone bedrock

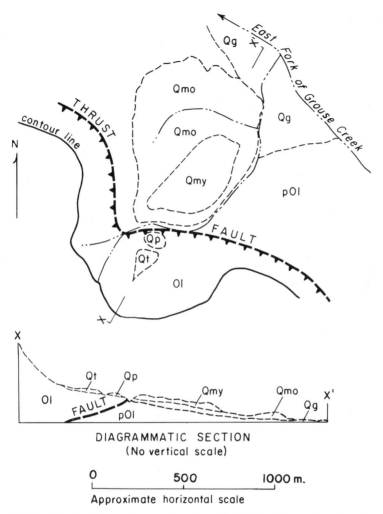

Fig. 11. Glacial deposits at the type locality of the Mint River glaciation. Location shown on Figure 1. *Qt*, talus; *Qp*, protalus rampart; and *Qg*, terrace gravel of Recent age. *Qmo*, older, and *Qmy*, younger moraines of Mint River glaciation. *Ol*, Ordovician limestone; *pOl*, pre-Ordovician slaty limestone.

exposed at the surface in cirque walls. Gravel terrace deposits overlie and bevel the drift of the York Glaciation but do not notch or bevel the terminal moraines and drift of the Mint River Glaciation. For these reasons, the Mint River drift is thought to record a glacial advance separate and distinct from recessional features of the York Glaciation.

Glacial deposits of the Mint River Glaciation are generally confined to valleys of north-flowing streams that head in well-defined cirques in the

York Mountains. End moraines of the Mint River Glaciation lie as much as 5 km downstream from some cirques but much nearer to others. End moraines were seen at the mouth of a cirque in the valley of Kanauguk River (Fig. 9) and in the central headwaters of the Mint River about 5 km north of the divide between the Kanauguk and Mint Rivers. Glaciers apparently were lacking in valleys draining the south front of the York Mountains during the Mint River Glaciation.

At least two ice advances during the Mint River Glaciation are recorded at the type locality on the East Fork of Grouse Creek (Fig. 11). The morainal complex there was deposited by a small glacier originating in a compound cirque carved mostly from medium- to thick-bedded limestone, but the moraines lie on thin-bedded argillaceous limestone. A hummocky and lobate outer moraine (Qmo on Fig. 11) encloses and is overlaid by an elongate inner moraine (Qmy) that is succeeded upstream by a protalus rampart (Qp) probably of Recent age. The older moraine consists almost entirely of fragments of medium- to thick-bedded limestone, badly broken by frost action and reduced at the surface to sizes less than 0.3 meters across; the younger moraine also consists of medium- to thick-bedded limestone from the cirque headwall, but contains blocks as much as 4 meters across. The compound cirque in which the Mint River–aged glaciers originated consists of a steep north-facing bowl containing only a moderate amount of talus and an east-facing bowl that is considerably modified by erosion and contains much weathered talus. The differences in the amount of frost weathering on the moraines and in the cirques suggest that the times of formation of the two moraines were separated by a considerable time interval, but both are thought to be of late Wisconsin age.

### MORAINES OF RECENT AGE

Small moraines at high altitudes in some cirques in the York Mountains evidently were formed by small glaciers during the Recent Epoch. The protalus rampart in the cirque at the type locality of the Mint River Glaciation on the East Fork of Grouse Creek (Fig. 11) is also probably of Recent age.

## Quaternary History

The Quaternary chronology inferred from the geomorphic and stratigraphic relationships described in this paper is summarized in Table 1.

The Pleistocene record begins with the cutting of the York terrace during the Yarmouth(?) Interglaciation. The position of sea level at that time cannot be established with certainty, but the lack of still higher terraces suggests that sea level had never stood higher during the Pleistocene Epoch. Flint

(1957, p. 270) concludes that the probable highest position of sea level during the last warm interglaciation was +30 meters, and that the as yet undetermined amplitudes of earlier fluctuations would be of the same order of magnitude as the last major change if they were commensurate with the size of the glaciated areas. Several authors cited by Woldstedt (1954, p. 292–293) calculated that sea level would rise 30–60 meters if all the present glacial ice were melted. Thus we may conclude that, during the cutting of the York terrace, sea level was no more than 30–60 meters above its present position.

The York terrace was uplifted some time after the Yarmouth(?) Interglaciation and before the Sangamon Interglaciation. The exact time of uplift is not known, but certain deductive evidence suggests that it took place late in Illinoian time:

1. No glacial deposits of Illinoian or older age are recognized in the York Mountains or on the York terrace, although elsewhere on the Seward Peninsula the Illinoian ice sheets were more extensive than those of Wisconsin age (Hopkins, 1959a, p. 1522). Possibly, the ice of Illinoian age extended far beyond the present shores, so that its deposits now lie in submerged areas, but it seems more likely that the absence of Illinoian drift indicates that the York Mountains were too low to support a glaciation as widespread as that of the Wisconsin Glaciation. If this is true, then the York terrace must have been uplifted after the peak of the Illinoian Glaciation.

2. If the York terrace had been uplifted early in Illinoian time, most streams probably would have eroded below the level of the Sangamon(?) terrace. It would be extremely coincidental for the minor streams to cut down only to the level of the Lost River terrace of Sangamon age if time alone had controlled the down-cutting.

The amount of uplift of the York terrace can be determined only within broad limits, because of the uncertainty of the position of sea level when it was cut. The highest part of the York terrace stands 215 meters above present sea level beneath the marine conglomerate 4 km west of Lost River and at a point 0.8 km seaward from the old sea cliffs. Assuming an original position of sea level between +30 and +60 meters, we conclude that this part of the terrace has been uplifted 155–185 meters. If the 150-meter terrace on Cape Mountain at the western tip of Seward Peninsula is correlative with the York terrace, the uplift there amounted to 90–120 meters.

How far west in the Bering Strait area the uplift occurred cannot be determined without information about Big and Little Diomede Islands, in Bering Strait some 40 km west of Cape Mountain. Fairway Rock, 29 km west of Cape Mountain, is a granitic pluton whose flat top suggests marine plantation. Although many faults cut the bedrock of the York terrace in the area discussed here, none displaces the planed rock surface of the York terrace; hence there is no reason to believe that the terrace is terminated by a fault in Bering Strait. That the uplift in post-Yarmouth(?) time continued into the central part of the Bering Strait is indicated by present evidence, but cannot be proved until additional fieldwork is done on the Diomede Islands.

After uplift of the York terrace, a widespread glaciation of early Wisconsin age covered a large part of the York Mountains and reached northward to the present shoreline of Chukchi Sea. Following the recession of glaciation, large alluvial fans were constructed in the glaciated valleys.

In late Wisconsin time, ice of the Mint River Glaciation filled north-facing cirques, and a pronounced series of terraces was formed in some valleys. At least two fluctuations occurred during the Mint River Glaciation.

In Recent time, a minor glaciation created ice-fillings of some high-level north-facing cirques, and during a later cold period, protalus ramparts were constructed in these same cirques. Modern streams are trenching all of these deposits.

## Implications for the History of the Bering Land Bridge

The Pleistocene history of the western Seward Peninsula, as outlined above, has strong implications with respect to the narrowest part of the Bering Land Bridge—the area adjacent to Bering Strait, which has been pointed out by all writers as the narrowest seaway between Alaska and Siberia. Hopkins (1959a, p. 1525) presented drawings showing the expansion of the land bridge at successive stages of lowered sea level, and concluded (p. 1524) that "A reduction in sea level of 300 feet (92 meters)—to the level recorded during early Wisconsin time more than 35,000 years ago—would result in the exposure of nearly all of the Bering-Chukchi platform, and Alaska and Siberia would be joined by an almost featureless plain extending nearly 1,000 miles (1,610 km) from the north shore of a shrunken Bering Sea to the south shore of the Arctic Ocean." He further stated (p. 1526), "However, even when the more extensive Illinoian glaciation was at its height, a broad ice-free corridor extended from central Alaska across the land bridge to eastern Siberia."

The Pleistocene history of the western Seward Peninsula, as outlined in this paper, is generally compatible with the conclusions of Hopkins and others regarding Wisconsin time (except for the extent of ice), but is clearly incompatible with an extension of these conclusions into early Illinoian time (and earlier), particularly regarding the eastern Bering Strait area of the land bridge. Prior to the uplift of the York terrace, the Bering Strait was probably a seaway during all parts of Pleistocene time, during which sea level lay within 100 meters of its present level. If sea level were eustatically lowered more than 100 meters, Bering Strait would be gripped in the most severe parts of a glacial stage, and much of eastern Siberia and the Bering Strait probably would have been covered by ice. It is probable, then, that any land migrations by animals and plants across Bering Strait, were restricted to Wisconsin time; even then the extreme climatic severity and large ice sheets may have prevented any migration of animals. During the earlier Pleistocene, the migrations by land, which are discussed in detail by Simpson (1947) and by several of the present contributors, most probably would have been along a route considerably south of Bering Strait, and could have been via St. Lawrence Island. It would seem desirable for students of paleontology and archaeology to search for evidence of land migration in early Pleistocene time in areas well south of Bering Strait.

## REFERENCES

Bradley, W. C. 1958. Submarine abrasion and wave-cut platforms (Calif.): Geol. Soc. America Bull., v. 69, p. 967–974.

Brooks, A. H., G. B. Richardson, and A. J. Collier. 1901. A reconnaissance of the Cape Nome and adjacent gold fields of Seward Peninsula, Alaska, in 1900: U.S. Geol. Survey Spec. Pub., p. 1–222.

Collier, A. J. 1902. A reconnaissance of the western portion of Seward Peninsula, Alaska: U.S. Geol. Survey Prof. Paper 2, 70 p.

Flint, R. F. 1957. Glacial and Pleistocene geology: Wiley, New York, 553 p.

Hopkins, D. M. 1959a. Cenozoic history of the Bering Land Bridge (Alaska): Science, v. 129, no. 3362, p. 1519–1528.

——— 1959b. History of Imuruk Lake, Seward Peninsula, Alaska: Geol. Soc. America Bull., v. 70, p. 1033–1046.

Hopkins, D. M., F. S. MacNeil, and E. B. Leopold. 1960. The coastal plain at Nome, Alaska—A late Cenozoic type section for the Bering Strait region: Rept. 21st Internat. Geol. Cong. (Copenhagen), 1960, pt. 4, p. 46–57.

Knopf, Adolph. 1910. The probable Tertiary land bridge between Asia and North America: Univ. Calif. Dept. Geol. Bull. 5, p. 413–420.

Scruton, P. C. 1960. Delta building and the deltaic sequence, p. 82–102 *in* F. P. Shepard *et al.*, eds., Recent sediments, northwest Gulf of Mexico: Am. Assoc. Petroleum Geologists.

Shepard, F. P. 1960. Gulf Coast barriers, p. 197–220 *in* F. P. Shepard *et al.*, eds., Recent sediments, northwest Gulf of Mexico: Am. Assoc. Petroleum Geologists.

Simpson, G. G. 1947. Holarctic mammalian faunas and continental relationships during the Cenozoic: Geol. Soc. America Bull., v. 58, no. 7, p. 613–668.

Steidtmann, E., and S. H. Cathcart. 1922. Geology of the York tin deposits, Alaska: U.S. Geol. Survey Bull. 733, 130 p.

Woldstedt, P. 1954. Das Eiszeitalter: Ferdinand Enke Verlag, Stuggart, 374 p.

# 7. Paleogeography of Chukotka During Late Neogene and Quaternary Time

O. M. PETROV

*Geological Institute, Academy of Sciences of the USSR*

The Chukotka Peninsula is a generally mountainous country constituting the strongly dissected eastern spur of the Anadyr Range. Here and there, the mountains closely approach the sea, forming steep cliffs as much as 300–400 meters high; these mountainous segments of the coast are due to faulting during Neogene and Quaternary time. More commonly, the mountains are separated from the sea by a narrow belt of lowlands underlain by unconsolidated Quaternary deposits; the most extensive of these is the Vankarem lowland on the Arctic coast (Fig. 1). The Vankarem lowland is continuous with an intermontane lowland, the Kolyuchinsk–Mechigmen depression, which extends southeastward to the shore of Bering Sea. Lowlands are also found on the south side of Chukotka along the shore of Kresta Bay.

The Quaternary paleogeography of this region has been discussed by Kiriushina (1939), Obruchev (1939), and Sachs (1946), who noted that Chukotka had undergone several glaciations and several marine transgressions. In 1953, Sachs included Chukotka in a discussion of all the information then available concerning the Quaternary deposits of Siberia. Geological investigations carried out from 1953 to 1961 now permit an expansion and a partial modification of these earlier concepts of the late Neogene and Anthropogene[a] history of Chukotka. The results of these studies are summarized in previous papers on the stratigraphy of the Quaternary deposits (Petrov, 1963) and on marine mollusk faunas (Merklin *et al.*, 1962).

Before we proceed to the discussion of the Quaternary history of Chu-

A translation by M. C. Blake of "Paleogeografiia Chukotskogo Poluostrova v pozdnem Neogene i Chetvertichnom Periode," p. 65–88 *in* Anthropogene Period in the Arctic and Subarctic: Trans. Res. Inst. Geology of the Arctic, 143, 359 p. (1965); in Russian with English abstract.
[a] "Anthropogene" in Soviet stratigraphic nomenclature corresponds approximately to "Quaternary" as used by most of the American and European contributors to this symposium. ED.

kotka, some preliminary remarks must be made. Special attention is given in this paper to the record of marine transgressions; other aspects of the paleogeography of the Quaternary period—in particular the late Quaternary glacial history—are presented more schematically. All the conclusions about changes in vegetation and climate are based on palynological data. The contemporary physical-geographic conditions of Chukotka are assumed to typify interglacial conditions, and the ancient spore-pollen complexes are compared with modern samples extracted from the fresh muds of the river bottoms and from the bottom sediments of lakes and lagoons. Discussion of the fossil marine faunas is based mostly on the bivalves; the fossil gastropods from the Quaternary deposits have not yet been thoroughly studied, and knowledge of the contemporary gastropod faunas of Bering Sea is still very limited.

## Late Neogene Time

Uncertainty about the position of the lower boundary of the Quaternary system necessitates an examination of the paleogeography of Chukotka during the last stages of Neogene time.

Undisturbed marine deposits of the Pestsov Suite apparently attest to the last Neogene transgression. These beds consist of sandstone with interbeds of conglomerate exposed in cliffs 8–15 meters high along the Pestsov River and along a number of other, unnamed streams that dissect the piedmont plain adjoining the northwestern slopes of the Zolotoy Range (Fig. 1). The coarse nature of these deposits and the character of the faunal remains that they contain indicate that the Pestsov Suite accumulated in a shallow marine basin.

The molluscan fauna in the Pestsov Suite includes several extinct species, several species now living much farther south, and a smaller number of species now living in the Gulf of Anadyr. The molluscan fauna in the Pestsov Suite has a generally south-boreal aspect and includes thermophilous species of the genera *Glycimeris, Arca,* and *Ostrea.* The hydrologic regime of the marine basin corresponded approximately to the present conditions in the northern part of the Japan Sea and the southern part of the Sea of Okhotsk. K. P. Yevseev, in determining the fossils collected by B. I. Vasil'ev in 1953, suggested that they may correlate with the faunal complex of the Kavran Suite (Late Miocene through Middle Pliocene [Il'ina, 1963]) of Kamchatka. These lines of evidence seem to indicate that the Pestsov Suite cannot be younger than Middle Pliocene. The data available at the present time favor an Early Pliocene or even a Late Miocene age for the Pestsov Suite.

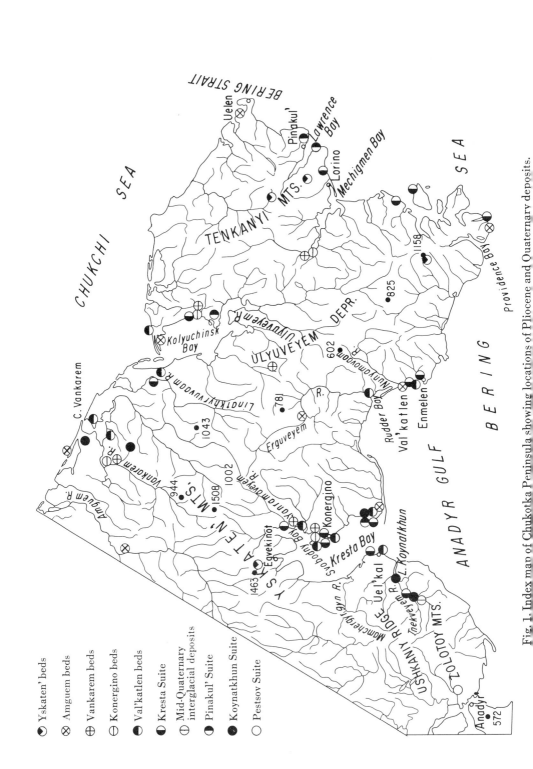

Fig. 1. Index map of Chukotka Peninsula showing locations of Pliocene and Quaternary deposits.

Palynological data indicate that during the time when the Pestsov Suite accumulated, the shores of the basin supported forests consisting chiefly of conifers (spruce, pine, hemlock, larch, fir) and narrow-leaved angiosperms (birch and alder), with an admixture of broad-leaved trees (hornbeam, oak, elm). At the time of the Pestsov transgression, all of the present lowlands seem to have been drowned, and the sea penetrated farther into the area of the Anadyr depression than at present. Climatic conditions were much more equable than at present and were characterized by mean annual air and sea temperatures above freezing.

Most previous publications touching upon the stratigraphy and paleogeography of the Quaternary Period in Chukotka correlate the preglacial deposits there with certain deposits of supposed marine origin in the Anadyr region. However, the Quaternary age of these deposits and also their marine origin are doubtful. Thus, in discussing an area that lies west of that shown in Fig. 1, Eliseev writes (1936), "traces of post-Pliocene marine transgressions were found by the author in the area of the Russian Range along the Shchechnaya River in the form of clayey sands with a fauna of *Corbicula fluminalis* Mull., nov. var., in *uncertain conditions of superposition* [italics mine—O.M.P.] on a basement of strongly disturbed volcanogenic strata of Tertiary Age. A curious fact is that these post-Pliocene sands are cut by dikes of basalt and are covered by a basaltic sheet, making up the highest points in the Russian Range. These sands were found at an altitude of 180 meters." Eliseev apparently considered these sands to be of post-Pliocene age solely on the basis of their position on top of deformed sediments. Furthermore, an objection is raised to his assumption of a marine origin for sediments containing *C. fluminalis*—a mollusk living primarily in fresh water and sometimes found in the brackish water of estuaries but not in salt water (Zhadin, 1952). In my opinion, the sands containing *C. fluminalis* probably are deltaic beds correlative with the marine Pestsov Suite—i.e., they are undoubtedly of pre-Quaternary age.

## Preglacial Time

Preglacial time, as used in this paper, denotes the time immediately preceding the great Quaternary glaciations; it corresponds to the later part of the Pliocene Epoch. The preglacial interval embraces a lengthy time interval in the continental development of the Chukotka Peninsula during which sea level was lower than at present. The tectonic regime was quiet during this interval, as is attested by the relatively fine-grained sediments that make up the lower part of the Koynatkhun Suite.

The Koynatkhun Suite consists of preglacial sediments that are exposed in both the northern and the southern parts of Chukotka. In the Lake Koynatkhun area and farther to the west, these deposits occupy considerable portions of the interstream areas and are exposed below glacial sediments in the river and lake bluffs. Exposed thicknesses reach 30–40 meters. In southwestern Chukotka the Koynatkhun Suite consists of brown and brownish-gray micaceous sand and sandy loam containing abundant coarse blocks of weakly coalified wood. Interbeds and lenses of lignite, 2–3 meters thick and consisting exclusively of coalified wood fragments, are a characteristic feature. To the north, in the Vankarem lowland, the preglacial deposits are made up of light-gray, fine- and coarse-grained arkosic sand, with interbeds of pebble gravel and masses of weakly coalified wood fragments. The minerals in the sand are greatly weathered.

Wood in the preglacial deposits is entirely derived from the conifers, spruce and pine. Cones of *Pinus monticola* Dougl., *P.* ex sect. *Strobus*, *Picea bilibinii* Vask., *P.* cf. *anadyrensis* Kryscht., and *Larix* cf. *sibirica* L. have been found. The spore-pollen spectra of these deposits indicate a coniferous forest, possibly containing some broad-leaved trees. The annual rings of the wood fragments are as much as 3–4 mm thick and have a large radius of curvature, suggesting that they are derived from well-developed trees one meter or more in diameter. Thus, the climate must have been characterized by mean annual temperatures above freezing when the Koynatkhun Suite was deposited.

The base of the Koynatkhun Suite lies below sea level along the shores of Kresta Bay (Figs. 2, 5), indicating a much lower position of sea level when these beds were deposited. The present continental shelf of the Bering and Chukchi Seas probably was drained, and northeastern Asia probably was connected with North America. The possibility of such a connection during the Pliocene Epoch is strengthened by the presence of cones of American western white pine (*Pinus monticola*) in the preglacial deposits of Chukotka.

A change in composition upward in the section of the Koynatkhun Suite, expressed by a disappearance of the lignite interbeds, by a diminution in the quantity of organic material, and by the replacement of fine-grained sediments by coarse sand and pebbly sand, evidently records several episodes of crustal movement and a change for the worse in climatic conditions. A deteriorating climate is also indicated by pollen spectra in a series of deposits developed on the periphery of the Koynatkhun depression and in the upper parts of exposures in the central part of the same depression. In these beds, arboreal pollen makes up only 30–40 per cent of the total pollen, and the

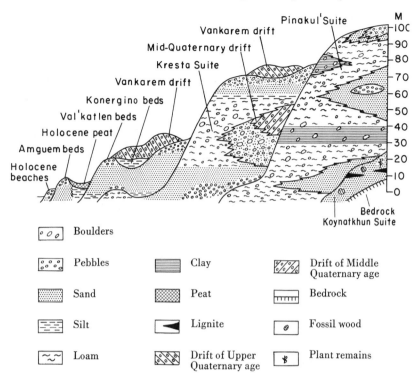

Fig. 2. Schematic cross section through the Quaternary deposits of coastal plain areas of Chukotka. Highest terrace is locally as high as 160 meters.

quantity of spores increases to 58 per cent or more, largely as a consequence of increase in abundance of *Sphagnum*. All of these facts attest to an overall cooling, to a thinning of the forest, and to the formation of open, swampy areas. The first glaciers may have formed in the mountainous regions of Chukotka at the very end of the time during which the Koynatkhun deposits were formed.

In my opinion, the lower boundary of the Quaternary Series on Chukotka may lie within the upper part of the Koynatkhun Suite.

## Early Quaternary Time

Against the background of a general deterioration in climatic conditions during Early Quaternary time, a comparative lowering of the land took place in the Chukotka region. The sea occupied the area of the continental shelf and the present lowlands as well, approaching the base of the mountain slopes. The deposits of this transgression—the Pinakul' Suite—overlie partly eroded deposits of the Koynatkhun Suite on the eastern shore of Kresta Bay.

The most complete section of the Pinakul' Suite is found near the village of Pinakul' on Lawrence Bay. Here, the section is made up of consolidated sandy loam and sandy clay containing pebbles and boulders. The sediments are gray with brownish hues, and have a total thickness of 85 meters. Interfingered with the clayey strata are sequences of thin-bedded sand with lenses and interbeds of gravel and pebbly sand; all of these strata are distinctly laminated, owing to the presence of horizontal or slightly undulating interbeds of sand several centimeters thick. Detailed examination of the more homogeneous sandy loam also discloses a horizontal lamination reflected in changes from gray to brown in the color of the sediment.

The sandy loam is characterized by the presence of small, sparsely distributed iron-manganese concretions. Here also are found cigar-shaped calcite concretions having a rhombic cross section; these have as a core either a calcite rhomb or a mass of mollusk shells or algal remains. Less commonly, loaf-shaped concretions occur conformably within the stratification.

Grit, pebbles, and boulders up to 1–2 meters across are irregularly disseminated through all the sandy-loam strata. In places, accumulations of coarse fragmental material comprise as much as 30–40 per cent of the volume of the sediments; these deposits greatly resemble ordinary glacial till. The coarse material has a varying form and degree of rounding, but angular and poorly rounded material predominates.

The sand beds of the Pinakul' Suite are well sorted and have a distinct horizontal lamination; they grade laterally into sandy loam. Well-rounded pebbles and boulders are found in the sand, but in much lesser quantity than in the sandy loam. The sand beds contain many mollusk shells, occurring in layers as unbroken valves and often as closed, paired valves. The shells are generally large and commonly enclose a tough, stony core. On the eastern shore of Kresta Bay, the faunal remains are generally found in ironmanganese concretions.

Among the faunal remains found in sandy beds in the outcrops near the village of Pinakul', *Macoma calcarea* (Gmelin), *Serripes groenlandicus* (Brug.), and *Nucula tenuis* (Mont.) predominate, and representatives of *Musculus nigra* (Gray), *Mya arenaria* L., *Natica russa* (Gould), and *Neptunea communis* (Midd.) are present. On the coast of Kresta Bay, *Cardium californiense* Deshayes, *Serripes groenlandicus* (Brug.), *Macoma inquinata* (Deshayes), and *Mya arenaria* L. predominate, and examples of *Mytilus edulis* L. and *Macoma incongrua* Martens are present. On the whole, the molluscan complex has a boreal-arctic aspect, characterizing a basin with normal salinity and depths of up to 50 meters. Shells are rare and occur mainly as isolated specimens in the till-like sandy loam that makes up the larger part of the Pinakul' exposures, but the accumulations of shells are found in isolated areas, especially in the lower parts of the exposures. The sandy loams have yielded remains of *Macoma calcarea* (Gmelin), *Astarte borealis borealis* (Schumacher), *A. borealis arctica* Gray, *A. montagui* (Dillw.), *A. alaskensis* Dall, *Mya arenaria* L., *M. truncata* L., *Nucula tenuis* (Mont.), *Yoldia hyperborea* (Loven), and *Portlandia arctica siliqua* (Reeve), as well as algae, sponges, and bryozoa. The fossil mollusks found in the Pinakul' Suite represent species living at the present time. The mollusk complex is characterized by the coexistence of North Atlantic and North Pacific species, with a distinct predominance of Atlantic forms. Evidently, representatives of the arctic mollusk fauna (e.g., *Portlandia arctica siliqua*) first penetrated from the Arctic Basin southward into Bering Sea at the time of the Pinakul' transgression.

The diatoms in the Pinakul' Suite also represent boreal and arctic species characteristic of the sublittoral zone and related to modern diatom complexes in Providence Bay (Jousé, 1962). Thus, the hydrologic regime of the sea during the Pinakul' transgression was similar to present conditions in the northwestern part of Bering Sea.

Granite is exposed at the shoreline, 1.5 km east of the village of Pinakul', and farther east the surface cut on the granite slopes upward to an altitude of 100–200 meters. The unconsolidated strata overlap this surface and ultimately wedge out on it. In all probability, the smooth surface on the granite is an erosion surface developed at the time of the Pinakul' transgression. Farther east (8 km east of Pinakul' village), well-sorted, horizontally laminated, fine gray sand containing fragments of *Mya arenaria* L. and *Hiatella arctica* (L.) rests on an irregular surface underlain by granitic gravel about 1 meter thick; the granitic gravel probably represents a colluvial deposit of ancient corestones of weathered granite. The preservation of these weathered cores and the presence of fine-grained sediments immediately above them suggest a quite rapid advance of the sea, as a result of which marine abrasion did not succeed in completely destroying the colluvium.

All of the marine deposits found at altitudes greater than 100 meters on the Chukotka Peninsula—for example, sandy loam and sand found at altitudes of 150–160 meters along the southern part of the east shore of Kolyuchinsk Bay—are evidently correlated with the Pinakul' transgression. The much higher position of the Pinakul' deposits on Kolyuchinsk Bay, as compared with those of Lawrence Bay, is evidently due to large later uplift. Thus, it is reasonable to suggest that during the Pinakul' transgression the Bering

and Chukchi Seas were joined not only through the widened Bering Strait
but also through the Kolyuchinsk–Mechigmen depression.

Palynological data apparently characterizing the lower part of the
Pinakul' deposits exposed on the eastern shore of Kresta Bay indicate tundra
vegetation. During separate stages of formation of these deposits, the vegetation composition is dominated by birch, alder, and bush pine (*Pinus
pumila*). The most characteristic of the Pinakul' deposits, however, have
somewhat different pollen spectra, which I believe can be considered the
glacial type of spectra in this region. They are characterized by a dominance
of spores of Bryales; arboreal, shrub, and herbaceous pollens are rarely
encountered. The spore-pollen spectra indicate more severe climatic conditions on land during the Pinakul' transgression than at the present time
(herbaceous pollen predominates in contemporary spore-pollen samples).

Evidently, there were local glaciers on the Chukotka Peninsula at the
time of the Pinakul' transgression—a conclusion also suggested by the
moraine-like aspect of the sandy-loam sediments. The relatively small quantity of coarsely fragmental material in the sandy loam of the Pinakul' Suite
compared with that in the Kresta Suite, described below, attests to the limited
transporting capacity of the glaciers, which in most cases probably did not
extend to the sea. The good sorting of the sand beds and the rounding of
pebbles in the littoral deposits within the Pinakul' Suite indicate considerable reworking of terrigenous sediments by waves; this would be possible
only if an ice blanket were lacking. Evidently, only small valley glaciers
were present in Chukotka; in any case, glaciation during the Pinakul' transgression undoubtedly was less extensive than the later glaciations.

## Middle Quaternary Time

The Middle Quaternary stage of development of Chukotka embraces the
time of Maximum Glaciation in northeastern Asia[b] and the interglacial
period that preceded it. The Middle Quaternary deposits include the marine
and glacial deposits of the Kresta Suite and, probably, the unnamed interglacial beds exposed below drift of the maximum glaciation along the
Tnekveyem River, southwest of Lake Koynatkhun. Partially eroded deposits
of the Pinakul' Suite are overlapped by the marine deposits of the Kresta
Suite. The presence in the Kresta beds of rounded, mollusk-bearing concretions reworked from the Pinakul' strata and, especially, differences in

---

[b] The term "Maximum Glaciation" is used in northern Siberia to denote an extensive
glacial episode thought to correspond to the Riss Glaciation of Europe and the Illinoian
Glaciation of North America. ED.

aspect and in degree of fossilization of faunal remains indicate an interruption in sedimentation between the Pinakul' and the Kresta Suites. The
base of the Middle Quaternary drift sheet lies considerably below sea level
in Chukotka, attesting to deep erosional cutting before it was deposited;
sea level at that time must have lain below present sea level. Evidently, the
first half of Middle Quaternary time was characterized by appreciable tectonic movements, as a result of which the principal topographic features of
present-day Chukotka, the position of the basic orographic units, and the
absolute and relative altitudes were formed.

Alluvial sediments exposed on the right bank of the Tnekveyem River
lying beneath drift of the Maximum Glaciation and resting on deposits of the
Koynatkhun Suite probably represent the oldest Middle Quaternary deposits
in Chukotka. Unfortunately, conclusive biostratigraphic data confirming
the Middle Quaternary age of these deposits have not yet been developed,
but they are tentatively referred to the first Middle Quaternary interglaciation on the basis of diatom analysis. These alluvial deposits consist of gritty
sand and pebbly sand containing interbeds as much as 25–30 cm thick of
clean sand, plant detritus, and peat. The total thickness is 10 meters. Pollen
of arboreal, shrub, and herbaceous plants is present in about equal quantities in the lower part of the section and spores are of subordinate importance. A slight increase in the quantity of herbaceous pollen is found at
higher levels. Among the arboreal-shrub pollen types, birch (47–74 per cent)
and alder (12–43 per cent) clearly predominate over pine (3–19 per cent).
Tree birch pollen is more abundant (35–52 per cent) than shrub birch
(14–23 per cent) in the lower part of the section, but shrub birch predominates (39–66 per cent) over tree birch (2–17 per cent) in the upper part.
Alder pollen is represented almost exclusively by shrub forms (up to 40
per cent). Among the herbaceous pollens, Cyperaceae (sedges) predominate
(up to 53 per cent) ; and Sphagnales dominate the spores (47–91 per cent).
Such a spore-pollen spectrum attests to a forest-tundra type of vegetation,
allowing one to suppose that climatic conditions at the time the alluvium
was deposited were somewhat better than at present. The differences between
these spore-pollen spectra and those from the Pinakul' Suite and also the
poor degree of fossilization of the peat and plant remains provide the basis
for assigning a Middle Quaternary age to the strata exposed beneath the
Middle Quaternary drift along the Tnekveyem River.

The second half of the Middle Quaternary stage of development of
Chukotka is recorded by the marine and glaciomarine beds of the Kresta
transgression. The Kresta Suite occupies a clear stratigraphic position; it
rests upon the Lower Quaternary deposits of the Pinakul' Suite or upon

much older rocks, and is covered by marine deposits of Upper Quaternary age that form lower terraces or, near the mountains, by glacial complexes of Upper Quaternary age. In areas that were glaciated again during Late Quaternary time, the Kresta deposits form hummocky residual terraces, commonly at altitudes of 50–60 meters and locally as high as 100 meters.

The Kresta Suite has a complex stratigraphy; three subsuites have been distinguished within it. The upper and lower subsuites are made up of laminated sandy gravel and sand, and the middle subsuite consists of till-like dark-blue-gray sandy loam grading toward the mountains into typical glacial till or into extremely coarse pebble-boulder deposits. The boundaries between the subsuites are generally gradational, but they are always distinct.

The sand beds of the lower subsuite accumulated at the beginning of the Kresta transgression. On the east shore of Kresta Bay, these beds contain massive accumulations of *Astarte* shells, often in life position, consisting predominately of *A. borealis borealis* (Schumacher) and *A. borealis placenta* Morch. Somewhat similar assemblages have been noted in sandy beds near the mouth of the Enmelen River, which contain large shells of *Serripes groenlandicus* and *Mya arenaria*, together with *Nucula tenuis*, and in an accumulation of slightly rounded pebbly material containing many broken shells exposed at the coast east of the village of Lorino. The shells in the lower subsuite contrast with those in the Pinakul' strata in having a fresh appearance, and are characterized by good preservation of the outer organic sheath, or periostracum. A considerable portion of the shells in the limey beds shows the effects of solution, and many specimens are represented only by excellently preserved periostraca, the calcareous parts having been completely dissolved away. This indicates the presence of free $CO_2$ in the aquifer, which is possible at low temperatures.

Good sorting, distinct lamination, and the presence of a boreal-arctic molluscan fauna indicate that the lower subsuite of the Kresta Suite consists of normal marine sediments that were deposited before the maximum glaciation had begun to influence sedimentation.

The middle subsuite accumulated later in the Kresta transgression under conditions of great water depths and intensive glacial transportation of terrigenous material. Typical sandy-loam till or coarse pebble-boulder gravel was laid down near the active ice front where the sediments were not substantially reworked by marine processes. Areas of this type were located immediately in front of the entrances to fjords and at the mouths of large rivers, such as the Enmelen and the Nunyamovaam (Fig. 3). Laterally, these coarse deposits quickly grade into horizontally laminated, dark-gray-blue loam containing irregularly distributed grit, pebbles, and boulders of various forms, sizes, and degrees of rounding. The grit and pebble-sized grains, as a rule, are angular or only slightly rounded, but the boulders are generally well rounded. The abundance of coarse material is extremely variable, ranging from isolated inclusions up to accumulations of pebbles and boulders in which almost no sandy-loam matrix can be found.

Fragments and isolated complete valves of mollusks are scattered throughout the sandy-loam beds of the middle subsuite. Mass accumulations of shells with closed, paired valves are also found; some of these consist of *Portlandia arctica siliqua* (Reeve) and *Bathyarca glacialis* (Gray), and others are of various species of *Astarte*, with a predominance of *A. alaskensis* Dall. The *Astarte* accumulations are sometimes found in thin sandy interbeds.

Numerous investigations indicate that a distinct diminution in abundance of mollusk shells coincides with an increase in the quantity of coarse fragmental material in the sandy loam or with the transition from sandy loam to pebble-boulder gravel, and that some sort of boundary exists beyond which no shells or even shell fragments are to be found, the sandy-loam strata acquiring the characteristics of a glacial till. Rarely, this transition is observed within a distance of several meters in a single outcrop. Occasionally, an absence of shells is characteristic of the glaciomarine sandy loams of an entire region; for example, no shells have been found in deposits of this type west of the Linatkhyrvuvaam River within the Vankarem lowland. Evidently, conditions were extremely unfavorable for molluscan life near the shore because of the glacial activity. Unfavorable factors probably included a freshening of the sea water due to the admixture of glacial meltwater, an absence of organic detritus in the terrigenous material delivered by the glaciers, and hindrance to the development of bottom vegetation due to the thick ice cover. One must conclude that during middle Kresta time the biocoenose of the littoral zone simply did not exist.

The mollusks collected in the bouldery sandy loam of the middle subsuite belong to the sublittoral zone of the sea. Arctic species such as *Portlandia arctica siliqua* (Reeve), *Yoldiella intermedia* (Sars), *Y. lenticula* (Moller), and *Bathyarca glacialis* (Gray) predominate, along with *Astarte alaskensis* Dall, *A. invocata* Merklin & Petrov, and *Macoma calcarea* (Gmelin). The presence of *Bathyarca glacialis* and of representatives of the genus *Yoldiella* indicates conditions characteristic of the lower sublittoral zone; apparently, some of the sediments of the middle subsuite accumulated in water depths as great as 100 meters. Negative water temperatures (°C.) prevailed at the sea bottom.

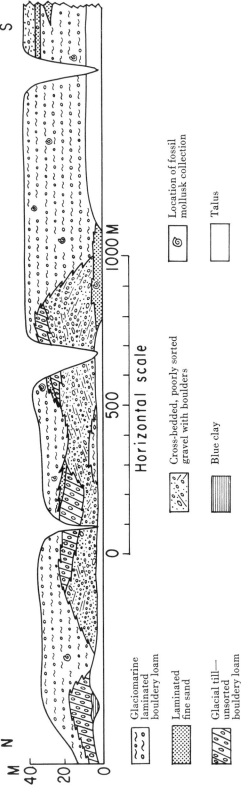

Fig. 3. Exposures of the Kresta Suite in the Middle Quaternary terrace southeast of the mouth of the Enmelen River.

**Legend:**

Glaciomarine laminated bouldery loam

Laminated fine sand

Glacial till—unsorted bouldery loam

Cross-bedded, poorly sorted gravel with boulders

Blue clay

Location of fossil mollusk collection

Talus

Horizontal scale

0    500    1000 M

N

M
40
20
0

S

While the middle subsuite of the Kresta Suite was accumulating, the Bering Sea was once again broadly connected with Chukchi Sea. Species of the genera *Portlandia, Bathyarca,* and *Yoldiella* penetrated from the north through Bering Strait and into Bering Sea. These mollusks apparently no longer live in Bering Sea; the arctic species of the genus *Yoldiella* are now found in the northern part of Chukchi Sea (Filatova, 1957).

The boundary between the middle and upper subsuites of the Kresta Suite is marked by a gradual transition from moraine-like sandy loam upward into silt and horizontally laminated fine silty sand that commonly possesses lamination of the ribbon type. Shells of *Yoldiella* and *Portlandia* persist in the transition zone. The changing conditions of sedimentation evidently reflect changes taking place in adjoining land areas; the deposition of "clean" marine sediments and the cessation of delivery of coarse fragmental materials to the sea herald the beginning of deglaciation. The increased melting of the glaciers resulted in some freshening of the coastal waters, which probably explains the insignificant quantities of faunal remains found in the upper Kresta sands. Distinct freshening during the regressive phase of the Kresta transgression is also probably responsible for the presence of shells of *Cyrtodaria kurriana* Dunker in muddy sands along the shores of Svobodny Bay.

The presence of pebbly sand and gravel in the uppermost deposits of the Kresta transgression indicates a subsequent regression of the sea. The bays were formed at this time, and sandy sediments were laid down in areas protected from direct wave action. Deposits of this type make up marine terrace remnants at altitudes of 30–40 meters in the estuary of the Yanramaveyem River on Svobodny Bay. Palynological analysis of these deposits (Fig. 6) seems to indicate that a climatic amelioration characteristic of the beginning stages of an interglaciation took place at the end of the Kresta transgression.

Spore-pollen and diatom analyses have been attempted on numerous specimens from other parts of the Kresta Suite in various regions, but the results are not diagnostic. This circumstance evidently is not due merely to bad luck; it reflects unfavorable conditions for the preservation of pollen, spores, and dinoflagellates at the time the sandy horizons were formed, as well as a deterioration of physical-geographical conditions, which limited the growth of vegetation and the development of diatom floras.

The geologic and paleontologic data all indicate that the Kresta Suite was formed in the extremely severe climatic conditions of a glacial epoch. In all probability, this was the most extensive glaciation that has ever affected Chukotka. Erratic boulders found by several investigators at alti-

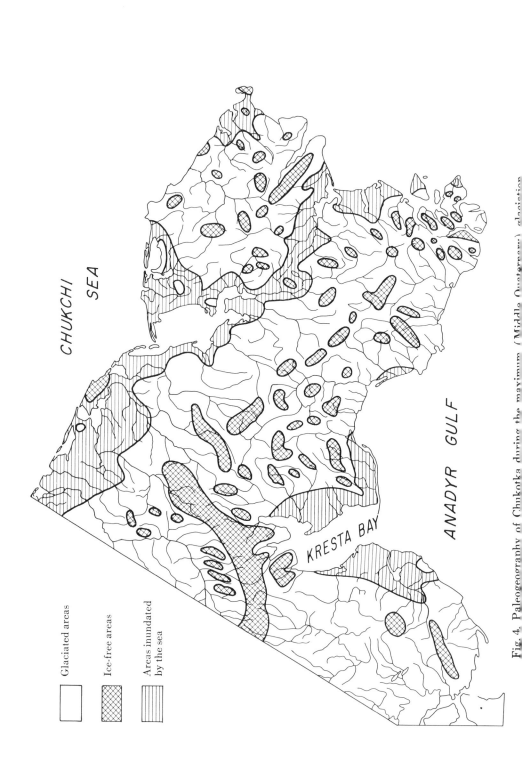

CHUKCHI SEA

ANADYR GULF

KRESTA BAY

Glaciated areas

Ice-free areas

Areas inundated by the sea

Fig. 4. Paleogeography of Chukotka during the maximum (Middle Quaternary) glaciation

tudes up to 500–600 meters and sometimes even higher on flat interfluve areas undoubtedly belong to this glacial epoch. Evidently, the glaciation that coincided with the Kresta transgression was of the ice-cap or semi-ice-cap type (Fig. 4).

## Late Quaternary Time

The Late Quaternary stage in the history of the development of Chukotka is divided into two large segments of time: one of these is the interglacial interval during which both the marine deposits of the Val'katlen transgression and the alluvial-lacustrine deposits of the Konergino Suite were formed; the other is the segment of time during which a twofold glaciation took place, the deposits of which are separated by interglacial alluvial beds and by the marine Amguem beds.

Several observations indicate that, at the end of the Kresta transgression, sea level lay below its present level. Thus, silty sand containing a lacustrine diatom complex underlies beds containing a marine diatom complex in the 20-to-30-meter terrace on the shores of Rudder Bay. And in exposures near the sod hut of Val'katlen, a steep-sided ravine filled with silty sand is incised in sandy loam of the Kresta Suite.

Sands of the Val'katlen transgression form well-developed terraces that are widespread at altitudes up to 30 meters everywhere on the shores of Chukotka. In some places, two sets of terraces are found at altitudes of 10–15 meters and 20–30 meters, perhaps representing different stages of regression of the Val'katlen Sea. An oscillation in sea level during Val'katlen time is also indicated by the presence of a layer of buried peat in the marine deposits making up the Val'katlen terrace on the shores of Vankarem Lagoon.

In some places, the Val'katlen terrace is erosional in nature and is underlain by Middle Quaternary glaciomarine bouldery-sandy loam. A terrace traceable for a considerable distance along the southern part of the east shore of Kresta Bay, for example, has a varying structure—accumulative in some places, abrasive-accumulative or abrasive in others (Fig. 5). I must emphasize that a comparable abrasive-accumulative surface is forming in the littoral zone along this part of the coast at the present time. Thus, the beach zone, which is more than 100 meters wide in some places, is made up of dark-blue-gray glaciomarine sandy loam of the Kresta Suite, covered only locally by a thin sheet of gritty and pebbly sand. Farther seaward, at a distance of 2–3 km from the shore, depths at low tide do not exceed 1 meter, and sediments are not accumulating there at the present time. Still farther from the shore there are submerged bars, composed mostly of gritty pebble

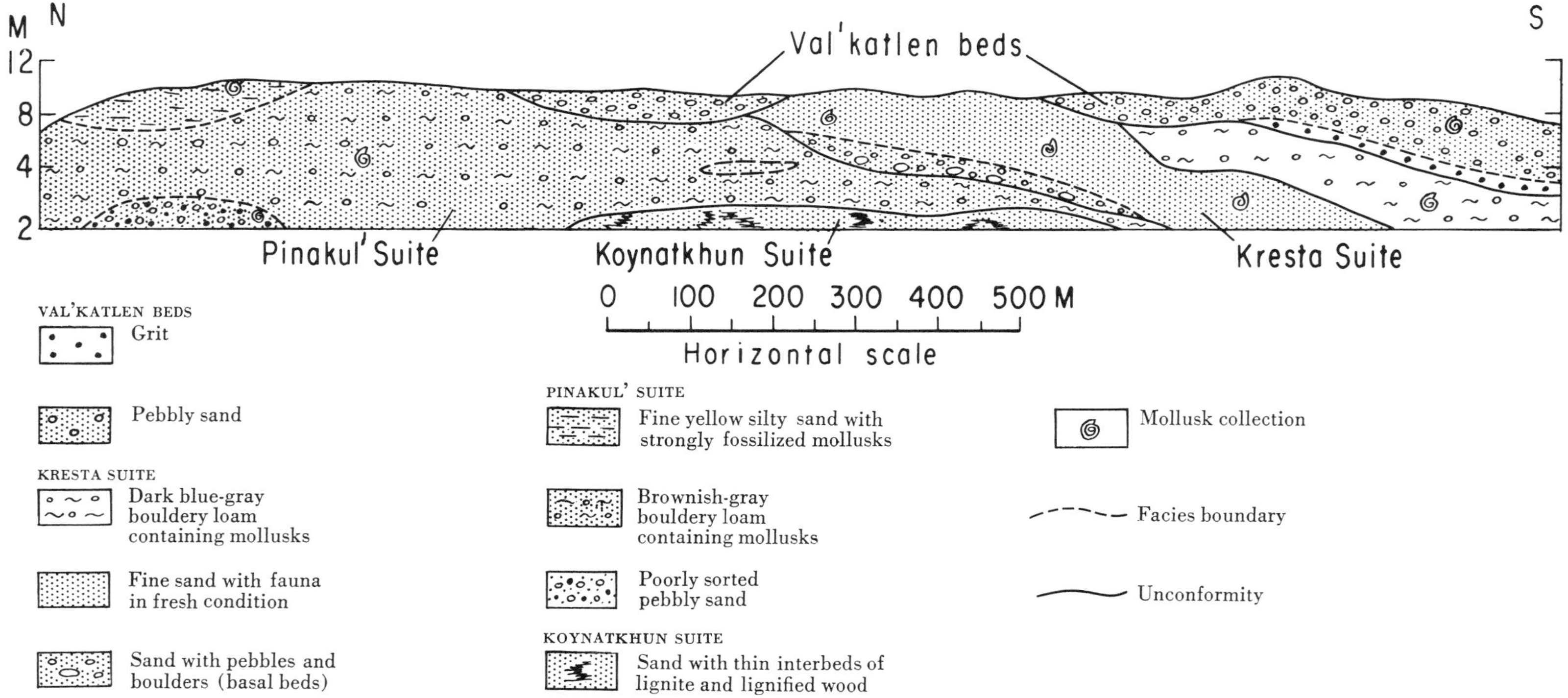

M N

M
12
8
4
2

S

Pinakul′ Suite

Val′katlen beds

Koynatkhun Suite

Kresta Suite

Horizontal scale

0   100   200   300   400   500 M

VAL′KATLEN BEDS

Grit

Pebbly sand

KRESTA SUITE

Dark blue-gray boulderly loam containing mollusks

Fine sand with fauna in fresh condition

Sand with pebbles and boulders (basal beds)

PINAKUL′ SUITE

Fine yellow silty sand with strongly fossilized mollusks

Brownish-gray bouldery loam containing mollusks

Poorly sorted pebbly sand

KOYNATKHUN SUITE

Sand with thin interbeds of lignite and lignified wood

Mollusk collection

Facies boundary

Unconformity

Fig. 5. Structure of the late Quaternary abrasive-accumulative terrace on the southern part of the west shore of Kresta Bay.

gravel, that are exposed at the very lowest tides. Only beyond the outer margins of the bars is there a sudden increase in depth. A relative subsidence is presently in progress on the eastern coast of Kresta Bay (Budanov and Ionin, 1956). It seems likely that similar littoral conditions existed locally at the time of the Val'katlen transgression. The development of abrasive-accumulative surfaces in unconsolidated sediments was evidently furthered by the subsidence of the land, shallow water depths, the argillaceous composition of the eroded rocks, the strong tidal currents, and finally, intensive wave action.

Boulder piles found along the southern part of the east coast of Kresta Bay apparently fixed the position of the shoreline at the time of the Val'katlen transgression. They are located at the rear of the Val'katlen terrace at altitudes of 30–35 meters, where the terrace meets the scarp to the 60-to-70-meter terrace, which is generally composed of glaciomarine bouldery-sandy loam. The altitude of 30–35 meters most likely represents the maximum level of the Val'katlen transgression on the shores of both Bering and Chukchi Seas. The sediments of this transgression consist almost entirely of sand; the sand is underlain by partly eroded sandy loam of the Kresta Suite, indicating that water depths were modest during Val'katlen time, because the bottom evidently lay above wave base. These observations also indicate a decrease in the ice cover on the sea, at least during the summer months, in comparison with conditions during the Kresta transgression.

Mollusk shells are widely distributed in the Val'katlen sands. Especially common are mass accumulations of various species of *Astarte,* with *A. borealis* predominating in some places and *A. alaskensis* in others. Mass accumulations of *A. invocata* are found in the sediments of the Val'katlen terrace in two places along the shores of Vankarem Lagoon. The wide distribution of *Astarte* indicates normal salinities in the sea water. *Portlandia arctica siliqua* and *Gomphina fluctuosa* commonly are also present in the lower part of the section. *Mytilus edulis* is found in the upper beds, indicating not only a shallowing of the basin but also a new penetration of this mollusk to the Chukotka coast, since its remains are not found in the Kresta deposits, though they are present in the much older beds of the Pinakul' Suite.

The species composition of the fauna during the Val'katlen transgression seems to have been very close to that of the present. This circumstance alone allows one to see the Val'katlen beds as interglacial deposits.

The marine sands contain little pollen and few dinoflagellates, probably because conditions are unfavorable for preservation in these comparatively coarse sediments. The coarse dinoflagellate *Melosira sulcata* is commonly the only diatom found. The palynological data suggests that tundra vegetation existed in the land areas during the Val'katlen transgression. This is especially clearly substantiated by the organic remains from the upper beds in the 30-meter terrace. The appearance there of interbeds of gritty gravel indicates that a shallowing and gradual lowering of the sea took place during the late stages of the Late Quaternary interglacial, apparently at the beginning of the last glaciation.

The climatic optimum during the first Late Quaternary interglaciation evidently coincided with the accumulation of the Konergino lacustrine and alluvial beds. Palynological data indicate that the lowlands were clothed in shrub tundra or even forest tundra when the Konergino beds were deposited.

An interglacial lacustrine deposit exposed in the coastal bluffs on the east shore of Kresta Bay, north of the village of Konergino, consists of sandy clay with peat interbeds occupying a depression on the eroded surface of glaciomarine bouldery-sandy loam of the middle Kresta subsuite and covered by moraines of the Late Quaternary glaciation. Arboreal-shrub pollen, consisting almost entirely of alder (78–98 per cent) predominates in the lacustrine deposits, but birch pollen (51–61 per cent) dominates over alder (34–43 per cent) in two samples from the lower part of the section (Fig. 7).

Somewhat to the northwest, in the middle part of a similar interglacial deposit, the following plant remains were obtained: Bryales, *Carex* sp., *Betula nana* L., *Alnus fruticosa* Rupr., *Alnus* sp., *Ranunculus aquatalis* L., *Rubus chamaemorus* L., *Rubus* sp., *Hyppurus vulgaris* L., *Empetrum nigrum* L., *Arctostaphylos uva-ursi* Spr., *Vaccinium uliginosum* L., and *Menyanthes trifoliata* L. According to T. D. Kolesinkovaya, all of these species are presently growing in the region or close by. This fact and the weak degree of fossilization of the plant remains—the material can hardly be distinguished from modern material—forces one to accept a comparatively young age for the enclosing deposits. Under no circumstances can they be older than the early part of the Late Pleistocene. Figure 8 gives the pollen diagram for the section from which the plant remains were obtained. One must note that alder no longer grows at all in the interstream areas of Chukotka; it is confined to river valleys, forming thickets in valleys in the Anadyr Range, well to the west of the fossil locality. Thus temperatures somewhat warmer than at present are indicated for the last interglaciation. The vegetation of the lowlands on Chukotka at that time probably was similar to the present vegetation along the lower course of the Anadyr River.

The first Late Quaternary glaciation of Chukotka—the Vankarem Glaciation—began and reached a maximum at a time when sea level was much lower than at present. Borehole data indicate that the lower courses of rivers in mountainous parts of the peninsula were excavated 25–30 meters below present sea level, and in the northern part of Kresta Bay the base of the Upper Quaternary glacial deposits lies below sea level. The extent to which the continental shelves of Bering and Chukchi Seas were laid bare can be established only indirectly; studies of diatom floras in bottom samples from Bering Sea indicate that a large part of the shelf was dry land during the last glaciation (Jousé, 1962).

The Vankarem Glaciation consisted of valley glaciers in the mountains and broad piedmont glaciers at the edges of the lowlands. Well-developed ice-scoured relief characterizes this glaciation in the mountains, and fresh accumulative forms composed of boulder loam, sandy loam, cross-bedded glaciofluvial pebbly sand and gravel, and glaciolacustrine sand and silt are found in coastal plain areas. The thickness of the glacial and glaciofluvial deposits amounts to not less than 30–50 meters, depending on the relative height of the morainal hillocks.

The extent and the retreatal history of glaciers during the Vankarem Glaciation can be determined by study of the composition of the glacial deposits and of the glacial and glaciofluvial relief forms. The most typical glacial relief is found within the Vankarem lowland, where end moraines are distributed 150 km from the source of supply; these seem to represent the longest glaciers on Chukotka during the Vankarem Glaciation. In the northern part of Kresta Bay, the limit of the Vankarem Glaciation extended approximately along a line connecting the mouth of the Mamchergyrgyn River and the village of Konergino, 80–90 km from the sources of the glaciers. Glaciers were somewhat shorter in the eastern parts of the peninsula.

Analysis of the extent, distribution, and orientation of the morainal forms indicates that at the culmination of the Vankarem Glaciation the mountain valleys were occupied by active glaciers that descended to the foothill plains, which, for the most part, were occupied by firn, snow, and stationary ice. Numerous recent investigators have noted that glaciofluvial sediments predominate among the deposits of the Vankarem Glaciation. Unsorted deposits predominate only very near the mountain massifs and in small intermontane depressions, such as the Ulyuveem depression. Evidently, the glaciers quickly stagnated after reaching the lowlands, so that the basic relief-forming role was assumed by meltwater.

A complete absence of megascopic remains of plants and animals is also a characteristic feature of the deposits of the Vankarem Glaciation. It is not possible to reconstruct the vegetation of that time from palynological data because of the scarcity of pollen and spores in the glacial and glaciofluvial deposits, and because of the difficulty in distinguishing reworked pollen

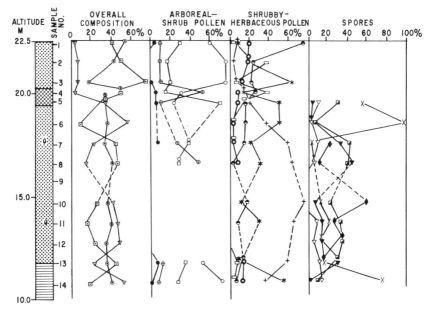

Fig. 6. Spore-pollen diagram of the upper subsuite of the Kresta Suite in the lower reaches of the Yanramaveyem River near Svobodny Bay.

## EXPLANATION FOR FIGURES 6–9

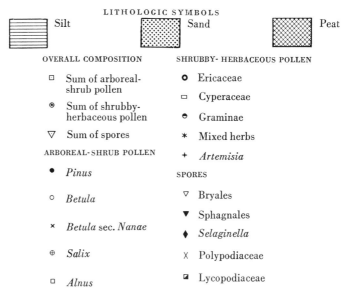

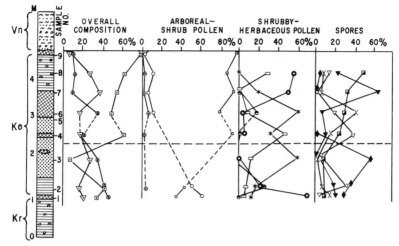

Fig. 7. Spore-pollen diagram of Upper Quaternary interglacial (Konergino) lacustrine deposits north of the village of Konergino.

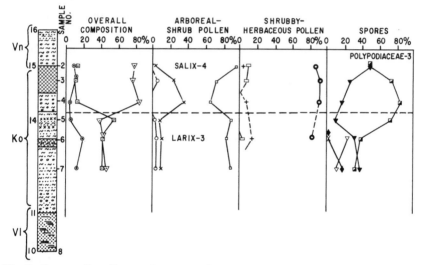

Fig. 8. Spore-pollen diagram for section from which plant remains were obtained from Konergino interglacial deposits northwest of section illustrated in Figure 7.

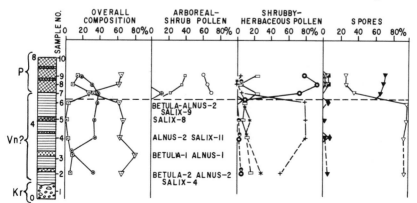

Fig. 9. Spore-pollen diagram of the lacustrine deposits correlated with the Vankarem glaciation north of the village of Uel'kal.

from pollen that is contemporary with the sediments. However, lacustrine deposits exposed near the village of Uel'kal, in my opinion, give information concerning the vegetation in unglaciated areas during the Vankarem Glaciation. The lacustrine clay near Uel'kal occupies a depression on the surface of a 15-meter terrace made up of Kresta loams; it is covered by peat, apparently of postglacial age. The major role in the spore-pollen complex of the lacustrine clay is played by pollen of *Artemisia* and spores of Bryales; the remainder of the spectrum is made up of individual grains of the forms found in older pollen complexes (Fig. 9). This spectrum may represent a periglacial vegetation association that grew in immediate proximity to an active glacier. It is interesting to note that comparable quantities of *Artemisia* pollen have not been detected in spectra from steppes and semideserts (Grichuk and Zaklinskaia, 1948).

Although, as I have noted, the geomorphological data allow one to propose the existence of several stages in the development of the last glaciation, only two main stages—evidently reflecting two independent glaciations—can be distinguished on the basis of stratigraphic and palynological data at present. The interglacial interval that preceded the second Late Quaternary glaciation is represented by the Amguem beds—the deposits of the second river-flood plain terrace and correlative sediments of the second marine terrace. The second flood plain terrace covers the bottoms of the ancient mountain valleys, lying 5–13 meters above river level and incised 20–30 meters below glaciofluvial deposits of the Vankarem Glaciation (Fig. 10). The terrace deposits characteristically contain interbeds and lenses of peaty mud, peat, and plant detritus. Fragments of alder and willow wood are common. Palynological analysis indicates that when the alluvium of the second terrace accumulated, the vegetation was either tundra similar to the present vegetation or somewhat more shrubby than at present. Shrub vegetation is clearly indicated by spore-pollen spectra in the deposits of the second flood-

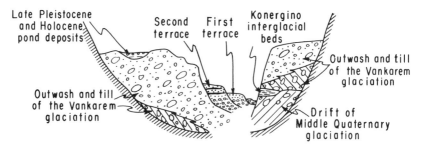

Fig. 10. Schematic cross section showing the stratigraphy of the Quaternary deposits in the river valleys of Chukotka Peninsula. Symbols same as in Figure 2.

plain terrace of the Amguem River; arboreal-shrub pollen predominates there (68–87 per cent), consisting chiefly of alder (41–90 per cent—probably shrub forms) and dwarf birch (10–54 per cent). Heaths and mixed herbs predominate among the herbaceous pollens. In the opinion of R. E. Giter·man, "the vegetation at the time these deposits formed consisted of shrub tundra or forest tundra with a dominance of *Alnus fruiticosa,* shrub birch, and heaths, and also quite a rich herbaceous cover. Vegetation of this type indicates climatic conditions more favorable than at present in this region." These facts suggest that the alluvium of the second terrace accumulated during an interglacial interval, when the glaciers apparently had completely disappeared.

A marine terrace correlated with the second floodplain terrace is found at altitudes of 8–10 meters in many parts of Chukotka, almost everywhere well represented. The marine terrace is made up chiefly of flat-bedded or cross-bedded sandy-gritty pebble gravel. Ancient barrier bars, composed of sand and pebbles, are commonly found on the surface of the terrace. The distribution of ancient lagoons is indicated in the terraces by sandy-muddy sediments generally covered by peat. An ancient barrier bar on the shore of Bering Sea can be traced more than 50 km eastward from Kresta Bay. The terrace surface slopes landward from the bar, apparently representing an ancient lagoon area. The modern offshore bars in areas protected from wave action have altitudes of 4–5 meters. If we assume that the ancient off-shore bars on the terraces extended only 4–5 meters above the sea level of the time, then we may suppose that the sea stood 5–6 meters higher than at present when the second marine terrace was formed. Evidently, a connection between Bering and Chukchi Seas was established through Bering Strait when the Amguem beds were deposited—a connection that has persisted without interruption until the present time.

Boreal-arctic mollusks have been collected from the deposits of the second marine terrace. Exposures on the shores of Bering Sea have yielded *Astarte borealis* (Ch.), *A. alaskensis* Dall, *Mytilus edulis* L., *Serripes groenlandicus* (Brug.), and others; from the shores of Chukchi Sea have come *Astarte invocata* Merklin & Petrov, *A. montagui* (Dillw.), *Mya arenaria* L., *Mytilus* sp., and *Macoma* sp.

The second Late Quaternary glaciation—the Yskaten' Glaciation—is recorded by large morainal ridges up to 30 meters high, which blockade headwater valleys at altitudes of 200–450 meters. These ridges are especially clearly developed in the Yskaten' Range north of Kresta Bay. Deposits of the Yskaten' Glaciation are distributed at similar altitudes near the bases of the largest mountain masses on the eastern part of the peninsula, and similar hypsometric indications are also seen on the floors of the largest and freshest cirques. It is necessary to note that the freshest glacially scoured relief observed in the summit areas of the large mountain massifs is connected exclusively with the Yskaten' Glaciation. The Yskaten' glaciers were mostly cirque glaciers. Small valley glaciers 5–10 km long were present only in isolated places—for example, in the Yskaten' Range.

## Recent Time

During the last phases of the Yskaten' Glaciation, the rivers incised themselves to form the first terrace (the high flood plain), and, concurrently, the deposits of the proluvial fans were formed. Postglacial time began at the moment of accumulation of the upper beds of the terrace, which generally are represented by the floodplain facies of alluvium. Palynological analysis of the upper beds of the terrace indicate that climatic conditions were better than at present and that Chukotka Peninsula supported forest-tundra vegetation.

An intensive development of peat bogs apparently took place in low-lying areas on the coastal lowlands, in the large intermontane depressions, and especially on the surface of the second marine terrace at this time. The bogs are underlain by peat 3–4 meters thick, but peat formation has essentially ceased at present. The occurrence of fragments of willow shrubs in the peat in areas where only dwarf willows now grow (shrub willows are found in the peat as far north as the vicinity of the village of Vankarem on the shores of Chukchi Sea) attests to a milder climate. In all probability, the first terrace was formed during the postglacial climatic optimum. Palynological analysis of the deposits of the lower floodplain indicates that the deposits accumulated after the present climatic conditions had begun.

The largest part of the coast of Chukotka is undergoing a relative submergence at the present time (Budanov and Ionin, 1956).

Obruchev reported that the present extent of glaciation on Chukotka consists of several small cirque glaciers in the Yskaten' Range, in the Pekuleney Range, and possibly in the main core of the Chukotka Range (Obruchev and Salishchev, 1936). However, similar small glaciers were discovered by M. T. Kiriushina (1939) in eastern Chukotka, as well. In 1957, I discovered traces of three small glaciers in the Yskaten' Range. They were noted on August 21, at the time of the greatest summer melting of snow, of permafrost, and thus of glaciers. All of the glaciers had a northern exposure, and all were located in cirques, the walls of which reach heights of 200–400 meters. The largest glacier, 650 meters long and 500 meters wide, lying at the foot of a 1,082-meter mountain, had a concave longitudinal profile, and its lower boundary was at an altitude of 700–750 meters. The central and lower parts were devoid of snow cover, and greenish-blue ice was exposed at the surface there. The lower margin was buried beneath a moraine 30–40 meters high. Another cirque glacier found 3.5 km to the northeast had a total length of 700 meters and a width of 350 meters. A lake whose surface is about 500 meters above sea level adjoins this glacier, held in by a large morainal ridge. Scattered bedrock outcrops exposed in the lower part of the glacier attested to the small thickness of ice, which hardly exceeded 10 meters. The third glacier was almost completely buried by fragmental material fallen from the walls of the cirque. All of these glaciers are related to the mountain massif on the southern slope of the Yskaten' Range, where summit altitudes reach 1,000–1,100 meters. In spite of greater altitude (1,300–1,400 meters), no glaciers are present in the central part of the range. The position of the observed glaciers indicates that their existence is dependent on the exposure and local orographic conditions, and that they are now in a state of degradation.

## Summary

The Chukotka Peninsula was subjected to repeated marine transgressions during the Quaternary Period; these were evidently produced by the combined action of several factors, with tectonic movements playing the major role.

Chukotka suffered four glaciations during the Quaternary Period; the second glaciation, which was of Middle Quaternary age, was the most extensive and may have had the character of an ice sheet or semi-ice sheet.

The first or Early Quaternary Glaciation and the second or Middle Quaternary Glaciation both coincided with marine transgressions. The Third (Vankarem) glaciation took place when sea level lay much lower than at present; and the fourth (Yskaten') took place when sea level was at its present position.

The development of the marine molluscan fauna of Bering Sea during Neogene and Quaternary time is characterized by successive increases in the role of cold-loving forms, culminating in the formation of an arctic faunal complex in Middle Quaternary time. The role of the arctic elements has decreased since then, the fauna has become boreal-arctic in character, and there has been a gradual approach in the composition of the faunas to that of the present time. In short, the history of the molluscan faunas records one wave of chilling that reached a peak in Middle Quaternary time.

Tundra vegetation appeared in Chukotka at the beginning of the Quaternary Period. Since then, there has been no forest but only forest tundra, during even the most optimal climatic conditions of the interglacials.

## REFERENCES

Budanov, V. I., and A. C. Ionin. 1956. Sovremennye vertikakal'nye dvizheniia zapadnykh beregov Beringova moria (Recent vertical movements on the western coast of Bering Sea): Okeanografich Komis., Trudy, v. 1, p. 65–72.

Eliseev, B. N. 1936. Materialy k geologii i poleznym iskopaemym Anadyrskogo kraya (Material on the geology and useful fossils of the Anadyr region): Vsesoiuznyi Arkticheskii Inst., Trudy, v. 48, p. 72–122.

Filatova, Z. A. 1957. Obshchii obzor fauny dvustvorchatykh molliuskov severnykh morei SSSR (General aspects of the marine bivalve fauna of the northern seas of the USSR): Akad. Nauk SSSR, Inst. Okeanologiya, Trudy, v. 20, p. 3–59.

Grichuk, V. P., and E. D. Zaklinskaia. 1948. Analiz iskopaemykh pil'tsy i spor i ego primenenie v paleogeografii (Analysis of fossil pollen and spores and their application in paleogeography): Geographgiz (Moscow), 223 p.

Il'ina, A. P. 1963. Molliuski Neogena Kamchatki (Neogene mollusks of Kamchatka): Vsesoyuznyi Neftianoi Nauchno-Issledovatel'skii Geologorazvedochnyi Inst. (VNIGRI), Trudy, v. 202, 242 p.

Jousé, A. P. 1962. Stratigraphicheskie i paleogeograficheskie issledovaniia v severo-zapodnoi chasti Tikhogo okeana (Stratigraphic and paleogeographic investigations in the northwestern part of the Pacific Ocean): Akad. Nauk SSSR, Inst. Okeanologiia (Moscow), 258 p.

Kiriushina, M. T. 1939. Geomorfologiia i chetvertichnye otlozheniia severo-vostochnoi chasti Chukotskogo poluostrova (Geomorphology and Quaternary deposits of the northeastern part of Chukotka Peninsula): Vsesoiuznyi Arkticheskii Inst., Trudy, v. 131, p. 7–47.

Krishtofovich, L. V. 1964. Molliuski Tretichnykh Otlozhenii Sakhalina (Mollusks of the Tertiary sediments of Sakhalin): Vsesoyuznyi Neftyanoi Nauchno-Issledovatel'skii Geologorazvedochnyi Inst. (VNIGRI), Trudy, v. 232, 228 p.

Merklin, R. L., O. M. Petrov, and O. V. Amitrov. 1962. Atlas-opredelitel' molliuskov chetvertichnykh otlozhenii Chukotskogo poluostrova (Atlas-guide to the mollusks of the Quaternary deposits of Chukotka Peninsula: Akad. Nauk SSSR, Kom. Izucheniia Chetvertichnogo Perioda (Moscow), 57 p. English translation available from U.S. Dept. of Commerce Clearinghouse for Fed. Sci. Tech. Inf.

Obruchev, S. V. 1939. Drevnee oledenenie i chetvertichnaia istoriia Chukotskogo (Ancient glaciations and the Quaternary history of the Chukotka Region): Akad. Nauk SSSR, Izv., Ser. Geogr. Geofiz., p. 129–146.

Obruchev, S. V., and K. A. Salishchev. 1936. Chukotskaia lëtniaia ekspeditsiia 1932–1933 gg (Air expedition to Chukotka, 1932–1933): Vsesiuznyi Arkticheskii Inst., Trudy, v. 54, 184 p.

Petrov, O. M. 1963. Stratigrafiya chetvertichnykh otlozhenii iuzhnoi i vostochnoi chastei Chukotskogo poluostrova (Stratigraphy of the Quaternary deposits of the southern and eastern parts of the Chukotka Peninsula): Akad. Nauk

SSSR, Kom. Izucheniya Chetvertichnogo Perioda, Bull. 28, p. 135–152. English translation available from U.S. Dept. of Commerce Clearinghouse for Fed. Sci. Tech. Info.

Sachs, V. N. 1946. Chetvertichnaia istoriia Chukotskogo okruga (Quaternary history of the Chukotka Region) : Problemy Arktiki, No. 3.

———— 1963. Osnovnye problemy izucheniia chetvertichnogo perioda (The principal problems of study of the Quaternary Period) : (no further information available).

Zhadin, V. I. 1952. Molliuski presnykh vod SSSR (Fresh-water mollusks of USSR) : Vsesoiuznyi Arkticheskii Inst., Trudy, v. 48, p. 72–122.

# 8. Radiocarbon Chronology in Siberia

N. V. KIND

*Geological Institute, Academy of Sciences of the USSR*

Important progress has been made in the study of Quaternary deposits in Siberia during the last decade. At first, efforts were devoted mainly to establishing local stratigraphic schemes for individual areas. More recently, attempts have been made to correlate these schemes. Correlations of this sort are, naturally, often rather difficult. This is explained not only by the vast areas involved and the consequent differences in the geological history of the separate parts, located as they are in essentially different structural-tectonic and physiographic provinces, but also by the differing extent of our knowledge of different areas.

## Existing Stratigraphic Schemes for Siberia

The stratigraphy of Anthropogene deposits has been studied most thoroughly in Western Siberia. There, a regional stratigraphic scheme has been created for the Quaternary deposits, and attempts have been made to correlate it with the stratigraphic scheme for the European part of the USSR. The scheme is based on the data of lengthy and detailed investigations, mostly in the glaciated regions of the northern part of Western Siberia and in the territory just to the east—in the lower valley of the Ob, around northern Yenisei, in the Taimyr lowland. At the end of the 1940's, V. N. Sachs (Sachs and Antonov, 1945; Sachs, 1951a) worked out a stratigraphic scheme for the Quaternary deposits of Northern Siberia, using the material provided by these investigations. His scheme is also reflected in the regional stratigraphic scheme for Western Siberia, and it is accepted for the northern Yenisei area almost without change (Table 1). Both schemes still contain many points of dispute, most of them arising within the lower part of the column—that is, within the Lower and Middle Pleistocene. One may men-

TABLE 1. *Stratigraphic Scheme for the Anthropogene of Western Siberia*

| *Division* | *Regional Subdivision (Stage)* | *Local Subdivisions for the Lower Yenisei* | | | |
|---|---|---|---|---|---|
| HOLOCENE | RECENT | Fluvial deposits of floodplain terraces; lacustrine and lacustrine-bog deposits | | | |
| UPPER PLEISTOCENE | SARTAN | SARTAN SUITE Glacial deposits | | Fluvial deposits of Terrace I | |
| | KARGINSKY | Fluvial deposits of the "Karginsky" Terrace | | | |
| | ZYRIANKA | ZYRIANKA SUITE Glacial and glaciofluvial deposits | | | |
| | KAZANTSEVO | KAZANTSEVO SUITE Fluvial and lacustrine deposits; marine deposits with *Cyprina islandica* | | Fluvial deposits of Terrace III | |
| MIDDLE PLEISTOCENE | TAZ OR SANCHUGOV | SANCHUGOV SUITE Marine and glacio-marine deposits | | TAZ SUITE Glacial and glacio-fluvial deposits | |
| | MESSOV | MESSOV SUITE Marine and fluvial deposits | | Lacustrine-fluvial deposits of Shirta river | |
| | SAMAROV | SAMAROV SUITE Glacial and glaciofluvial deposits | | | |
| | TOBOL | Fluvial, lacustrine, and marine(?) deposits in the buried ancestral Yenisei valley | | | |
| LOWER PLEISTOCENE | DEMYANKA | Fluvial, lacustrine, and moraine-like (glacial?) deposits in the buried ancestral Yenisei valley | | | |
| | PRECLACIAL | Fluvial deposits of deep horizons in the buried canyon | | | |

(BAKHTIAN SERIES spans the Middle Pleistocene regional subdivisions; UST-YENISEI SERIES spans the local subdivisions from Kazantsevo through Messov.)

tion here such debatable questions as the existence of pre-Samarov (pre-Maximum) glaciations and of the Taz Glaciation, which supposedly followed the Maximum Glaciation but preceded the Zyrianka Glaciation, and which is conditionally correlated with the Moscow Glaciation of the Russian plain.

Leaving aside these questions and such general problems as the synchroneity of glaciations and marine transgressions, let us analyze the upper part of the column in detail, inasmuch as it bears directly on the subject under discussion. It should first be said that there is almost no controversy regarding this part of the column. Its large divisions seem to reflect correctly the main climatic events in the Late Pleistocene and are consequently of regional importance.

Different genetic facies of correlative age can be found not only in glacial but also in nonglacial regions of Siberia. Thus, a pronounced cold interval that followed the last large interglacial, the Kazantsevo Interglacial, was recorded in the north by a complex of glacial and glaciofluvial deposits connected with the Zyrianka Glaciation. Owing to its fresh morphological features, the boundaries of this glaciation can be traced in greater detail than the boundaries of the Maximum Glaciation. Zyrianka glacial deposits extend across the vast interfluve area in the northwestern part of Western Siberia, the Taimyr lowland, and a considerable part of the central Siberian Plateau; and they fringe the base of the Polar Urals and the Verkhoyansk Mountains. The main centers of the Zyrianka Glaciation, as well as of the Maximum Glaciation, were in the Polar Urals, the mountains of the northern part of the Taimyr Peninsula, the middle Siberian highland, and the Verkhoyansk Range.

In the nonglacial regions, Zyrianka deposits include a complex of lacustrine-alluvial and alluvial deposits, as well as the loesslike sediments of the interfluves. All these are characterized by cold-loving mammalian faunas and spore-pollen spectra typical of periglacial steppe landscapes. The same can also be said of the second important Upper Pleistocene cold interval corresponding to the Sartan Glaciation, a glacial event that was confined to the mountain regions, such as the central Siberian Plateau (the Putoran Plateau). The Karginsky warm period that preceded the Sartan Glaciation is well characterized by floristic complexes; Karginsky alluvial deposits are characterized by forest spore-pollen spectra in the northern area that is now tundra. Within postglacial deposits, two time intervals with differing climates can be distinguished equally distinctly on the basis of plant remains; these include a warm spell during the first half of the Holocene Epoch and a colder interval during the second half.

Although this scheme of Upper Pleistocene stratigraphy has been adopted as a whole, there remain a number of unsettled problems. One of these concerns the number of substages in the Zyrianka Glaciation. Two pronounced belts of marginal glacial forms left by the Zyrianka glaciers are recorded in the glaciated regions; a third is less distinct. Recognition of the presence of these end moraines, which actually reflect recessional stages of the Zyrianka glaciers, does not solve the problem of whether or not interstades existed within the Zyrianka interval. Investigations in nonglacial or periglacial regions of Siberia are important to a solution of this problem, because there one may find a more detailed sequence of Upper Pleistocene deposits containing evidence of relatively small climatic changes. Thus, recent investigations have distinguished a fossil soil horizon, within the Zyrianka complex, reflecting the existence of a warmer interval that may be regarded as an interstadial within the Zyrianka Glaciation (Ravskii *et al.*, 1964; Zeitlin, 1964, 1965). Similar traces of a warmer time are also established within sequences of deposits of Sartan age.

The problem of the independence of the Sartan Glaciation also remains obscure. There has been a tendency recently to regard it as an independent glaciation, and not as the concluding stage of the Zyrianka Glaciation. In my view, this problem is not of fundamental importance at the present level of knowledge. The correlation of Siberian glaciations with those of other countries seems much more important.

Absolute geochronology is the only reliable method for establishing correlations of Upper Pleistocene deposits across vast areas. The radiocarbon method must be depended upon primarily; it has proved successful in correlating geological events between Europe and North America during the last glaciation and during the postglacial interval.

## Stratigraphy and Radiocarbon Datings

Until recent years, only single radiocarbon datings were available for the territory of Siberia, and they were quite insufficient for the reconstruction of the absolute chronology of geological events and climatic fluctuations during Late Pleistocene and Holocene time. They were even less adequate for any large-scale correlations between the Late Quaternary glaciations in Siberia and glaciations in other parts of the world. The first data on the absolute chronology of the Holocene were based on measurements of a series of samples from the Indigirka basin (Vinogradov *et al.*, 1962). These determinations made it possible to establish the synchroneity of the principal Holocene stages recognized in northeastern USSR, Western Europe, and North America (Lavrushin *et al.*, 1963). In 1962, systematic studies aimed at working out an absolute time scale for the upper part of the Quaternary system in Siberia began in the Absolute Age Laboratory of the Geological

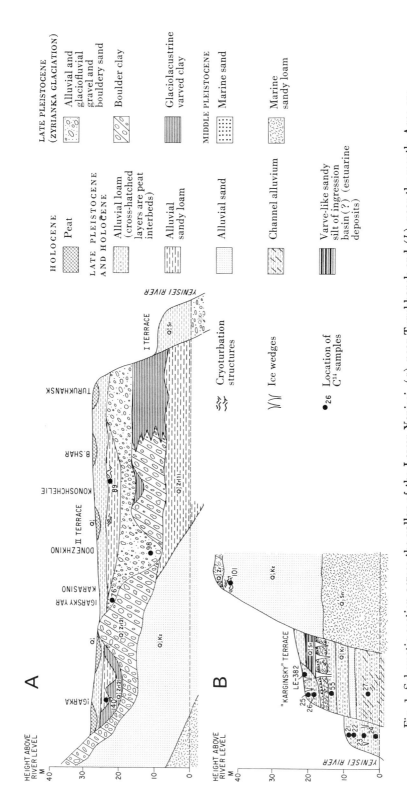

Fig. 1. Schematic section across the valley of the Lower Yenisei, (a) near Turukhansk and (b) near the mouth. Ages are indicated by letter symbols: $Q^2_4$—Upper Holocene; $Q^1_4$—Lower Holocene; $Q^3_3Sr$—Sartan; $Q^3_3Kr$—Karginsky; $Q^2_3Zr$—Late Zyrianka; $Q^3_3Zr(1)$—Early Zyrianka; $Q^3_3Zr(2)$—Middle Zyrianka (maximum phase); $Q^1_3Kz$—Kazantsevo; $Q_2Sn$—Sanchugov.

Institute (GIN) of the Academy of Sciences of the USSR under the direction of Professor V. V. Cherdyntsev. The first region chosen for investigation was the Lower Yenisei basin. The Quaternary stratigraphy had been worked out thoroughly enough there that this region could be regarded as a reference area; its intermediate position between western and eastern Siberia is also of importance.

The proposed scheme given here for the absolute chronology of the Upper Pleistocene and Holocene is still very preliminary and is based both on the results of age determinations of samples from the northern Yenisei area (Cherdyntsev *et al.*, 1964, 1965; Alekseev *et al.*, 1965; Kind, 1965) and on datings from other Siberian regions obtained in the laboratory of the Geological Institute and in other laboratories in the USSR. The correlation scheme for the Upper Quaternary deposits in the Lower Yenisei area and the position of $C^{14}$ samples in these deposits is given in Fig. 1.

The northernmost region around the mouth of the Yenisei is an undulating plain with a discontinuous cover of Zyrianka glacial and glaciofluvial deposits, overlying various pre-Zyrianka deposits consisting mainly of Kazantsevo marine sand. The post-Zyrianka or late Zyrianka valley of the Yenisei is incised into the plain, forming a number of large expanses occupied by the wide steps of the 20-to-25-meter terrace (the so-called "Karginsky terrace") and by an upper flood plain.

Borehole data indicate that the deposits of the "Karginsky terrace" are about 50–60 meters thick and that they can be divided into several stratigraphic units (Fig. 1*b*). The lowest unit consists of channel alluvium more than 30 meters thick, lying entirely below river level. This is overlain by floodplain deposits, 25–30 meters thick, consisting of silt with bands of peat; the upper part extends above river level and forms the basal part of the visible section of the "Karginsky terrace." In the outcrops along the Malaya-Kheta River, there is a layer of coarse sandy loam with pebbles and boulders, overlain by gray varved clay (estuarine deposits) and yellowish-gray sandy loam (Alekseev *et al.*, 1965). The underlying deposits bear traces of cryoturbation and are split by ice wedges (Fig. 1*b*). Finally, sandy loam with peat interbeds occupies depressions on the surface of the terrace at a level of about 16–20 meters above the river; these are also locally split by ice wedges. A spore-pollen diagram of this sequence compiled by O. V. Matveeva shows two distinct climatic optima separated by a cold spell (lower part of Fig. 2). These coincide with the three sedimentary sequences of different age distinguished by S. L. Troitskii, the lowermost part being of Karginsky age, the middle of Sartan age, and the upper of Lower Holocene age (Alekseev *et al.*, 1965).

A section through Lower Holocene deposits formed during the post-

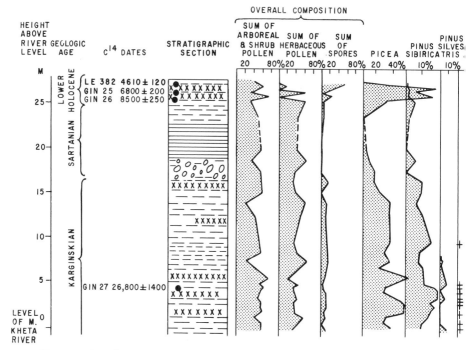

Fig. 2. Pollen diagram by O. V. Matveeva of Karginsky, Sartan, and Lower Holocene deposits in the "Karginsky" Terrace, lower Malaya Kheta River.

glacial climatic optimum is also provided by exposures through the peat
bogs on the surface of the "Karginsky terrace," along the lower reaches
of the Malaya Kheta River. Spore-pollen spectra from these deposits are
characterized by an abundance of spruce (*Picea*) in the lower part of the
section, and by an increasing prominence of tree birches (*Betula*) in the
upper part (Barkova, 1960; Matveeva in Alekseev *et al.*, 1965) (Fig. 2,
upper part). Among the macroscopic plant remains, cones and seeds of fir
(*Abies*) and bark of *Betula* sec. *Albae* are abundant. The plant remains
indicate the development of forests in an area that is now tundra, with a
climate warmer than at present.

Quite different relations are observed among the Upper Quaternary
deposits exposed along the Yenisei between Igarka and Turukhansk (Fig.
1*a*). Alluvium of Terrace II (25–35 meters) occurs there on a high step
formed by Zyrianka glacial and glaciofluvial deposits. In the Igarka region,
a mass of sandy loam identified as Karginsky deposits by V. N. Sachs
(1951b) rests on a strongly eroded surface underlain by varved clay and
till of the Zyrianka Glaciation; these rest in turn on Kazantsevo marine sand.
This sand, together with older Quaternary strata underlying it, drops southward and upstream below the river edge (Arkhipov, 1960; Arkhipov and

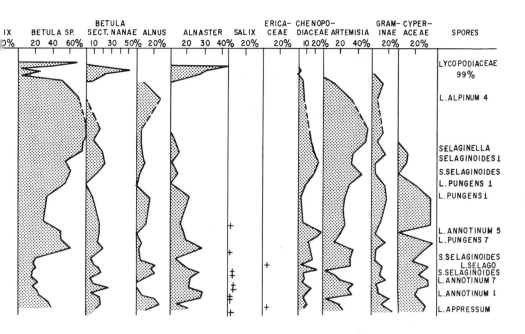

others, 1960; Zubakov, 1960; Lavrushin, 1961), and the base of the visible part of the terrace section farther upstream consists of early Zyrianka alluvium. Traces of cryoturbation are observed in the upper part of this mass. The irregular surface of the early Zyrianka alluvium is covered by a moraine that is replaced farther upstream, beyond the outer boundary of the Zyrianka Glaciation, by a thick sequence of varved clay representing the sediments of a proglacial lake. The Zyrianka glacial and glaciolacustrine deposits are overlain with erosional unconformity by cross-bedded sand containing boulders and pebbles. This grades upward into a sequence of laminated ox-bow lake deposits of late Zyrianka age that contain abundant plant remains. Near the top of the section is a horizon showing the effects of cryoturbation. The sequence is mantled by a thin mass of yellowish-gray sandy loam. Holocene peat bogs occur in depressions on the terrace surface.

These two sequences of Upper Pleistocene and Holocene deposits in the Lower Yenisei area are summarized in Table 2. Radiocarbon age determinations are now available for many of the stratigraphic units represented there.

For the late Kazantsevo marine sands, an infinite date was obtained (>45,000 GIN-101) (Cherdyntsev *et al.*, 1965). For the late Zyrianka glaciofluvial and alluvial deposits there are the following dates: 36,900 ± 400

**TABLE 2.** *Absolute Chronology of the Upper Pleistocene and Holocene of Siberia*

Thousands of Years (left scale: 0, 5, 10, 15, 20, 25, 30, 35, 40)

| Stratigraphic Division | | | Climate | Geological Processes | Radiocarbon dates (Years B.P.) |
|---|---|---|---|---|---|
| HOLOCENE | UPPER HOLOCENE | | Moderately Warm (Recent) to Moderately Cold | Aggradation of lower and upper (late phase) floodplain terraces; ice-wedge formation in the north | 220 ± 140 (GIN-21)<br>765 ± 85 (GIN-22)<br>2,100 ± 160 (MO-228)<br>2,285 ± 160 (MO-231)<br>3,470 ± 170 (MO-230)<br>3,700 ± 100 (GIN-23)<br>4,125 ± 180 (MO-227)<br>4,330 ± 160 (GIN-24)<br>4,610 ± 190 (LE-382)<br>4,770 ± 280 (MO-229) |
| | LOWER HOLOCENE | WARM PERIOD | Warm (Warmer than Recent) | High moorland formation; upper floodplain terrace aggradation (early phase) in the north | 6,800 ± 200 (GIN-25)<br>6,850 ± 225 (MO-245)<br>7,820 ± 210 (MO-233)<br>7,850 ± 250 (MO-234)<br>8,500 ± 250 (GIN-26)<br>8,670 ± 270 (MO-232) |
| | | EARLY HOLOCENE | Moderately Cold | Erosion | 8,960 ± 60 (GIN-96) |
| UPPER PLEISTOCENE | SARTAN GLACIATION | | Cold | Loesslike alluvium aggradation on I (II) terraces (cold phase); cryoturbation | |
| | | | Moderately Warm | Formation of two converging horizons of buried soils | 11,450 ± 250 (T-297)<br>11,700 ± 300 (MO-3) |
| | | | Cold (Glacial) | Glaciation in mountain regions and on high plateaus; transgression in north (?); fluviatile aggradation of I (II) terraces (late phase) in periglacial regions; cryoturbation | 12,940 ± 270 (LE-526)<br>13,300 ± 50 (GIN-91)<br>13,330 ± 100 (GIN-90)<br>14,320 ± 330 (LE-469)<br>15,460 ± 320 (LE-540)<br>20,900 ± 300 (GIN-117)<br>21,700 ± 1,700 (GIN-55) |
| | KARGINSKY INTERGLACIAL (INTERSTADIAL?) | | Moderately Cold (Similar to Recent) | Fluviatile aggradation of "Karginsky" terrace on northern Yenisei; erosion; fluviatile aggradation of I (II) terraces (early phase) farther south | 26,000 ± 1,600 (MO-215)<br>26,800 ± 1,400 (GIN-27)<br>30,700 ± 300 (GIN-126) |
| | ZYRIANKAN GLACIATION | LATE PHASES | Cold (Glacial) | Oscillation of late Zyriankan glacier in north (?); fluviatile aggradation of II (III) terraces (late phase); formation of lacustrine-fluviatile plain in Lower Yenisei | 32,500 ± 700 (GIN-99)<br>>24,000 (MO-4)<br>35,400 ± 300 (GIN-140) |
| | | | Temperate Warm (?) | Glacial retreat; ice-melting of preceding phase of Zyriankan Glaciation; erosion; fluviatile and glaciofluvial aggradation of II (III) terraces (early phase) | 35,800 ± 600 (GIN-76)<br>36,900 ± 400 (GIN-98)<br>37,000 ± 1,900 (GIN-61)<br>40,760 ± 580 (GIN-149) |
| | | | ? | ? | |
| | | MAXIMUM PHASES | Cold (Glacial) | Maximum advance of Zyriankan glacier; formation of proglacial lakes; development of glaciolacustrine sediments | |
| | | EARLY PHASES | Cold | Advance of early Zyriankan glacier; earliest fluvial accumulation of II (III) terraces | |
| | KAZANTSEVO INTERGLACIAL | | Warm (Interglacial) | Boreal transgression in north; formation of III (IV) terraces | >45,000 (GIN-101) |

(GIN-98), obtained on wood from cross-bedded sand with pebbles and large boulders filling a deep scour in the moraine near the back suture of Terrace II in the vicinity of Donezhkino village, and 35,800 ± 600 (GIN-76) for wood fragments from the base of glaciofluvial gravel in the section of Terrace II exposed near Igarka Yar (Fig. 1a). A similar date of 35,400 ± 300 (GIN-140) was obtained as a result of redetermination of wood from deposits in the Igarka region considered to be of Karginsky age by V. N. Sachs (1951b). An earlier determination in the laboratory of the Vernadsky Institute gave an age of more than 24,500 (MO-4) for this wood. On the basis of this new determination, however, we are inclined to regard the enclosing deposits as being older than Karginsky and to refer them to late Zyriankan time. All of these deposits indicate a warm spell of late Zyriankan age, a corresponding recession of the glacier, and an episode of deep channeling. The top of the late Zyriankan alluvium is dated as 32,500 ± 700 years old (GIN-99). For the climatic optimum represented by the true Karginsky deposits, there is a dating of 26,800 ± 1,400 (GIN-27) from the mouth of the Yenisei. The beginning of the Sartan cold interval is represented by an age determination of 21,700 ± 1,700 (GIN-55).

Dating of the Early Holocene climatic optimum is based on wood obtained from peat bogs along the lower Malaya Kheta River, represented by the upper part of the pollen diagram in Fig. 2. The following age determinations were obtained: 8,500 ± 250 (GIN-26) for the lower part; 6,800 ± 200 (GIN-25) for the middle part; and 4,610 ± 190 (LE-382) for the upper part. Thus, the Holocene climatic optimum falls in a period between approximately 8,500 and 4,500 years ago.

For the Upper Holocene deposits, there is a series of age determinations on wood from the visible part of the sediments making up the high flood plain of the Yenisei River—a mass 7 meters thick of clayey silt with bands of peat and stems of driftwood. The specimens were obtained on the Malaya Kheta River, 6 km above the mouth; their ages are 220 ± 140 years at a depth of 0.25 meters (GIN-21); 765 ± 85 years at a depth of 1.3 meters (GIN-22); 3,700 ± 100 years at a depth of 3.5 meters (GIN-23); and 4,330 ± 160 years at a depth of 6.8 meters (GIN-24). These figures show that the lower layers accumulated at the very end of the postglacial warm interval. This is confirmed by the presence of a considerable admixture of tree taxa (e.g., *Picea* up to 28 per cent) in spore-pollen spectra from these layers (Barkova, 1961).

These radiocarbon dates permit us to outline the general age boundaries of the principal stratigraphic subdivisions of the Upper Quaternary deposits of the Lower Yenisei area. Before drawing further conclusions, however, let us analyze some additional datings for geological specimens from other

Siberian regions and for archaeological specimens from the Yenisei region, which make it possible to distinguish more details and even to expand our ideas on the absolute chronology of Late Pleistocene and Holocene events in Siberia.

A date of 37,000 ± 1,900 years (GIN-61) was obtained for the alluvium of Terrace III of the Lower Tunguska River, which corresponds to Terrace II of the Yenisei. The alluvium there rests on a channeled surface of Zyrianka glacial and glaciofluvial deposits; because of the composition of its seed flora, spores, and pollen (Zeitlin, 1964), it is assigned to the late Zyrianka warm interval. This date agrees well with the dates GIN-61 and GIN-98 for analogous deposits on Terrace II of the Yenisei. It may be that the same late Zyrianka warm interval is represented by the date of 40,760 ± 580 years (GIN-149), which was obtained on wood from a peat layer included in lacustrine-fluvial deposits of the Nyimingde River (one of the Lena's right tributaries near the Verkhoyansk Range); these are overlain by glaciofluvial deposits.

Mention should also be made of the Chekurovka mammoth carcass from the lower valley of the Lena River, which has been determined to be 26,000 ± 1,600 years old (MO-215) (Korzhuev and Federova, 1962). This carcass was found in a landslip derived from the thin alluvial cover on Terrace I (17–20 meters), which is composed mostly of bedrock. Palynological studies of adjacent and overlying sediments show that the mammoth died when taiga flora was widely developed; yet the sediments underlying the dead body are characterized by a predominance of spores and pollen, indicative of meadow and bog vegetation. One may conclude from these data that the mammoth died during the first half of the Karginsky warm interval, when the forest was advancing and nutritive foods for mammoths were becoming scarce.

A date of 30,700 ±300 (GIN-126) was obtained for a buried soil with a forest bed at the base of the alluvium of Terrace II of the Irtysh River near the village of Lipovskayia.

An age of 20,900 ± 300 (GIN-117) was obtained for charcoal from the culture layer of Afontova Gora II, an Upper Paleolithic site in the Krasnoyarsk region along the middle Yenisei. The culture layer lies within the lowermost beds of the deposits mantling the surface of Terrace II of the Yenisei. The composition of the mammalian fauna indicates cold periglacial conditions during this time interval (Gromov, 1948), which probably corresponds to the maximum of the Sartan Glaciation.

A series of age determinations for Upper Paleolithic sites near the village of Kokarevo in the middle course of the Yenisei are of great im-

portance for drawing the upper age boundary of the Sartan Glaciation. The
culture layers at these sites belong to the upper part of the loesslike alluvial
deposits of Terrace II, which is 10–14 meters high there (Ravskii and Zeitlin, 1965; Zeitlin, 1965). Cryoturbation structures are recorded in these
sediments that are syngenetic with the accumulation of the alluvium. They
are overlain by cover loam containing two closely spaced buried soil horizons. The uppermost part of the cover loam also bears traces of cryoturbation. The soils, which are younger than the culture layers, are referred to
relatively warm interstadials of late glacial age by Ravskii and Zeitlin
(1965). Charcoal from these sites is dated as follows: Telezhny Ravine
(Kokarevo site II), 13,330 ± 100 (GIN-90); Zabochka (Kokarevo site I),
13,300 ± 50 (GIN-91) and 12,940 ± 270 (LE-526); Kiperny Ravine (Kokarevo site IV), 14,320 ± 330 (LE-469), and 15,460 ± 320 (LE-540) for
Excavation 4 at the same site.

The Taimyr mammoth also provides dating for events during the late
glacial period. The mammoth was originally buried in the fluvial sediments
of Terrace II of the Mamontova River, but was later reworked into Terrace I, according to Popov (1959). The carcass has been determined to be
11,450 ± 250 years old (T-297) (Heintz and Garutt, 1964), and wood from
the sediments that originally contained the mammoth has been determined
to be 11,700 ± 300 years old (MO-3) (Vinogradov *et al.*, 1951). The
plant remains in these deposits indicate that tundra vegetation prevailed,
that the landscape was generally similar to the present landscape, but that
the climate was somewhat warmer, according to Tikhomirov (1950, 1951)
and Zaklinskaia (1959). According to these researchers, the younger floodplain facies of Terrace I accumulated during the Holocene climatic optimum,
when the climate was considerably warmer than at present. Thus, the mammoth died during a climatic warming that preceded the Holocene warm
period. The rapid burial of the carcass during an increasingly colder spell,
accompanied by the growth of ground ice (Popov, 1959), may possibly indicate that the animal died at the very end of this warm phase or at the beginning of the final cold spell of the late glacial interval. This late Sartanian
warm spell apparently corresponds to the above-mentioned two continuous
horizons of buried soil in the sequence near the village of Kokorevo, and
the final cold spell to the upper horizon of cryogenic disturbances (Zeitlin,
1965).

For the beginning of the Holocene, a date of 8,960 ± 60 (GIN-96) was
obtained for fossil bones from a culture layer of "Ustye Belaya," an early
Mesolithic site at the mouth of the Belaya River, a tributary of the Angara
River (Cherdyntsev *et al.*, 1965). The culture layer belongs to the upper

part of the floodplain deposits of Terrace I. In the immediately underlying sediments, cryoturbation structures are recorded (Logachev *et al.*, 1964; Ravskii and Zeitlin, 1965).

The absolute chronology of Holocene climatic fluctuations in eastern Siberia can be reconstructed from an extensive series of radiocarbon determinations and spore-pollen analyses of Holocene deposits in the Indigirka River basin (Vinogradov *et al.*, 1962). Spore-pollen spectra studied by Lavrushin and Giterman indicate that the lacustrine and lacustrine-meadow deposits determined to be 8,670 ± 270 years old in the basal layers to 4,770 ± 280 years old at the top (MO-232, MO-234, MO-233, MO-245, MO-229 in Table 2) accumulated during the Holocene climatic optimum (Lavrushin *et al.*, 1963). Trees, now absent from these latitudes, were present during that interval. A rather cold spell began approximately 4,500 years ago, during which the alluvium accumulated that now makes up the visible part of the high floodplain of the Indigirka. Age determinations on these deposits range from 4,125 ± 180 to 2,100 ± 160 years old (MO-227, MO-230, MO-231, MO-228 in Table 2).

## Late Quaternary Chronology

The data furnished above suggest the following scheme for the absolute chronology of the principal geologic events and climatic fluctuations during Late Quaternary time in Siberia (Table 3).

The maximum southward advance of the Zyrianka glaciers was undoubtedly in progress more than 40,000 years ago. The interstadial warming during late Zyrianka time began approximately 38,000 years ago and ended about 35,000 years ago. During this interval, the Zyrianka glacier receded northward, and a deep channeling took place, followed by the deposition of the lower alluvial horizons of Terrace II in the Lower Yenisei and Terrace III in more southern and more mountainous regions. A new cold spell evidently followed this warming, as is shown by the spore-pollen spectra for the upper part of the alluvium of Terrace II and by the traces of cryoturbation. For these deposits, there is a date of 32,500 ± 700 years (GIN-99).

Next came the true Karginsky warm interval, characterized by a climate that at its optimum, about 26,000 or 27,000 years ago, seems to have been somewhat warmer than at present, judging from the presence of forests in northern latitudes where tundra now prevails. The Karginsky warm interval lasted about 6,000 to 8,000 years, extending from approximately 30,000 to 24,000–22,000 years ago. Consequently, the duration of the Sartan cold interval is between 14,000 and 12,000 years, having extended from 24,000–

**TABLE 3. Correlation of Climatic and Glacial Events in the Northern Hemisphere**

| Thousands of years | NORTH AMERICA — Dreimanis, 1960a, b Great Lakes Area | WESTERN AND CENTRAL EUROPE — Woldstedt, 1960 / Gross, 1964 | WESTERN AND CENTRAL EUROPE — Zagwijn, 1961, 1963 | WESTERN AND CENTRAL EUROPE — Fink, 1964 / Ložek, 1964 | EUROPEAN PART OF USSR — Velichko, 1965 | EUROPEAN PART OF USSR — Moskvitin, 1954* / Neustadt, 1962 | SIBERIA |
|---|---|---|---|---|---|---|---|
| — | POSTGLACIAL | POSTGLACIAL | POSTGLACIAL | POSTGLACIAL | HOLOCENE | HOLOCENE | WARM PERIOD / HOLOCENE |
| 10 | LAST STAGES OF RETREAT | ALLERØD INT. / BØLLING INT. | | | | | |
| | TWO CREEKS INT. | LATE WÜRM | LATE GLACIAL | LATE GLACIAL | LATE GLACIAL | LATE GLACIAL | WARM PERIOD |
| | MAIN OR CLASSICAL WISCONSIN — GLACIAL STAGES | MAIN WÜRM / MAXIMUM ADVANCE STAGES (W₃) | PLENIGLACIAL B | MAIN STAGES OF WÜRM | VALDAI GLACIATION / SECOND MAIN PHASE | OSTASHKOV GLACIATION | SARTAN GLACIATION |
| 20 | | | | | | | |
| | MID-WISCONSIN (SIDNEY) INTERSTADIAL — PLUM POINT INTERSTADIAL | MIDDLE WÜRM / PAUDORF (and Arcy?) INTERSTADIALS | PAUDORF INTERSTADIAL | STILLFRIED B (PK-1) | BRIANSK INTERSTADIAL | MOLOGA-SHEKSNA INTERGLACIAL (INTERSTADIAL) | KARGINSKY INTERGLACIAL (INTERSTADIAL) |
| 30 | RETREAT AND OSCILLATION OF THE GLACIER | RETREAT AND OSCILLATION OF THE GLACIER | INTERSTADIAL? | | | | |
| | MID-WISCONSIN ADVANCE STAGE | INTERSTADIAL | | | | | LATE PHASES — INTERSTADIAL |
| 40 | | PLENIGLACIAL A / INTERSTADIAL? | EARLY WÜRM | VALDAI GLACIATION / FIRST MAIN PHASE | | | |
| 50 | PORT TALBOT INTERSTADIAL | MAXIMUM STAGE (W₁) | | | KALININ GLACIATION | | ZYRIANKAN GLACIATION / MAXIMUM PHASE |
| | EARLY WISCONSIN | EARLY WÜRM / BROERUP INTERSTADIAL | BROERUP INTERSTADIAL | BROERUP INTERSTADIAL (PK-2) | | TRANSITION PHASE | |
| 60 | ST. PIERRE INTERSTADIAL | STADIAL (W₁) | | STILLFRIED A | MOSCOW-VALDAI INTERGLACIAL | UPPER VOLGA INTERSTADIAL | EARLY PHASES (?) |
| | | AMERSFOORT INTERSTADIAL / STADIAL (W₁) | AMERSFOORT INTERSTADIAL | AMERSFOORT INTERSTADIAL (PK-3) | | | |
| 70 | SANGAMON INTERGLACIAL | RW / EEM INTERGLACIAL | RW / EEM INTERGLACIAL | | | MIKULINO INTERGLACIAL | KAZANTSEVO INTERGLACIAL |

\* The stratigraphic scheme of Moskvitin for the Pleistocene is interpreted chronologically by the author.

22,000 years ago until about 10,000 years ago. One or possibly two warm spells are recorded within the interval between 13,000 and 11,000 years ago; the dating of the Taimyr mammoth (11,700 ± 300 and 11,450 ± 250 B.P.) probably refers to the latest of these.

A climatic optimum extending from approximately 8,500 to 4,500–4,300 years ago is recorded within the Holocene. The climate of this interval was considerably warmer than the present climate; coniferous taiga was developed far to the north of its present limits. The lower beds making up the visible part of the high floodplains along Siberian rivers accumulated at the end of the Holocene warm interval, and the rest of the section was formed under considerably more severe climatic conditions during the second half of the Holocene. To this recent cold interval, which began at the end of the fifth millennium B.P., belongs the development of permafrost in the north and the formation of ice wedges in the floodplain alluvium and in Lower Holocene peat bogs.

## Correlations with Other Regions

These data on the absolute age of the main time boundaries in the Late Quaternary history of Siberia permit certain comparisons to be made with distant parts of the Northern Hemisphere, notably Western Europe and North America. The absolute geochronological time scale for the last glaciation and the postglacial period is worked out comparatively fully in these areas, and Table 3 shows that the principal climatic changes in the upper part of the geological column are synchronous.

The comparatively great age of the Karginsky warm interval should be stressed first of all; this excludes the possibility of referring it to the Alleroed of Western Europe, as some researchers had suggested (Zubakov, 1963). Instead, the Karginsky warm period appears to be simultaneous with the Paudorf Interstadial of Europe, the age boundaries of which have been established by a considerable number of radiocarbon dates as 30,000–29,000 B.P. and 25,000–24,000 B.P. (Gross, 1958, 1959, 1960a, 1960b, 1964; Movius, 1960, 1963; Woldstedt, 1960). The Karginsky interval is also correlative with the Plum Point Interstadial of the Great Lakes area of North America, which occupies the interval from 30,000 to 25,000–24,000 years ago (Dreimanis, 1960a, 1960b; Flint, 1957, 1963; Flint and Brandtner, 1961; Frye and Willman, 1963).

The Zyrianka Glaciation is correlative with Würm I and II of Europe and with the Early Wisconsin Glaciation of North America. This confirms the opinion of some researchers, who suggest the existence of several sub-stades within the Zyrianka Glaciation. The late Zyrianka warm interval, 38,000 to 35,000 years ago, appears to correspond to a warm interval established by spore-pollen analyses in France (Leroi-Gourhan, 1960) and in Holland (Zagwijn, 1963).

The Sartan Glaciation, which extended from approximately 22,000–21,000 B.P. until 11,000–10,000 B.P. is evidently correlative with Würm III and the Late Glacial period of Western Europe—the Young Würm of Woldstedt (1960). The late Sartan warm spell about 11,500 years ago coincides almost exactly with the Alleroed of Europe and the Two Creeks Interval in North America. The coincidence between the postglacial climatic optimum from approximately 9,000 to 4,500 years ago in Siberia and Western Europe (Godwin *et al.*, 1957; Godwin and Willis, 1959; Krog, 1960; Godwin, 1961; West, 1963) should also be pointed out; it includes the second half of the Boreal Period and the whole of the Atlantic Period of the Blytt-Sernander scheme. This warm interval is also well defined and has the same age boundaries in the European part of the USSR (Neushtadt *et al.*, 1962; Vinogradov *et al.*, 1962). The corresponding warm spell in Alaska began somewhat earlier (9,000–10,000 years ago) and lasted only about 2,000–3,000 years, according to McCulloch and Hopkins (1966).

There is still much that is obscure in these correlations, because of the insufficient data on the absolute chronology of the Pleistocene in Siberia. The chronology of the Zyrianka remains incompletely known, and the time limits of the Sartan Glaciation are also not quite exact. A more important problem also arises from these correlations. The maximum Late Pleistocene expansion of north European and North American glaciers is known to have occurred during the last glacial episode, that is, during Würm III in Europe and during the classical Wisconsin in North America. The Sartan Glaciation, which is correlated with these events, covered, on the contrary, an extremely limited territory and cannot be compared in scale with the much more extensive Zyrianka Glaciation. I believe the contradiction is more apparent than real. There is no doubt that scale differences for the last glaciation between Northern Siberia and other regions were caused by the special climatic conditions of Northern Siberia—by the greater aridity in that region, which inhibited the growth of the glacial cover. All this proves once again that in spite of the obvious synchroneity of the climatic fluctuations in different parts of the Northern Hemisphere, geologic processes can proceed quite differently under the actual physical-geographic conditions of individual regions.

# REFERENCES

Alekseev, V. A., N. V. Kind, O. V. Matveeva, and S. L. Troitskii. 1965. Novye dannye no absoliutnoi khronologii Verkhnego Pleistotsena i Holotsena Sibiri (New data on the absolute chronology of the Upper Pleistocene and Holocene of Siberia) : Akad. Nauk SSSR, Dokl. 160, p. 1147–1150. In Russian; English translation p. 69–72 in running transl. pub. by Am. Geol. Inst.

Arkhipov, S. A. 1960. Stratigrafiia chetvertichnykh otlozhenii, voprosy neotektoniki i paleogeografii basseina srednego techeniia Eniseia (Stratigraphy of the Quaternary deposits and problems of the neotectonics and paleogeography of the basin of the middle course of the Yenisei) : Akad. Nauk SSSR, Tr. Geol. Inst., Bull. 30, 171 p.

Arkhipov, S. A., E. V. Koreneva, and Yu. A. Lavrushin. 1960. Stratigrafiia chetvertichnykh otlozhenii prieniseiskogo rayona mezhdy ust'iami rek Bakhta i Turukhan (Stratigraphy of the Quaternary deposits in the vicinity of the Yenisei region between the mouths of the Bakhta and Turukhan Rivers) : Akad. Nauk SSSR, Tr. Geol. Inst., Bull. 26, p. 248–280.

Barkova, M. V. 1960. Palinologicheskie spektry iz torfianika korginskoi terrasy v rayona poselka Malaia Kheta (Palynological spectra from peaty low terraces in the settled region of Malaya Kheta) : Tr. In-ta. Geol. Arktiki, Vyp. 20, p. 65–71.

———— 1961. Palinologicheskaia kharakteristika chetvertichnykh otlozhenii rayona Ust'Porta (Palynological characteristics of the Quaternary deposits of the Ust'-Port region) : Tr. In-ta. Goel. Arktiki, 124, Bull. 2, p. 177–187.

Cherdyntsev, V. V., V. A. Alekseev, N. V. Kind, V. S. Forova, and L. D. Sulerzhitskii. 1964. Radiouglerodnye daty laboratorii Geologicheskogo Instituta (GIN) AN SSSR (Radiocarbon dates, Laboratory of the Geological Institute, Academy of Sciences of the USSR) : Geokhimiia, no. 4, p. 315–324. In Russian with English abstract.

Cherdyntsev, V. V., V. A. Alekseev, N. V. Kind, V. S. Forova, F. C. Zavelskii, L. D. Sulerzhitskii, and I. V. Churikova. 1965. Radiouglerodnye daty laboratorii Geologicheskogo Instituta (GIN) AN SSSR (Radiocarbon dates, Laboratory of the Geological Institute, Academy of Sciences of the USSR) : Geokhimiia, no. 12, p. 1410–1422.

Dreimanis, A. 1960a. The early Wisconsin in the eastern Great Lakes region, North America: Deutsche Akad. Wissenschaften, Abh. 3, p. 196–205.

———— 1960b. Pre-classical Wisconsin in the eastern portion of the Great Lakes region, North America: Rept., 21st Int. Geol. Congress (Copenhagen), 1960, pt. 4, p. 108–119.

Fink, J. 1961. Die Gliederung des Jungpleistozäns in Österreich: Mitteil. Geol. Gesellschaft Wien, v. 54, p. 1–25.

———— 1964. Die Gliederung der Würm-liszeit in Österreich: Rept., Internat. Assoc. Quaternary Res. (INQUA), 6th Cong. (Warsaw), v. 4 (Symposium on Loess), p. 451–462.

Flint, R. F. 1957. Glacial and Pleistocene geology: Wiley, New York, 553 p.

———— 1963. Status of the Pleistocene Wisconsin Stage in central North America: Science, v. 139, p. 402–404.

Flint, R. F., and F. Brandtner. 1961. Climatic changes since the last interglacial: Am. Jour. Sci., v. 259, p. 321–328.

Frye, J. C., and H. B. Willman. 1963. Development of Wisconsin classification in Illinois related to radiocarbon chronology: Geol. Soc. America Bull., v. 74, p. 501–506.

Godwin, H. 1961. Radiocarbon dating and Quaternary history in Britain: Roy. Soc. London Proc., v. 153, p. 287–320.

Godwin, H., D. Walker, and E. H. Willis. 1957. Radiocarbon dating and post-glacial vegetational history: Scaleby Moss: Roy. Soc. London Proc., v. 147, p. 352–366.

Godwin, H., and E. H. Willis. 1959. Radiocarbon dating of the late-glacial period in Britain: Roy. Soc. London Proc., v. 150, p. 199–215.

Gromov, V. I. 1948. Paleontologicheskoe i arkheologicheskoe obosnovanie strati-grafii kontinental'nykh otlozhenii chetvertichnogo perioda na territorii SSSR (Mlekopitaiuschchie, Paleolit) (Paleontological and archaeological basis of the stratigraphy of continental deposits of the Quaternary period in the territory of the USSR [mammals, paleoths]): Akad. Nauk SSSR, Tr. Geol. Inst., Vyp. 64, Geol. Ser. 17, 521 p.

Gross, H. 1958. Die bisherigen Ergebnisse von C14-Messungen und paläolithis-chen Untersuchungen für die Gliedurung und Chronologie des Jungpleisto-zäns in Mitteleuropa und den Nachbargebieten: Eiszeitalter und Gegenwart, v. 9, p. 155–187.

——— 1959. Zur Frage der Gliederung und Chronologie der letzten Eiszeit (Würm oder Weichsel) in Mitteleuropa: Forschungen und Fortschrifte, v. 33, p. 332–336.

——— 1960a. Die lösung des Problems der Gliederung und Chronologie der letzten Eiszeit in Mitteleuropa: Forschungen und Fortschrifte, Jg. 34, p. 297–301.

——— 1960b. Die Beteutung des Göttweiger Interstadials im Ablauf der Würm-Eiszeit: Eiszeitalter und Gegenwart, v. 11, p. 99–106.

——— 1964. Das Mittelwürm in Mitteleuropa und angrenzenden Gebieten: Eiszeitalter und Gegenwart, v. 15, p. 187–198.

Heintz, A. E., and V. E. Garutt. 1964. Opredelenie absolyutnogo vozrasta isko-paemykh ostatkov mamonta i sherstistogo nosoroga iz vechnoi merzloty Sibiri pri pomoshchi radioactivnogo ugleroda (C14) (Determination of absolute age of remains of mammoth and woolly rhinoceros from permafrost in Siberia by use of radioactive carbon [C14]): Akad. Nauk SSSR, Dokl., v. 154, p. 1367–1370. In Russian; English translation p. 168–170 in running transl. pub. by Am. Geol. Inst.

Kind, N. V. 1965. Absoliutnaia khronologiia osnovnykh etapov istorii poslednego oledeneniia i poslelednikov'ia (Absolute chronology of the fundamental stages of the history of the final glaciation and the postglacial): p. 157–175 *in* Chetvertichnyi period i ego istoriia (Quaternary period and its history): Izd-vo Nauka, Moscow (published for the VII INQUA Congress).

Korzhuiev, S. S., and R. V. Fedorova. 1962. Chekurovskii mamont i usloviia ego obitaniia (The Chekurovka Mammoth and its food supply): Dokl. Akad. Nauk SSSR, v. 143, p. 181–183. In Russian; English translation p. 16–18 in running transl. pub. by Am. Geol. Inst.

Krog, H. 1960. Post-glacial submergence of the Great Belt dated by pollen-analy-sis and radiocarbon: Rept., 21st Internat. Geol. Congress (Copenhagen), 1960, pt. 4, p. 127–133.

Lavrushin, Iu. A. 1961. Tipy chetvertichnogo alliuviia Nizhnego Eniseia (Types of Quaternary alluvium of Lower Yenisei): Akad. Nauk SSSR, Tr. Geol. Inst., Bull. 47, 94 p.

Lavrushin, Iu. A., A. L. Devirts, R. E. Giterman, and N. G. Markova. 1963. Pervye dannye po absoliutnoi khronologii osnovnikh sobytii golotsena Severo-Vostoka SSSR (First data on absolute Holocene dates, Northeastern USSR): Akad. Nauk SSSR, Kom. Izucheniiu Chetvertichnogo Perioda Bull. 28, p. 112–126.

Leroi-Gourhan, A. 1960. Flores et climat du Paléolithique recent: Congr. Prehist. de France à Monaco, 1959, Le Mans, Comptes Rendu.

Logachev, N. A., T. K. Lomonosova, and V. M. Klimanova. 1964. Kainozoiskie otlozheniia Irkutskogo amfiteatra (Cenozoic deposits of the Irkutsk amphitheatre): Izd-vo "Nauka," Moscow, 194 p.

Ložek, V. 1964. Die Umwelt der urgeschichtlichen Gesellschaft nach neuen Ergebnissen der Quartärgeologie in der Tschechoslowakei: Jahrschr. mitteld. Vorgeschichte, no. 48, p. 7–24.

McCulloch, D. S., and D. M. Hopkins. 1966. Evidence for a warm interval 10,000 to 8,300 years ago in Northwestern Alaska: Geol. Soc. America Bull. v. 77, p. 1089–1108.

Moskvitin, A. I. 1954. Stratigraficheskaia skhema chetvertichnogo perioda v. SSSR: Akad. Nauk SSSR, Izvestia, Geol. Ser., no. 3, p. 20–50.

Movius, H. L. 1960. Radiocarbon dates and Upper Paleolithic archaeology in central and western Europe: Current Anthropology, v. 1, p. 355–391.

——— 1963. L'âge Perigordien de l'Aurignacien et du Proto-Magdalenien en France sur le base des datations au carbon-14: Bull. Soc. Merid. Spéleol. et Préhist. no. 6–9, p. 131–142.

Neushtadt, M. I., A. L. Devirts, N. G. Markova, E. N. Dobkina, and N. A. Khotinskii. 1962. Datirovka Geolotsenovykh otlozhenii radiouglerodnym metodom i dannymi pyl'tsevogo analiza (Radiocarbon and pollen dating of Holocene strata): Dokl. Akad. Nauk SSSR, v. 144, p. 1129–1131. In Russian; English translation p. 117–121 in running transl. by Am. Geol. Inst.

Popov, A. J. 1959. Taimyrskii mamont i problema sokhraneniia ostatkov mamontovoi fauny v chetvertichnykh otlozheniiakh Sibiri (The Taimyr mammoth and the problem of conservation of the mammoth fauna in the Quaternary deposits of Siberia), p. 259–275 in K. K. Markov and A. J. Popov, eds., Lednikovyi Period na territorii Evropeiskoi chasti SSSR i Sibiri (Ice age in the European section of the USSR and in Siberia): Moskva Gosudarstv. Univ., Geogr. Fak., 560 p.

Ravskii, E. N., L. P. Aleksandrova, E. A. Vangengeim, V. G. Gerbova, and L. V. Golubeva. 1964. Antropogenovye otlozheniia iuga Vostochnoi Sibiri (Anthropogene deposits in the south of Eastern Siberia): Akad. Nauk SSSR, Tr. Geol. Inst., Vyp. 105, 277 p.

Ravskii, E. N., and S. M. Zeitlin. 1965. Geologicheskaia periodizatsiia pamiatnikov paleolita Sibiri (The geological periodization of the Siberian Paleolithic), p. 387–393, in Osnovnye problemy izucheniia Chetvertichnogo Perioda (Principal problems of the study of the Quaternary Period): Izd. "Nauka," Moscow, 496 p. Published for the VII INQUA Congress.

Sachs, V. N. 1951a. Chetvertichnye otlozheniia severnoi chasti Zapadno-Sibirskoi nizmennosti i Taimyrskoi depressii (Quaternary deposits of the northern part of the West Siberian lowland and the Taimyr Depression): Tr. In-ta Geol. Arktiki, v. 14, 114 p.

——— 1951b. Geologicheskii ocherk raiona goroda Igarki (Geological sketch of Igarka village): Tr. In-ta. Geol. Arktiki, v. 19, p. 3–13.

Sachs, V. N., and K. V. Antonov. 1945. Chetvertichnye otlozheniia i geomorfologiia raiona Ust'-Eniseiskogo porta (Quaternary deposits and geomorphology of the region around the harbor of Ust'-Enisei) : Tr. Gorno-geol. Upravleniia, p. 65–117.

Tikhomirov, B. A. 1950. K kharakteristike rastitel'nogo pokrova epokhi mamonta na Taimyre (On the characteristics of the vegetation of the beds of the mammoth epoch in Taimyr) : Botanich. Zhurn., v. 35, p. 428–497.

——— 1951. O rastitel'nosti epokhi mamonta na severe Sibiri (On the vegetation of the mammoth epoch in northern Siberia) : Priroda, no. 1, p. 33–40.

Velichko, A., and T. Morozova. 1965. Glavnye osobennosti razvitiia perigliatsial'-noi oblasti v epokhi valdaiskogo oledeneniia (Principal characteristics of the distribution of periglacial areas during the epoch of the Valdai Glaciation), p. 59–65 *in* Section "Less, pogrebennye pochvy i kriogennye fenomeny" (Loess, buried soils, and cryogenic phenomena) in Poslednii evropeyskii lednikovyi pokrov (End of the European Ice Sheet): Izd-vo "Nauka," Moscow. Published for the VII INQUA Congress.

Velichko, A., Iu. Moiskii, and G. Richter. [1965.] Perigliatsial'nye oblasti Vosto-chnoi i Srednei Evropy (Periglacial districts of Eastern and Central Europe), p. 162–166 *in* "Less, pogrebennye pochvy i kriogennye fenomeny" (Loess, buried soils, and cryogenic phenomena) in Poslednii evropeyskii lednikovyi pokrov (End of the European Ice Sheet): Izd-vo "Nauka," Moscow. Published for the VII INQUA Congress.

Vinogradov, A. P., A. L. Devirts, E. I. Dobkina, and N. G. Markova. 1962. Opredelenie absoliutnogo vosrasta po $C^{14}$: So. 3 (Determination of absolute age by the $C^{14}$ method: Pt. 3) : Geokhimiia, 1962, no. 5, p. 387–402. In Russian; English translation p. 439–457 in running trans. pub. by Geochem. Soc.

Vinogradov, A. P., A. L. Devirts, E. I. Dobkina, N. G. Markova, and L. G. Martish-chenko. 1951. Opredelenie absoliutnogo vozrasta po $C^{14}$ pri pomoshchi pro-portsional'nogo schetchika (Determinations of absolute age by $C^{14}$ with the assistance of the proportional counter) : Izd-vo Akad. Nauk SSSR, Moscow, 1961, 59 p.

West, R. G. 1963. Problems of the British Quaternary: Geologists' Assn., Proc., v. 74, p. 147–186.

Woldstedt, P. 1960. Die letzte Eiszeit in Nordamerica und Europa: Eiszeitalter und Gegenwart, v. 11, p. 148–165.

Zagwijn, W. H. 1961. Vegetation, climate, and radiocarbon datings in the late Pleistocene of the Netherlands; Pt. 1, Eemian and early Weichselian: Geol. Stichting, Med., n.s., no. 14, p. 15–16.

——— 1963. Pleistocene stratigraphy in the Netherlands based on changes in vegetation and climate: Verh. Kon. Nederl. Geol. Mijnb. Genootschap., Geol. Ser., v. 21/2, p. 173–196.

Zaklinskaia, E. D. 1959. Sporovo-pyl'tsevye spektry chetvertichnykh otlozhenii raiona nakhodki Taimyrskogo mamonta (Spore-pollen spectra of the Quaternary deposits in the vicinity of the Taimyr mammoth locality), p. 276–300 *in* K. K. Markov and A. J. Popov, eds., Lednikovyi period na territorii Evropei-skoi chasti SSSR i Sibiri (Ice age in the European section of the USSR and in Siberia): Moskva, Gosudarstv. Univ., Geogr. Fak., 560 p.

Zeitlin, S. M. 1964. Sopostavlenie chetvertichnykh otlozhenii lednikovoi i vneled-nikovoi zon Tsentral'noi Sibiri (bassein Nizhnei Tunguski) (A comparison of the Quaternary deposits of the glaciated and nonglaciated zones of Central Siberia [Lower Tunguska basin]) : Akad. Nauk SSSR, Tr. Geol. Inst., Vyp. 100, 181 p.

——— 1965. O raschlenenii poslednego lednikov'ia Sibiri (On the subdivision of the last glaciation in Siberia), p. 175–183 *in* Chetvertichnyi Period i ego istoriia (Quaternary period and its history) : Izd-vo "Nauka." Published for VII INQUA Congress.

Zubakov, V. A. 1960. Stratigrafiia i paleogeografiia pleistotsena Prieniseiskoi Sibiri (Pleistocene stratigraphy and paleogeography of the Pri-Yenisei area of Siberia) : Vsesoiuzn. Nauchn.-Issled. Geol. In-t., Tr., n.s., v. 66, p. 135–150.

———— 1963. Lednikovyi vek Sibiri (Glacial age in Siberia) : Rept., 6th Internat. Cong. on Quaternary (Warsaw), 1961, v. 3 (Geomorphology), p. 411–423.

# 9. Neogene and Early Quaternary Vegetation of Northwestern North America and Northeastern Asia

JACK A. WOLFE
U.S. Geological Survey, Menlo Park, California

ESTELLA B. LEOPOLD
U.S. Geological Survey, Denver

In contrast to the Neogene floral record at middle latitudes, the Neogene floras at high latitudes are poorly known. The gradual cooling of the non-marine environment during the Neogene and the effects of this cooling upon the subtropical and warm-temperate vegetation that then existed have been discussed by several authors (Tanai, 1961; MacGinitie, 1962; Axelrod, 1964; Wolfe, 1964). However, most authors have considered all the warm-temperate Tertiary plants known from high latitudes to belong to the Paleogene. Thus, Krystofovich (1935) and Chaney (1936) considered the Turgai or arcto-Tertiary flora at high northern latitudes to be of Paleocene or Eocene age; this has consequently confused the history of Tertiary vegetation, because floras of this type are no older than Oligocene. Recent work in Northeastern Siberia (various workers cited in Vangengeim, 1961) and in Alaska (Wolfe *et al.*, 1966; Wolfe, 1966) has somewhat clarified the floral history in these high-latitude regions.

At present, the Neogene floral sequence is more completely known in Alaska than in Siberia, and our discussion will draw data primarily from the Alaskan Neogene. What is known of Tertiary vegetation in Siberia, however, indicates that the floristic histories of Northeastern Siberia and Alaska are essentially one and the same.

We have established three provincial stages for the Neogene of Alaska (Wolfe *et al.*, 1966). The relationship of these stages to the traditional Tertiary series is not precisely known. Our Seldovian Stage probably represents the early and middle Miocene though some early Seldovian beds may be of Oligocene age; the upper part of this stage, which is of primary concern in this discussion, is certainly of Miocene age. The Homerian Stage is mostly

Publication authorized by the Director, U.S. Geological Survey.

or entirely of late Miocene age, and the Clamgulchian Stage represents at least part of the Pliocene. The upper part of the Clamgulchian may be equivalent to beds designated as Eopleistocene by Vangengeim (1961) and Alekseev *et al.* (1962).

The most complete floral sequences in the Alaskan Neogene are in the Cook Inlet basin (lat. 60–62° N.; Fig. 1) and in the central part of the Alaska Range (lat. 64° N.). These plant sequences have been studied from both microfossils and megafossils. We have also studied the data from microfossils in the late Cenozoic on Seward Peninsula and at Kivalina, and from megafossils in the Wrangell Mountains.

## The Floral Records

*Seldovian Stage.* Within the Seldovian Stage, we recognize two different assemblages of plants. The younger assemblage, which we will be concerned with here, is composed largely of broad-leaved plants now exotic to Alaska and Northeastern Siberia. Over 40 genera of woody dicotyledons have been recognized, including *Carya, Juglans, Pterocarya, Comptonia, Alnus, Carpinus, Corylus, Fagus, Quercus, Ulmus, Zelkova, Cocculus, Cercidiphyllum, Liquidambar, Platanus, Acer, Ilex, Tilia,* and *Nyssa.* Except for members of Taxodiaceae, megafossil remains of conifers are rare. One microfossil assemblage from the Cook Inlet basin, however, contains a significant amount of *Picea* (28 per cent), and we therefore feel that spruce was represented locally in the forests of the coastal lowlands during Seldovian time. It is significant that the sample with the high representation of *Picea* also contains *Abies* and *Tsuga.* Pinaceae, except for *Pinus,* are consistently poorly represented in the Seldovian samples from the Alaska Range area; there were probably no high mountains there during Seldovian time (C. Wahrhaftig, unpublished data).

Late Seldovian floras in the Wrangell Mountains grew in a region of considerable topographic diversity and at a higher altitude (possibly 800 to 1,000 meters) than either the Cook Inlet or Alaska Range Seldovian floras. The Wrangell floras contain some of the now-exotic dicotyledons—for example, *Pterocarya, Fagus,* and *Ulmus*—but conifers such as *Chamaecyparis, Thuja, Abies, Picea, Pinus,* and *Tsuga* are also common as megafossils.

During Seldovian time in Alaska, therefore, it appears that coniferous forest was largely restricted to the uplands, although conifers were common locally near the coast. Most of the coastal lowland and all the interior lowland areas of southern Alaska were occupied by a diversified broad-leaved

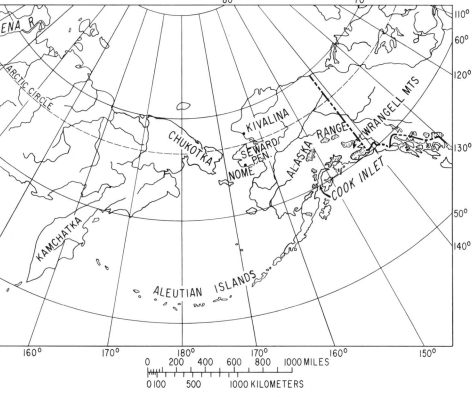

Fig. 1. Map of Alaska and Northeastern Siberia showing important Neogene and early Quaternary fossil plant localities.

deciduous forest, containing only such conifers as Taxodiaceae and *Pinus* that are today commonly associated with such an assemblage.

The similarity of the Alaskan Seldovian floras to the early and middle Miocene floras of Japan and Oregon is remarkable. On the generic level, of course, the resemblance is especially obvious, but the resemblance is almost as great on the specific level. Many extinct species are found in both Alaska and Oregon; many are found in Alaska and Japan; and some are found in all three areas. Among the widely distributed dicotyledon species are *Populus kenaiana, P. reniformis, Comptonia naumani, Carya bendirei, Pterocarya nigella, Alnus fairi, Fagus antipofi, Quercus bretzi, Q. axelrodi, Ulmus newberryi, Zelkova oregoniana, Cercidiphyllum crenatum, Cocculus auriculata, Liquidambar mioformosana, Platanus bendirei, Cladrastis japonica, Acer ezoanum, A. fatisiaefolia, A. chaneyi, A. glabroides, A. macrophyllum,*

*Aesuculus majus,* and *Viburnum latanifolium.* Even more significant is the fact that some of these species have only a narrowly restricted stratigraphic range. It is apparent that a broad-leaved deciduous forest was continuous around the northern Pacific during early and middle Miocene time.

As in Alaska, the coniferous forests in Japan were restricted to the uplands, if indeed purely coniferous forests existed there at all. Tanai and Suzuki (1963) have presented evidence to indicate that conifers such as *Abies, Picea,* and *Tsuga* were generally restricted to slopes above altitudes of 500 meters, even in Hokkaido. Similarly, in Oregon, members of these genera are poorly represented in floras that grew at altitudes as high as 700 meters. It was not until late in Seldovian time that a mixed deciduous and broad-leaved conifer forest was present at altitudes down to 700 meters in Oregon. It is probable, therefore, that the areas that were occupied by coniferous forests were highly disjunct during the Seldovian in northwestern North America and northeastern Asia.

*Homerian Stage.* In contrast to the Seldovian floras, the Homerian floras in Alaska contain only a few deciduous broad-leaved genera now exotic to Alaska. Except for the fluviatile *Salix* and *Alnus,* none of the broad-leaved tree genera are common either as megafossils or microfossils. Presumed shrubs belonging to *Rhododendron, Vaccinium,* and *Spiraea* are present at most localities and typically are abundant. The pollen floras contain *Picea* and *Tsuga* at almost all localities. In swampy lowland basins, such as the Cook Inlet area, the forest was probably composed of the moisture-tolerant mesophytic plants, and the conifers and rare trees of *Carya* and *Ulmus* were confined to the drier interfluves.

In more southern regions, such as Oregon, the Queen Charlotte Islands off British Columbia (Martin and Rouse, 1966), and Japan, the lowland vegetation was still similar to the Seldovian flora; that is, the forests were of warm-temperate character, with Juglandaceae, Fagaceae, Ulmaceae, and Aceraceae as the dominant tree forms. The change from a Seldovian to a Homerian type of flora in Alaska resulted in a disjunction between the broad-leaved deciduous forests of northwestern North America and northeastern Asia.

By the late Miocene, the coniferous forests apparently began to occupy large areas in the uplands. Floras found at intermediate elevations (above 300 meters) throughout western United States contain moderate to large amounts of *Abies, Picea,* and *Tsuga,* both as megafossils and as microfossils. It was probably in the late Miocene that a continuous coniferous forest extended, for the first time, from the uplands of Oregon northward through British Columbia and into Alaska.

*Clamgulchian Stage.* The Cook Inlet Clamgulchian floras all contain *Picea* and *Tsuga,* but *Abies* is lacking. The only now-exotic genera found are *Glyptostrobus* and *Rhus,* both known as megafossils. In the microfossil record, we have encountered a few grains of *Pterocarya* and *Ulmus,* but we think these may have been redeposited from older beds. The only plants commonly represented as megafossils or microfossils are *Pinus, Populus, Salix* (including *S. crassijulis*), *Alnus, Betula,* and *Spiraea.* In addition, the herbaceous flora is better represented (*Epilobium,* Caryophyllaceae, *Valeriana*). The reduced representation of *Picea, Tsuga,* and Ericales in comparison with the Homerian may be due to the lack of drier habitats in the rapidly subsiding Cook Inlet basin, but climatic factors may be involved as well.

In the Homerian or lower Clamgulchian rocks of the Alaska Range, we have a flora preserved *in situ* by an ash fall. The dominates in this forest were *Pinus monticola* and *Betula papyrifera.* Also represented are *Abies grandis, Larix laricina, Picea glauca, Tsuga heterophylla, Salix richardsoni, S. glauca, Populus tacamahacca,* and *Alnus* sp. On the specific level, this flora contains only two now-exotic trees or shrubs, the fir and the pine. The only exotic dicotyledon known from the Alaska Range Clamgulchian is *Trapa,* whose fruits occur stratigraphically above the ash-fall forest. It should be noted that in the stratigraphically higher part of the Alaska Range Clamgulchian, the pollen floras from some localities contain large amounts of exotic dicotyledons that apparently have been redeposited from older Tertiary or Cretaceous(?) rocks nearby.

Pollen floras from the lower member of the Kougarok Gravel on Seward Peninsula are probably of Homerian (late Miocene) age. These floras are dominated by *Picea, Pinus, Tsuga,* and Betulaceae. Minor elements include *Abies,* Cupressaceae, *Carya, Pterocarya, Myrica, Ulmus, Ilex,* Ericales, Gramineae, and Cyperaceae (Benninghoff, written communication, 1962; Wolfe, unpublished data). These floras appear to represent a rich Boreal Forest,[a] as in the Cook Inlet Homerian. Floras from the upper member of the Kougarok Gravel lack the regionally exotic dicotyledon genera, but still contain several types of Pinaceae.

Pollen from Submarine Beach at Nome probably represents the late Clamgulchian, that is, the late Pliocene or early Pleistocene (the Eopleistocene of Vangengeim, 1961). The pollen evidence is, however, inconsistent. Some samples lack any regionally exotic genera; in all these samples, *Picea* is poorly represented, but Gramineae, *Betula,* and Ericales are well represented. Some other samples from Inner Submarine Beach, however, contain

---

[a] The capitalized vegetational terms used in this report are those of Wang (1961).

abundant *Pinus* and several grains of *Abies* and *Carya*(?) and a few grains of *Ilex* and *Pterocarya*. In addition, there is pollen from two species elsewhere unknown above the Paleogene. We therefore consider that the latter samples are contaminated by redeposited pollen and are hence unreliable. The remaining samples thus indicate that the vegetation near Nome during the time of Submarine Beach was primarily shrubby, and that the coniferous forest was absent locally. At the same time, some spruce forest occurred at Kivalina.

The interpretations of the fossil plant record in Northeastern Siberia have been continually modified during the past two decades. As in Alaska, the well-dated records of the Mixed Mesophytic Forest are no older than Oligocene.

In the valleys of the Lena and its tributaries—where the coldest temperatures for the Northern Hemisphere have been recorded—the late Oligocene or early Miocene Tandia Suite of Mamantov Mountain contains a flora consistent with a Mixed Mesophytic Forest. The cool Pinaceae (*Abies* and *Picea*) are poorly represented, but Juglandaceae, Fagaceae, and other broad-leaved plants are characteristic members of the pollen spectra (Alekseev *et al.*, 1962). The next younger Miocene beds contain reduced representation of the warm broad-leaved plants and increasing proportions of *Picea* and *Tsuga*. By the "Eopleistocene," the pollen samples indicate that the vegetation was dominated by Pinaceae and Betulaceae, with rare occurrences of *Ulmus* and *Ilex*; the nuts of *Juglans cinerea*, however, occur commonly. The middle Pleistocene flora of the Lena River valley is not significantly different from the extant flora.

The late Miocene or early Pliocene Pestsov Suite in southern Chukotka contains pollen floras that indicate the presence of a rich Boreal Forest. *Picea, Pinus, Tsuga*, and Betulaceae dominate floras containing small quantities of *Larix, Abies, Corylus, Fagus, Quercus, Ulmus*, and Ericales (Petrov, 1963).

The late Neogene and early Quaternary sequence on Kamchatka indicates that the vegetation was similar to that now present in the maritime province of southeastern Siberia in the Amur River valley (the Boreal Forest of the northeastern provinces, described by Wang, 1961). This particular region has a forest dominated by one or more genera of Pinaceae, but with several broad-leaved tree genera in small amounts. The late Pliocene Ermanov Suite of western Kamchatka contains *Tsuga, Salix* (ten species), *Betula, Alnus, Corylus, Ilex, Acer, Quercus, Juglans*, and *Vaccinium* (Vas'kovskii, 1959). Probably equivalent beds in northern Kamchatka have *Pinus* and *Juglans cinerea*, but *Metasequoia* is also present. The flora of the lower

Pleistocene beds indicates that the regionally exotic genera were eliminated, and that the vegetation was dominated by species of *Picea, Pinus,* and *Tsuga.* It is of interest that *Juglans cinerea* and *Metasequoia* lingered until about the end of the Pliocene from Japan north to Kamchatka. Thus far, *Juglans cinerea* has not been found in Alaskan Cenozoic rocks, and *Metasequoia* became regionally extinct by the end of the Miocene (Homerian).

The Northeastern Siberian record thus indicates that by the late Miocene or early Pliocene the Mixed Mesophytic Forest had become regionally extinct, and the vegetation was dominated by coniferous trees, with minor amounts of hardwoods now associated with the Boreal Forest of the Amur River valley. The Pliocene coniferous forests consisted of either extinct species or extant Asian species, except for *Pinus monticola. Larix occidentalis* was thought by Vas'kovskii (1959) to occur in the Pliocene of Kamchatka, but the illustration of the cone assigned to this species is not convincing (the cone figured lacks the characteristic bracts of *Larix*). The Siberian records of *Tsuga* cf. *heterophylla* and *Pinus radiata* have not been documented. The history of *P. radiata* and related species in California (Mason, 1932) makes its occurrence highly improbable in Siberia. The occurrence of *P. monticola* in the Siberian Pliocene is not surprising, because this species has a long history in the Mixed Mesophytic Forest and has about the same phytogeographic significance as its associated hardwoods. In general, therefore, the northeastern Asian floras are similar in the Neogene to the floras of northwestern North America. The maritime regions supported a rich broad-leaved deciduous forest throughout most of the Miocene. By the late Miocene the broad-leaved forest was restricted to middle latitudes, and the high latitudes were occupied by a predominantly coniferous forest with a few hardwoods. By the early Pliocene, the flora of Beringia, although somewhat richer than the extant Boreal Forest in Alaska, contained few or no regionally exotic dicotyledons. The late Pliocene flora was probably a taiga, i.e., tundra with sparse groves of *Picea* and *Pinus.* The rich conifer forest persisted into the late Pliocene in areas such as Kamchatka. It should be noted that the Mixed Mesophytic Forest of the early and middle Miocene included several species that ranged across Beringia. The Boreal vegetation of the late Miocene, Pliocene, and Quaternary had or has no tree species (except for the relictual *Pinus monticola*) that ranged or now range across Beringia. Instead, it is the shrubby flora, e.g., *Salix, Rubus, Spiraea,* that indicates strong floristic connection. If the Boreal Forests of Asia and North America were contiguous during the later Neogene, it was under conditions suboptimal for the growth of tree species.

In Oregon, the early Pliocene floras west of the Cascade Range contain exotic broad-leaved genera. The continuing presence of *Carya, Pterocarya, Ulmus*, and *Platanus* indicates the existence of an impoverished broadleaved deciduous forest. There is no evidence that the coniferous forest then occupied the coastal lowlands and valleys west of the Cascades. By the early Pleistocene, however, the flora was essentially modern; pollen floras from the lower Pleistocene marine deposits at Cape Blanco in southwestern Oregon contain *Picea, Pinus, Pseudotsuga, Tsuga, Alnus, Godetia*, and Compositae, but lack pollen of the broad-leaved deciduous exotics. The lastrecorded now-exotic genera in western United States are *Trapa, Persea*, and *Ilex* in the upper Pliocene of the San Francisco area (Axelrod, 1944).

In Japan, the broad-leaved deciduous forest has persisted to the present day. Indeed, except for the presence of a few species and genera now restricted to the Asian mainland, the Japanese Pliocene floras are similar to the extant Japanese flora.

## Climatic Interpretations

Although the climatic factors that control plant distribution are diverse and not well understood, certain types of climatic data appear to be significant. In general, temperature is the most significant factor in the replacement of one forest type by another. Although the effects of precipitation are perhaps more obvious in regions of little rainfall, in the area we are here concerned with there appears to have been ample precipitation throughout the Neogene; the floristic changes we have outlined, therefore, are probably not related to precipitation, except locally on the leeward side of mountain ranges. Some writers have emphasized the effect of the seasonal distribution of precipitation on the distribution of forest types, but the coastal region of southern Alaska receives heavy precipitation throughout the year. In Oregon, on the other hand, both temperature and seasonality of precipitation are important factors to plant distribution.

It is therefore in the realm of temperature that we must seek the cause for the vegetational changes outlined here. Leopold (1958) has pointed out that summer temperature, as expressed by the July mean, appears significantly related to the southern boundary of *Picea* in eastern North America. It should be pointed out that the approximate coincidences of ranges of genera with particular isotherms do not indicate that the isotherms, which are abstractions, control the ranges of genera, which are at least secondorder abstractions. The coincidence does, however, indicate (1) that the various species and their individual members within a genus such as *Picea* have a similar function that is limited by the same environmental factor;

and (2) that this factor in the environment is involved in the concept of average July temperature. The various other genera of Pinaceae appear to have their southern limits determined by the same factor-function. The southern limits of North American species of *Abies*, *Picea*, and *Tsuga* (except *T. caroliniana*) all coincide approximately with the 21° C. July isotherm. Exceptions to this coincidence are primarily in areas where low precipitation precludes the existence of a forest. The northern limits of the three genera differ. The most northern—*Picea* (Fig. 2)— has a northern limit approximately coincidental with the 10° C. July isotherm. *Abies* is apparently less tolerant of cool summers, because its northern limit is approximately coincidental with the 13° C. July isotherm. The northern limit of *Tsuga* parallels the minus 12° C. January isotherm more nearly than any other isotherm. *Pinus* is the most tolerant of warm temperatures of any genera of Pinaceae, and its southern limit is not relevant here; the northern limit of *Pinus* is approximately coincidental with the 12° C. July isotherm. We think that these data, which are based on several species within each genus, can be reliably used in the interpretation of Neogene climates.

Among the broad-leaved tree genera, the data are less trustworthy. There is evidence from the fossil record that the tolerances of some dicotyledon genera have changed significantly during the Tertiary (see MacGinitie, 1962, p. 90–91). Particularly when considering obvious monotypic relicts, such as *Platycarya*, *Cercidiphyllum*, and *Trochodendron*, it would be dangerous to apply the tolerances of the living species to interpretations of paleoclimates, especially in the Paleogene. However, genera that are still diverse today—for example, *Fagus*, *Quercus*, *Ulmus*, and *Acer*—appear in Neogene associations that are consistent with their present tolerances, and hence these genera may be reliable indicators of the Neogene climates in which they lived. A few genera that are still diverse must be utilized with caution, nevertheless; for example, the wide geographic range of *Juglans cinerea* during the late Tertiary— sometimes in association with a predominantly coniferous forest— indicates that the tolerance of some species and thus of the genus was different in the late Tertiary than it is today. Based on their Neogene associations, we think that the following genera did not have significantly different tolerances in the Neogene than today: *Liquidambar*, *Nyssa* (see Fig. 3), *Fagus*, *Quercus*, *Tilia*, and *Ulmus*. Less reliable are *Carya* and *Juglans*.

The presence in the coastal Seldovian floras in Alaska of *Liquidambar* and *Nyssa* indicates that the average July temperature was at least 20–21° C. The local presence of *Picea*, on the other hand, indicates that locally the July average was not in excess of 20–21° C. In the central Alaskan area,

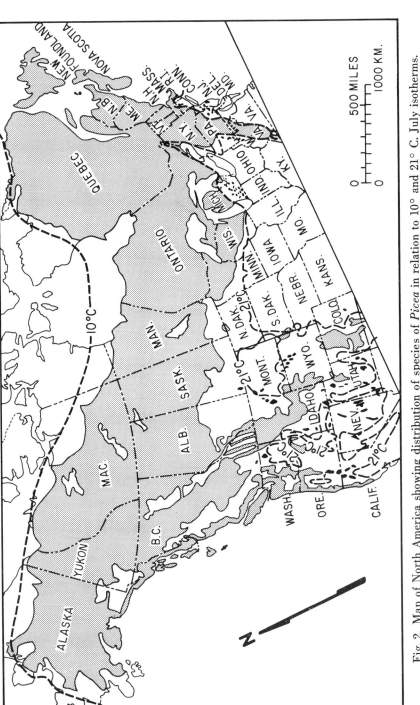

Fig. 2. Map of North America showing distribution of species of *Picea* in relation to 10° and 21° C. July isotherms.

the July temperature was apparently warmer, because *Abies, Picea,* and *Tsuga* are rare, whereas *Nyssa* and *Liquidambar* are present. In areas such as the Wrangell Mountains, where *Liquidambar* and *Nyssa* are absent and the conifers abundant, the July temperature was probably less than 20° C.

During the late Seldovian (middle Miocene), a group of genera (*Juglans, Fagus, Quercus, Liquidambar, Nyssa*) that have summer temperature requirements of about 18–21° C., as expressed by the July average, became extinct. In the younger Homerian, both *Carya* (18° C.) and *Ulmus* (16° C.) drop out of the record. The local extinction of *Abies* (13° C.) in the Cook Inlet basin also apparently took place in the Homerian, which indicates a very rapid decline in summer temperatures there. The persistence of *Abies* in the Homerian or lower Clamgulchian of central Alaska indicates a slightly warmer summer than in the Cook Inlet region, where only *Pinus* (12° C.) was present.

We think, therefore, that the major climatic deterioration in the Neogene of Alaska took place in the later half of the Miocene, when a drop in the July average of about 7° C. took place within a time span of about 4 million years.

The rapid climatic deterioration in the late Miocene is further supported by the fact that the vegetation changed from a Mixed Mesophytic to a Boreal Forest, apparently going through the transitional Northern Hardwood Forest, in a short period of time. In fact, we have no clear evidence of such a Northern Hardwood Forest in the Neogene of the Cook Inlet area or the central Alaska Range. *Nyssa, Liquidambar,* and other warm hardwoods disappeared and were replaced by members of Pinaceae.

This interpretation of the fossil record and its climatic indications is in agreement with the biogeographic implications of Recent floras. The close resemblance between vicarious species in the warm-temperate broad-leaved deciduous vegetation (the Mixed Mesophytic Forest) is best explained by floristic continuity between eastern North America and eastern Asia as late as the middle part of the Miocene, that is, about 15 million years ago. The Arcto-Tertiary concept, which envisions a disjunction as long ago as the early or middle Oligocene (30–35 million years ago), does not seem plausible in light of either biological processes or the fossil record (Mason, 1947).

Diversified coniferous forests expanded over large areas in Siberia and northern North America as the northern limits of various mesophytic hardwoods retreated southward. One might expect that the Siberian and North American coniferous forests would have been floristically continuous, forming a single gene pool, just as the Mixed Mesophytic Forests of the two continents had at an earlier time. However, the complete lack of conspecificity,

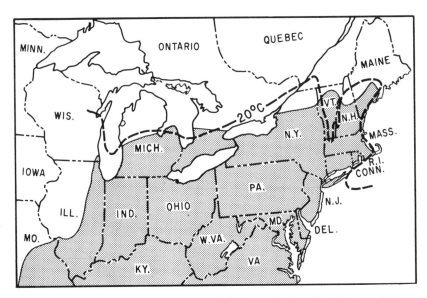

Fig. 3. Map of eastern United States and Canada showing distribution of *Nyssa sylvatica* in relation to 20° C. July isotherm.

in both the modern and the fossil floras, seems to indicate that the coniferous forests of Siberia and North America have never been continuous. It should be pointed out that our interpretation indicates that only during the Home-rian (late Miocene) was it possible for an interchange to have taken place between the coniferous forests of the two continents; but it was during part or all of the late Miocene that evidence from marine organisms (Durham and MacNeil, this volume) indicates that the Bering Strait was open, which would inhibit floristic interchange.

During the Pliocene, although some of the area was forested, it was pri-marily the shrubby plants (*Salix*, *Alnus*, Ericales) that occupied the regions adjacent to Bering Strait, and it is among the shrubby and herbaceous flora that the flora of northeastern Asia and northwestern North America have their greatest resemblance. It was, for example, probably during the Pliocene that shrubby species such as *Salix crassijulis*, S. *glauca*, S. *richardsoni*, *Alnus crispa*, *A. incana*, and *Spiraea beauvardiana* were distributed across Beringia. All these species or their ancestors occur in the Alaskan Pliocene.

The present disjunct distribution of *Salix crassijulis* and *Spiraea beau-vardiana* indicates a pre-Pleistocene distribution, which is in agreement with the fossil record. It is noteworthy that the dicotyledonous trees that were present in the Clamgulchian, for example, *Populus tacamahacca* and *Betula papyrifera*, do not today have a trans-Bering distribution.

# Summary

The Mixed Mesophytic Forest of the early and middle Miocene was continuous from Japan through Alaska and into conterminous northwestern United States.

At high altitudes, mixed conifer–broad-leaved-deciduous forests were present.

In the late Miocene, a severe decline in summer temperatures in Alaska resulted in the disjunction of the Mixed Mesophytic Forest.

In Alaska and adjacent Siberia, the late Miocene vegetation was a rich Boreal Forest, i.e., conifer-dominated, with a few hardwood tree species.

The rich Boreal Forest was gradually depauperated in the north during the Pliocene, but survives today in regions such as the Amur River valley and coastal Oregon and Washington.

By the early Pleistocene, the forests around the North Pacific basin were similar to the extant forests of the same region. A few exotic plants persisted locally, but mixed forest-tundra was present in Beringia.

The Boreal Forest was effectively partitioned during the late Miocene—probably by edaphic factors and by the opening of Bering Strait—and during the Pliocene by both edaphic and climatic factors.

## REFERENCES

Alekseev, M. N., N. P. Kuprina, A. I. Medyanntsev, and I. M. Khoreva. 1962. Stratigrafiia i korreliatsiia Neogenovykh i Chetvertichnykh otlozenii severo-vostochnoi chasti Sibirskoi platformy i ee vostochnogo skladchatogo obramleniia (Stratigraphy and correlation of the Neogene and Quaternary deposits of the northeastern part of the Siberian platform and its folded framework) : Akad. Nauk SSSR, Tr. Geol. Inst., Vyp. 66, 126 p.

Axelrod, D. I. 1944. The Sonoma flora: Carnegie Inst. Washington Pub. 553, p. 167–203.

—— 1964. The Miocene Trapper Creek flora of southern Idaho: Univ. Calif. Pub. Geol. Sci., v. 51, 148 p.

Chaney, R. W. 1936. The succession and distribution of Cenozoic floras around the northern Pacific Basin, *in* Essays in geobotany in honor of William Setchell: Univ. Calif. Press, Berkeley, p. 55–85.

Krystofovich, A. N. 1935. A final link between the Tertiary floras of Asia and Europe: New Phytologist, v. 34, p. 339–344.

Leopold, Estella B. 1958. Some aspects of late-glacial climate in eastern United States: Zurich Geobot. Inst. Veröffent., n. 34, p. 80–85.

MacGinitie, H. D. 1962. The Kilgore flora: Univ. Calif. Pub. Geol. Sci., v. 35, p. 67–158.

Martin, H. A., and G. E. Rouse. 1966. Palynology of late Tertiary sediments from Queen Charlotte Islands, British Columbia: Canadian Jour. Botany, v. 44, p. 171–208.

Mason, H. L. 1932. A phylogenetic series of the California closed cone pines suggested by the fossil record: Madroño, v. 2, p. 49–55.

——— 1947. Evolution of certain floristic associations in western North America: Ecol. Monogr., v. 17, p. 201–210.

Petrov, O. M. 1963. Stratigrafiia chetvertichnykh otlozhenii iuzhnoi i vostochnoi chastei Chukotskogo poluostrova (The stratigraphy of the Quaternary deposits of the southern and eastern parts of Chukotka Peninsula): Akad. Nauk SSSR, Biul. Komm. po Izucheniyu Chetvertichnogo Perioda 28, p. 135–152. In Russian; English translation available from Am. Geol. Inst.

Tanai, T. 1961. Neogene floral change in Japan: Hokkaido Univ. Jour. Fac. Sci., ser. 4, v. 11, p. 119–398.

Tanai, T. and N. Suzuki. 1963. Miocene floras of southwestern Hokkaido, Japan, in Tertiary floras of Japan. I. Miocene floras: Geol. Survey Japan, p. 9–149.

Vangengeim, E. A. 1961. Paleontologicheskoe obosnovanie stratigrafiia Antropogenovykh otlozhenii severa vostochnoy Sibiri (Paleontological basis of the stratigraphy of the Anthropogene deposits of Northeast Siberia): Akad. Nauk SSSR, Tr. Geol. Inst., Vyp. 48, 182 p.

Vas'kovskii, A. P. 1959. Kratkii ocherk rastitel'nosti, klimata i khronologii chetvertichnogo perioda v verkov'iakh rek Kolymy, Indigirki i na severnom poberezh'e Okhotskogo moria (A brief essay on the vegetation, climate, and chronology of the Quaternary Period in the upper Kolyma and Indigirka Rivers and on the northern coast of the Sea of Okhotsk): p. 510–556 in K. K. Markov, and A. J. Popov, eds., Lednikovyi period na territorii Evropeiskoi chasti SSSR i Sibiri (Ice age in the European section of the USSR and in Siberia): Moskva Gosudarstv. Univ., Geogr. Fak., 560 p. In Russian; English translation, p. 464–512 in H. N. Michael, ed. (1964), The archaeology and geomorphology of northern Asia: Arctic Inst. North America, Anthropology of the North: Translations from Russian Sources, no. 5, 512 p.

Wang, Chi-Wu. 1961. The forests of China: Harvard Univ. Bot. Mus., Maria Moors Cabot Found., Pub. 5, 313 p.

Wolfe, J. A. 1964. Miocene floras from Fingerrock Wash, southwestern Nevada: U.S. Geol. Survey Prof. Paper 454-N, p. N1–N36.

——— 1966. Tertiary plants from the Cook Inlet Region, Alaska: U.S. Geol. Survey Prof. Paper 398-B, p. B1–B32.

Wolfe, J. A., D. M. Hopkins, and Estella B. Leopold. 1966. Tertiary stratigraphy and paleobotany of the Cook Inlet region, Alaska: U.S. Geol. Survey Prof. Paper 398-A, p. A1–A29.

# 10. Quaternary Vegetational History of Arctic Alaska

PAUL A. COLINVAUX
*Ohio State University*

The lowlands of arctic Alaska and of the islands in Bering Sea have never been completely glaciated, and thus have supported vegetation throughout the Pleistocene Epoch. In the absence of glaciers, polliniferous lacustrine and peat deposits have survived for long periods, leaving pollen records of ancient Beringian plant communities. These records, and similar records from Eastern Siberia and the Lena valley, should enable us to reconstruct the history of the vegetation of northern regions by means of pollen analysis.

## The Pollen Record

The arctic lowlands are now occupied by tundra. Megafossil and biogeographic evidence suggests that the lowlands have supported only tundra throughout Pleistocene and late Tertiary times (Hultén, 1937; Hustich, 1953; Wolfe and Leopold, this volume). Since tundra communities have probably always been present in the region, the arctic pollen analyst must interpret pollen records of tundra. This presents special problems not encountered by pollen analysts of other regions, who may rely rather heavily on the pollen record of forest trees. A certain amount of tree pollen is to be found blowing about over a modern tundra, but this consists of pollen blown from distant sources or pollen from shrub species of genera otherwise represented by trees. Pollen of spruce trees (*Picea*) and alder bushes (*Alnus*) from the subarctic forests may represent 30 per cent or more of the total pollen collected at a tundra site. Much birch (*Betula*) pollen is produced in some tundra communities by the dwarf birch, *Betula nana*, but the other common tundra shrubs—the willows (*Salix*) and the heaths (Ericaceae)— produce proportionately less pollen, and therefore may be represented by only a few per cent in the pollen rain. About half, and sometimes nearly all, of the pollen collecting at a tundra site consists of herb pollen, particularly pollen of grasses and sedges.

Grass (Gramineae) and sedge (Cyperaceae) pollen can only be identified to family, so that these pollen taxa embrace many species, possibly representing many different habitats and environments. The lesser amounts of *Artemisia* pollen can be identified to genus. The pollen of other arctic herbs usually occurs only in trace amounts, but when present it may often be identified with a precision not usually expected of pollen analysis, perhaps even to species. This is because the limited arctic species list allows one to eliminate many of the pollen types with which pollen from an arctic plant might be confused. Unfortunately, the utility of this tactic is impaired by the fact that most arctic plants have wide distributions and broad environmental tolerances, so that the precise record of a species having been present rarely allows one to deduce much about the community of which it was a member.

The pollen analyst of tundra regions, then, must interpret pollen diagrams that usually record only a few broadly defined taxa of tundra plants. Of the true tundra plants, perhaps only the dwarf birch, grass, and sedge will be represented in significant amounts in the pollen diagram. In addition, the analyst may have to weigh the importance of quantities of spruce and alder pollen that he suspects have been blown in from outside his local area, and he will also encounter small amounts of willow and heath pollen, and of the pollen and spores of a variety of other tundra plants. A tundra pollen diagram is thus much simpler and less informative than some of the splendid forest records of temperate regions. In the Arctic, pollen analysis is a rather blunt instrument.

LIVINGSTONE'S THREE-ZONE SEQUENCE IN THE BROOKS RANGE

Our understanding of the modern pollen rain and of the pollen stratigraphy of arctic Alaska stems from the work of D. A. Livingstone, who raised a series of sediment cores from lakes in the Brooks Range and published pollen diagrams for the cores (1955). The cores were not dated by radiocarbon, but Livingstone was able to construct a chronology for his diagrams by comparing the sedimentary record with the known glacial history of the region. Later, Livingstone (1957) was able to support his chronology by pollen analysis of a radiocarbon-dated core of valley fill from Umiat (Fig. 1). Livingstone's conclusions may be understood by reference to his pollen diagram for a 5-meter core from Chandler Lake (see Fig. 1), the major parts of which are presented as Fig. 2.

The Chandler Lake diagram is thought to span most of postglacial time, so that the bottom of the core represents the period some 7,000 or 8,000

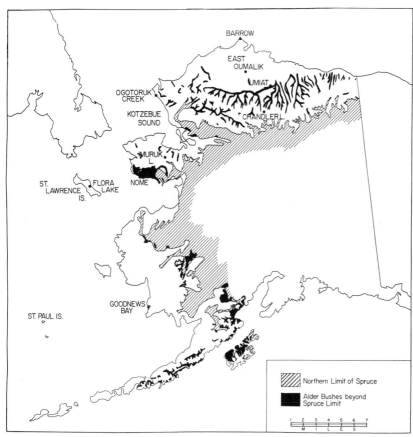

Fig. 1. Map showing sampled sites and extent of forests in Alaska. Note that the spruce line is not the limit of trees; poplar and birch trees are found locally beyond spruce timberline. The vegetation boundaries are taken from L. A. Spetzman's (1963) map of the vegetation of Alaska. In this map, alder is plotted, together with shrub willows and poplar, as "high brush." It may be, therefore, that alder is not locally present in some of the places where it is indicated, but the limits shown are approximately correct. Small shrub or prostrate willows occur in nearly all parts of the Alaskan tundra.

years ago (Livingstone, 1955; Porter, 1964). The diagram falls naturally into three zones, an herbaceous zone, a birch zone, and an alder zone, which Livingstone labeled zones I, II, and III. Each of the zones was found to be comparable to modern spectra from various parts of arctic Alaska. Two spectra similar to those of zone I were obtained from modern deposits at Barrow, 400 km to the north, on the shore of the Arctic Ocean. A modern spectrum comparable to zone II was obtained from East Oumalik in the northern foothills of the Brooks Range. Several samples of the surface mud

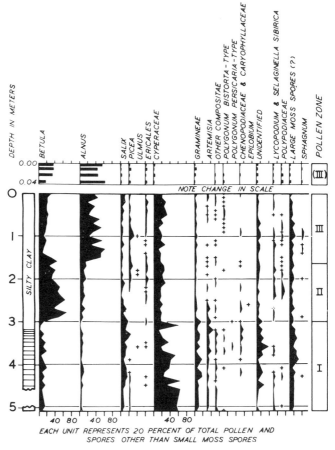

Fig. 2. Pollen diagram from Chandler Lake. The diagram is part of that published by Livingstone (1955). Livingstone's division of the pollen zones into subzones is not followed.

of Chandler Lake yielded pollen spectra assignable to zone III. Livingstone explained the three-zone sequence as follows.

*Zone I.* During the zone I period, an herbaceous tundra comparable to that of modern Barrow occupied the region. Dwarf birches did not grow locally, the small amount of birch pollen found in the zone having been blown from the nearest birches many kilometers away. That only a trace of alder and spruce pollen was found suggests that the nearest alder and spruce trees were perhaps hundreds of kilometers away. Of the pollen present in large amounts, only that for the sedge, grass, and willow was produced locally. This picture closely mirrors the present vegetation at Barrow, where there is a sedge and grass tundra devoid of true shrubs. Creeping willows are present, but the nearest dwarf birch is many kilometers inland. The nearest alder bushes are over 100 km away, and the nearest spruce tree is some 300 to 400 km to the south.

*Zone II.* In zone II times, dwarf birches had colonized the site, and were a prominent feature of the local tundra. Alder bushes and spruce trees were still too far away for their pollen to have been incorporated in significant amounts. The tundra was still largely made up of sedges and grasses. The birch-pollen maximum resulted from the dwarf birches, having produced relatively more pollen than the herbs, so that the inclusion of even a minority of birch plants in the neighborhood had a dramatic effect on the pollen spectra. A comparable tundra, with its associated pollen spectrum, was found by Livingstone at East Oumalik, 150 km north of Chandler Lake.

*Zone III.* This zone reflects the present vegetation of the Chandler Lake region. The area still supports only tundra, but the nearest alder bush is now within 10 km of the lake, and there are spruce trees within about 50 km. Apparently, the pollen spectra of zone III are produced when a local pollen rain similar to that of the zone II period is mixed with tree pollen blown from sources only a few kilometers away.

Livingstone's matching of the three zones with surface spectra demonstrated the sort of vegetation shifts within the Alaskan tundra that could be interpreted from the pollen data. A tundra with dwarf birches can be distinguished, on the basis of pollen data alone, from one in which dwarf birches are absent. The proportions of spruce and alder pollen included in the local pollen rain provide useful indicators of the position of the tree line, knowledge of which can be of great value to the ecologist. But the pollen clues to the detailed composition of the tundra are much less clear. Various investigations subsequent to Livingstone's have shown that there can be wide fluctuations in the proportions of such taxa as Gramineae, Cyperaceae, *Artemisia*, and Ericales (Colinvaux, 1964a; Heusser, 1963; Hopkins *et al.*, 1960) that cannot yet be so certainly explained. Some of these problems are discussed below. Part of the difficulty is the present paucity of published modern pollen spectra for Alaska. Eventually, work that carefully relates modern pollen production to tundra descriptions should allow refinement of Livingstone's zones,[a] but at present the Livingstone three-zone sequence forms the only safe basis for interpreting fossil pollen spectra from arctic Alaska.

[a] I collected a series of 36 surface samples during the summer of 1965 from Barrow, Ocean Point on the Colville River, Cape Beaufort, Cape Thompson, and various sites on Seward Peninsula; analysis of these should aid in interpreting the details of fossil spectra.

OGOTORUK CREEK, CAPE THOMPSON

A pollen record from peat deposits at Ogotoruk Creek (Heusser, 1963) extends Livingstone's Brooks Range history more than 500 km westward to the Bering Sea in the region of Cape Thompson. The glaciers of the Brooks Range did not reach the Ogotoruk Creek site, but were thought to have come to within 100–150 km. There is no radiocarbon control of the sections, and there is reason to believe that the sections are truncated at the top. In spite of these difficulties, pollen diagrams from the three sections (Fig. 3) are valuable in that they show clearly a change from Livingstone's herbaceous zone I at the bottom of the sections to the birch zone II higher up. Heusser concludes that zone I represents the late glacial period, or a time soon after, when the local vegetation and climate reflected the existence of glaciers on the Brooks Range to the east. Zone II records the inclusion of dwarf birches in the local tundra during the postglacial period.

The Ogotoruk Creek record supports Livingstone's view (1955, 1957) that the postglacial period has seen the vegetation of the latitude of the Brooks Range progress from one similar to that of modern Barrow, through various forms of tundra comparable with those found at intermediate latitudes, to the final condition of the present day. For the student of the Bering Land Bridge, this means that the latitudes of the Brooks Range supported tundra equivalent to that of modern Barrow in the period immediately after the severing of the last land link.

ARCTIC COAST AT BARROW

A pollen record spanning the last 25,000 years is available for the region near Barrow. The record has been constructed from a series of discrete samples of peat, preserved in permanently frozen ground, which were unearthed by Dr. Jerry Brown and his colleagues at the Cold Regions Research and Engineering Laboratory in the course of their permafrost drilling program. There is usually no stratigraphic evidence for the age of the samples, but each has been dated by means of radiocarbon, allowing the pollen spectra from them to be arranged in chronological order. The samples come from borings within an area a few kilometers square. They have been described and discussed by Brown (1965). Results of pollen analysis from some of the samples have already been published (Colinvaux, 1964b, 1965). The pollen data are here summarized in Table 1. Two modern pollen spectra, taken from Livingstone's 1955 paper, have been included in Table 1 to complete the pollen sequence.

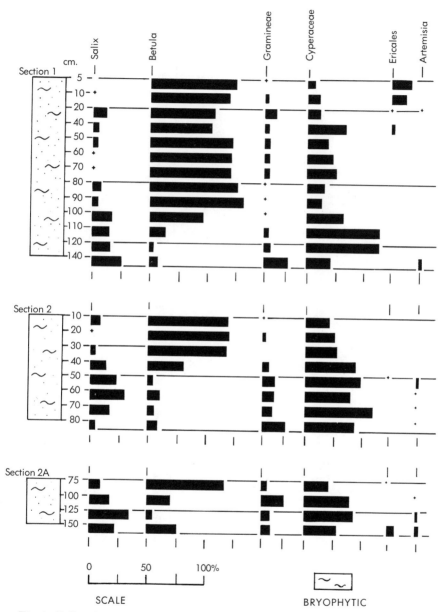

Fig. 3. Pollen diagrams from Ogotoruk Creek Cape Thompson. These diagrams are parts of those published by Heusser (1963). Scale is per cent total pollen and fern spores less *Sparganium* pollen; + represents 2 per cent or less.

TABLE 1. *Pollen Spectra from Barrow, Alaska*[1]

| Specimen | Livingstone: Footprint Lake | Livingstone: Unnamed Lake | Laboratory Serial Numbers and Age B.P. | | | | | | |
|---|---|---|---|---|---|---|---|---|---|
| | | | I-699 1,750 ±120 | I-1544 3,200 ±230 | I-1545 5,010 ±320 | I-700 9,550 ±240 | I-701 10,525 ±280 | I-1171 14,000 ±500 | I-1384 25,300 ±2,300 |
| % *Betula* | 7 | 5 | 5 | 6 | 6 | 3 | – | + | + |
| % *Alnus* | 5 | 10 | 12 | 6 | 6 | – | + | – | + |
| % *Picea* | + | + | + | + | + | + | + | + | + |
| % *Salix* | 3 | 4 | 3 | 3 | 3 | 16 | 41 | 16 | + |
| % *Ericales* | + | – | 6 | 8 | 10 | – | – | + | + |
| % *Cyperaceae* | 36 | 24 | 13 | 28 | 46 | 23 | 5 | 31 | 71 |
| % *Gramineae* | 41 | 49 | 57 | 44 | 25 | 49 | 41 | 44 | 14 |
| % *Artemisia* | – | – | + | + | + | – | + | + | 5 |

[1] The two modern samples from Footprint Lake and an unnamed lake are taken from Livingstone's 1955 paper. The remaining samples represent units of peat or organic matter buried in the frozen ground. They are identified by the serial numbers of Isotopes, Inc. The figures represent per cent total pollen and pteridophyte spores. The plus sign represents 2 per cent or less; the minus sign indicates that no pollen grains were found in the analysis. The minor elements have not been included.

The two modern Barrow samples yield herbaceous pollen spectra comparable to the zone I spectra of the Brooks Range sequence. The small quantity of birch pollen can be explained as the result of wind transport from dwarf birches some tens of kilometers away. A similar quantity of alder (which was not present in zone I of the Brooks Range) records the presence of the nearest alder groves on the arctic slope of the Brooks Range, perhaps 100 km away. The 60–80 per cent of grass and sedge pollen in the two spectra are produced in the flat, marshy, meadow-like tundra of the coastal plain at Barrow.

The three samples spanning the period of 1,700–5,000 years ago have pollen spectra essentially the same as those of the present day, differing only in the inclusion of a small amount of heath pollen. It seems likely that the rise in heath content of these samples reflects local site differences rather than broad changes in the vegetation of the area. Heath and willow pollen tend to be locally overrepresented in pollen diagrams, and it must be remembered that these samples represent a series of slightly different sites within a general area. It seems, then, that no broad change in the vegetation of the Barrow area is detectable in the pollen record during the last 5,000 years.

The sample from 9,500 years ago includes no alder pollen. At this date, there was apparently no alder in the arctic foothills. Older samples, from as far back as 25,000 years ago, have no significant alder content, and also lack birch. These spectra are even more "herbaceous" than those of the Chandler Lake zone I. They represent an herbaceous tundra that was far more remote from the nearest alders and dwarf birches than was Chandler Lake at any time in Livingstone's record. The radiocarbon dates show that such a tundra prevailed on the arctic coastal plain during the time of the "classical" Wisconsin Glaciation.

SEWARD PENINSULA

Two pollen records are available from Seward Peninsula: Estella B. Leopold's record from Nome (Hopkins *et al.*, 1960) and my long record from Imuruk Lake (Colinvaux, 1964a).

*Nome.* The Nome record comes from a peat deposit exposed in a swale, and from specimens of buried peat dated by radiocarbon, associated fossils, and stratigraphy. An abbreviated form of Leopold's pollen diagram is presented in Fig. 4. A peat sample with a radiocarbon age of about 13,000 years yielded an herbaceous pollen spectrum comparable to those of zone I at Chandler Lake and modern Barrow. A second sample, with a radiocarbon age of about 10,000 years, yielded a birch-pollen spectrum comparable to

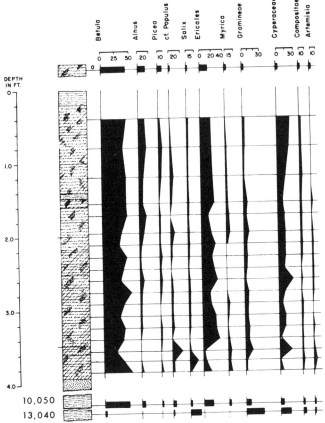

Fig. 4. Pollen diagram from Nome. This diagram is part of the one published by Leopold (in Hopkins *et al.*, 1960). The minor elements are not included in this version. The figures represent per cent total pollen. The radiocarbon dates on the two peat samples at the bottom of the diagram are as follows: 13,040 ± 200 (W-463), and 10,050 ± 270 (W-461). The taxon *Populus* is now believed to be wrongly identified. Most probably these grains should be assigned to Cyperaceae.

that of the Chandler Lake zone II. The pollen sequence from the swale peat begins with spectra similar to that from the 10,000-year-old sample and ends with a pollen zone including alder that may be compared with the Chandler Lake zone III. This 13,000-year record thus shows that events at Nome have paralleled those in the Brooks Range, but that the sequence started much earlier. This is understandable when it is considered that Nome is 400 km south of Chandler Lake, that Nome is on the coast, and that the glacial histories of the two places have been different.

Pollen spectra in three peat samples from the Sangamon deposits at Nome have been determined by Miss Leopold (Table 2). The relative ages

TABLE 2. *Sangamon Pollen Spectra from Nome*[1]

| Pollen Taxa | Stratigraphic Position of Sample and U.S. Geological Survey Paleobotany Number | | |
| --- | --- | --- | --- |
| | Early Sangamon D-1213-2 | Early Sangamon D-1213-4 | Sangamon Maximum D-1478 |
| % *Betula* | 30.0 | 28.2 | 9.6 |
| % *Alnus* | 8.6 | — | 67.3 |
| % *Picea* | 4.0 | — | 10.6 |
| % *Salix* | 1.6 | — | — |
| % *Ericales* | 22.1 | 49.7 | 1.9 |
| % Cyperaceae | 9.5 | 0.3 | 1.9 |
| % Gramineae | 6.2 | 16.5 | — |
| % *Artemisia* | 1.6 | — | — |
| % Juniperus (?) | 1.6 | — | — |
| % *Sagittaria* type | — | 4.0 | — |
| % Compositae | 0.8 | 0.2 | — |
| % Undetermined dicots | 4.0 | 0.5 | — |
| % Undetermined monocots | 10.0 | 0.6 | 5.8 |
| % *Polygonum bistorta* | — | — | 2.9 |
| Pollen sum | 127 | 370 | 104 |

[1] Samples taken from the Second Beach at Nome and analyzed for pollen by E. B. Leopold. Sample D-1478 was associated with spruce wood, and is considered to date from the time of the Sangamon maximum (Hopkins, personal communication). Pollen results from this sample are published here for the first time. The other two samples were associated with marine shells and are considered to be of early Sangamon age. Pollen results on these two samples have already been published (Hopkins *et al.*, 1960), though the sample D-1213-4 was then erroneously shown as containing alder pollen, like D-1213-2. The two samples were close together, apparently part of the same deposit. Figures represent per cent total pollen. Minus signs indicate that the pollen type was not seen during analysis. Pteriodophyte spores were not included in the pollen sum and have been omitted from the table.

of the three samples are not certainly known, but all three are firmly assigned to the Sangamon Interglaciation (Hopkins *et al.*, 1960, and personal communication). All three samples yielded pollen spectra high in birch, and two had considerable amounts of alder. One spectrum includes 70 per cent alder pollen as well as 11 per cent spruce pollen. Heath pollen appears in different amounts in the three spectra, but this is attributable to local variations in the populations of heath plants in the parent vegetation mat. The spectra suggest that during part of the Sangamon period the vegetation at and near Nome was comparable to that of the present day. Locally, there was tundra with dwarf birches, and there were groves of alder bushes no more than a few tens of kilometers away. At one time during the Sangamon Interglaciation, alders were much closer than now, perhaps actually growing at the site. At this period, too, there may have been spruce trees quite close to Nome.

*Imuruk Lake.* The pollen record from an 8-meter core of the sediments of Imuruk Lake has been fully discussed elsewhere (Colinvaux, 1964a). An abbreviated form of the pollen diagram is reproduced as Fig. 5. Radio-

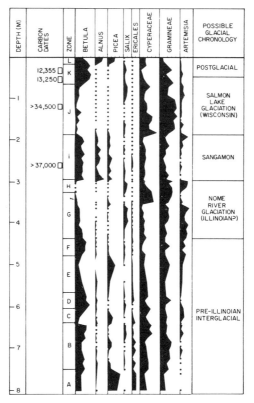

Fig. 5. Pollen diagram from Imuruk Lake. This diagram is part of that published by Colinvaux (1964a). Each division represents 10 per cent total pollen grains, pteridophyte spores, and *Sphagnum* spores. Dots represent 2 per cent or less. Radiocarbon numbers are Y-1144 (12,355 ± 160), I-588 (13,250 ± 700), Y-1142 ( > 34,500), Y-1143 ( > 37,000), all in years before present.

carbon dating suggests that the record is an extremely long one, with the top meter spanning something like 30,000 years and the rest of the core being correspondingly older. The geologic history of the Imuruk Lake basin is consistent with the assignment of great age to the sediments (Hopkins, 1959). Some slightly positive radiocarbon determinations on material from the bottom of the core are not thought to reflect true ages (Colinvaux, 1964a).

The Imuruk Lake pollen diagram was divided into 12 zones, A–L, by Colinvaux. At the top of the diagram is a three-zone sequence (J-L) spanning a period from beyond the limits of radiocarbon dating to the present. The three zones are an herb zone, J, a birch zone, K, and a zone with alder and spruce, L. Zones J, K, and L parallel zones I, II, and III at Chandler Lake, though they span a much longer period of time. It is reassuring that a similarly prolonged three-zone sequence is suggested for Nome by Leopold's discontinuous record (Hopkins *et al.*, 1960). Apparently, an herbaceous tundra, in which dwarf birches were absent or scarce, occupied central Seward Peninsula from a time early in the Wisconsin Glaciation until its close.

Dwarf birches then became common. Later, alder advanced to its present limit, some 10 km east of Imuruk Lake, and spruce reached its present position some 50 km away (Fig. 1).

A notable feature of the Imuruk Lake pollen diagram is that spruce maxima generally accompany the alder maxima in a way that is not so apparent in pollen diagrams from farther north. Undoubtedly, the spruce maxima record the advance westward of spruce on Seward Peninsula to something like its present position, for modern pollen samples from Imuruk Lake include 15 to 20 per cent of spruce pollen, which must have been transported by wind from trees more than 50 km away.

Zone i[b] is an alder and spruce zone comparable to zone L and to surface spectra for the region. It lies below, and is therefore older than, the herbaceous zone J of Wisconsin age. Zone i is thought to be of Sangamon age. It is reassuring that the Sangamon samples from Nome yielded pollen spectra generally similar to those of zone i.

Underlying zone i are the two herbaceous zones G and H. Zone G may be compared with the Wisconsin zone J and with the modern spectra from Barrow. Zone H spectra are similar only to spectra from the glacial period at Barrow, suggesting that not only alder and spruce but also dwarf birches receded to remote distances, perhaps some hundreds of kilometers from the lake. It is reasonable to equate zones G and H with the Illinoian Glaciation, locally represented by the Nome River Glaciation (Hopkins, 1963). The suggestion conveyed by the pollen spectra of zone H—that local tundras were similar to those of Barrow during the glacial period—implies a glacial climate more severe than that experienced during the Wisconsin episode of zone J. This is in accord with other evidence that the Illinoian Glaciation was the more severe in the region (Hopkins, 1963).

Zones A–F all contain varying amounts of alder and spruce pollen. Since they underlie zones assigned to the Illinoian Glaciation, they are considered to represent a pre-Illinoian interglaciation.

It should be emphasized that the correlation of zones A–i with glacial events is based on extrapolation and is not radiometrically controlled. It is

---

[b] Zone "i" is written in lower case to avoid confusion with Livingstone's zone "I."

possible to construct other chronologies using different extrapolations (Colinvaux, 1964a), and the conflicting radiocarbon determinations from the bottom of the core, though discredited, have not been completely explained. Yet the chronology suggested here seems the most satisfactory. It is consistent with the known glacial history of the area, and is therefore cautiously accepted.

This long pollen record from Imuruk Lake, then, is thought to record the vegetation of the area from some time in the pre-Illinoian interglaciation, through Illinoian, Sangamon, Wisconsin, and Recent times. Throughout this long period, the vegetation of central and western Seward Peninsula was tundra. In glacial periods, alder and spruce retreated far to the east of their present position, and dwarf birch was scarce or absent from the Imuruk Lake region. In interglacial periods, dwarf birches were locally abundant, and alder and spruce reached about their present limits.

KOTZEBUE SOUND

A pollen record of late Wisconsin age is provided by a 7-meter core of sediments from Kotzebue Sound taken in approximately 39.5 meters of water. The stratigraphy and importance of the core have been discussed by Creager and McManus (1965). A pollen diagram for the core has been published (Colinvaux, 1964a), and an abbreviated form of the diagram is reproduced as Fig. 6. On the basis of radiocarbon dating and fossil evidence, Creager and McManus concluded that the bottom 5½ meters of the core represent deltaic deposits that were deposited rapidly and continuously.

The delta is thought to have been covered by only 5 or 6 meters of water, suggesting that sea level was then some 35 meters lower than now. The approximate mean of five radiocarbon dates from the deltaic material is 14,000 years B.P., but Creager and McManus believe the true age to be about 12,000 years B.P. Whichever age is correct, it is apparent that the deltaic deposits were laid down toward the close of the Wisconsin Glaciation, when the last portions of the Bering Land Bridge were being flooded by the rising sea.

The bottom six spectra in Fig. 6 are from the ancient deltaic deposits. They constitute a birch zone comparable to zone K at Imuruk Lake. Zone K, dated independently at the two sites as having been in existence some 12,000 years ago, is shown by the Kotzebue Sound core to record the vegetation of the land bridge lowlands just before they were flooded. A tundra with dwarf birches had developed, but the western limits of spruce and alder still lay east of their present positions. This is direct evidence for a tundra environment on the Bering Land Bridge.

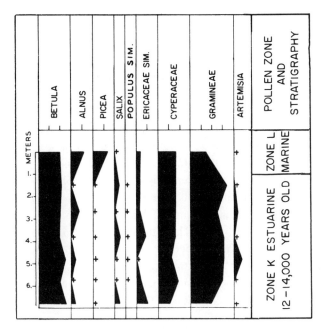

Fig. 6. Pollen diagram from Kotzebue Sound. This diagram is part of that published in Colinvaux (1964a). Each unit represents 10 per cent total pollen grains, pteridophyte spores, and *Sphagnum* spores. Plus signs represent 2 per cent or less. Dating of the estuarine deposits is from Creager and McManus (1965).

The top one meter of the core was shown by Creager and McManus to be marine sediment of postglacial age. A pollen spectrum from the very top contains alder and spruce. It may be assigned to the Imuruk Lake zone L. This spectrum confirms that alder and spruce have advanced to their present position in western Alaska during postglacial time.

The position of the Kotzebue Sound record is close enough to Ogotoruk Creek (Fig. 1) to permit valid comparison of the development of pollen zones at the two sites. The Ogotoruk Creek diagrams (Fig. 3) span the period when the birch maximum developed. In Kotzebue Sound, the birch zone begins before 12,000 years ago, suggesting an even older date for the start of the Ogotoruk Creek sequence. On the other hand, comparison with Livingstone's sequences from Umiat and Chandler Lake (Livingstone, 1955, 1957) suggests that the oldest part of the Ogotoruk Creek pollen sequence is about 8,000 years old. This shows that it is unsafe to use pollen zones as time-stratigraphic units in arctic Alaska, except in the most general way or when the sites are close together.

ST. LAWRENCE ISLAND

A pollen diagram for a 2-meter core of sediments from Flora Lake on St. Lawrence Island (Fig. 1) is presented as Fig. 7. A stratigraphic break in the core at the one-meter level separates the pebbly bottom half from the top half, which is composed of silty gyttja. Both stratigraphic members are polliniferous. Only the upper member is thought to represent the deposits of open Flora Lake, the older deposit having preceded the establishment of a permanent lake. At the very bottom of the core is a thin stratum of different material that is thought to be discontinuously overlain by the pebbly stratum. Pollen spectra from this stratum are not included in Fig. 7.

At the bottom of the pollen diagram is an herb zone closely comparable to zone J of the Imuruk Lake sequence. A short interval of nonpolliniferous sediment overlies the herb zone. There follows a birch zone, but this is interrupted by spectra containing alder or spruce. This interrupted birch zone is shown by radiocarbon dating to have been in existence 5,650 years ago. A second nonpolliniferous section separates the interrupted birch zone from a single surface spectrum containing little birch. The explanation of the

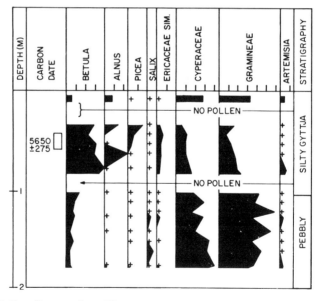

Fig. 7. Pollen diagram from Flora Lake, St. Lawrence Island. This is a hitherto unpublished diagram. It represents the top 1.8 meters of a 2-meter core. The minor elements have not been included. Each division of the diagram represents 10 per cent pollen grains, pteridophyte spores, and *Sphagnum* spores. Plus signs represent 2 per cent or less. The carbon date of 5650 ± 275 is I-993.

sudden changes in the pollen spectra throughout the top meter of the core is probably to be found in the fact that Flora Lake lies on an island. There are no spruce trees or alder bushes on St. Lawrence Island, so that spruce and alder pollen must be blown from sites on the mainland. The incidence of spruce and alder pollen is thus influenced by the vagaries of the wind over the sea. In view of this, the pollen diagram may be thought to conform reasonably well to that from the top meter of the Imuruk Lake core, and to the general trend of pollen diagrams from the adjacent mainland.

GOODNEWS BAY

Three pollen spectra for the Goodnews Bay area are presented as Table 3. The samples were obtained by D. M. Hopkins in 1957 from peat interbedded with lake sediments in a drained kettle in an end moraine of Late Wisconsin age exposed in coastal bluffs 1.5 km north of the mouth of the Salmon River. The samples were analyzed for pollen by E. B. Leopold, and are published here for the first time. The oldest sample has a radiocarbon age of 11,500 years ± 500 years (I-426).

The spectra present some peculiarities, such as the high content of pollen of *Valeriana* in the oldest sample, which may be considered as overrepresentation of plants growing in the peat deposit as it formed. These peculiarities do not mask the essential characteristics of the spectra. The oldest is an herb spectrum, the second a birch spectrum, and the youngest an alder spectrum. The three spectra thus parallel the three-zone sequence at Chandler Lake and zones J, K, and L of the Imuruk Lake sequence. The herb zone was still in existence 11,500 years ago, by which time it had been superseded by birch spectra at Imuruk Lake, in spite of the fact that Goodnews Bay is 600 km south of Imuruk Lake. It is apparent that purely local factors overrode the effects of latitude in determining when dwarf birches reached the Goodnews Bay region. Similar local factors must be responsible for the fact that the high alder content of the youngest sample is not accompanied by spruce, as it is in zone L at Imuruk Lake. The edge of the spruce forest is still 200 km east of Goodnews Bay (Fig. 1), whereas it has approached within 50 km of Imuruk Lake.

The Goodnews Bay spectra take the vegetation record close to the southern margin of the Bering Land Bridge. Apparently, the vegetation near the southern land-bridge coast in late Wisconsin time was comparable to that then prevailing at the latitude of Bering Strait. Alder bushes and spruce trees were remote from both places. Tundra prevailed on the southern coast as well as on the land bridge to the north.

TABLE 3. *Pollen Spectra from Goodnews Bay*[1]

| | Stratigraphic Position of Sample and U.S. Geological Survey Paleobotany Number | | |
| Pollen Taxa | Lowest Sample (11,500 ± 250 Years B.P.; I-426) D-1354-1 | Middle Sample D-1354-2 | Highest Sample D-1354-3 |
|---|---|---|---|
| % *Betula* | 1.6 | 41.2 | 9.1 |
| % *Alnus* | — | — | 17.5 |
| % *Salix* | — | 2.5 | 0.9 |
| % Ericales | 0.5 | 0.8 | 2.7 |
| % Cyperaceae | 1.1 | — | — |
| % Cyperaceae (?) | 13.3 | 5.0 | 4.3 |
| % Gramineae | 16.0 | 6.6 | 57.2 |
| % *Artemisia* | 1.6 | 3.3 | 0.9 |
| % *Dryas* cf. *octopetala* | — | 0.8 | — |
| % Compositae | 2.1 | 2.5 | 1.4 |
| % *Polemonium* | 4.8 | 4.2 | 0.4 |
| % *Polygonum bistorta* | — | 0.8 | — |
| % Verbenaceae | — | 1.7 | — |
| % *Valeriana* cf. *officionalis* | 37.2 | 3.3 | 0.4 |
| % Caryophyllaceae | 6.9 | 2.5 | 0.4 |
| % Boraginaceae | 2.1 | 0.8 | — |
| % *Saxifraga* (?) | 4.8 | 3.3 | — |
| % Onagraceae | 0.5 | 2.5 | 0.4 |
| % Cruciferae | 1.1 | 2.5 | — |
| % *Sagittaria* type | 1.1 | — | — |
| % Rubiaceae | — | 2.5 | — |
| % Ranunculaceae (?) | — | 3.3 | — |
| % Umbelliferae | — | 2.5 | — |
| % Caprifoliaceae (?) | 2.1 | — | — |
| % Undetermined monocots | — | 1.7 | 1.3 |
| % Undetermined dicots | 3.2 | 5.7 | 2.7 |
| Pollen sum | 201 | 136 | 222 |

[1] Results of pollen analysis performed by E. B. Leopold on three samples of interbedded lake and peat sediments in a drained kettle lake. The taxon here identified as Cyperaceae (?) was originally identified as *Populus*, but this now seems to have been in error; the grains most likely belong to Cyperaceae. Figures represent per cent total pollen. Minus signs indicate that the taxon was not observed during the analysis. Pteridophyte spores were not included in the pollen sum and have been omitted from the table.

THE PRIBILOF ISLANDS

Analyses are in progress of a number of lake and bog cores from the Pribilof Islands. These islands are set on the southern edge of the Bering-Chukchi platform, so that they may have been part of the land-bridge coast during glacial maxima. The longest continuous record comes from St. Paul Island, and is in the form of a 14-meter core of sediment that reaches the rock basin of the lake in the crater of the small extinct volcano known as Lake Hill. The lake is 200 meters in diameter, and has no outlet.

Figure 8 is a partial pollen diagram for the top 5.5 meters of the core; radiocarbon dating shows that the pollen diagram spans the last 9,000 years.

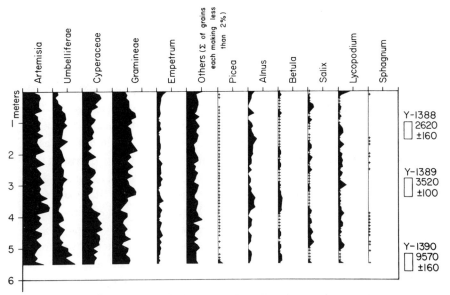

Fig. 8. Pollen diagram from the lake in Lake Hill cone, St. Paul Island, Pribilofs. This is a hitherto unpublished diagram of the results of pollen analysis of the top 5.5 meters of a 14-meter core of lake sediment. The diagram is complete, the minor elements having been added together and represented as a sum in the diagram. Each division represents 10 per cent pollen grains, pteridophyte spores, and *Sphagnum* spores. Plus signs represent 2 per cent or less.

The remaining 8 meters of the core are thought to carry the record back until before the maximum of the Wisconsin Glaciation, but problems of stratigraphy and dating of this part of the core are still under investigation. The pollen record of the bottom 8 meters shows that only tundra was present on St. Paul Island throughout the period recorded.

The pollen spectra of the last 9,000 years fall into a single herb zone, which shows important differences from those of mainland sites. Large amounts of Umbelliferae and *Artemisia* pollen are present in all the spectra. The minor elements—that is, pollen taxa present in amounts of 2 per cent or less—always make up a significant proportion of the total pollen. There are variable though relatively small amounts of alder and birch throughout the diagram.

The spectrum from the surface of the lake bottom conforms to spectra deeper in the core, showing that the pollen sequence can be explained in terms of the modern vegetation of the island. The flora of the Pribilofs is rather well known (Hultén, 1941–50 and personal communication). No species of alder or birch is found on the islands; the alder and birch pollen in the modern deposits has been blown there from continental sites as much as 500 km away. That such long-distance transport should produce a notable proportion of the pollen rain suggests that pollen production by the island vegetation is low. The large amount of minor-element pollen confirms this view.

The Pribilof tundra of mesic sites forms a thick rough turf, of which Umbelliferae (particularly *Angelica*) and *Artemisia* (particularly *A. tilesii* and *A. arctica*) are prominent components. Lupines, valerians, *Rubus*, many saxifrages, and other plants are also important parts of the turf. Grasses and sedges are abundant, but do not dominate the turf to quite the extent that they do on arctic mainland sites. On areas of thin soil there are communities in which *Empetrum nigrum* is prominent. The inside walls of the crater of Lake Hill support both the thick turf and the thin soil communities.

The surface pollen spectra are consistent with the known composition of the St. Paul Island tundra, in general, and with the vegetation of the crater that holds the lake, in particular. The pollen record shows that similar vegetation has been present on St. Paul Island for the last 9,000 years. The peculiar vegetation of the Pribilofs may be attributed to the cold, wet, windy, oceanic climate of the islands. A similar climate for the whole of the last 9,000 years is suggested by the pollen record.

## *Artemisia* Maxima in Arctic Pollen Profiles

A feature of the St. Paul Island pollen diagram is the content of 20 or even 30 per cent of *Artemisia* pollen. This could be explained by the presence of *A. tilesii* and *A. arctica* as common constituents of the mesic tundras of the islands. These species grow in association with umbellifers, and their pollen is accompanied by notable amounts of umbellifer pollen. The climate associated with this *Artemisia*-umbellifer complex is cold and wet.

*Artemisia* maxima are a common feature of the ancient herb spectra of mainland sites. Up to 10 per cent *Artemisia* pollen is present in zone I at Chandler Lake (Fig. 2) ; 5 per cent appears in the oldest samples from Barrow (Table 1) ; over 20 per cent is recorded in zones G and H (Illinoian) of the Imuruk Lake sequence (Fig. 5). These *Artemisia* maxima always appear in herb zones, zones comparable to Livingstone's zone I, indicating that the tree line was remote at the time. The spectra containing the maxima do not include notable amounts of umbellifer pollen as do the modern samples from the Pribilofs, and they also differ in subtler ways from the Pribilof spectra. They have not yet been matched by surface spectra from anywhere in arctic Alaska. Even the modern spectra from Barrow, those on which the interpretation of the herb zones is based, include little or no *Artemisia* pollen. An explanation for the ancient *Artemisia* maxima would add considerably to our understanding of the land-bridge vegetation.

Colinvaux (1964a) concluded that the high *Artemisia* pollen content of certain spectra in the Imuruk Lake core was probably produced by artemisias growing on bare ground produced by ablation or frost action. This conclusion follows the general belief that artemisias are plants of dry sites. The occurrence of notable amounts of *Artemisia* pollen on the Pribilofs under conditions of moist climate does not necessarily invalidate the conclusion, because the pollen spectra of the mainland *Artemisia* maxima are in other ways so different from those of the Pribilofs. The high *Artemisia* content in certain fossil pollen spectra in arctic Siberia have been thought, partly on the basis of other evidence, to reflect a large admixture of steppe elements in certain tundra floras of the past (Giterman and Golubeva, this volume). It still seems most likely that the ancient *Artemisia* pollen maxima in Alaska represent an abundance of dry sites.

## Megafossil Evidence for Postglacial Warming

McCulloch and Hopkins (1966; McCulloch, this volume) have collected evidence for an advance of the tree line on Seward Peninsula and to the north of Kotzebue Sound in the period 10,000 to 8,300 years ago. Dated fossil logs of alder, birch, and poplar; fossil beaver dams; and beaver-chewed wood have been found on Seward Peninsula and the Baldwin Peninsula well to the west of the present position of the tree line. McCulloch and Hopkins conclude that these finds record a westward expansion of the forest that took place in response to the warming that ended the Wisconsin Glaciation. They suggest that the continental climate that prevailed until the final flooding of the Bering Land Bridge allowed the trees to advance beyond their present limits. Once the rising sea had approached its modern level, it produced a cool maritime climate in the area, and the tree line fell back to its present position.

It is of interest to note that this temporary westward extension of the tree line has not so far been detected in pollen diagrams from the region, though the evidence for its occurrence seems conclusive. This may be partly because there is as yet no good detailed postglacial pollen diagram of the Seward Peninsula area (the Imuruk Lake diagram, for instance, covers the interval in only four or five spectra), but it is likely to be partly due to the bluntness of the pollen instrument. The pollen of poplar, which is the tree line on Seward Peninsula, does not appear in pollen diagrams. Tree-birch pollen is masked by shrub-birch pollen. The vagaries of the wind do not allow one to extrapolate the distance of trees from a pollen-collection site with much accuracy. To these difficulties are added the uncertainties of radiocarbon dating of the silty sediment of Alaskan lake deposits. Reliance on megafossils, when they are available, overcomes many of these uncertainties and difficulties.

Another instance in which megafossils have given clearer results than pollen diagrams is Tedrow and Walton's (1964) dating of an advance of alder north of the Brooks Range, 5,650 ± 230 years ago. Such an advance, attributable to the hypsithermal interval, seemed to have been indicated by Livingstone's pollen diagrams (1955), but the pollen evidence had not been decisive. Plant megafossils are widely and well preserved in the frozen sediments of arctic Alaska. More study of the megafossils should strengthen our knowledge of the vegetation history of the region.

## Conclusion

The pollen record suggests that the Bering Strait region has never supported forest, from some time in the interglacial interval that preceded the Illinoian Glaciation until the present. During glacial periods, the tree line receded far to the west and south of its present position.

In the pre-Illinoian interglacial interval, Seward Peninsula supported vegetation similar to that of the present day. The western end of the peninsula supported a shrub tundra, with dwarf birches, and there were alder bushes and spruce trees on the peninsula's eastern base. The tree line fluctuated somewhat. In the succeeding Illinoian Glaciation, all trees were removed far to the east, and dwarf birches were absent or rare. An herbaceous tundra remained, much like the tundra of the modern Alaskan arctic coast at Barrow. However, the high *Artemisia* content is also suggestive of the arid steppe-tundra of Siberia. It is probable that a continuous steppe occupied the land-bridge plains from somewhere east of Seward Peninsula to somewhere in Central Siberia to the west.

These conclusions about Illinoian and pre-Illinoian vegetation derive from the single Imuruk Lake core, and suffer the disadvantage of being unsupported by radiometric dating. For the Sangamon and Wisconsin intervals, we are on surer ground. Sangamon records from Nome and Imuruk Lake agree in suggesting that the vegetation of Seward Peninsula was similar to that of modern times, with a shrub tundra on the western part of the peninsula and trees along its eastern base.

By the time of the Wisconsin maximum, trees had again retreated far to the east of Seward Peninsula, and dwarf birches had become rare. There was an herbaceous tundra like that of modern Barrow; however, it may have had some of the steppe characteristics of Illinoian times. At Barrow itself, a tundra even more arctic than that of the present is indicated. There were probably no dwarf birches or alder bushes north of the Brooks Range.

From late Wisconsin time to the present, pollen diagrams from many parts of arctic Alaska record parallel developments of vegetation. In glacial times, spruce, alder, and dwarf birch were all far to the south or east of their present limits. At each site, the vegetation was comparable to that now found several hundred kilometers farther north. With the end of the Wisconsin Glaciation, there was an advance of trees and shrubs, with dwarf birches always well in front of the alder and spruce lines. This advance was completed some 10,000 years ago on Seward Peninsula, when the tree line had reached or slightly surpassed its present limits. The rise of the sea to within a few meters of its present level then brought a maritime climate to what is now the Alaskan coastal strip, and local adjustments were made to the tree line, which brought it back to about its present position. In mountainous regions, the development of modern vegetation was delayed by the presence of mountain glaciers. There were local small advances of alder during the hypsithermal period.

The whole eastern edge of the Bering-Chukchi platform is still fronted by tundra, which forms a broad strip down the Alaskan coast. Dwarf birches have reached the sea everywhere except on the arctic coastal plain, but alder and spruce are still usually tens or hundreds of kilometers away to the east or south.

The withdrawal of dwarf birches and the tree line in glacial times strongly suggests that the climate was colder than now. Since it is the warmth and duration of the summer rather than the low temperature of winter that is likely to affect the distribution of arctic plants, it may be further inferred that ice-age summers were cooler and shorter than those of modern times. A continental region, as this then was, may be expected to experience seasonal extremes of climate. For summers to have been cooler under continental conditions than under the present maritime regimen seems to suggest an absolute fall in the energy received by the region during the continental period. It seems reasonable to conclude, therefore, that there was a major reduction in the total annual heat budget of the region during glacial times.

The colder climate of glacial times was experienced on the arctic coast no less than on Seward Peninsula or the southern land-bridge plains. It follows that the Arctic Ocean was ice-covered in glacial as well as in interglacial intervals, and hence that the conditions required for the Ewing and Donn theory of ice ages (Ewing and Donn, 1956, 1958; Livingstone, 1959; Colinvaux, 1964b, 1965; Donn and Ewing, 1965) cannot be met.

The vegetation record also shows that the Bering Land Bridge of Illinoian and Wisconsin times had a cold arctic environment.

## REFERENCES

Brown, J. 1965. Radiocarbon dating, Barrow, Alaska: Arctic, v. 18, p. 36–48.

Colinvaux, P. A. 1964a. The environment of the Bering Land Bridge: Ecol. Monogr., v. 34, p. 297–329.

———— 1964b. Origin of ice ages: Pollen evidence from arctic Alaska: Science, v. 145, p. 707–708.

———— 1965. Pollen from Alaska and the origin of ice ages: Science, v. 147, p. 633.

Creager, J. S., and D. A. McManus. 1965. Pleistocene drainage patterns on the floor of the Chukchi Sea: Marine Geology, v. 3, p. 279–290.

Donn, W. L., and M. Ewing. 1965. Pollen from Alaska and the origin of ice ages: Science, v. 147, p. 632–633.

Ewing, M., and W. L. Donn. 1965. A theory of ice ages: Science, v. 123, p. 1061–1066.

———— 1958. A theory of ice ages, II. Science, v. 127, p. 1159–1162.

Heusser, C. J. 1963. Pollen diagrams from Ogotoruk Creek, Cape Thompson, Alaska: Grana Palynologica, v. 4, p. 149–159.

Hopkins, D. M. 1959. History of Imuruk Lake, Seward Peninsula, Alaska: Geol. Soc. America Bull., v. 70, p. 1033–1046.

———— 1963. Geology of the Imuruk Lake area, Seward Peninsula, Alaska: U.S. Geol. Survey Bull. 1141-C, 98 p.

Hopkins, D. M., F. S. MacNeil, and Estella B. Leopold. 1960. The coastal plain at Nome, Alaska: a Late Cenozoic type section for the Bering Strait region: Rept., 21st Internat. Geol. Cong. (Copenhagen), 1960, pt. 4, p. 46–57.

Hultén, E. 1937. Outline of the history of arctic and boreal biota during the Quaternary Period: Bokförlags Aktiebolaget Thule, Stockholm, 168 p.

Hultén, E. 1941–50. Flora of Alaska and Yukon, part 1: Lunds Univ. Arssk., N.F. avd. 2, Band 37, no. 1, p. 1–127; part 2, Band 38, no. 1, p. 132–412; part 3, Band 39, no. 1, p. 415–567; part 4, Band 40, no. 1, p. 571–795; part 5, Band 41, no. 1, p. 799–978; part 6, Band 42, no. 1, p. 981–1066; part 7, Band 43, no. 1, p. 1069–1200; part 8, Band 44, no. 1, p. 1203–1341; part 9, Band 45, no. 1, p. 1342–1482; part 10, Band 46, no. 1, p. 1483–1902.

Hustich, I. 1953. The boreal limits of conifers: Arctic, v. 6, p. 149–162.

Livingstone, D. A. 1955. Some pollen profiles from arctic Alaska: Ecology, v. 36, p. 587–600.

———— 1957. Pollen analysis of valley fill near Umiat, Alaska: Am. Jour. Sci., v. 255, p. 254–260.

———— 1959. Theory of ice ages. Science, v. 129, p. 463–465.

McCulloch, D. S., and D. M. Hopkins. 1966. Evidence for a warm interval 10,000 to 8,300 years ago in northwestern Alaska: Geol. Soc. America Bull., v. 77, p. 1089–1108.

Porter, S. C. 1964. Late Pleistocene glacial chronology of north-central Brooks Range, Alaska: Am. Jour. Sci., v. 262, p. 446–460.

Spetzman, L. A. 1963. Terrain study of Alaska. Part 5. Vegetation: U.S. Army, Office of Chief of Engineers, Engineer Intelligence Study 301.

Tedrow, J. C. F., and G. F. Walton. 1964. Some Quaternary events of northern Alaska: Arctic, v. 17, p. 268–271.

Tikhomirov, B. A. 1962. The treelessness of the tundra. Polar Record, v. 11, p. 24–30.

## 11. Vegetation of Eastern Siberia During the Anthropogene Period

R. E. GITERMAN and L. V. GOLUBEVA
*Geological Institute, Academy of Sciences of the USSR*

This paper reports the results of palynological studies of the Quaternary deposits of Eastern Siberia. Pollen in the Quaternary deposits of Yakutia was analyzed by Giterman, and pollen in the Quaternary deposits of the Baikal region and the southern part of the Central Siberian Plateau was studied by Golubeva. Earlier discussions of the developmental history of the vegetation of Eastern Siberia during the Anthropogene Period include articles by Karavaev (1955), Popova (1955), Puminov (1957, 1959), Baskovich (1959), Vas'kovskii (1959), Grichuk (1959), and Boiarskaia (1961).

Three stages—the Eopleistocene, the Pleistocene, and the Holocene—are distinguished in the Anthropogene developmental history of the vegetation of Eastern Siberia. The Eopleistocene stage is characterized by a wide distribution of coniferous forests containing *Tsuga* (hemlock), exotic species of *Pinus* (pine) and *Picea* (spruce), and several broad-leaved tree species. The Pleistocene stage is characterized by repeated changes of climate—alternations of warm and cold phases—and by alternations between forest and forest-steppe landscapes on the one hand and the periglacial vegetation of the tundra and tundra-steppe landscape on the other. The contemporary vegetation was formed during the Holocene stage.

### The Eopleistocene Stage

Eopleistocene deposits consisting of strongly ferruginous, coarse clastic deposits are exposed in many places in Yakutia. Alluvial deposits of the high terraces of the Lena, Vilyui, and Aldan Rivers are also referred to the Eopleistocene.

Pollen from these deposits indicates that during the first half of the Eopleistocene, Yakutia supported coniferous forests dominated by extant species of *Picea, Pinus,* and *Abies* (fir), but also composed of such Neogene relicts as *Pinus* sect. *strobus, Picea* sect. *omorica, Tsuga,* and *Juglans* (walnut). Thus the first half of the Eopleistocene was a transition period during which the typical Neogene plant formations and their relict species gradually disappeared, while the typical plant formations of the Anthropogene Period had not yet been formed.

Great changes in the vegetation took place in Yakutia during the second half of the Eopleistocene. Light-coniferous taiga, dominated by *Larix* (larch) and *Pinus* and containing the broad-leaved trees *Tilia* (linden) and *Quercus* (oak), grew in the middle valley of the Lena River, but by the end of the Eopleistocene, this vegetation was altered to dark-coniferous taiga dominated by *Picea* and *Abies* and containing only scattered *Tilia.* Light-coniferous forests of *Pinus* and *Larix* became widely distributed in northern Yakutia. The climate during the second half of the Eeopleistocene was warmer than at present; the vegetation zonation was similar to that of the present time, but the forest boundaries were shifted northward.

Paleobotanical material has been obtained in the southern part of Eastern Siberia from Eopleistocene deposits that are widely distributed in the intermontane depressions of the Tunkin region. At the beginning of the Eopleistocene, these deposits record a coniferous forest dominated by *Picea* and containing *Tsuga* and *Pinus. Tsuga* was represented by at least three species and occupied an important position in the vegetation. The pines played a smaller role, but species of the sections *Eupitys, Cembrae,* and *Strobus* were present. *Juglans* was present in small quantities, as were the broad-leaved trees *Ulmus* (elm), *Tilia,* and *Quercus. Myrica* is also represented. Judging by the composition of the vegetation, we would conclude that the climate of the southern part of Eastern Siberia was warm and humid during early Eopleistocene time.

The coniferous forests with prominent *Tsuga* in the south of Eastern Siberia were altered to birch forests at the end of the first half of the Eopleistocene. *Tsuga* was still present, but only in considerably diminished abundance; *Ulmus, Tilia, Quercus,* and *Carpinus* (hornbeam) persisted in the most favorable sites. The birch forests were sparse in character, and alternated with steppes composed of *Artemisia* (sage), Chenopodiaceae, grasses, and various herbs. These changes in the vegetation indicate a shift toward a colder and dryer climate.

During the second half of the Eopleistocene, dark-coniferous forests containing relict species of pine and the broad-leaved *Quercus, Tilia,* and *Ulmus* again became widely distributed in the southern part of Eastern Siberia, including the Central Siberian Plateau, the Baikal region, and the region of the Tunkin depression. The character of the vegetation was different in western Transbaikalia;[a] birch and pine forest with an oak component predominated in the Chikoy River basin, and forest-steppe and steppe predominated in the Selenga River basin.

The paleobotanical data indicate that large ice sheets did not exist during the Eopleistocene. The Eopleistocene glaciation of Eastern Siberia indicated by some authors (Kuprina, 1958; Vas'kovskii, 1959) evidently was only of alpine character, and its influence on the vegetation was insignificant.

## The Pleistocene Stage

In contrast to the Eopleistocene, the Pleistocene stage is characterized by repeated changes of climate and vegetation. Four cold phases and three warm phases are distinguished in the developmental history of the vegetation. The cold phases correspond to the Samarov, Tazov, Zyrianka, and Sartan Glaciations, and the warm phases to the Messov, Kazantsevo, and Karginsky Interglaciations. The vegetation of the glacial and interglacial intervals will be discussed separately.

### VEGETATION OF GLACIAL INTERVALS

Deposits of glacial age are exposed in many sections in Eastern Siberia. Moraines, glaciofluvial deposits, and lacustrine sediments are found in the glaciated areas, and the periglacial zone contains alluvium recording aggradation and an enlargement of areas of floodplain and channel facies. Lenticular lake deposits are widely distributed. The periglacial deposits characteristically contain various forms of ground ice, the melting out of which has resulted in cryogenic structures.

Three phases in the development of the vegetation can be distinguished within a typical glacial interval. During the first phase, when the glaciers are growing, tundra and forest-tundra are developed. Dwarf birch becomes rather widely distributed, and there is a wide development of grass and herb assemblages and of *Sphagnum* and sedge bogs. During the second phase, the xerophytes—*Artemisia*, Chenopodiaceae, and *Ephedra*—become widely developed. *Selaginella sibirica* (Milde) Hieron. is characteristic of this phase in the eastern regions of Siberia. The climate is cold and dry.

---

[a] Siberian regions are often described according to their location relative to the European part of the USSR. Thus, Transbaikalia is the region east of Lake Baikal, and Pribaikalia is the region west of the lake.

During the third phase, which coincides with the melting of the glaciers, dwarf birch increases in abundance, as do *Sphagnum* and the green mosses. The role of forest vegetation becomes greater. The climate becomes more humid during this third phase than during the second, but remains drier than during the first phase.

Pollen diagrams from most geological sections contain only the first and second phases, evidently because of the erosion that took place at the end of glacial episodes and at the beginnings of interglaciations. Consequently, the third phase can be found only in sections where there are deposits representing the transition from a glaciation to an interglaciation.

The Far East is a special paleobotanical province in which these phases cannot be distinguished. A dry phase, assigning strong roles to xerophytic assemblages, is lacking in the vegetation sequences of glacial periods in this province, evidently because of the monsoonal maritime climate.

Abundant paleobotanical material representing the Pleistocene stage has now accumulated in Siberia, permitting a detailed description of the history of vegetation during the Samarov and Zyrianka Glaciations. In the non-glacial area, the deposits of the Samarov (Maximum) Glaciation include the alluvium and lacustrine-alluvial sediments making up Terrace IV and sometimes Terrace V of the great rivers, the loess on the surface of the high terraces, and the lacustrine deposits in the intermontane depressions. The Khazar mammalian faunal complex has been found in deposits that correspond to the Samarov Glaciation. The deposits correlated with the Zyrianka Glaciation consist chiefly of alluvial sediments of Terrace II of the great rivers, but locally include the upper part of the alluvium of Terrace III and loess on the surface of the high terraces.

During the first phase of the Samarov Glaciation, the vegetation zonation (Fig. 1) differed from the zonation during the pre-Samarov period in that forests were widely distributed. The tundra and forest-tundra boundaries were shifted southward. Open, parklike woodland and muskegs became widely distributed in areas previously occupied by forests. The role of the steppes decreased.

During the maximum development of the Samarov Glaciation—the second phase in the vegetation development (Fig. 2)—the climate became dryer and the role of xerophytes increased, as they shifted northward to regions previously occupied by tundra and forest-tundra. The unique periglacial vegetation was formed as *Artemisia* assemblages containing Chenopodiaceae, *Ephedra*, and *Selaginella sibirica* became widely distributed in tundra regions. Periglacial steppes were also widely distributed. It is possible that isolated areas of light-coniferous forest were present along the river

## EXPLANATION FOR FIGURES 1-3

**VEGETATION**

Arctic tundra

Arctic tundra with xerophytes

Tundra-steppe

Alpine tundra

Interspersed alpine tundra and valley glaciers

Interspersed alpine tundra and open woodland in mountainous areas

Interspersed alpine tundra and open *Larix* woodland in mountainous areas

Forest tundra

Open woodland and muskeg

Open dark-coniferous woodland

Mixed coniferous and birch forest

Interspersed periglacial steppe and open woodland

Interspersed open forest and steppe in mountainous areas

Cold steppe and periglacial steppe

- - - - - - - Shoreline

Political boundary

Limit of continental glaciation

**PLANT SYMBOLS**

*Betula* sec. *Albae*

*B.* sec. *Nanae*

*Alnaster*

*Larix*

*Picea*

*Pinus silvestris*

*Pinus pumila*

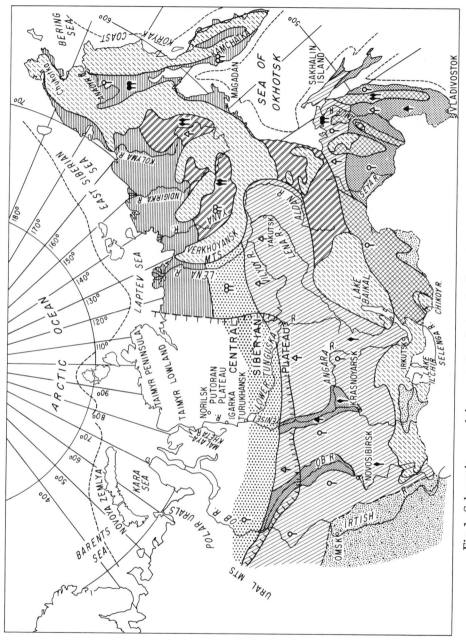

Fig. 1. Schematic map of the vegetation during the first phase of the Samarov Glaciation. See explanation of keys on p. 236.

237

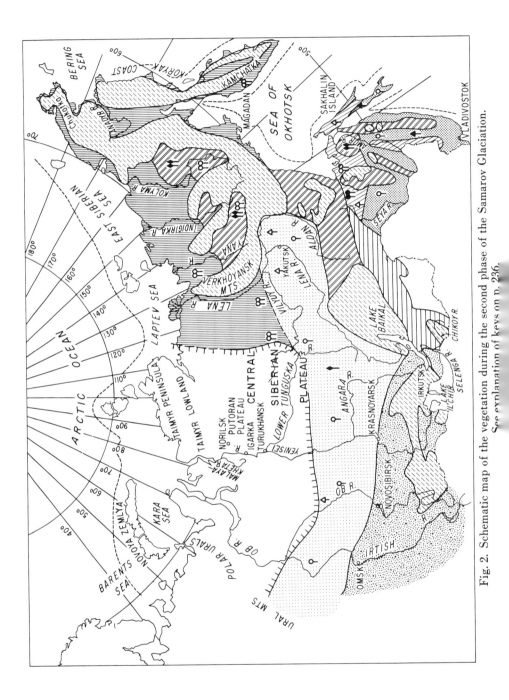

Fig. 2. Schematic map of the vegetation during the second phase of the Samarov Glaciation. See explanation of keys on p. 236.

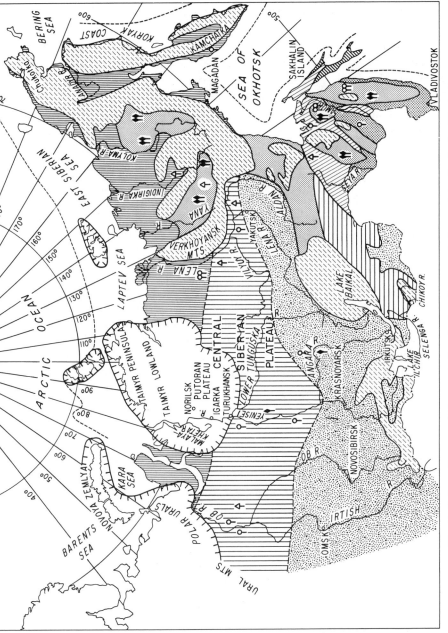

Fig. 3. Schematic map of the vegetation during the second phase of the Zyrianka Glaciation. See explanation of keys on p. 236.

239

valleys. In the Far East—the coastal region, Priamurya, and Sakhalin—the more arid climate resulted only in a somewhat restricted distribution of forests and in an increase in the role of birch.

Analogous phases in the history of the vegetation cover have been established for the Tazov and Zyrianka Glaciations (Golubeva and Ravskii, 1964; Golubeva, 1964). But the vegetation during the Late Pleistocene Zyrianka Glaciation differed somewhat from the vegetation during the older glaciations in that there was a more general merging of species. As a consequence of an increase in continentality, the climate became dryer (Zeitlin, 1964). The forest vegetation underwent considerable degradation, but the xerophytes reached their widest distribution, especially during the second phase.

During the first phase of the Zyrianka Glaciation, as during Samarov time, the boundaries of tundra and forest interfingered far to the south, and still farther south there were only islands of coniferous forest and open, parklike woodland.

Tundra-like landscapes continued to exist in the north and northeast during the second phase (Fig. 3), which began at the time the Zyrianka glaciers reached maximum distribution, but they contained a considerable admixture of xerophytes. The central regions were occupied by tundra-steppes; concentrations of xerophytes predominated, but tundra elements were prominent. The distribution of forest vegetation was reduced to a minimum; birch and larch grew only in the most favorable ecological niches. *Selaginella sibirica* was prominent in the vegetative cover (Giterman, 1963).

Periglacial steppes were widely developed in the southern regions, and their boundaries were shifted northward from those of the present-day steppes, especially in Central Siberia, where periglacial steppes invaded the headwaters of the Lower Tunguska River. *Artemisia* assemblages containing Chenopodiaceae, *Ephedra*, Plumbaginaceae, and *Selaginella sibirica* were widely developed. In the Far East, birch forests and open, parklike woodland dominated during the second phase of the Zyrianka Glaciation, as they had during the second phase of Samarov time.

The Sartan Glaciation—the last Late Pleistocene glaciation—was of alpine character and restricted distribution; it is represented in unglaciated regions by the alluvial sediments of Terrace I of the rivers in East Siberia. In contrast with the preceding glacial intervals, phases in the development of the vegetation were not established during the Sartan Glaciation; instead, the vegetative cover of different regions showed a marked provincialism. Periglacial steppes were developed in the south. Xerophytic conditions existed in the intermontane depressions along the middle course of the

Chikoya River, and the vegetative cover there consisted mainly of Cheno-podiaceae, *Artemisia*, and *Ephedra*. *Ephedra* was represented by no fewer than five species, although at the present time only one species (*E. mono-sperma* C. A. M.) is present there. Among the Chenopodiaceae, *Kochia prostrata* (L.) Schrad. and *Eurotia ceratoides* (L.) C. A. M. were present. Periglacial steppes were also developed on the eastern bank of Lake Baikal (Vipper, 1962) ; they extended northward and merged with the cold steppes of Yakutia.

VEGETATION OF THE INTERGLACIAL INTERVALS

Three phases are distinguished in the developmental history of a typical interglacial interval. The climate of the first phase is moderately cold and relatively dry; the second is moderately warm and moist; and the third phase is moderately cold and moist. The optimal climates during the inter-glacial intervals of East Siberia were the moderately warm and moist cli-mates of the second phase, during which taiga became widely distributed. Unfortunately, one almost never finds stratigraphic sections providing a complete record of an interglacial interval, because there were erosional episodes at the ends of the glaciations and during the beginnings of the interglaciations. The sediments of the interglacial alluvial terraces were de-posited during the second half of the interglaciations, beginning at the end of the second climatic phase and extending through the third phase.

Paleobotanical materials characterizing the vegetation during the Ka-zantsevo Interglaciation are now abundant. This seems to have been the warmest of the Pleistocene interglaciations, and characteristic phases and floristic provinces can be distinguished for it. During the first phase of the Kazantsevo Interglaciation, swamp-tundra and forest-tundra developed in northern regions. Open, parklike pine-birch and spruce-birch forests with some larch grew in the central regions. Steppes predominated in the south, but pine-spruce forests with some birch were present in the valleys.

During the second, optimal phase of the Kazantsevo Interglaciation, three floristic provinces can be distinguished. The first was represented by an area of dark-coniferous forests distributed across Western and Central Siberia. The second province, Eastern Siberia, was characterized by light-coniferous forests dominated by larch and pine. In the south, however, the complexion of the vegetation varied according to relief and slope exposure; dark-coniferous forest grew on the northern slopes, whereas the southern slopes supported larch forest and steppe. The Far East, including Sakhalin Island, constituted a third floristic province characterized by coniferous forests containing broad-leaved species; the mixed coniferous and broad-leaved forests of eastern Transbaikalia were part of this province.

The paleobotany of the other interglaciations of Eastern Siberia has received little study. Deposits of the Messov Interglaciation are found in the Baikal region and in central Yakutia. Forests of larch containing birch flourished on slopes, and dark-coniferous taiga was present in the valleys in these regions during the second half of the Messov Interglaciation. Forests of larch, birch, and pine were distributed in Central Siberia, with spruce in the river valleys. Forest tundra and open forests flourished in the northern regions.

Deposits of the Karginsky Interglaciation are exposed in several places in the Baikal region and in central Yakutia; these are characterized by light-coniferous forests of larch and pine and by birch forests.

## Holocene Stage

The Holocene stage is represented by the deposits of the upper and lower floodplain terraces of the great rivers of Eastern Siberia. Four phases are distinguished in the developmental history of the vegetation during the Holocene. The first is transitional between the vegetation of the late Sartan cold period and that of the climatic optimum; forests of larch, birch, and pine were widely distributed, and pine-spruce forests grew in the valleys. The second and third phases represent the climatic optimum, during which the climate was warmer than that of the present day. During the climatic optimum, the boundary of the forest zone extended well north of the present forest boundaries, and fingers of forest penetrated far into the unforested tundra. Coniferous forests with some *Abies*, *Pinus sibirica*, and *Picea* developed in the central regions of Eastern Siberia. Timberline was higher in mountainous areas; for example, pine and larch forest including considerable Siberian pine (*Pinus sibirica*) was present in the region of Lake Il'chir in eastern Sayanakh west of the south end of Lake Baikal, where only alpine tundra grows today. Pollen evidence indicates that the broadleaved species *Quercus mongolica* Fisch. and *Ulmus pumila* L. migrated into Pribaikalia and western Transbaikalia from the east and southeast at this time. Oak is now completely absent from this region, and elm occurs only in isolated forests in the valleys of western Transbaikalia.

The fourth phase represents the period during which the contemporary vegetation was formed.

## Conclusions

A comparison of our paleobotanical data on the periglacial vegetation in Siberia with descriptions of the vegetation during Pleistocene glaciations in Europe and America shows certain common features. The periglacial landscapes that prevailed during glacial episodes were very similar across great distances. The principal difference is that forest vegetation was more widely developed during glacial episodes in Europe and North America than in Siberia, evidently because of the intense continentality of the glacial climates in Siberia.

The vegetation during interglacial epochs in Eastern Siberia differed considerably both in its floristic composition and in the character of the vegetation formations from the interglacial vegetation of Europe and perhaps of America (paleobotanical material representing interglaciations is very rare in America). Broad-leaved deciduous tree species played only an inconsequential role in the interglacial vegetation formations of Eastern Siberia, and that only in the southern regions, such as Transbaikalia and Pribaikalia.

The stages and phases that have been distinguished in the history of the development of vegetation during the Anthropogene Period in Eastern Siberia, when coupled with geological and faunal data, are very useful in the stratigraphic subdivision of Quaternary deposits.

## REFERENCES

Baskovich, R. A. 1959. Sporovo-pyl'tsevye kompleksy chetvertichnykh otlozhenii severo-vostoka SSSR (Spore-pollen complexes in the Quaternary deposits of northeastern USSR), p. 434–450 in Tr. Mezvedomstv. soveshchaniia po razrabotke unifitsirovannykh stratigraficheskikh skhem severo-vostoka SSSR, 1957g (Proc. Interdepartmental conference for the development of a unified stratigraphic scheme for northeastern USSR, 1957) : Magadan, 483 p.
Boiarskaia, T. D. 1961. K voprosu o razvitii rastitel'nosti basseina Angary v chetvertichnom periode (On the question of the development of the vegetation of the Angara River basin during the Quaternary Period), p. 160–173 *in* Paleogeografiia chetvertichnogo perioda SSSR (Paleogeography of the Quaternary Period in the USSR) : Izd.-vo. Moskva, Gosudarstv. Univ., 213 p.
Giterman, R. E. 1963. Etapy razvitiia chetvertichnoi rastitel'nosti Yakutii i ikh znachenie dlya stratigrafii (Stages in the development of the Quaternary vegetation of Yakutia, and its significance for stratigraphy): Akad. Nauk SSSR, Tr. Geol. Inst., Vyp. 78, 192 p.
Golubeva, L. V. 1964. O tipakh perigliatsial'noi rastitel'nosti pleistotsena Vostochnoi Sibiri (Types of Pleistocene periglacial vegetation in Eastern Siberia) : Akad. Nauk SSSR, Doklady v. 155, no. 4, p. 810–813. In Russian; English translation p. 74–77 in running transl. pub. by Am. Geol. Inst.
Golubeva, L. V., and E. N. Ravskii. 1964. O klimaticheskikh fazakh vremeni zyrianskogo oledeneniia Vostochnoi Sibiri (On the climatic phases of the time of the Zyrianka Glaciation in Eastern Siberia) : Akad. Nauk SSSR, Kom. Izucheniia Chetvertichnogo Perioda, Bull. 29, p. 132–148.
Grichuk, M. P. 1959. Rezul'taty paleobotanicheskogo izucheniia chetvertichnykh otlozhenii Priangar'ia (The results of paleobotanical study of the Quaternary

deposits in Priangarie), p. 442–497 *in* K. K. Markov and A. J. Popov, eds., Lednikovyi period na territorii Evropeiskoi chasti SSSR i Sibiri (Ice age in the European section of the USSR and in Siberia): Moskva Gosudarstv. Univ., Geogr. Fak., 560 p.

Karavaev, M. I. 1955. Paleogeograficheskaia rekonstruktsiia landshchaftov Tsentral'no-Iakutskoi ravniny v kainozoe (Paleogeographic reconstruction of the landscape of central Yakutian lowland in the Cenozoic): Dokl. Akad. Nauk SSSR, v. 102, no. 4, p. 797–799.

Kuprina, N. P. 1958. Novye dannye ob oledenenii Zapadnogo Verkhoian'ia (New data on the glaciation of western Verkhoiansk): Akad. Nauk SSSR, Doklady, v. 121, p. 1071–1074. In Russian; English translation p. 673–676 in running transl. of Geol. Sci. Sect. pub. by Am. Geol. Inst.

Popova, A. I. 1955. Sporovo-pyl'tsevye spektry chetvertichnykh otlozhenii Tsentral'noi Yakutii v sviazi s istoriei razvitiia rastitel'nosti v posletretichnoe vremia (Spore-pollen complexes from the Quaternary deposits of central Yakutia, in connection with the history of the development of vegetation in post-Tertiary time): Akad. Nauk SSSR, Tr. In-ta Biol. Yakutsk Fil., Bull. 1, p. 136–146.

Puminov, A. P. 1957. Osnovnye etapy istorii rastitel-nosti v verkhov'iakh rek Olenek i Markhi v chetvertichnom periode (Fundamental stages in the history of the vegetation in the upper Olenek and Markhi drainage in the Quaternary Period), p. 76–80 *in* Sbornik statei po paleontologii i biostratigrafii NIIGA (Symposium articles on the paleontology and biostratigraphy of NIIGA), Bull. 2, 80 p. (Leningrad).

———— 1959. K istorii razvitiia rastitel'nosti na Severo-Vostoke Srednesibirskogo ploskogor'ia v poslezyrianskoe vremia (On the history of the development of vegetation of the northeastern Central Siberia Plateau in post-Zyriansk time): Trudy Nauchno-Issle. Inst. Geol. Arktiki, v. 102, Vyp. 10, p. 138–151.

Vas'kovskii, A. P. 1959. Kratkii ocherk rastitel'nosti, klimata i khronologii chetvertichnogo perioda v verkhov'iakh rek Kolymy, Indigirki i na severnom poberezh'e Okhotskogo moria (A brief essay on the vegetation, climate, and chronology of the Quaternary Period in the upper Kolyma and Indigirka Rivers and on the northern coast of the Sea of Okhotsk), p. 510– 556 *in* K. K. Markov and A. J. Popov, eds., Lednikovyi period na territorii Evropeiskoi chasti SSSR i Sibiri (Ice age in the European section of the USSR and in Siberia): Moskva Gosudarstv. Univ. Geogr. Fak., 560 p.

Vipper, P. B. 1962. Poslelednikovaia istoriia landschaftov v Zabaikal'e (Postglacial history of the landscape of Transbaikalia): Akad. Nauk SSSR, Doklady, v. 145, no. 4, p. 871–874. In Russian; English translation p. 82–84 in running transl. pub. by Am. Geol. Inst.

Zeitlin, S. M. 1964. Sopostavlenie chetvertichnykh otlozhenii lednikovoi i vnelednikovoi zon Tsentral'noi Sibiri (bassein Nizhnei Tunguski) (A comparison of the Quaternary deposits of the glaciated and nonglaciated zones of Central Siberia [Lower Tunguska Basin]): Akad. Nauk SSSR, Tr. Geol. Inst., vyp. 100, 187 p.

## 12. Distribution, Ecology, and Cytology of the Ogotoruk Creek Flora and the History of Beringia

ALBERT W. JOHNSON
*San Diego State College*

JOHN G. PACKER
*University of Alberta*

We are concerned here with the problem of the extent to which our present knowledge of the western arctic angiospermous flora can contribute to an understanding of the history of Beringia and, more specifically, to an understanding of the kinds of conditions that existed when the land bridge between Asia and North America allowed migrations of plants and animals between the two continents. We emphasize an approach that makes use of inferences based on the genetic and evolutionary history of modern species by analysis of their chromosome numbers. By combining this information with other data on the geographical and ecological distributions of species in the angiospermous flora of the Ogotoruk Creek–Cape Thompson area of Alaska, we reach conclusions that can be reconciled with the more direct evidence on the Cenozoic history of Beringia available from the physical sciences.

During the past 30 to 40 years, research in the cytology of arctic plants and in chromosome geography has resulted in the discovery that a considerable number of species consist of two or more chromosome races; i.e., they have populations with different chromosome numbers. As yet, taxonomists have not reached a consensus on how these entities should be handled, but regardless of their taxonomic status, the phytogeographer can hardly ignore the fact that they exist and that they may have had very different histories, reflected at least partially in their different distributions at the present time. Unfortunately, relatively little information is available on the distribution of most of these chromosome races, primarily because of the regional imbalance in cytological knowledge of arctic floras (Packer, 1964). This is particularly unfortunate, in the context of this paper, because the majority of these taxa are probably of recent origin; and their distribution, especially in relation to the Bering Strait, could tell us more about the environment of the land bridge in recent times. Because distribution data are unavailable, we have omitted from consideration 37 taxa having two or more chromosome races, at least one of which occurs in the Ogotoruk Creek flora. We believe, however, that sufficient other data are available to provide insight into the history of the Beringian flora and to allow an evaluation of the importance of this type of phytogeographical study.

## Origin and History of the Flora

Before examining the Ogotoruk Creek flora in detail, we shall consider those aspects of the origin and history of the arctic-alpine tundra that are relevant to our analysis of its modern features.

Although the fossil history of the arctic tundra flora is relatively poorly known, there are a number of reasons for supposing that it originated during the Tertiary. Wolfe and Leopold (this volume) describe sea-level boreal forest floras at 60° N. during the late Miocene. Since latitudinal and altitudinal zonation of climates, and presumably vegetation, have been characteristic of the earth for as long as paleoclimatic conditions can be detected, it is reasonable to assume that colder climates must have prevailed at higher latitudes and altitudes. Axelrod (1964), for example, shows altitudinal stratification in the Miocene of the western United States. He describes a montane coniferous forest and infers alpine habitats with increasing altitude. The distribution of the extant arctic-alpine flora (Hultén, 1937, 1958; Weber, 1965), when considered along with modern concepts of rates of migration and evolution in the angiosperms, is best explained by supposing an essentially circumpolar arctic-alpine tundra flora during the Tertiary upon which Pleistocene events have been superimposed.

That tundra plants have not been found in Tertiary deposits is negative and therefore insubstantial evidence for concluding that they did not exist at that time. To us it seems more reasonable to argue that the herbaceous nature and upland existence of the early tundra plants did not favor their fossilization.

Because the fossil record is of so little help in discovering the origin of the tundra flora, it is necessary to use the modern flora, especially its distribution and ecology, to formulate some general hypotheses. On such grounds, Tolmatchev (1959) thinks that while much of what is now arctic tundra was still forest-covered, some herbaceous species had probably become adapted to specialized habitats, such as uplands, bogs, snow beds, and coastal strands. In a sense, these species were preadapted to full arctic conditions, and Tolmatchev believes they became widespread following the retreat of the forest. Löve (1959) suggests that another group of species adjusted to changing environments *in situ*, and became components of the arctic flora. Species of genera whose major distribution is still associated with the forest—e.g., *Ledum, Empetrum, Rubus, Linnaea,* and *Adoxa*—are probably included in that group. Last, alpine species finding congenial conditions at lower altitudes on mountain slopes migrated into the Arctic along the north-south-trending mountain ranges.

The precise timing of these events is admittedly speculative, and in all likelihood was spread over millions of years. Nevertheless, there is consensus among arctic botanists that, by the beginning of the Pleistocene, floras occupying tundra habitats were essentially modern, and the evidence from distribution of modern tundra plants supports the idea of a diverse origin for the tundra flora.

## The Beringian Flora and the Pleistocene

If the present arctic flora is essentially Tertiary in origin, then some of its species must have survived the Pleistocene glaciations in high-latitude refugia. Evidence for this conclusion is based on palynological and paleobotanical data, on distributional data of the present arctic flora, and on cytological data.

### PALEOBOTANICAL EVIDENCE FOR SURVIVAL

The history of late Tertiary and early Quaternary arctic floras is relatively unknown for several reasons. First, the predominantly herbaceous tundra vegetation did not leave a significant fossil record, or, if it did, there has not yet been enough palynological work in the Arctic to describe it; second, Quaternary glaciations eliminated deposits in which much of the record may have been preserved; and third, fossil pollens of this kind of vegetation are difficult to interpret, for the reasons discussed below. As a result, the only data that pertain to the Quaternary are from relatively late bog deposits, most of which are of Wisconsin and post-Wisconsin age. This is especially true of palynological data from the western Arctic (see Colinvaux, this volume).

In our opinion, conclusions based on nonarboreal pollens in arctic palynological spectra must still be regarded as highly speculative. The ecological conditions that supported the predominantly herbaceous tundra flora cannot be reconstructed from palynological data alone, because pollens of sedges, grasses, willows, tundra forbs, and other plants whose identifications are usually generic or familial are not reliable ecological indicators, except in a most general way. What these data suggest and what is relevant to our discussion here is that climatic changes associated with glacial and interglacial periods have apparently never been of sufficient amplitude to eliminate higher plants from unglaciated parts of Beringia.

## DISTRIBUTIONAL EVIDENCE FOR SURVIVAL

In 1937, Hultén proposed his theory of progressive equiformal areas to explain the distribution of northern plants. He showed that when modern plant distributions are mapped, the ranges of individual species overlap in areas that may represent refugia to which these species were restricted during glacial advances and from which they migrated when climates ameliorated. The Bering Strait area figures most importantly in Hultén's work because much of it is unglaciated and because his analyses show that "most arctic and numerous boreal plants radiated from the Bering Sea area" in interglacial and postglacial periods. Hultén concluded that the precursors of many modern arctic species already existed at high latitudes during the Tertiary period, although Quaternary events have modified their distribution patterns in various ways. Distributional information that has become available since 1937 (Hultén, 1958, 1962) apparently has not changed Hultén's basic conclusions. Subsequent work by arctic taxonomists and phytogeographers has modified some details of Hultén's specific conclusions (Raup, 1947; Porsild, 1955), but his general theory is accepted by most arctic biologists.

Ideas about the persistence of plants and animals in unglaciated refugia during the Pleistocene have occupied an important place in northern biogeography for several decades (Fernald, 1925; Hultén, 1937; Raup, 1941; Dahl, 1946, 1955; Heusser, 1954; Porsild, 1955; Löve and Löve, 1956; Tolmatchev, 1959; Savile, 1960; Packer, 1963; and Ball, 1963). Survival of at least some higher plants and animals in refugial areas during the Pleistocene glaciations is accepted by the vast majority of biogeographers, and disjunct distribution patterns and the distribution of endemics provide a substantial basis for this opinion.

## CYTOLOGICAL EVIDENCE FOR SURVIVAL

During the past decade, the importance of cytological data in studies of historical plant geography has been clearly demonstrated (Favarger, 1954, 1958, 1961; Favarger and Contandriopoulos, 1961). Early studies of the geographical distribution of polyploids seemed to indicate a proportional relationship between the severity of the environment and the frequency of polyploidy. It now appears, however, that the level of polyploidy in floras is not directly related to the "severity" of the environment but to the historical development of the flora.

This change of thought is largely the result of studies of floras that appear to have been relatively unaffected by Pleistocene climatic changes and that apparently represent the lineal descendents of Tertiary floras that existed in the same area. Examples are the floras of the Canary Islands (Larsen, 1960), of the Algerian Sahara (Reese, 1957), of the island of Sakhalin (Sokolovskaia, 1960), and of the unglaciated parts of the Alps in Switzerland (Favarger, 1954). All these floras possess relatively low frequencies of polyploidy. The correlation between largely unmodified Tertiary floras and relatively low frequencies of polyploidy is so great as to be beyond coincidence. Thus, an arctic flora occupying an unglaciated area relatively unaffected by Quaternary events and possessing a relatively low frequency of polyploidy may represent an extant Tertiary flora. The flora of Ogotoruk Creek in unglaciated northwestern Alaska fits these criteria (Johnson and Packer, 1965), and it must have existed throughout the Pleistocene much as it is today; it offers excellent support for the view that many arctic species existed at high latitudes throughout the Pleistocene.

## The Ogotoruk Creek Flora

We now consider the angiospermous flora of the Ogotoruk Creek–Cape Thompson area of northwestern Alaska (lat. 68° 06′ N.: long. 165° 46′ W.) and its relationship to the cytological and phytogeographical problems raised here. This flora is more or less typical of the western Arctic, and lies well within Hultén's Beringian refugium (Hultén, 1937). The Ogotoruk Creek–Cape Thompson flora consists of about 300 species found within an area of 40 square miles. The flora and vegetation of the Ogotoruk Creek–Cape Thompson area have been described by Johnson *et al.* (1966), and plant distribution has been correlated with environmental gradients. Species diversity is highest on well-drained slopes and uplands at Ogotoruk Creek, although plant cover is most complete on bottomlands near sea level where several species of *Carex* and *Eriophorum* dominate.

Johnson and Packer (1965) have reported on the chromosome numbers of the angiosperm flora at Ogotoruk Creek (Table 1). Of particular interest is that the frequency of polyploidy among the flowering plants is lower than has been determined for floras from comparable latitudes in Europe. We relate this to the difference in the glacial history of the two areas.

The problem of the age of floras can also be approached through the relationship between polyploidy and ecology. Recently, Pignatti (1960) and

TABLE 1. *Polyploid Frequencies in the Ogotoruk Creek Flora*[1]

| Group | Species (all sources) | Species Known Cytologically Total (all sources) | From Ogotoruk Creek | Diploids[2] No. | Per Cent | Polyploids[2] No. | Per Cent |
|---|---|---|---|---|---|---|---|
| Monocotyledons | 52 | 52 | 40 | 8(9) | 15.4(17.3) | 44(43) | 84.6(82.7) |
| Dicotyledons | 214 | 206 | 194 | 97(105) | 47.1(51.0) | 109(101) | 52.9(49.0) |
| Total | 266 | 258 | 234 | 105(114) | 40.7(44.2) | 153(144) | 59.3(55.8) |

[1] Values in parentheses calculated by considering ambiguous cases to be diploids.
[2] Frequency calculations of diploids and polyploids include non–Ogotoruk Creek data.

Johnson and Packer (1965) have demonstrated that polyploid plants are not distributed randomly where there is significant habitat diversity. At Ogotoruk Creek, for example, the highest frequency of diploid species occurs on the slopes and uplands, which are the warmest, driest, and most stable of all plant habitats in the area. We also regard these as the oldest habitats, or those least modified by the events of the Pleistocene. Apparently, the diploid species survived in these relatively stable upland habitats to which they became adapted in Tertiary or Quaternary times, and few have been able to invade the colder, wetter, and more highly disturbed habitats of the lowlands, which are of more recent origin and where polyploid species occur in their highest frequencies.

### THE FAVARGER POLYPLOID SPECTRUM METHOD

Favarger's method of using polyploid frequencies in studies of the history and evolution of floras is only one application of such data to these problems. In an important paper, Favarger (1961) points out that while comparisons of polyploid frequencies in different floras are useful, this method fails to take into account the fact that polyploids are of varying age. Favarger's "polyploid spectrum" method attempts to correct this deficiency by establishing criteria for distinguishing, at least in a general way, the relative ages of certain types of cytological categories. In descending order from oldest to youngest, these categories are diploids and paleopolyploids, mesopolyploids, neopolyploids, and chromosome races. These he defines as follows.

Paleopolyploids are species with high chromosome number, suggesting polyploidy, but in which the diploid ancestors have apparently become extinct. Examples are monotypic polyploid genera lacking close affinity with other genera; certain genera in which all the species are polyploid and lack close relatives; or polyploid species forming a particular section of a genus in which the diploid species, if they occur, are classified in another section and differ too markedly to be considered ancestors of the polyploids.

Mesopolyploids are "good" polyploid species in which the diploid ancestral types or diploid types derived from the ancestral diploids may be found in the same genus as the polyploid or in a closely related genus.

Neopolyploids are polyploid forms or races of a "Linnaean" species that in a given territory have supplanted the diploid form. Favarger states by way of explanation that polyploid races of a species tend to eliminate the ancestral diploids from the territory they occupy, so that in a given area the species will eventually be represented only by the polyploid taxon.

Chromosome races are races of a species not possessing any significant degree of morphological divergence. The important difference between chromosome races and neopolyploids is that, in the former, diploids and polyploids occur together in the same area. Because the chromosome race has not yet eliminated its diploid ancestor, it is regarded as being of more recent origin than the neopolyploid.

Comparisons of the frequencies of these five cytological categories should provide information on differences in the evolution of different floras. Favarger (1961) has provided an example (Table 2) in his comparison of the flora of certain parts of the Swiss Alps with that of southwestern Greenland. Favarger interprets the high frequency in the Alps of diploid and paleopolyploid species, which he regards as ancient taxa that probably originated in the early Tertiary, as evidence of great antiquity of the Alpine flora. The contrast in the frequency of these ancient elements and of mesopolyploids in the Alpine and Greenland floras is interpreted by Favarger as indicating that the Alpine flora has enjoyed greater environmental stability through the Tertiary, and that evolution has occurred gradually and often at the diploid level. The flora of southwestern Greenland is much higher in mesopolyploids, presumed to date from middle to late Tertiary time; this reflects a much less stable Tertiary environment, one

TABLE 2. *Comparison of the Polyploid Spectra of the Angiosperm Floras of Certain Swiss Alps and of Southwest Greenland*[1]

|  | Swiss Alps |  | Southwest Greenland |  |
| --- | --- | --- | --- | --- |
| Number of species ................ | 220 | | 259 | |
| Number of species in study ......... | 204 | | 238 | |
| Diploids ........................ | 53% | 57% | 36% | 38% |
| Paleopolyploids ................... | 4% | | 2% | |
| Mesopolyploids ................... | 31% | | 48% | |
| Neopolyploids .................... | 7% | 12% | 10% | 14% |
| Chromosome races ................ | 5% | | 4% | |

[1] After Favarger (1961).

in which Favarger envisions frequent mingling of floras as a result of climatic change and variation in other environmental factors. Both floras reflect Pleistocene environmental upheavals in their similar frequencies of neopolyploids and chromosome races, the most recently formed polyploid taxa.

Some comment is required on Favarger's definition of the polyploid groups and their interpretation in his polyploid spectrum approach. Favarger states that the separation of diploid and paleopolyploid taxa is the most difficult to make. In *Saxifraga*, for example, Hamel (1953) and others have inferred from the basic numbers occurring in related genera that the basic number $x = 13$ is a polyploid derivative of diploid taxa, with $x = 6$ and $x = 7$.[a] Favarger, however, treats them as diploids, just as he does basic numbers from 12 to 18 in *Minuartia* and 18 in *Cerastium*. His reason is that these stand as diploid numbers in relation to polyploid series subsequently evolving in these genera. Basically, we agree with Favarger, though so far as we can see there could be little objection to treating them as paleopolyploids. Like Favarger, we think that both diploids and paleopolyploids can be taken as indications of antiquity in a flora, perhaps dating from the early to middle Tertiary, as he suggests.

Although the separation of diploids and paleopolyploids may present problems, we believe that the separation of neopolyploids and chromosome races is even more difficult. Favarger's prime criterion for distinguishing them is that, in time, polyploid races tend to eliminate their diploid ancestors from the area they occupy. These polyploid taxa are the neopolyploids. As long as the ancestral diploids and derived polyploids persist in the same area, however, they are called chromosome races. A difficulty here, as Favarger himself notes, is that polyploids often exploit habitats different from those of their ancestors, and so could not eliminate the diploids through competition. Still another problem arises in cases where the polyploid migrates into areas never occupied by its diploid ancestors. It is clear that there is no objective way to decide when derived polyploids are sufficiently sympatric with their diploid ancestors to call them chromosome races or sufficiently allopatric to call them neopolyploids.

According to Favarger, neopolyploids and chromosome races are of Quaternary age, largely contemporaneous with the glaciations. Thus, these infraspecific polyploids can be regarded as an indication of youth in floras in which they occur. In practice, Favarger treats neopolyploids and chromo-

---

[a] The recent discovery of the chromosome number of $2n = 12$ in *Saxifraga eschscholtzii* (Johnson and Packer, unpublished) confirms this. In this paper, however, we regard species of *Saxifraga* with counts of $2n = 26$ as diploids.

some races collectively, the joint total being taken as evidence of Pleistocene disturbance. No attempt is made to separate them in age because the method is not sufficiently precise. For all of these reasons then, we include all infraspecific polyploids in the neopolyploid category.

Finally, in classifying taxa as chromosome races, Favarger includes not only the sympatric polyploid or polyploids but the diploid as well. Logically, the status of the diploid should not be changed in these circumstances; the recently evolved races cannot have had any effect upon the age or earlier history of the diploid. At Ogotoruk Creek, *Thalictrum alpinum* is represented by the diploid with $2n = 14$ and also a chromosome race with $2n = 21$. The $2n = 14$ taxon we classify as a diploid—the taxon with $2n = 21$ is the chromosome race.

We now consider the mesopolyploids, which are thought by Favarger to have originated in the middle to late Tertiary. In some ways, this is the easiest group to deal with, because those polyploids that do not fall into other categories must belong here. In our opinion, however, the separation of mesopolyploids from neopolyploids is the most difficult problem of all.

Favarger assumes that mesopolyploids can be distinguished from their diploid ancestors by obvious morphological differences resulting from differentiation over a long time. Many mesopolyploids are placed here with no difficulty, but because evolutionary rates are not the same for all plant groups, Favarger's criteria must not be applied blindly. Genera in which evolution seems to have occurred rapidly must be contrasted with those in which all available evidence suggests slower rates of evolution. In other words, the age assignments applied to polyploid categories are relative only within any clearly defined phylogenetic line. Thus, although some mesopolyploids may have originated in the Tertiary as Favarger suggests, others have probably evolved more recently; and this is where problems arise in attempting to separate mesopolyploids and neopolyploids. Within a line where evolutionary change proceeds slowly, for example, mesopolyploids that have not diverged markedly from the parental diploids would be called neopolyploids. As more information accumulates on relationships and rates of evolution within lines, the treatment of the cytological categories should become more uniform.

## THE POLYPLOID SPECTRUM OF THE OGOTORUK CREEK FLORA

Using the criteria described above, we have calculated the polyploid spectrum of the Ogotoruk Creek flora (Table 3). We have omitted the family Cyperaceae from our calculations because of the difficulty of in-

TABLE 3. *The Polyploid Spectrum of the Ogotoruk Creek Flora*

| Category | Number | Per Cent |
|---|---|---|
| Diploid/paleopolyploid .................. | 96 | 43 |
| Mesopolyploid ......................... | 101 | 46 |
| Neopolyploid/chromosome race .......... | 24 | 11 |
| Total ............................ | 221 | 100 |

terpreting polyploidy in this family. Also, as indicated previously, we have not included 37 taxa for which cytological and distributional data are insufficient for analysis.

About 90 per cent of the flora is divided evenly between two major categories—the combined diploid-paleopolyploid group and the mesopolyploid group. The remainder consists of the more recently evolved neopolyploid-chromosome race group. Our interpretation of the ages of each of these groups follows Favarger, at least in broad outline.

PHYTOGEOGRAPHY AND CYTOLOGY

The taxa discussed by Hultén (1937) in his analysis of refugial centers in Beringia include most of those at Ogotoruk Creek, and of the 40 distribution patterns he proposes, 22 apply to the Ogotoruk Creek flora. Distributional data collected since 1937 require that some taxa be reassigned to categories other than those originally proposed by Hultén. His 22 categories are here combined into four more inclusive ones; namely, oceanic plants (here primarily strand species) (Fig. 1), arctic plants (those that occur only in the Arctic) (Fig. 2), montane plants (those with one or more alpine wings to their distributions) (Fig. 3), and boreal plants (belonging primarily to the boreal forest, but entering at least the low Arctic) (Fig. 4). Using this system, about 50 per cent of the Ogotoruk Creek angiosperms belong to the montane group, between 30 and 35 per cent are arctic species, about 10 per cent are boreal, and the remaining five per cent are strand species (Table 4).

TABLE 4. *Number of Diploid and Polyploid Species in Phytogeographical Groups*

| Group | Diploids | Paleo-polyploids | Meso-polyploids | Neo-polyploids | Cyper-aceae | Unknown | Unassigned Local Chromo-some Races | Total |
|---|---|---|---|---|---|---|---|---|
| Oceanic .......... | 10 | 1 | 3 | | 1 | 2 | | 17 |
| Arctic ........... | 26 | 4 | 37 | | 8 | 6 | 3 | 84 |
| Montane ......... | 44 | 2 | 33 | 1 | 15 | 14 | 22 | 131 |
| Boreal ........... | 6 | 5 | 4 | 1 | 1 | 4 | 10 | 31 |
| Total ........ | 86 | 12 | 77 | 2 | 25 | 26 | 35 | 263 |

We have excluded two (not entirely mutually exclusive) groups of plants from this analysis. The first of these consists of the 37 local chromosome races to which we have referred previously. In this case, the excluded taxa consist of two or more chromosome races, and since the separate distributions of both races are not known completely, we prefer not to assign either to a geographical group. For example, *Saxifraga hirculis* is known to consist of a diploid race with $2n = 16$, and a polyploid race with $2n = 32$. The $2n = 16$ race is apparently arctic-montane, being known from the Colorado Rockies, from the Pamirs, and from Ogotoruk Creek. The $2n = 32$ race occurs in the Eurasian Arctic and from the Canadian Arctic and Greenland. From what is known, we could assign the diploid race to the montane category and the tetraploid race to the arctic group. But relatively few counts of either race have been made, so we do not feel justified in placing them in any category.

The other group excluded from Table 4 consists of 11 species of mixed geographical relationships that Hultén (1937) includes in his Group 12, "Plastic American Plants Radiating From the Yukon Valley." Some of the plants are Alaskan endemics, some range rather widely and enter the northern Rockies, whereas still others seem to be primarily boreal plants. Four of the 11 species are diploids, five are mesopolyploids, and two are included in the unassigned chromosome race group.

In Table 5 we have calculated the frequency of diploids and polyploid types for each geographic category. The frequencies differ among the four groups, as would be expected if they represent elements in the flora that have different histories.

The data for the oceanic/strand group, which has a very high percentage of diploids, suggest that this is probably a rather ancient element and that the strand habitat retains its basic characteristics irrespective of climatic changes. The essential similarity of coastal strands in widely different climatic zones is reflected in the widespread distribution of many strand species. Environmental constancy and the ease of migration—such that different physiological races of strand taxa at different latitudes could always "keep pace" with climatic fluctuations — have effectively minimized the

TABLE 5. *Percentages of Diploids and Polyploid Types in Phytogeographical Groups*

| Group | Diploids | Paleo-polyploids | Meso-polyploids | Neo-polyploids |
|---|---|---|---|---|
| Oceanic | 71 | 7 | 21 | 0 |
| Arctic | 39 | 6 | 55 | 0 |
| Montane | 55 | 2 | 41 | 1 |
| Boreal | 38 | 31 | 25 | 6 |

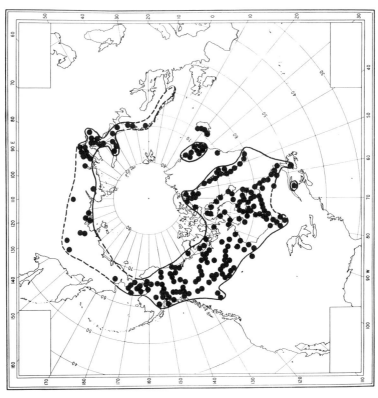

Fig. 2. Distribution of *Pyrola grandiflora* Radius as an example of an arctic distribution pattern (after Hultén, 1958, p. 143).

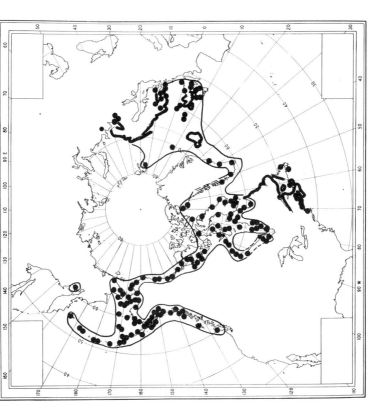

Fig. 1. Distribution of *Mertensia maritima* (L.) S. F. Gray as an example of a maritime distribution pattern (after Hultén, 1958, p. 295).

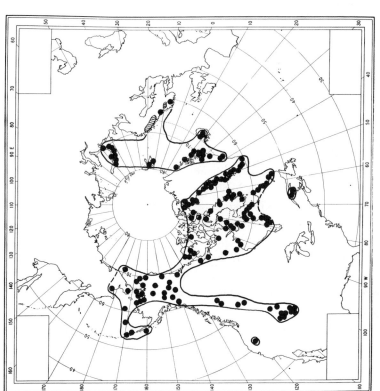

Fig. 3. Distribution of *Campanula uniflora* L. as an example of an arctic montane distribution pattern (after Hultén, 1958, p. 193).

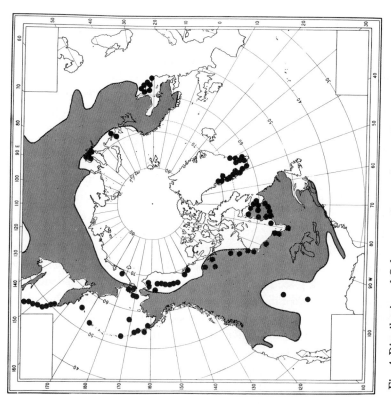

Fig. 4. Distribution of *Calamagrostis canadensis* (Michx.) Nutt. sens. lat., as example of boreal distribution pattern (after Hultén, 1962, p. 189).

adaptive advantages associated with polyploidy. We suspect that the strand species were components of the Arcto-Tertiary flora that adjusted to the changing climatic conditions *in situ*, edaphic and biotic conditions being very little altered. It is interesting to note that Hultén (1937), on the basis of distributional evidence alone, suggests that *Montia lamprosperma*, a component of the strand flora at Ogotoruk Creek, which is widely distributed in the Northern Hemisphere, is probably a very ancient taxon which must have experienced the vicissitudes of earlier periods. Our data entirely support this statement with regard to *Montia lamprosperma* and for the strand flora generally.

The interpretation of the arctic and montane groups is best undertaken on a comparative basis. The almost reversed frequencies of diploids and mesopolyploids in these two elements suggest that the development of the arctic element took place in circumstances of greater climatic and environmental disturbance than did the montane. If this is correct, it is difficult to see how these two elements could have coexisted during their early development. It is our tentative conclusion that they did not.

The arctic element (see Fig. 2) is largely restricted to the Arctic, and we suggest that it probably represents that component of the Arcto-Tertiary geoflora which adapted in the Arctic to the Tertiary climatic deterioration that eventually gave rise to tundra environments. The taxa of the montane element (see Fig. 3), in contrast to those of the arctic element, occur also in southern mountains. The high diploid and lower mesopolyploid frequency suggest that this element differentiated in rather more stable circumstances than the Arctic. Favarger (1961), on the basis of his comparison of the flora of southwest Greenland with that of the alpines in the Swiss Alps, suggests that the middle to late Tertiary was a time of greater climatic fluctuation and environmental disturbance in the Arctic than in the Alps of Europe. If this is correct—and we have no evidence to contradict it—the frequencies of polyploids in the montane element could be accounted for by its differentiation in southerly mountains and its subsequent migration into the Arctic, as appropriate habitats developed or became accessible. On the basis of our present information, this is the most satisfactory explanation we can furnish to account for the differences in the arctic and montane elements.

It is perhaps significant that the frequencies in the different polyploid categories in our montane element and in the alpine flora investigated by Favarger (1961) are rather similar. The somewhat higher frequency of mesopolyploids in the montane element is consistent with the greater instability of environmental conditions at higher latitudes in the later Tertiary, and would be expected in north–south-trending mountain ranges running up into the Arctic.

The boreal element in the Ogotoruk Creek flora is really too small for meaningful analysis. Historically, its closest affinity is with the arctic element. We suspect that most species of this element adapted *in situ* to the changing Tertiary conditions and that it represents a nonmigratory element in the arctic flora. What cytological evidence there is tends to support this broad conclusion, the frequency of diploids being about the same as in the arctic element. The high frequency of paleopolyploids in this element and the higher frequency of paleopolyploids in the arctic as compared with the montane are not inconsistent with our suggestion that they adapted from the preexisting arctic flora.

Despite the impossibility of assigning absolute ages to individual species or groups of species, the cytological, taxonomic, and geographic data on these species lead to the conclusion that the modern arctic flora is a composite of species of various ages and origins. The generalizations that speciation is most likely to occur during periods of climatic and geological instability, and that hybridization and polyploidy are important evolutionary mechanisms during such periods, are established axioms of higher-plant biology. While the fossil record provides the best kind of evidence for the origin and evolution of species, it is of little help in understanding the modern arctic flora.

In the absence of any evidence to the contrary, arctic-plant biologists are agreed that the processes that led to the formation of arctic species began during the Tertiary at high altitudes and latitudes in the Northern Hemisphere. The recent evidence (Wolfe and Leopold, this volume) that Tertiary temperate floras in Alaska are younger than was formerly believed implies that habitats and climates that might be considered tundra-like probably did not appear in Alaska until later than had been thought, probably during the late Miocene or early Pliocene. It is suggested that the diploid and paleopolyploid species discussed here evolved during this time, and although they have no doubt changed genetically since that time, evolution by polyploidy has not occurred. Mesopolyploids and neopolyploids have evolved more recently. Favarger thinks that most mesopolyploids originated in the Tertiary, whereas neopolyploids are of Quaternary age. It must be admitted that there is no direct evidence to support any age assignment to these groups, and, for at least the present, these suggestions are rather tentative.

We now return to the question posed earlier: Do the phytogeographical relationships of the Ogotoruk Creek taxa give us any clue to the effectiveness of the Bering Land Bridge as a barrier to plant migration, thereby allowing, from a knowledge of the normal habitats of successful and unsuccessful migrants, conclusions to be drawn regarding land bridge environments? For at least 90 per cent of the Ogotoruk Creek angiosperm flora, the answer to this question seems to be No. Most species in this flora have an amphi-Beringian or circumpolar distribution; one can only suggest that the barrier was crossed, but little can be said about how, when, or where. Species with a markedly asymmetrical amphi-Beringian distribution, *Koeleria asiatica*, for example, may indicate recent crossing, but alternatively may indicate the near-elimination of previously widespread taxa on one side or the other of the Bering Strait. In fact, disjunct distribution patterns are most often interpreted in this way. In the absence of good evidence for long-distance dispersal (Dahl, 1963; Löve, 1963), the ecological diversity of taxa with these distributions suggests that the Bering Land Bridge has, at one time or other (not necessarily contemporaneously), supported a wide range of tundra habitats. Further evidence for this conclusion is to be found in the diverse ecological preferences of the taxa that Hultén (1937) thinks radiated from Beringia during interglacial periods and in postglacial times.

About 25 species that occur on the American side of the Chukchi Sea are not known on the Asian side. These taxa, all angiosperms, and all recorded from the Ogotoruk Creek–Cape Thompson area, are as follows:

*Festuca baffinensis* Polunin
*Poa brachyanthera* Hultén
*Puccinellia vaginata* (Lge.) Fern. & Weatherby
*Carex microchaeta* Holm
*Carex ramenskii* Kom.
*Kobresia hyperborea* Porsild
*Zygadenus elegans* Pursh
*Salix brachycarpa* Nutt. subsp. *niphoclada* Argus
*Anemone multiceps* (Greene) Standley
*Smelowskia borealis* Drury and Rollins var. *jordalii* Drury and Rollins
*Saxifraga reflexa* Hook.
*Saxifraga tricuspidata* Rottb.
*Boykinia richardsonii* Rothr.
*Astragalus polaris* (Seem.) Benth.
*Lupinus arcticus* S. Wats.
*Hedysarum* sp.
*Bupleurum americanum* Coult. and Rose
*Gentiana propinqua* Richards.
*Eritrichium splendens* Kearney
*Castilleja pallida* (L.) Spreng. subsp. *caudata* Pennell
*Artemisia arctica* Less. subsp. *comata* (Rydb.) Hultén
*Erigeron hyperboreus* Greene
*Senecio lugens* Richards.
*Senecio conterminus* Greenm.
*Taraxacum phymatocarpum* J. Vahl

A note of caution is entered here, however, because some of these have close relatives in Asia that may, upon closer scrutiny, turn out to be indistinguishable taxa. There is a similar group of about equal size of species on the Asian side that do not occur in North America, but these are not considered here.

There are several possible explanations for the absence of these species in Asia. First, it is conceivable that they have reached the Alaskan coast only recently and have not had an opportunity to reach Asia; second, they may have been eliminated on the Asian side; and third, the nature of the land bridge may have prevented their crossing it. Regardless of which of the three explanations is most likely, it is probable that each of them applies in specific cases. Regarding the last of the three possibilities, it is interesting to note that nearly all of the taxa listed in the tabulation are plants of the slopes and upland. Taxa such as *Poa brachyanthera* and *Anemone multiceps* are never found on bottomland soils. If the Quaternary Bering Land Bridge had the same monotonous topography that the submerged Bering and Chukchi shelf has today, one might expect that it would have supported the kinds of species that occupy the coastal lowlands today. The contemporary coastal lowlands contain wet meadow and tussock plants that tend to form a nearly closed vegetation, and upland plants do not occur in such habitats. On the other hand, none of the species growing in wet meadow and tussock habitats at Ogotoruk Creek is restricted to the American side of the Chukchi Sea. This suggests that there may have been an ecological control over the occupants of the land bridge, during at least the later part of its subaerial history.

## Summary and Conclusions

The arctic flora came into existence as a characteristic part of the world's flora during the late Tertiary, and by the beginning of the Quaternary a predominantly herbaceous perennial flora similar in composition to modern arctic floras existed throughout the North. Available fossil evidence and inferences based on the distribution and ecology of modern species suggest that the arctic flora was derived from species that had been associated with the high-latitude Arcto-Tertiary forest. They adapted themselves to local arctic-like conditions before they became widespread, from species that evolved *in situ* as the climate deteriorated during the late Tertiary, and from species that migrated into the Arctic, which were essentially pre-adapted to arctic habitats by virtue of alpine adaptations.

The influences of Quaternary events on this flora were not uniform.

Although great portions of northern regions were glaciated, many areas were not, and palynological, phytogeographical, and cytological evidence points to the persistence of a flora much like that of today in unglaciated refugia, such as parts of Beringia. In the broader context of Beringian history, the significance of these observations resides in the fact that they indicate that the land adjacent to and including much of the Bering Land Bridge supported a tundra flora growing under the influence of tundra climates throughout the Quaternary.

Our use of cytological data to supplement the distributional and fossil information on the arctic flora is based on the realization that, in all biological interpretations and reconstructions, precise identification of the organisms involved is essential. The use of chromosome data, like other properties of plants, constitutes one more characteristic that can be used in identification procedures. But chromosome statistics may offer greater precision than other characteristics in interpreting the *history* of species, particularly where polyploidy is involved. In most cases of polyploidy examined thus far, it is believed that the high chromosome numbers in a polyploid series are derived from the lower numbers. Moreover, it becomes increasingly clear that polyploid species are especially likely to arise and persist during periods of climatic and physiographic disturbance, when new habitats are available for occupation. Thus, the diploid/polyploid ratio in a flora gives us a crude measure of the conditions to which species of that flora have been exposed during its history. According to this reasoning, a flora containing a relatively large proportion of diploid species suggests a history of evolution in stable or slowly changing conditions, while the opposite is true for floras consisting predominantly of polyploid species.

Also, polyploids themselves have arisen at different times, and the degree to which they have diverged from their parental species and the extent to which they are distributed geographically provide measures of their relative age. Favarger (1961) has used these criteria for establishing several age categories of polyploids. A polyploid that has diverged very little from its ancestors and occupies a restricted geographical area must be relatively young, as compared, for example, with one that is clearly separated from all other related species and has spread to a circumpolar distribution. Favarger's method is the most objective yet proposed to analyze the age of polyploids, and its use here is a test of its applicability to arctic problems.

Our analyses of the Ogotoruk Creek–Cape Thompson flora suggest that it consists of four major geographical elements: an oceanic element that is, in general, very widespread along northern strands; a montane element that may have evolved in mountainous regions far to the south, in North America or Eurasia; an arctic element that probably evolved and remained in the North; and a boreal element whose primary distribution is still associated with the boreal forest. On the basis of the polyploid spectrum within each of these elements, it appears that the oceanic and montane elements, both high in diploids, have evolved under relatively stable conditions, whereas the high frequency of polyploid species in the arctic and boreal elements suggests conditions of relative instability favoring the survival of newly formed polyploid species.

Nearly all of the data available to us show that the Bering Strait has not been an effective barrier to arctic tundra plant dispersal. More than 90 per cent of the 300 species of angiospermous plants at Cape Thompson occur in Asia as well as in North America, although the incompleteness of the cytological data does not allow us to conclude that every one of these species is uniform cytologically throughout its range. In some cases, we are sure that they are not. On the other hand, the dissimilarity between the arboreal floras of Asia and western North America suggests that the Bering Strait has prevented a commingling of forest species between the two continents for a relatively long time, probably since the Miocene or early Pliocene. In view of the frequent occurrence of a land bridge in the region during Quaternary times, it must be assumed that the barrier that prevented forest species from occupying the land-bridge surface was primarily climatic in nature.

Based on evidence that tundra climates have persisted in the amphi-Beringian area during all of the Quaternary Period, and on our understanding of the plant ecology of the western Arctic, supplemented by the physical data that may be used to reconstruct the Quaternary Bering Land Bridge surface, we envision a flat-to-slightly-undulating tundra plain whose primary plant constituents were lowland species—cottongrass, sedges, grasses, and moisture-tolerant herbs and low shrubs. The fact that the relatively small group of species (about 25) restricted to the American side of the Chukchi Sea is made up of upland species that never occur in meadowy habitats could be interpreted as supporting this reconstruction.

The historical plant geographer must work with what remains from the past. We have attempted here to illustrate how a synthesis of physical and biological data may lead to meaningful interpretations of past events in the area around the Bering Land Bridge. We conclude that the present terrestrial habitats and their biota in this area are reasonably representative of Quaternary Beringian landscapes.

## REFERENCES

Axelrod, D. I. 1964. The Miocene Trapper Creek flora of southern Idaho: Univ. Calif. Pub. Geol. Sci., v. 51, 181 p.

Ball, G. E. 1963. The distribution of the species of the subgenus *Cryobius* (Coleoptera, Carabidae, *Pterostichus*), in J. L. Gressitt, ed., Pacific Basin biogeography: Bishop Mus. Press, Honolulu, p. 133–151.

Dahl, E. 1946. On different types of unglaciated areas during the ice ages and their significance to phytogeography: New Phytologist, v. 45, p. 225–242.

———— 1955. Biogeographic and geologic indications of unglaciated areas in Scandinavia during the glacial ages: Geol. Soc. America Bull., v. 66, p. 1499–1519.

Dahl, E. 1963. Plant migrations across the North Atlantic Ocean and their importance for the paleogeography of the region, p. 173–188 *in* A. Löve and D. Löve, eds., North Atlantic biota and their history: Pergamon Press Ltd., Oxford; Macmillan Co., New York.

Favarger, C. 1954. Sur le pourcentage des polyploides dans la flore de l'étage nival des Alpes Suisses: 8th Internat. Bot. Cong. (Paris), Rap. et Commun. Sects. 9 and 10, p. 51–56.

———— 1958. Contribution à l'étude cytologique des genres *Androsace* et *Gregoria*: Veröffentlichungen Geobot. Inst. Rübel Zurich, v. 33, p. 59–80.

———— 1961. Sur l'emploi des nombres de chromosomes en géographie botanique historique: Zurich Geobot. Inst. Eidgenoessische Tech. Hochschule, Stiftung Rübel, Bericht, v. 32, p. 119–146.

Favarger, C., and J. Contandriopoulos. 1961. Essai sur l'endémisme: Bull. Soc. Bot. Suisse, v. 71, p. 384–408.

Fernald, M. L. 1925. Persistence of plants in unglaciated areas of boreal America: Mem. Am. Acad. Arts & Sci., v. 15, 237–342.

Hamel, J. L. 1953. Contribution à l'étude cytotaxinomique des Saxifragacées: Rev. Cytologie Biol. Végétales, v. 14, p. 113–313.

Heusser, C. J. 1954. Nunatak flora of the Juneau ice field, Alaska: Torrey Bot. Club Bull., v. 81, p. 236–250.

Hultén, E. 1937. Outline of the history of arctic and boreal biota during the Quaternary period: Stockholm, Bokförlags Aktiebolaget Thule, 168 p.

———— 1958. The amphi-atlantic plants and their phytogeographical connection: Kgl. Svenska Vetenskapsakademiens Handlingar, Fjärde Ser. 4, v. 7, no. 1, p. 1–340.

———— 1962. The circumpolar plants. I. Vascular cryptogams, conifers, monocotyledons: Svenska Veteskapsakadamiens Handlingar, Ser. 4, bd. 8, nr. 5, 275 p.

Johnson, A. W., and J. G. Packer. 1965. Polyploidy and environment in arctic Alaska: Science, v. 148, p. 237–239.

Johnson, A. W., L. A. Viereck, R. E. Johnson, and H. Melchior. 1966. Vegetation and flora, p. 277–354 *in* N. J. Wilimovsky, ed., Environment of the Cape Thompson region, Alaska: U.S. Atomic Energy Comm. Pub. PNE-481.

Larsen, J. 1960. Cytological and experimental studies on the flowering plants of the Canary Islands: Kgl. Danski Vidensk. Selsk. Biol. Skr., v. 11, p. 1–60.

Löve, A. 1959. Origin of the Arctic flora, *in* Problems of the Pleistocene and Arctic: McGill Univ. Mus. Pubs. no. 1, p. 82–95.

Löve, A., and D. Löve. 1956. Cytotaxonomical conspectus of the Icelandic flora: Acta Horti Gotoburgensis, v. 20, p. 65–291.

Löve, D. 1963. Dispersal and survival of plants, p. 189–205 *in* A. Löve and D. Löve, eds., North Atlantic biota and their history: Pergamon Press Ltd., Oxford; Macmillan Co., New York.

Packer, J. G. 1963. The taxonomy of some North American species of *Chryso-splenium* L. Section *Alternifolia* Franchet, Canadian Jour. Bot., v. 41, p. 85–103.

———— 1964. Chromosome numbers and taxonomic notes on western Canadian and arctic plants: Canadian Jour. Bot., v. 42, p. 473–494.

Pignatti, S. 1960. Il significato delle specie poliploidi nelle associazioni vegetali: Ist. Veneto, Cl. Sci. Nat., Atti, no. 118, p. 75–98.

Porsild, A. E. 1955. Vascular plants of the western Canadian arctic archipelago: Nat. Mus. Canada Bull, no. 135, 226 p.

Raup, H. M. 1941. Botanical problems in boreal America: Bot. Rev., v. 7, p. 147–248.

———— 1947. The botany of southwestern Mackenzie: Sargentia, v. 6, 275 p.

Reese, G. 1957. Über die polyploidiespektren in der nordsaharischen Wüsten-flora: Flora, v. 144, p. 598–634.

Savile, D. B. A. 1960. Limitations of the competitive exclusion principle: Science, v. 132, p. 1761.

Sokolovskaia, A. P. 1960. Geograficheskoie rasprostranenie poliploidnykh vidov rastenii (issledovanie flory o. Sakhalina): Vestnik Leningrad Univ., Ser. Biol. 4, no. 21, p. 42–58.

Tolmatchev, A. I. 1959. Der autochthone Grundstock der arktischen Flora und ihre Beziehung zu den Hochgebirgsfloren Nord- und Zentralasiens: Bot. Tidsskr., v. 55, p. 269–276.

Weber, W. A. 1965. Plant geography in the southern Rocky Mountains, p. 453–468 *in* H. E. Wright, Jr., and D. G. Frey, eds., The Quaternary of the United States: Princeton Univ. Press, Princeton, N.J.

# 13. Mammal Remains of Pre-Wisconsin Age in Alaska

T. L. PÉWÉ
*Department of Geology, Arizona State University, and*
*U.S. Geological Survey, Tempe, Arizona*

D. M. HOPKINS
*U.S. Geological Survey, Menlo Park, California*

Alaska, like northern Siberia, has long been famous for the abundant remains of extinct Pleistocene mammals that are found in frozen unconsolidated deposits along major rivers and in the valleys of many minor streams. The earliest account of these fossils seems to be that of Kotzebue (1821, p. 218–220), who had found abundant vertebrate remains at Elephant Point in Eschscholtz Bay during his expedition to Chukchi Sea in 1816. Because early explorers reported a great abundance of bones, several expeditions (Maddren, 1905; Gilmore, 1908; Quackenbush, 1909) were conducted in Alaska in the hope of finding complete skeletons or perhaps even frozen carcasses comparable with those that had been found in Siberia. When large-scale gold-mining operations were initiated in the Fairbanks district in 1928, extensive fossil collecting was undertaken there and elsewhere in Alaska by the late O. W. Geist and other collectors on behalf of the American Museum of Natural History. Part of this material has been described by Frick (1937) and by Skinner and Kaisen (1947). More recently, Geist and other workers have collected vertebrate fossils from the Fairbanks area and northern Alaska for the Museum of the University of Alaska, but systematic descriptions of this material have not yet been prepared.

Unfortunately, most of the Alaskan Pleistocene vertebrate material now in museums has no known stratigraphic context. Fossils are most readily seen when the enclosing matrix has been washed away, and consequently most of the vertebrate fossils have been found lying loose on ocean and lake beaches, on river banks, and on the floors of placer-mining excavations. Furthermore, detailed studies of Quaternary stratigraphy were not initiated in Alaska until the late 1940's, so that even when bones were col-

Publication authorized by the Director, U.S. Geological Survey.

lected *in situ* in earlier years, no conclusions could be drawn concerning the age of the enclosing matrix. Skinner and Kaisen's thorough study of fossil *Bison* is based upon Alaskan material whose stratigraphic context is not known; their speculations on *Bison* evolution are therefore based entirely upon the morphology of the skeletal material and upon assumptions of probable trends with time of changes in such critical dimensions as horn width and tooth size.

The important finds of Siberian animals such as *Bos* (yak) and *Saiga* (steppe antelope) near Fairbanks (Frick, 1937) cannot be related with certainty to any particular stratigraphic unit, and consequently we cannot be sure when, within middle or late Pleistocene time, these animals ranged eastward into Alaska. However, Péwé concludes from a study of mining-company records that most of the *Bos* and *Saiga* material described by Frick came from silt of Wisconsin age.

During the two decades since detailed studies of Quaternary stratigraphy began in Alaska, a modest but growing number of vertebrate fossils have been found in significant stratigraphic contexts. Most of these have been found in the Fairbanks area, but a few have been found in other parts of Alaska. Most were in beds of Wisconsin and Recent age, but a significant few were in beds of Sangamon, Illinoian, and possibly pre-Illinoian age. The pre-Wisconsin fossils are of special interest, because some of these may represent taxa that arrived in Alaska long before their earliest appearances in the conterminous United States. The following checklist contains all of the mammalian fossils found thus far in Alaska in beds recognized as being of pre-Wisconsin age. The evidence upon which stratigraphic dating is based is discussed as follows: Fairbanks area—Péwé (1952, 1965); Nenana Valley—Wahrhaftig (1958); Chandalar Valley—Williams (1962); Nome area—Hopkins *et al.* (1960); Kotzebue Sound area—McCulloch *et al.* (1965); Alaska generally—Péwé *et al.* (1965).

*Checklist of Mammals Found in Beds of Pre-Wisconsin Age in Alaska*

| Order and Species | Age and Area Found |
|---|---|
| Order Rodentia: | |
| *Castor* sp. (beaver) | Illinoian, Fairbanks[a] |
| *Ondrata zibethecus* (muskrat) | Late Illinoian or Sangamon, Kotzebue Sound[b] |
| *Dicrostonyx torquatus* (lemming) | Pre-Illinoian interglaciation (?), Kotzebue Sound[c] |
| Order Carnivora: | |
| *Xenocyon* sp. (hunting dog) | Illinoian, Fairbanks[d] |
| *Canis* sp. (wolf) | Illinoian, Fairbanks[e] |
| *Vulpes* sp. (fox) | Illinoian, Fairbanks[e] |
| *Felis* sp. (large cat) | Illinoian, Fairbanks[e] |
| Order Proboscidea: | |
| *Mammut* sp. (mastodon) | Illinoian, Fairbanks[e] |

| Order and Species | Age and Area Found |
|---|---|
| Order Proboscidea (cont.): | |
| Mammuthus sp. (mammoth) | Illinoian, Nenana Valley;[f] Illinoian, Fairbanks; [e,g] pre-Illinoian interglaciation (?), Fairbanks[h] |
| Order Artiodactyla: | |
| Cervus sp. (elk) | Illinoian, Fairbanks[e] |
| Cervalces sp. (giant elk) | Illinoian, Fairbanks[e] |
| Alces sp. (moose) | Illinoian, Fairbanks[e] |
| Rangifer sp. (caribou) | Illinoian, Fairbanks;[e,g] pre-Illinoian interglaciation (?), Fairbanks;[h] pre-Illinoian interglaciation(?), Kotzebue Sound[c] |
| Bison (Superbison) sp. (large bison) | Sangamon, Tofty mining district;[i] Illinoian, Fairbanks;[e,g] pre-Illinoian interglaciation(?), Fairbanks;[g] pre-Illinoian interglaciation(?), Kotzebue Sound[c] |
| Bootherium sp. (extinct musk-ox) | Illinoian, Fairbanks[e] |
| Ovibos moschatus (modern musk-ox) | Illinoian, Nome[j] |
| Ovibos sp. (musk-ox) | Illinoian, Fairbanks[e] |
| Ovis sp. (mountain sheep) | Illinoian, Fairbanks[e] |
| cf. Ovis sp. (mountain sheep) | Illinoian, Nenana Valley[f] |
| Order Perissodactyla: | |
| Equus lambei (horse) | Illinoian(?), Yukon Flats[k] |
| Equus sp. (horse) | Illinoian, Fairbanks[e,g] |
| Order Pinnipedia: | |
| Odobenus sp. (walrus) | Sangamon, Nome[l] |

[a] Beaver dams and beaver-cut wood have been recognized by Péwé in redeposited loess of Illinoian age in the valley of Sheep Creek in the Fairbanks area.

[b] The nearly complete skeleton of a muskrat (identified by C. A. Repenning) was recovered by D. S. McCulloch in 1962 from bedded sand and silt banked against an Illinoian end moraine on the east shore of Baldwin Peninsula facing Hotham Inlet in the Kotzebue Sound area. The bedded sand and silt represent ancient floodplain deposits of the Kobuk River of late Illinoian or possibly Sangamon age (McCulloch et al., 1965).

[c] In 1961, D. M. Hopkins collected the roof of a skull, the horn cores, and one horn of Bison (Superbison) sp., the pelvis of a cervid, probably Rangifer, and a mandible with $M_{1-3}$ of Dicrostonyx torquatus (identified by C. A. Repenning). The bones were found on the floor of a steep, newly eroded gully cut in the coastal bluffs of the Baldwin Peninsula along the shore of Kotzebue Sound about 4.5 km north from the Arctic Circle. The bones were lying loose in a part of the gully cut into glacially deformed marine clay of the pre-Illinoian Kotzebuan transgression, but when collected, they were assumed to have been washed down from loess and peat of Illinoian and post-Illinoian age, which lay concealed beneath turf at the head of the gully. Upon cleaning, the interior of the bison skull was found to contain dried but well-preserved maggots encased in peat, silt, and sand. Sediment removed from the interior of the skull contained the following poorly preserved pollen and spores (identified by E. B. Leopold):

| | Grains | | Grains |
|---|---|---|---|
| Pinus | 3 | Scirpus | 1 |
| Pseudotsuga | 1 | Undetermined monocots | 19 |
| Picea | 14 | Ericales | 2 |
| Tsuga(?) | 1 | Artemisia | 1 |
| cf. Juniperus | 1 | Caryophyllaceae | 1 |
| cf. Ilex | 1 | Undetermined dicots | 8 |
| Graminae | 1 | Sphagnum | 32 |
| Alnus | 3 | Lycopodium | 2 |
| Betula | 7 | Undetermined spores | 3 |

The preservation of maggots in the skull indicates that the polliniferous matrix was washed into the skull soon after the death of the animal. The presence of *Pinus* and *Pseudotsuga* and the possible presence of *Tsuga* and *Ilex* distinguish this small pollen flora from any Illinoian or post-Illinoian pollen floras yet studied in northwestern Alaska. However, wood of *Pinus* has been identified by R. A. Scott in deltaic sediments of the Kotzebuan transgression on Baldwin Peninsula; and *Pinus*, *Pseudotsuga*, and *Tsuga* are present in older interglacial marine deposits in western Alaska. It is possible, though improbable, that these exotic pollen grains were reworked from Tertiary or older Pleistocene beds somewhere in the drainage basins of the Kobuk or Noatak River; even if this were the case, reworked pollen would be most likely to be present in the deltaic deposits of the Kotzebuan transgression and would be very unlikely to be present in appreciable quantities in late Illinoian and post-Illinoian loess. Hopkins therefore concludes that the *Bison* skull is almost certainly derived from underlying beds of the pre-Illinoian Kotzebuan transgression and that the cervid and the *Dicrostonyx* are also probably derived from the Kotzebuan beds.

[d] A mandible of a carnivore collected by Geist in the Illinoian sediments of the "Cripple Creek sump" has been identified by M. C. McKenna as *Xenocyon* (written communication, McKenna to C. A. Repenning, Feb. 9, 1966).

[e] The richest assemblage of fossil mammals found thus far in Alaskan beds of pre-Wisconsin age was in the "Cripple Creek sump," a mass of loess (possibly rebedded) of Illinoian age that was deposited on a down-warped or down-faulted surface of auriferous gravel beneath the present valley of Cripple Creek in the Fairbanks district. The deposit probably was first exposed in 1945 and 1946; excavations were still in progress in 1947 and 1948, during preparations for gold dredging. Vertebrate fossils were collected by O. W. Geist and T. L. Péwé and tentatively identified by Geist; they were then placed in the Frick collection of the American Museum of Natural History, and generic identifications were confirmed by the staff of the American Museum.

[f] Worn and abraded mammal remains were found in 1963 in a borrow pit in outwash of the Healy (= Illinoian) Glaciation about two and a half miles north-northwest of Ferry in E½ SW¼ Sec. 6 of T10S, R8W, Fairbanks Base and Meridian in the Fairbanks A-5 quadrangle. The fossils were discovered during the construction of the Nenana–Healy Highway through the northern foothills of the Alaska Range. The remains included a mammoth tusk found by P. E. Plack and identified by C. Wahrhaftig, and an Artiodactyl metatarsal found by Dan Couch. The Artiodactyl material was compared with *Odocoileus*, *Antilocapra*, *Rangifer*, and *Ovis* by Patricia Vickers and D. E. Savage of the University of California (Berkeley). The specimen most closely resembles *Ovis*. There was no saiga antelope material available for comparison. The Artiodactyl bone has been given to the geological museum of the University of Alaska.

[g] Mammoth, Bison (*Superbison*) sp., horse, and caribou have been collected by Péwé and Geist and identified by Geist and the staff of the American Museum of Natural History from many other exposures of loess of Illinoian age in the Fairbanks area.

[h] Mammoth and bison remains have been collected by Péwé and identified by Geist and the staff of the American Museum of Natural History in auriferous gravel beneath Illinoian loess in the Fairbanks area; Péwé considers the gravel to have been deposited during one or more pre-Illinoian glacial intervals.

[i] The right mandible with $M_{1-3}$ of a *Bison* was collected by D. M. Hopkins (identified by C. A. Repenning) in a forest bed of Sangamon age in the McGee placer mine on Dalton Creek, Tofty mining district, 85 miles west of Fairbanks. The Sangamon age assignment is based upon the position of the forest bed below loess of Wisconsin age and above loess containing an ash bed similar to a prominent ash near the top of the Illinoian deposits in the Fairbanks district.

[j] A skull identified by G. E. Lewis as *Ovibos moschatus* (Zimmerman) was recovered by Andrew Nerland from the buckets of a gold dredge working inland from Third Beach on the coastal plain at Nome in 1948. The dredge was digging in outwash of Illinoian age, 16 meters below the level of the dredge pond at the time the skull came up in the buckets.

[k] J. R. Williams (1962, p. 312) reports a discovery on the bank of the Chandalar River at the village of Venetie in the Yukon Flats. He found the symphysis and greater part

of the right ramus of a mandible with teeth of a horse identified by Jean Hough as *Equus lambei* Hay. Miss Hough is quoted as stating that she considered *E. lambei* to be a subspecies of *E. caballus*. The jaw was evidently derived from adjoining terrace deposits consisting of outwash graded to the outermost recognizable moraines along the Chandalar River in the southern foothills of the Brooks Range. Péwé *et al.* correlate these moraines with the Illinoian Glaciation (1965, Fig. 3).

l Walrus tusks have been found on several occasions during drift-mining of Second Beach, an auriferous beach deposit of Sangamon age at Nome, which form the type deposits of the Pelukian transgression of Hopkins (this volume). In 1956, walrus tusks were exposed and collected during stripping of estuarine deposits of Pelukian (Sangamon) age at Nome in preparation for gold dredging at deeper levels; these specimens were given to the University of Alaska by Carl Glavinovitz.

## REFERENCES

Frick, C. 1937. Horned ruminants of North America: Am. Mus. Nat. Hist. Bull., v. 69, 669 p.

Gilmore, C. W. 1908. Smithsonian exploration in Alaska in 1907 in search of Pleistocene fossil vertebrates: Smithsonian Misc. Coll., v. 51 (Publ. 1807), 38 p.

Hopkins, D. M., F. S. MacNeil, and Estella B. Leopold. 1960. The coastal plain at Nome, Alaska: a late Cenozoic type section for the Bering Strait region: Repts., 21st Internat. Geol. Cong. (Copenhagen), 1960, pt. 4, p. 46–57.

Kotzebue, O. von. 1821. A voyage of discovery into the South Sea and Beering's Straits, for the purpose of exploring a northeast passage, undertaken in the years 1815–1818: Longman, Hurst, Rees, Orme, and Brown, London, v. 1, 358 p.

Maddren, A. G. 1905. Smithsonian exploration in Alaska in 1904, in search of mammoth and other fossil remains: Smithsonian Misc. Coll., v. 49 (Publ. 1584), 117 p.

McCulloch, D. S., D. W. Taylor, and M. Rubin. 1965. Stratigraphy, non-marine mollusks, and radiometric dates from Quaternary deposits in the Kotzebue Sound area, western Alaska: Jour. Geology, v. 73, p. 442–453.

Péwé, T. L. 1952. Geomorphology of the Fairbanks area, Alaska: Ph.D. thesis, Stanford University, Stanford, Calif., 220 p.

——— 1965. Fairbanks area, p. 6–36 in T. L. Péwé, O. J. Ferrians, Jr., D. R. Nichols, and T. N. V. Karlstrom, Guidebook, Field Conf. F, Central and South Central Alaska: Internat. Assoc. Quaternary Res. (INQUA), 7th Cong. (Boulder), 141 p.

Péwé, T. L., D. M. Hopkins, and J. L. Giddings, Jr. 1965. The Quaternary geology and archaeology of Alaska, p. 355–374 in H. E. Wright, Jr., and D. G. Frey, eds., The Quaternary of the United States: Princeton Univ. Press, Princeton, N.J., 922 p.

Quackenbush, L. S. 1909. Notes on Alaskan mammoth expeditions in 1907 and 1908: Am. Mus. Nat. Hist. Bull., v. 26, p. 87–130.

Skinner, M. F., and O. C. Kaisen. 1947. The fossil bison of Alaska and preliminary revision of the genus: Am. Mus. Nat. Hist. Bull., v. 89, p. 123–256.

Wahrhaftig, C. 1958. Quaternary geology of the Nenana River Valley and adjacent parts of the Alaska Range, p. 1–78 in C. Wahrhaftig and R. F. Black, Quaternary and engineering geology in the central part of the Alaska Range: U.S. Geol. Prof. Paper 293, 118 p.

Williams, J.R. 1962. Geologic reconnaissance of the Yukon Flats District, Alaska: U.S. Geol. Survey Bull. 1111-H, p. 289–330.

## 14. On the Origin of the Mammalian Fauna of Canada

C. C. FLEROW

*Paleontological Institute, Academy of Sciences of the USSR*

It has been known for quite a long time that the faunal composition of the Palearctic is very similar to that of the Nearctic. This is especially true with respect to the mammalian faunas of Northeastern Siberia and Canada, as many zoogeographers and paleogeographers have noted (Baker, 1920; Romer, 1933; Colbert, 1937; Stirton, 1951; Jelinek, 1957; Zeuner, 1959). With the exception of a few purely American endemics, such as American porcupines, raccoons, and American deer, most of which are descended from South American groups, the mammal species in Canada are of Asiatic origin. Recent paleontological research has demonstrated not only similarity of the present Palearctic and Nearctic faunas but also their very great similarity throughout a considerable part of the Anthropogene Period.

The Pleistocene mammals of Siberia and northwestern North America are well known (Scott, 1937; Frick, 1937; Simpson, 1947; Gromov, 1948; Hibbard, 1958; Vangengeim, 1961). Therefore, after a few general remarks, I shall pass on to a review of the factors that determined the formation of the present-day mammalian faunas of North America.

The skeleton of a vertebrate animal gives a very complete idea of the organism, and permits one to reconstruct the functional importance of individual parts of the animal as well as its general structure and biology. Because they are represented in abundance and variety in modern faunas and in many fossil faunas, mammals provide extremely rich material for the understanding and reconstruction of extinct communities. Many mammal species are strictly associated with definite environmental conditions, so that they are very important for interpreting landscapes óf the past. The great variety of biological types represented, and of ecological niches occupied, make mammals perfect indicators of landscapes. The intimate interrelationships that exist between herbivorous mammals and the flora permit one to make a more reliable judgment concerning the vegetation that existed when a certain species was living. One must not forget that the slightest change in the composition of the vegetation serving as food, or in the method of food procurement or mode of movement of a mammal species, is reflected by evolutionary changes in the bony structure. All of these factors make mammals, as compared with other animals, among the best objects of study in paleogeographic reconstructions. For this reason, mammals provide extensive and abundant data concerning the role of Beringia in the faunistic history of North America and eastern Asia, and serve as an excellent basis for the interpretation of the factors in the development of the modern faunal complexes in these two regions.

A determination of routes, directions, and rates of dispersal is essential to an analysis of penetrative migrations of complete faunas. In establishing the routes, we simultaneously distinguish the obstacles, many of which result from geological transformations, that hinder migration of either the entire fauna or of some of its components. It is necessary, therefore, to determine the tempo or rate of migration of separate species as well as of entire faunas. It is quite obvious that dispersals across Beringia at times of intercontinental connection between Asia and America differed for different species, as each overcame the obstacles in its individual way. For example, elephants, bison, musk-oxen, sheep, and many species of deer crossed from Asia into North America comparatively easily. Saiga and yak, on the other hand, were unable to disperse past Alaska, the obstacles beyond proving to be insuperable for them; these inhabitants of open country did not enter the mountainous and forested areas of Canada. Woolly rhinoceros, musk deer, and squirrels were also unable to overcome the barriers they found on their way; since they were forest dwellers, the open, treeless country was impassable.

## The Bering Land Bridge

Recent studies lead us to assume that the Bering Land Bridge has appeared and vanished repeatedly during the Cenozoic Era (Hopkins, 1959 and this volume; Flint and Brandtner, 1961; Saidova, 1961). The earliest Cenozoic connection across this bridge took place during the Paleocene, at a time when there was a double link with America—eastward from Asia and westward from Europe. However, the animal migrations that took place at that remote time are of little importance for the understanding of the later history of the fauna and of the formation of the present-day faunal complexes. It is now definitely established that the Indo-Malayan, African, and Neotropical faunas had acquired their present character by the end of the Pliocene Epoch, while the Palearctic, Nearctic, and Alpine faunas were formed only during the Anthropogene. Consequently, my discussion will be confined to tracing Anthropogene events that affected mammals in Eastern Asia and North America.

Repeated migrations of mammals from Asia into North America took place during the Pleistocene across a Beringian isthmus that formed and disappeared several times. Saidova's study of benthonic foraminifera in the bottom sediments of Bering Sea and the North Pacific indicates that synchronous uplift of the ocean floor and lowering of sea level have brought the Beringian isthmus into existence at least three times during the Pleistocene; the earliest of these events corresponds to the Kansan or Mindel Glaciation (Saidova, 1961, and this volume).

During times of Anthropogene emergence, it is assumed that the Bering Land Bridge was an area of wide plains with low hills and sparse woodlands, affording an opportunity for numerous mammals to migrate from Asia to North America (Flerow and Zablozkii, 1961). This environment was such that most northern species of mammals could migrate across the land bridge, with the exception of those animals that inhabit forested mountains. The musk deer, whose ancestors lived in warm, humid, forested areas and probably migrated north and became adapted to cold climates and forested mountains as early as the second half of the Pliocene, is an example of a mammal that did not succeed in crossing the Bering Land Bridge. Neither the musk deer nor the squirrel (*Sciurus*) has migrated even to Kamchatka, since Kamchatka is separated from their main area of development by an expanse of open, treeless country. Conversely, the fact that the Rocky Mountain goat (*Oreamnos*) is derived from Asiatic ancestors suggests that those ancestors were adaptable to the plains of the Bering Land Bridge and that this goat was transformed into a true Alpine form only recently, at the beginning of the Pleistocene. The absence in Alaska of the woolly rhinoceros (*Coelodonta antiquitatis*) is also explained by the nature of the Bering Land Bridge. Rhinoceroses were confined to forest-steppe and forest areas that had a shrubby undergrowth. The woolly rhinoceros apparently did not inhabit open steppes; it would seem that this animal could not feed on harsh steppe grass, but browsed on trees, shrubs, and other soft fodder. Their remains are much rarer in the north of Eastern Siberia than in the south, where they are very numerous. Evidently in the northern part of their range in Eastern Siberia, they found the necessary food only in narrow belts along the rivers.

## Nature of Faunal Dispersal

As already mentioned, when speaking of the migration of entire faunas, one must realize that it is not the fauna that is moving; instead, we are observing a movement of biocoenotically related complexes of organisms. Both now and in the past, orographic, ecological, and morphological barriers played a major role affecting the dispersal of terrestrial animals. Some species of a faunal complex can expand into new territories and adjust themselves to new conditions relatively easily; others move slowly and have to reconstruct their ecology and morphology quite appreciably. Forms that have lived for a long period in the area of their origin can change little and only slowly when physical geographic conditions change slightly. On the other hand, that part of the population that migrates early into areas of contrasting conditions often begins to change rapidly as it adjusts itself to the new environment. The resulting morphological lability—the ability of the organism to acquire new adaptations rapidly—is a decisive factor in determining potentialities for migration and expansion into new areas. In other words, the more easily the organism is able to adjust itself to varying conditions, the more readily it can expand its range.

The prior presence in a new territory of ecological analogues is of great importance. For example, the numerous forms of Antilocapridae that developed in North America during the Miocene and the greater part of the Pliocene were full analogues of the deer of Europe and Asia, and occupied biotopes in North America that corresponded to those occupied by deer in the Old World. With the extinction of the majority of genera and species of Antilocapridae during the second half of the Pliocene came an explosive development and dispersal of deer in America (Flerow, 1950).

The main mass of migrants moved from Asia to America, and only a few moved from America to Asia. Most of the "Americans" achieved but limited penetrations into northeastern Asia, and only reindeer, musk-oxen, and a few others established large territories in northern Asia and Europe. The explanation for this east-west imbalance seems to lie in the fact that the principal ecological niches in Asia were already occupied by an abundant and varied mammalian fauna. The thermophile forms that had inhabited the forests of North America at the beginning of the Pleistocene were pushed southward with the appearance of the continental ice sheet in Canada. When the ice sheet disappeared, the American endemics were unable to adjust rapidly enough to reoccupy the newly deglaciated area. This permitted Asiatic immigrants to disperse explosively.

## Composition and Relationships of the
## North American Fauna

The present fauna of North America has a dual affinity to the Palearctic fauna. One component consists of species very similar to or even conspecific with species in Eastern Siberia; in some cases identical subspecies are present in both regions. Another component consists of taxa undoubtedly closely related to Asiatic forms but differing specifically or even generically.

Close study shows that a substantial part of the first group consists of taxa characteristic of northern latitudes in North America, including Alaska and part of Canada. The northern group includes the forest bison (*Bison priscus athabascae*), snow sheep (*Ovis nivicola*), musk-ox (*Ovibos moschatus*), moose (*Alces alces americanus*), New World elk or wapiti (*Cervus elaphus canadensis*), brown bear (*Ursus arctos middendorffi*), ermine (*Mustela erminea richardsoni*), weasel (*Mustela nivalis rixosa*), wolverine (*Gulo gulo luscus*), wolf (*Canis lupus*), red fox (*Vulpes vulpes*), lynx (*Lynx lynx*), arctic hare (*Lepus timidus*), lemmings (*Dicrostonyx torquatus, Lemmus sibiricus*), and voles (*Microtus oeconomys, Clethrionomys rutilus*).

The second and less closely related group, on the other hand, consists of taxa distributed chiefly in the southern part of North America. The ranges of the species included in this group extend either across southern Canada and the United States or through the Rocky Mountains south of Alaska. The second group includes the plains bison (*Bison bison*), bighorn sheep (*Ovis canadensis*), Rocky Mountain goat (*Oreamnos montanus*), grizzly bear (*Ursus horribilis*), black bear (*Ursus americanus*), American badger (*Taxidea taxus*), skunks (*Mephites* and others), coyote (*Canis latrans*); swift and kit fox (*Vulpes velox, macrotis*), bobcat (*Lynx rufus*).

As I suggested earlier, this dual similarity is explainable by differences in the glacial history in North America and Northeastern Siberia. Glaciation in Northeastern Siberia has consisted only of relatively small ice sheets and local alpine glaciers, and vast areas remained that were inhabited by a varied mammalian fauna, including the bisons. On the other hand, enormous areas in North America, including all of Canada, were subjected to glaciation for a lengthy period of time (Flint, 1947), and these areas were unsuitable for mammal life. Asiatic mammals that had migrated to North America before the maximum glaciation were cut off from their original homeland and lost all connection with it. Furthermore, they were pushed southward to the area of the present United States and were forced into new environments quite different from the northern landscapes. This resulted in rapid evolution and the acquisition of specific adaptations to the new conditions. During the retreat and disappearance of the continental glaciers in Canada, a new migration of Asiatic mammals across the Beringian isthmus took place. This history of immigration, evolution in isolation, and renewed immigration is illustrated by the history of sheep and bison, and explains the twofold degree of similarity to Asiatic mammalian faunas.

The sheep that migrated to America prior to the maximum glaciation were closely related to the living Asiatic argali (*Ovis ammon*). During the glaciation, they were forced southward and there evolved into the distinctly different American bighorn sheep (*Ovis canadensis*), which retains, however, many features of its Asiatic ancestor. Later, another Asiatic species, the Siberian snow sheep (*Ovis nivicola*), entered North America and dispersed southward along the Rocky Mountains to northern British Columbia. Its modern range includes areas on both sides of Bering Strait but does not merge with that of *Ovis canadensis*.

An even more striking illustration of the effect of the differences between Siberian and North American glaciation is provided by the history of the bison (Skinner and Kaisen, 1947; Flerow and Zablozkii, 1961). The early stages in the history of the bison are not yet clearly understood. However, an Asiatic origin is indicated by the fact that the earliest records of the genus are represented by *Bison sivalensis* (Falconer) Lidekker in the late Pliocene Siwalik deposits of India and by *B. paleosinensis* Chardin and Piveteau in the Late Pliocene Nihowan fauna of China.[a] During early Pleistocene time, bison dispersed into the temperate zone of Asia and Europe. At this time, short-horned forms, including first *B. tamanensis* Vereshchagin and the later *B. priscus schoetensaki* Freudenberg,[b] are found in southern Europe as far north as the north Caucasus (Predkavkazie). By middle Pleistocene time, bison apparently disappeared completely in southern Asia but became widely distributed across Europe, northern Asia, and northern North America as a large, long-horned form, *B. p. crassicornis* Richardson. During this long period, extending from the end of the Mindel–

---

[a] The *Bison*-bearing levels in the Siwalik Series and the Nihowan fauna are correlated by other authors in this volume with the Villafranchian Stage of Europe and are assigned by them to the Lower Pleistocene. ED.

[b] *B. cesaris* Schlottheim, 1821, has priority over *B. priscus* Bojanus, 1827, and *B. pallasi* Baer, 1823. However, both *B. cesaris* and *B. pallasi* are *nomen oblitum*, and I retain *B. priscus*.

Riss Interglaciation into the beginning of the Riss–Würm Interglaciation, landscape conditions were more or less uniform, the range of bison was continuous across Holarctica, and no substantially different species appeared.

*Bison priscus crassicornis* migrated from Asia to America before the Riss (or Illinoian) Glaciation. During this glaciation they were forced southward out of Canada to the present area of the United States, where they evolved into the giant species *B. latifrons* Harlan. This southern population was completely isolated from the northern group, which continued to live in Asia, Beringia, and Alaska. The ancestor of the modern plains bison, *B. alleni* Marsh then developed in the southern part of the United States, and this form evolved into the steppe bison of the Wisconsin Glaciation (*B. bison antiquus* Leidy) and finally into the living post-Pleistocene *B. bison bison* Linnaeus.

To the north, in Alaska and Siberia, the long-horned bison (*B. priscus crassicornis*) persisted through much of the late Pleistocene, but by the end of the Wisconsin (or Würm) Glaciation, bison began gradually to become smaller throughout Holarctica. Their horns became shorter, as well, and the long-horned *B. p. crassicornis* was replaced by the short-horned *B. p. priscus* Bojanus, which was equally holarctic in distribution and which also failed to produce distinct local forms.

With the withdrawal of the Wisconsin ice sheet, bison again penetrated into Canada from Alaska and Beringia to the northwest. However, like other late Pleistocene Asiatic immigrants, the northern bison dispersed only into the MacKenzie River basin and failed to reach the Hudson Bay region and eastern Canada. These areas probably were cleared of ice later than other parts of Canada and therefore remained unsuitable for such mammals as bison, moose, and wapiti. The presence of Hudson Bay also affected the climate and created very severe environments.

At the end of the Würm (or Wisconsin) Glaciation, the range of bison began to shrink and disintegrate, and by the beginning of the Holocene was completely disrupted. Bison became extinct over enormous areas and persisted only in parts of Europe, Eastern Siberia, Alaska, Canada, and the United States. The isolated populations began to differentiate, resulting in the wisent (*Bison bonasus*) of Europe, the large wood bison (*B. priscus athabascae*) of Eastern Siberia, Alaska, and Canada, and the plains bison (*B. bison bison*) of the United States. Further isolation in Europe produced a reduction from the moderate horn length of the ancestral wisent (*B. bonasus major* Hilzheimer) to the short-horned Lithuanian bison (*B. b.*

*bonasus* Linnaeus) and Caucasus bison (*B. b. caucasicus* Satunin). *B. b. major* was once present in the Caucasus and in Transcaucasia, but later its range diminished, and *B. b. caucasicus* persists only on the western half of the main ridge of the Caucasus.

Eastern Siberia, Alaska, and Canada were inhabited by the big *B. priscus athabascae* Rhoads, but a very small, short-horned endemic race evolved in an extreme northern part of Eastern Siberia isolated from the general range of the species by mountain ranges. This subspecies occupied lowlands adjoining the Arctic Ocean in the Lena, Indigirka, and Kolyma River basins and became extinct during the post-Würm thermal maximum.

In Canada, the post-Pleistocene history of bison has been one of range reduction and extinction similar to that in Eurasia. This history follows the same pattern seen in musk-ox (*Ovibos moschatus*), which became extinct throughout Palearctica and were preserved only in North America and Greenland, where the periglacial conditions to which they were adapted are still present. Even in North America, the musk-ox gradually disappeared west of the MacKenzie River and retreated eastward nearer to the glaciated areas, a process that has continued into historic time. Similarly, the range of forest bison shrank eastward, and the species finds its last refuge in the forests of the Great Slave Lake region of Canada (Seton, 1886, 1912, 1927; Flerow in Adlerberg *et al.*, 1935; Raup, 1935; Soper, 1939, 1941; Fuller, 1962) in an area of severe climatic conditions in which small patches of prairies are present in the taiga forest. These prairie patches apparently approach the character of the cold forest-steppes that were once widely developed in Siberia, Alaska, and Canada at the end of the Pleistocene and during the early Holocene. All features characterizing this big Canadian wood bison indicate its very close morphological affinity with the extinct Pleistocene and early Holocene species of the so-called "primary bison," *Bison priscus* (Rhoads, 1898).

We can say confidently that the typical Pleistocene mammals of northeastern Asia were unable to survive the period of the Holocene thermal maximum and the resulting change in the vegetation. The formation of *Sphagnum* associations (muskegs) in northern regions deprived most species of large herbivorous animals (including horses, oxen, and many species of deer and antelope) of their main source of forage. Rhinoceroses and mammoths became extinct at the same time and probably for the same reason. Only the musk-ox and the wood bison survived this crisis in America, persisting in areas where remnants of Pleistocene landscapes were preserved. These "living fossils," once contemporaries of the mammoth, survived to the present time as the last representatives of the Pleistocene fauna.

## REFERENCES

Adlerberg, G. P., B. S. Vinogradov, N. A. Smirnov, and C. C. Flerow. 1935. Zveri Arktiki (Arctic mammals) : Leningrad, 579 p.

Baker, F. C. 1920. The life of the Pleistocene or Glacial Period: Univ. Ill. Bull., v. 17, no. 41, 476 p.

Colbert, E. H. 1937. The Pleistocene mammals of North America and their relations to Eurasian forms, p. 173–184 *in* G. G. MacCurdy, ed., Early Man: Lippincott, New York, 363 p.

Flerow, C. C. 1950. Morfologiia i ekologiia oleneobraznykh v protsesse ikh evoliutsii (Morphology and ecology of deerlike animals and processes of their evolution): Akad. Nauk SSSR, Materialy po chetvertichnomu periodu SSSR (Data on the Quaternary Period) Bull. 2, p. 50–69.

Flerow, C. C., and M. A. Zablozkii. 1961. O prichinakh izmeneniia areala bizonov (On the reasons for the change in the range of bison) : Bull. Moskva Ob-va Ispytatelei Prirody Otd. Biologii, v. 46, p. 99–109.

Flint, R.F. 1947. Glacial geology and the Pleistocene Epoch: Wiley, New York, 589 p.

Flint, R. F., and F. Brandtner. 1961. Climatic changes since the last Interglacial: Am. Jour. Sci., v. 259, p. 321–328.

Frick, C. 1937. Horned ruminants of North America: Am. Mus. Nat. Hist. Bull., v. 69, 669 p.

Fuller, W. A. 1962. The biology and management of the bison of Wood Buffalo National Park: Canadian Wildlife Serv., Wildlife Mgmt. Bull. Ser. 1, no. 16, 52 p.

Gromov, V. I. 1948. Paleontologicheskoe i arkheologicheskoe obosnovanie stratigrafii kontinental'nykh otlozhenii chetvertichnogo perioda na territorii SSSR (Paleontological and archaeological bases of the stratigraphy of continental deposits of the Quaternary Period in the territory of the USSR) : Akad. Nauk SSSR, Tr. Geol. Inst., Bull. 64, Ser. Geol. No. 17, 519 p.

Hibbard, C. W. 1958. Summary of North American Pleistocene mammalian local faunas: Papers Michigan Acad. Sci., Arts, and Letters, v. 43, p. 3–32.

Hopkins, D. M. 1959. Cenozoic history of the Bering Land Bridge: Science, v. 129, p. 1519–1528.

Jelinek, A. J. 1957. Pleistocene faunas and early man: Papers Mich. Acad. Sci., Arts, and Letters, v. 42, p. 225–237.

Raup, H. M. 1935. Botanical investigations in Wood Buffalo Park: Nat. Mus. Canada Bull. 74, Biol. Ser. 20, 174 p.

Rhoads, S. N. 1898. Notes on living and extinct species of North American Bovidae: Proc. Acad. Nat. Sci., Philadelphia, v. 49, p. 483–502.

Romer, A. S. 1933. Pleistocene vertebrates and their bearing on the problem of human antiquity in North America, p. 49–81 *in* D. Jenness, ed., The American aborigines, their origin, and antiquity: Toronto Univ. Press, 396 p.

Saidova, H. M. 1961. Ekologiia foraminifer i paleogeografiia dal'nevostochnykh morei SSSR i severo-zapadnoi chasti Tikhogo Okeana (Ecology of foraminifera and paleogeography of the Far East Seas of the USSR and the northwestern part of the Pacific Ocean) : Akad. Nauk SSSR, Inst. Okeanologii, 232 p.

Scott, W. B. 1937. A history of land mammals in the Western Hemisphere, 2d ed.: Macmillan, New York, 786 p.

Seton, E. T. 1886. The ruminants of the Northwest: Proc. Can. Inst., Ser. 3, vol. 21, no. 4, fasc. 3, p. 113–117.

———— 1912. The Arctic prairies: Constable, London, 415 p.

———— 1927. Lives of game animals, v. 3. Doubleday, Doran, New York, 780 p.

Simpson, G. G. 1947. Holarctic mammalian faunas and continental relationships during the Cenozoic: Geol. Soc. America Bull., v. 58, p. 613–688.

Skinner, M. F., and O. C. Kaisen. 1947. The fossil *Bison* of Alaska and preliminary revision of the genus: Am. Mus. Nat. Hist. Bull., v. 89, Art. 3, p. 123–356.

Soper, J. D. 1939. Wood Buffalo Park; notes on the physical geography of the Park and its vicinity: Geogr. Rev., v. 29, p. 383–399.

———— 1941. History, range, and home life of the northern bison: Ecol. Monogr., v. 11, p. 347–412.

Stirton, R. A. 1951. Principles in correlation and their application to later Cenozoic holarctic continental mammalian faunas: Rept., 18th Internat. Geol. Cong. (London), 1948, pt. 11, p. 74–84.

Vangengeim, E. A. 1961. Paleontologicheskoe obosnovanie stratigrafii Antropogenovykh otlozhenii severo vostochnoi Sibiri (Paleontological basis of the stratigraphy of the Anthropogene deposits of Northeastern Siberia): Akad. Nauk SSSR, Tr. Geol. Inst., Vyp. 48, 182 p. In Russian; English translation of parts available from Am. Geol. Inst.

Zeuner, F. E. 1959. The Pleistocene Period; its climate, chronology and faunal successions: Hutchinson, London, 447 p.

## 15. The Effect of the Bering Land Bridge on the Quaternary Mammalian Faunas of Siberia and North America

E. A. VANGENGEIM
*Geological Institute, Academy of Sciences of the USSR*

The history of land connections between Asia and North America, and the effects of such connections on the faunal history of both continents, have occupied the minds of zoogeographers for a long time. And yet many questions remained unanswered because of insufficient knowledge of the fossil faunas of Eastern Siberia, especially of those representing the earlier part of the Anthropogene. Work by the Geological Institute of the Academy of Sciences of the USSR during the past ten years has greatly enlarged our knowledge of the Anthropogene fauna of Eastern Siberia. It appears now that there were at least three periods of intercontinental faunal exchange; at the beginning of the Anthropogene, during the Maximum (Samarov) Glaciation, and, most recently, during the Zyrianka and Sartan Glaciations. The distinction between the last two periods of continental connection is vague, and it may be that the intervening period of separation was short-lived.

### Early Anthropogene

According to the Anthropogene faunal record, the earliest connection across Beringia took place in the Villafranchian. Apparently, faunal exchange took place at this time in both directions, although our record seems to indicate that most of the migrating mammals moved from Asia to North America. Introduced to Asia from North America was a horse that was ancestral to the Asian *Equus sanmeniensis-sivalensis* lineage (horses with a long protocone differing from, but contemporary with, *E. stenonis*). We know of no horses of *E. stenonis* type in Asia; they were present only in Europe and Africa, possibly extending eastward as far as Kazakhstan. Apparently, there have been two routes by which horses penetrated the Old

World: the ancestor of *E. sanmeniensis* entered from the east; and that of *E. stenonis* from the west.[a]

During this period, elephants of the *Archidiskodon* line may have migrated from Asia into North America, although there is no record of their presence there until Middle Pleistocene time. This group occupied virtually the entire territory of Eurasia, Africa, and North America. The migration from Asia to North America of the Oviboninae also probably occurred at this time, although the earliest forms found in North America are of Middle Pleistocene age, and even these forms are only questionably referred to the family.

There are no representatives of "antiquoid" elephants of the genus *Paleoloxodon* in North America. At present, the reason for their absence in North America is not clear. The time when this group appeared in the Old World is not established precisely. The predominant opinion in the West is that the genus *Paleoloxodon* had separated from the *Archidiskodon* lineage no earlier than the end of the Villafranchian, and that it subsequently developed parallel to *Archidiskodon*. If such is the case, the absence of "antiquoid" elephants in America can be explained by the lack of a Bering land bridge after the genus *Paleoloxodon* had come into existence. There is, however, another point of view on the origin of *Paleoloxodon* (Garutt, 1965). *Paleoloxodon* and *Archidiskodon* originated simultaneously, the former in southern Asia and the latter in Africa. At the outset of its existence, the genus *Paleoloxodon* apparently split up into two ecological groups: forest elephants of the *P. antiquus* type and elephants of the *P. namadicus* type associated with the arid, pronouncedly continental climate of forest-steppes and steppes. The first group dispersed westward into Europe, while the second occupied mainly the central regions of Asia. The most northerly finds of elephants belonging to this group are in the upper valley of the Lena, in the Aldan region, and in Transbaikalia. Apparently, they could not penetrate farther north into the dark-coniferous taiga.[b] If this viewpoint is accepted, it provides a reason for the absence of the "antiquoid" elephants in the New World.

Similar controls by climatic zonation may account for the absence of rhinoceroses in North America. During the Villafranchian, representatives

---

[a] It should be noted here that there is some lack of agreement about the classification of caballine and zebrine horses, and also, therefore, on the necessity of two migration routes between North America and Eurasia. Viret (1954) has pointed out that *Equus sanmeniensis* from China is a zebrine horse similar to *E. stenonis*, and Hooijer (1949) feels that *E. sivalensis* from India is a caballine horse. As many people have pointed out, the length of the protocone is variable. ED.

[b] Soviet authors distinguish between dark-coniferous taiga (northern forests dominated by spruce) and light-coniferous taiga (northern forests dominated by larch). These terms are defined more fully by Giterman and Golubeva, this volume. ED.

of the genus *Dicerorhinus* ranged widely over all of northern Eurasia; but Villafranchian rhinoceroses have not yet been found in Northeastern Siberia, though we know of several later forms. Thus it is difficult to decide at present whether the absence of *Dicerorhinus* in Eastern Siberia is to be explained by incomplete geological records or by unfavorable forest conditions.

The early Villafranchian fauna of Eastern Siberia is known only from isolated finds. In the lower valley of the Aldan River, remains of an early form of *Equus* (*E.* ex gr. *sanmeniensis*) have been found, and remains of *Archidiskodon* cf. *meridionalis* are known in the valley of the Vilyui River (Vangengeim, 1961). A more or less complete faunal complex is recorded only in the extreme south of Eastern Siberia—in the western Transbaikal area. This is the so-called Chikoi complex (Ravskii *et al.*, 1964), similar in composition to the Nihowan fauna in northern China and belonging to the central Asiatic paleozoogeographical province.

The scanty information we possess on the fauna of Eastern Siberia during somewhat later intervals of the Anthropogene (Günz–Mindel Interglaciation and Mindel Glaciation in the Alpine time scale) does not warrant any judgment on the connections between the Asiatic and American faunas of this time. A fauna of Günz–Mindel age that includes forms unknown in the New World was obtained in the Aldan Valley in the southern part of the central Siberian Plateau. We find here *Trogontherium cuvieri* (widely developed all over northern Eurasia), a boreal form of *Alces latifrons* that is not known south of the latitude of Lake Baikal, and a number of central Asiatic species such as *Equus* cf. *sanmeniensis*, *Paleoloxodon* cf. *namadicus*, and *Canis* cf. *variabilis*.

Mammalian faunas of the next younger age are also of exclusively Old World composition. Their remains are found in the valley of the Vilyui River (*Archidiskodon* cf. *wüsti*, *Equus* cf. *mosbachensis*, *Dicerorhinus mercki*) and in the headwaters of the Lena River (*Paleoloxodon* ex gr. *namadicus*, *Dicerorhinus* cf. *mercki*, *Equus* cf. *mosbachensis*, *Sinocastor* sp.). The Itanta and Tologoi complexes of the Transbaikal region in the southern part of Eastern Siberia are also of central Asiatic aspect (E. A. Vangengeim *in* Ravskii *et al.*, 1964).

## Middle and Late Anthropogene

A prolonged connection between the Asiatic and American continents apparently existed from the time immediately preceding the Samarov (or Riss or Illinoian) Glaciation to the beginning of the Kazantsevo (or Riss–Würm or Sangamon) Interglaciation. During this time, faunal dispersal across Eurasia and Beringia into North America was greatly influenced by the circumpolar development of cold, treeless periglacial landscapes. In eastern Asia, the periglacial zone included vast areas extending from the southern limits of the continental ice to northern China. The zone of forests was greatly reduced, and northern tundra-steppes were in juxtaposition with arid central Asiatic steppes and semideserts (Vangengeim and Ravskii, 1965).

This floral and climatic condition was reflected in specific features of the mammalian fauna. Throughout an extremely extensive area, the bulk of the periglacial fauna consisted of mobile and ecologically plastic animals —especially big-hoofed forms characteristic of open country. A distinct "steppization" of the fauna took place as typical central-Asiatic species found an opportunity to disperse northward. At the same time, highly boreal and arctic animals extended their ranges southward. Because of these conditions, we find in single localities such seemingly ecologically incompatible pairs as arctic fox and the wild ass, *Equus hemionus*, or saiga antelope and reindeer. However, despite a certain homogenization of the environments, a definite zonation can be seen in the geographical distribution of the mammalian faunas. Southward across this area, occurrences of mammoth and other arctic species grow less common, and central-Asiatic elements increase in number, apparently because of greater aridity in the south. The periglacial faunal complex, known as the "mammoth" or Late Paleolithic complex, includes the mammoth (*Mammuthus primigenius*), woolly rhinoceros (*Coelodonta antiquitatis*), a horse (*Equus caballus*), and bison (*Bison priscus*). The arctic fox (*Alopex lagopus*) and lemmings (*Dicrostonyx torquatus* and *Lemmus obensis*) were present in the north, and wild ass (*Equus hemionus*) and yak (*Poephagus baicalensis*) were present in the south.

Asiatic mammoths, bison, elk, and a number of other forms were unhindered in dispersal from Siberia to the American continent, spreading first to Alaska and then expanding their range over the territory freed by the retreating glacial ice.

Apparently, reindeer (*Rangifer tarandus*) and musk-ox (*Ovibos moschatus*) returned to Asia from North America at this time; their remains are common in the deposits belonging to the time of the Maximum Glaciation in the Asiatic part of the USSR, but are unknown in older deposits.[c]

---

[c] There seem to be records of *Ovibos* in western Europe that are earlier than any found thus far in either Siberia or North America. Kahlke (1964) reports several occurrences of *Ovibos* and of the related but not ancestral *Praeovibos* in deposits of early Middle Pleistocene (Mindel or Elster) age. ED.

Although the earliest North American records of Ovibovinae in North America are from deposits of Illinoian (or Samarov) age near Fairbanks and Nome, Alaska (Péwé and Hopkins, this volume), I visualize their having descended from an ancestral form that migrated to North America at the end of the Pliocene. An adaptive radiation in the New World produced *Bootherium, Symbos*, and perhaps *Eucatherium*, which are unknown in the Old World, and *Ovibos*, which dispersed into the Old World at this time.

Beginning with Samarov (or Illinoian) time, Eastern Siberia and Alaska were virtually a single faunal province for such animals as mammoth and bison. A remarkable similarity is observed in the development of these forms in Eastern Siberia and Alaska during the second half of the Pleistocene: a gradual dwarfing resulted in dwarf mammoth and smaller-sized bison with shorter horns.

Mammoth, bison, reindeer, musk-ox, elk, brown bear, and a great number of other species ranged over the entire northern part of the Northern Hemisphere during the second half of the Pleistocene. There appears to be no explanation as to why certain forms that in the Old World were inseparable companions of the mammoth, reindeer, and other Holarctic forms did not migrate with them into the New World. Forms that did not migrate include the woolly rhinoceros and the saiga.

During the second half of the Pleistocene, the rhinoceros *Coelodonta antiquitatis* was present over the entire periglacial zone of Eurasia, extending eastward to Chukotka and the islands of the Arctic Ocean and southward to central Asia. Remains of this animal have also been found on Kamchatka. A great ecological plasticity is indicated for this species. However, northward, the frequency of discovery of this species decreases substantially, and the maximum number of specimens is recorded in the Transbaikal region. This region was an area of cold, very arid steppes similar to semideserts. Consequently, optimum conditions for the existence of woolly rhinoceros appear to have been pronouncedly arid environments, and it was less abundant in the more humid north.

During the late Anthropogene, including the Zyrianka and Sartan (or early and late Wisconsin or Würm) Glaciations, the lands of Beringia again united North America and Eastern Siberia. At this time, saiga had a range extending from eastern England to Alaska, but did not disperse beyond Alaska. In Eastern Siberia, saiga were probably associated with the dry phases of the second half of the Zyrianka and Sartan Glaciations, and the Alaskan occurrence of saiga is, therefore, also probably of this age.

The development of yaks is somewhat mysterious. Their main range was in the cold arid steppes and semideserts of central Asia; they did not extend northward beyond the Transbaikal region. Stray individuals occasionally penetrated farther north, but climatic and landscape conditions there were evidently unfavorable for this highly specialized animal. At present, only one record is known in the territory of Eastern Siberia, in the Vilyui River valley. Thus the discovery by Frick (1937) of yak remains in Alaska seems especially unexpected.

## Summary

Analysis of fossil faunas from the Anthropogene of Eastern Siberia indicates a possible connection between the American and Asiatic continents during the Villafranchian, when the migration of animals most probably took place across the area of the present continental shelf. Forests that developed farther south were probably avoided by open-country animals, such as horses. A close similarity between the faunas of Eastern Siberia and those of North America existed from the beginning of the Samarov (Illinoian) until the end of Sartan (Late Wisconsin) time. During this time interval, there could have been interruptions in the connection between the continents, as indicated by the presence of transgressive marine sediments in the extreme northeastern part of Asia. However, these interruptions need not have been lengthy, because neither in Asia nor in Alaska did substantially different mammals evolve.

A detailed study of the Quaternary deposits, fauna, and vegetation of Eastern and Northeastern Siberia, as well as preparation of detailed paleogeographical and vegetation maps for the various stages of the Anthropogene, will be helpful in enlarging and correcting these historical interpretations.

## REFERENCES

Frick, C. 1937. Horned ruminants of North America: Am. Mus. Nat. Hist. Bull. v. 69, 669 p.

Garutt, V. E. 1965. Iskopaemye slony Sibiri (Fossil elephants from Siberia), p. 106–130 in Antropogenovyi Period v arktike i subarktike (Anthropogene Period in the Arctic and Subarctic): Nauchno-Issled. Inst. Geol. Arktiki, Tr. 143, 359 p. In Russian with English abstract.

Hooijer, D. A. 1949. Observations on a calvarium of *Equus sivalensis* Falconer et Cautley from the Siwaliks of the Punjab, with craniometrical notes on recent Equidae: Archives Nederlandaises de Zoologie, v. 8, no. 3, p. 1–24.

Kahlke, H. D. 1964. Early middle Pleistocene (Mindel/Elster) *Praeovibos* and *Ovibos*: Soc. Sci. Fennica, Comm. Biol. v. 26, no. 5, 17 p.

Ravskii, E. I., L. P. Aleksandrova, E. A. Vangengeim, V. G. Gerbova, and L. V. Golubeva. 1964. Antropogenovye otlozheniia yugo-vostochnoi Sibiri (Anthropogene deposits of Southeast Siberia) : Akad. Nauk SSSR, Tr. Geol. Inst., Vyp. 105, 280 p.

Vangengeim, E. A. 1961. Paleontologicheskoe obosnovanie stratigrafii Antropogenovykh otlozhenii severo vostochnoi Sibiri (Paleontological basis of the stratigraphy of the Anthropogene deposits of Northeast Siberia) : Akad. Nauk SSSR, Tr. Geol. Inst., Vyp. 48, 182 p. In Russian; English translation of parts available from Am. Geol. Inst.

Vangengeim, E. A., and E. I. Ravskii. 1965. O vnutrikontinental'nom tipe prirodnoi zonal'nosti v chetvertichnom periode (Antropogene) (On the intracontinental type of natural zonality in Eurasia during the Quaternary Period (Anthropogene), p. 128–141 *in* Problemy stratigrafii kainozoia (Problems of Cenozoic stratigraphy) : Izd-vo "Nedra," Moscow, 143 p.

Viret, J. 1954. Le loess à banks durcis de Saint-Vallier (Drôme) et sa faune de mammiferes villafranchiens: Lyon. Mus. Hist. Nat., Nouv. Arch., v. 4, p. 1–200.

# 16. Palearctic-Nearctic Mammalian Dispersal in the Late Cenozoic

CHARLES A. REPENNING
*U.S. Geological Survey, Menlo Park, California*

"Mammalian ages" of continental scope tend to begin with an episode of intercontinental faunal dispersal. Initially conceived by vertebrate paleontologists as periods in the evolutionary history of the continental mammalian fauna, and bounded by hiatuses in the fossil record, mammalian ages become subject to redefinition as their limiting inter-age hiatuses are reduced or eliminated and as additional discoveries refine the record of evolutionary history into a more gradational series. Then age-delimiting concepts are focused on immigrants, and age boundaries are adjusted to conform to times of immigration—often without realization that the initial standards have been changed. The boundaries between the North American mammalian ages Hemphillian, Blancan, Irvingtonian, and Rancholabrean (conventionally representing middle Pliocene, late Pliocene and early Pleistocene, middle Pleistocene, and late Pleistocene, in North America) have long since evolved to the point where age-transitional faunas must be defined arbitrarily on the basis of the presence or absence of immigrants, such as the first cervid (Blancan), the first *Mammuthus* (Irvingtonian), or the first *Bison* (Rancholabrean).

This is not to say that the concept of a stage of either faunal or zoogeographic evolution is no longer useful, but rather that evolution is a continuum well suited to defining faunal character but lacking the abrupt changes now sought in definitions of interval boundaries on a continental scale. Extinction of taxa is an important historical consideration but is not chronometrically precise, because, like faunal and intracontinental zoogeographic evolution, it is not synchronous on a continental scale. Exotic faunal elements have become or are becoming the most respected criteria in

Publication authorized by the Director, U.S. Geological Survey.

biostratigraphic age assignment; when a correlation has been based on the introduction of exotic faunal elements, a new interpretation based on either evolution or extinction must be on strong ground to be accepted as a valid replacement.

The effect of this, in the published record, is to make intercontinental mammalian dispersal appear to have taken place in spasms corresponding to the beginnings of mammalian ages. This may be partly because knowledge of both phylogeny and ecology is not yet strong enough to place all local fossil faunas in a chronologic sequence within a given mammalian age. In addition, the fossil sample is often too incomplete to allow one to place significance on the absences of certain taxa. Thus, an exotic animal whose first known occurrence in North America is in a local fauna placed late in the time span of a mammalian age commonly must be counted as an immigrant not merely of the late part of the age but of the entire age.

The conclusion that intercontinental dispersal of mammals actually has been intermittent is supported by the contemporaneous arrival of several exotics from the same source area. It is reasonable to assume that the major faunal exchanges that mark the beginning of Blancan, Irvingtonian, and Rancholabrean mammalian ages are real, but it is questionable whether it should also be assumed that there were no minor exchanges within these ages, or that there was not a continuous exchange at varying rates, simply because lesser dispersals cannot be defended with present evidence.

In the following pages, the late Cenozoic mammalian fauna of North America is compared with that of the Old World for two purposes: first, to establish the reliability of temporal correlations based both upon the fauna and upon the sparse but impressive data of radiometric geochronology; and second, to determine as clearly as possible what mammals were involved in these intercontinental exchanges and what significance these may have for the history of the Bering Land Bridge.

I wish to emphasize that temporal correlation of Old and New World mammalian faunas is strongest when it is based on mutual intercontinental exchange of taxa that have rather precise chronometric significance in their native continents. The North American genus *Canis* differs little from its ancestor *Tomarctus*, and is not particularly useful in dating faunas in North America. Thus the appearance of *Canis* in the late Pliocene of the Old World yields only one piece of information—that there was late Pliocene immigration from the New World. By itself, the fact of this immigration does little to assist correlation. By contrast, the appearance of the North American zebra *Plesippus* in the late Pliocene of the Old World (China) has more significance. This horse is clearly more advanced than its North

American ancestor *Pliohippus*, and Old World faunas containing *Plesippus* can be no older than the earliest North American records of the genus. Similarly, a record in North America of the Old World genus *Cervus*, which could be no older than late Pliocene in Europe, would indicate that the North American fauna that contains it can also be no older than late Pliocene. If these two dispersants are found together in both Old and New World faunas that represent the earliest record of the native genus, then, within the sensitivity of our records, neither fauna can be older than the other and they are contemporaneous.

The sensitivity of our record of fossil history does not always permit conclusive decisions, nor does the certainty of our interpretations of phylogeny. However, the uncertainties of record and interpretation are reduced in inverse proportion to the number of chronologically significant taxa involved in mutual intercontinental faunal exchange.

This review, which naturally began with Simpson's study in 1947 of intercontinental dispersals, attempts to incorporate all significant publications since that time. Of particular usefulness have been the many works of Hibbard, Kurtén, Teilhard, Pei, Pilgrim, Vangengeim, Schaub, and Zdansky.

## Correlation of Faunas

The following section discusses the temporal correlations of the mammalian faunas. The taxa involved in intercontinental dispersal are tabulated, and numbered notes briefly explain new taxonomic, phyletic, and faunal data and interpretations used to support correlations.

### HEMPHILLIAN FAUNA

The Hemphillian North American mammalian age (Wood *et al.*, 1941, p. 12) is customarily considered to be of middle Pliocene age. However, available radiometric dates (Evernden *et al.*, 1964) suggest that the North American nonmarine Pliocene lasted from about 11.5 to 3.5 million years B.P., and that the Hemphillian lasted from about 10 to 4 million years B.P. and thus spans all except earliest and latest Pliocene time. The similarity of the North American Hemphillian faunas to those Old World faunas generally correlative to the Pannonian (or Hipparion or "Pontian"[a]) is one of the strongest in the Tertiary; this may reflect a relatively long period of

[a] The Pontian and Astian Stages are defined on the basis of marine strata and faunas. Their application to nonmarine strata and faunas has been widespread but inconsistent with the definition. Hence, they are shown here in quotes.

faunal exchange. About 24 genera are known to have been Holarctic at this time, of which nine appear to represent dispersal to Nearctica (Table 1) and five to Palearctica (Table 2). The remaining ten genera (Table 3) are those whose continent of origin is uncertain or that are known to have been Holarctic in distribution before the Hemphillian age.

Most of the genera shown in Tables 1, 2, and 3 suggest forested or aquatic environments, and none are clearly plains or steppe forms. The otters *Lutra* and *Lutravus* and beavers *Castor* and *Dipoides* are clearly

TABLE 1. *Hemphillian Species Representing the First Nearctic Record of a Palearctic Lineage*

*Ochotona spanglei* Shotwell,[1] pika
*Microtoscoptes disjunctus* (Wilson),[2] microtine rodent
*Promimomys mimus* (Shotwell), an ancestral meadow mouse[3]
*Castor* sp., modern beaver
*Simocyon marshi* (Thorp), specialized wolf
*Indarctos oregonensis* Merriam, Stock, & Moody, bearlike carnivore
*Agriotherium gregoryi* (Frick), bearlike carnivore
*Plionarctos edensis* Frick, primitive bear
*Lutravus halli* Furlong, otter-like carnivore

[1] Although ochotonids have a long North American record, the first true North American *Ochotona* appears to be from McKay, Oregon (Shotwell, 1956, p. 727), and its ancestry is clearly not in the New World.

[2] The ancestor of the microtine rodent *Microtoscoptes* is uncertain, but it appears to have lived in the Old World.

[3] New material representing all cheek teeth in several stages of wear shows that the primitive microtine *Prosomys* Shotwell (1956) is the junior synonym of *Promimomys* Kretzoi (1955). *Promimomys mimus* (Shotwell), from two Hemphillian localities in Oregon, is nearly identical to *P. cor* Kretzoi from the Csarnota fauna of south Hungary. Slightly greater complexity of the enamel border on the anterior loop of $M_1$ in little-worn specimens of *P. mimus* from Oregon suggests the more advanced structure of *Promimomys moldavicus* (Kormos) from Moldavia. Both central European species are of Csarnotan (or Rousillon or "Astian") age and thus are younger than the North American species. However, several Old World cricetine rodents appear to be morphologically suitable as an ancestral form. Kretzoi (1955, p. 91–93) discussed the similarities of these cricetines to the primitive microtine *Promimomys*. Also, one can hardly doubt that *Promimomys* must have had an earlier "Pontian" history when it is noted that the Csarnotan records of *Promimomys* are associated with much more progressive microtines. Some of these more progressive European microtines are of a specialization not seen in North America until late Blancan time.

TABLE 2. *Pannonian (or Equivalent) Species Representing the First Palearctic Record of a Nearctic Lineage*

Possibly *Anourosorex inexpectatus* (Schlosser), mole-shrew[1]
*Hypolagus schreuderi* Teilhard (and other species), rabbit
*Eutamias orlovi* Sulimski (and other species), chipmunk
*Dipoides majori* Schlosser (and other species), extinct beavers
*Sinohippus zitteli* (Schlosser), browsing horse

[1] Shrews apparently ancestral to *Anourosorex* have been found in the late Miocene and early Pliocene of North America. However, the lineage leading to *Anourosorex* is an Old World one, and whether *Anourosorex* is derived from its earlier relative in the North American Clarendonian or from an unknown form in Asia cannot be determined.

TABLE 3. *Hemphillian-Pannonian Genera of Debatable Ancestry or of
Earlier Holarctic Distribution*

---

*Sorex*, shrew
*Pseudaplodon*, mountain "beaver"[1]
*Pliozapus-Sminthozapus*, jumping mouse[2]
*Eomellivora*, wolverine-like carnivore
*Plesiogulo*, wolverine-like carnivore
*Lutra*(?), otter[3]
*Machairodus*, saber-toothed cat[4]
*Tapirus*, tapir
*Hipparion*, three-toed horse
*Gomphotherium-"Serridentinus,"* mastodon

---

[1] Shotwell (1958, p. 458) has stressed the fact that the middle Pliocene *Pseudaplodon asiatica* of China is less advanced than *P. occidentale* from the early Pliocene of North America, and that, therefore, the Asiatic species must represent dispersal to Asia of an earlier date followed by parallel evolution in Old and New Worlds.

[2] The jumping mice *Pliozapus* from the Hemphillian and *Sminthozapus* from the earliest part of the Weże breccia are both closely related to living *Eozapus*. Although it seems somewhat improbable that two forms could have developed so similarly through parallel evolution, it is even more improbable that the hemisphere of their ancestry can be determined from present records.

[3] A primitive otter questionably assigned to *Lutra* is known from Hemphillian and early Pliocene (Clarendonian) faunas of North America. *Lutravus*, from the Hemphillian, appears to be so distinct from *Lutra* that it is here included in Table 1 as a new lutrine immigrant from Asia.

[4] The scarcity of machairodonts in the middle and late Miocene and early Pliocene of North America, and the abundance of *Machairodus* in the Hemphillian, have suggested to many a new immigration of saber-toothed cats into North America from Asia.

aquatic specializations. In North America, the voles *Microtoscoptes* and *Promimomys* are found only in association with *Dipoides* and presumably also were closely associated with water; indeed, distribution of fossils within the Hemphillian lake beds near Rome, Oregon, suggests that *Microtoscoptes* was as well adapted to life in water as was *Dipoides*, possibly occupying the ecologic niche now filled by muskrats. The remainder are best considered to be forest forms but do not preclude broken forest with some grassland.

BLANCAN FAUNA

On the basis of faunal exchange, it seems fairly certain that the North American faunas here included in the Blancan mammalian age of Wood and others (1941, p. 12–13) are correlative in time to both the late Pliocene (Csarnotan or Rousillon or "Astian"[b]) and early Pleistocene (Villafranchian) faunas of the Old World.

[b] The day before this report was delivered to the publisher, I had the opportunity to discuss, at length, the correlation of late Cenozoic continental faunas of Europe with Professor Miklós Kretzoi of Budapest. During this discussion it developed that I had misinterpreted his Csarnotan age because of an error in Fig. 5 of Kretzoi and Vértes (1965, p. 78). Kretzoi clearly distinguishes between the older faunas comparable to those from Rousillon and the younger faunas that he places in his Csarnotan age. Both are considered younger than the middle Pliocene Pannonian faunas and are older than

A correlation of the Blancan faunas to both Csarnotan and Villafranchian faunas is also strongly suggested by recent potassium-argon age determinations. Curtis (1965) has published a K/Ar date of "over 3 million years," which has subsequently been calculated at 3.3 million years (Curtis, written communication, 1965), for the Etouaires fauna of earliest Villafranchian age at Perrier, and he has also assigned a date of 2.5 million years B.P. to the later Villafranchian Roccaneyra fauna at Perrier. The Hagerman fauna is the oldest Blancan fauna for which a K/Ar date is available; this fauna has a date of 3.5 million year B.P. (Evernden and others, 1964, p. 191), and is clearly more advanced than several earlier Blancan faunas, which, consequently, appear to be pre-Villafranchian on the basis of radiometric dates.

According to Kurtén (1963a, p. 12), recent collecting in the type locality of the Villafranchian stage has produced mammals of a very early aspect comparable to those of the Etouaires fauna; this fauna is, presumably, the most logical basis for a definition of the Villafranchian. The definition of the Pliocene-Pleistocene boundary as the base of the "Calabrian Formation (marine) together with its terrestrial (continental) equivalent the Villafranchian" (I.G.C. Commission [as approved by the I.G.C. Council on Sept. 1, 1948], 1950, p. 6) thus appears to place this time boundary within the Blancan mammalian age of North America.

Although the Blancan is defined as an interval in the history of North American mammals and not as a subdivision of either the Pliocene or the Pleistocene, some dissatisfaction has been expressed because neither limit of this definition coincides with the International Geological Congress's definition of the Pliocene-Pleistocene boundary. For this reason, some workers have suggested that the Pliocene part of the Blancan be placed in the Hemphillian mammalian age. Others have suggested separating those Blancan faunas of Pliocene age from both accepted mammalian ages and placing them in a new "age," intermediate between the Blancan and Hemphillian ages. Neither suggestion has been popular, largely because, for either definition, a particular part of the original Blancan age of North America must be correlated to the base of the Villafranchian of Europe—a requirement that cannot at present be met.

It does appear possible to subdivide the Blancan mammalian age into older and younger parts, largely on the basis of the apparent phylogeny of

---

faunas of earliest Pleistocene age here referred to as Villafranchian and equal to those Kretzoi places in his Villányian age. However, because of my misinterpretation, Csarnotan is applied throughout this report as an age term equal to both Rousillion and Csarnotan ages as used by Kretzoi.

some native North American lineages (Table 4) but partly on the basis of a few immigrants from the Old World first reported in later Blancan faunas (*Enhydriodon* and possibly *Synaptomys* and *Canimartes*). The distinction is vague, however, and cannot be correlated with Old World faunal units.

Some variation from conventional assignment of local faunas to the Hemphillian or Blancan ages occurs in this report. Transitional faunas, such as Goleta, Buis Ranch, and Saw Rock Canyon, are placed in the latest Hemphillian by most authorities. Such faunas are here placed in the earliest Blancan, because they, or apparently contemporaneous faunas, include Old World forms that are otherwise first recorded in the Blancan of North America. This decision derives the philosophy that immigrants make the most practical basis for establishing limits to the North American mammalian ages. The alternative is to retain these faunas in the latest Hemphillian, to describe the "Blancan" wave of Old World immigrants as beginning in the latest Hemphillian, and to correlate the latest Hemphillian with an early part of the Csarnotan age of Europe. If this is done, the definition of the Hemphillian-Blancan faunal boundary would appear to be more uncertain than in the usage adopted here. However, the change in assignment of local

TABLE 4. *Possible Phylogenetic Series Showing Differences Between Early[1] and Late Blancan Faunas*

| Hemphillian Ancestor | Early Blancan Forms | Late Blancan Forms |
|---|---|---|
| *Promimomys* | *Ogmodontomys, Pliophena-comys primaevus* | *Cosomys, Pliopotomys, Nebraskomys, Ondatra, Pliophenacomys parvus* |
| (Old World cricetines) | *Sigmodon* (*intermedius* stage) | *Sigmodon* (*curtisi* stage) |
| *Rhynchotherium* | *Stegomastodon* (*rexroadensis* stage) | *Stegomastodon* (*mirificus* stage) |
| *Pliozapus*(?) | *Zapus rexroadensis* | *Zapus sandersi* |
| *Pliohippus* (*mexicanus* stage) | *Plesippus* (Rexroad stage) | *Plesippus simplicidens* and other species |
| *Nannippus* | *Nannippus* (Rexroad-Benson stage) | *Nannippus phlegon* and other species |
| *Osteoborus* | *Borophagus* | *Borophagus diversidens* and other species |

1 As subdivided in this report, the following represent the best-known early Blancan faunas: Goleta, Buis Ranch, Saw Rock, Rexroad, Miñaca, Benson, and Red Corral. The Hagerman fauna, and its correlative, the San Joaquin fauna, are often placed in the early Blancan, partly because they appear to be Pliocene rather than Pleistocene. These faunas are here placed in the late Blancan because they contain several forms listed in this table that appear to be more advanced than those listed from early Blancan faunas.

faunas here adopted is entirely for convenience in North American usage. As Hibbard (1964, p. 116) states, in reference to the age assignment of the Saw Rock Canyon fauna, "it is of no great concern whether the fauna is placed in the late Hemphillian or early Blancan, as long as it is understood that early Blancan implies early upper Pliocene."

The Blancan faunas of North America include many genera that represent new immigration of Palearctic lineages. Some genera that have been cited by others in support of this dispersal, such as the Old World genus *Mimomys*, are not shown in Table 5, because there is valid reason for doubting that they entered North America in Blancan time.

In addition to those genera in Table 5 representing Palearctic immigrants in Blancan time, the Blancan faunas also contain many genera shared with contemporary faunas of the Old World. These may be grouped into three categories, as shown in Table 6: (1) Holarctic stock present in earlier Pliocene faunas, (2) emigrants from Nearctica, and (3) genera of uncertain zoogeographic significance.

In Table 6 no attempt has been made to reevaluate Old World correlations. In China, the "Villafranchian" of Teilhard and Leroy is accepted, but "Pontian" is enlarged to include their "Pontian" and their "middle Pliocene." In India, the upper Siwalik fauna is considered as Villafranchian. In Europe, use of the Villafranchian follows Kurtén (1963a) and is divided into early and late parts by his standards. The term "Csarnotan age" of Kretzoi (see Kretzoi and Vértes, 1965, Fig. 5) is used, as it is used in central Europe, for those faunas that are transitional between Pannonian and Villafranchian faunas and that are correlative to the marine Astian and Plaisancian Stages; faunas of this age have also been referred to as being of Rousillon type or as earliest Anthropogene.

From Tables 5 and 6 it can be seen that as many as 29 Blancan genera are identical or closely related to Old World forms of Csarnotan or Villafranchian age. As many as 19 of these appear to have dispersed between the Old and New Worlds during Blancan time. This is a record of even greater faunal similarity and of more extensive faunal exchange than is evidenced by Hemphillian faunas, and represents perhaps one-fifth of the time involved in the Hemphillian, according to available radiometric dates.

One obvious difference between the intercontinental faunal exchange of the Hemphillian and that of the Blancan is the directional imbalance of the Blancan dispersals. In the Hemphillian as many as nine genera arrived in the New World as immigrants from the Palearctic faunas, and as many as five genera left as emigrants from Nearctica. In the Blancan, as many as

TABLE 5. *Blancan Genera Representing the First Nearctic Record of Their Lineage*

| Blancan Genus | Closest Old-World Form and/or Earliest Record of Lineage | Age of Earliest Old-World Records of Genus |
|---|---|---|
| Notiosorex[1] | Chodsigoa bohlini | Middle Pleistocene, China |
| Sigmodon[2] Bensonomys Symmetrodontomys | Cricetine complex of Old-World "Pontian" | "Pontian" |
| Probably Synaptomys[3] | Unknown origin, possibly Old-World microtine complex | Unknown |
| Lynx | Lynx issiodorensis | Csarnotan of Europe and USSR |
| Trigonictis[4] | Pannonictis pliocaenica | Villafranchian of Europe |
| Canimartes(?)[5] | Unknown origin, but apparently Old World; possibly Trochichtis | Possibly late Miocene and Pliocene of Europe |
| Probably Enhydra[6] | Probably from Asiatic "Pontian" otters | Probably "Pontian" |
| Enhydriodon[7] | Enhydriodon lluecai | Pannonian and Csarnotan of Europe; "Pontian" and "Villafranchian" of India |
| Ursus | Ursus minimus | Csarnotan and early Villafranchian |
| Chasmaporthetes[8] | Euryboas lunensis | Villafranchian |
| Cervus[9] Odocoileus | Cervus | Csarnotan of Europe and USSR; "Villafranchian" of China |

1 *Notiosorex* belongs in an Old-World subfamily of shrews; the subfamily is also represented earlier in North America, but by the lineage leading to *Anourosorex*, not to *Notiosorex* and *Chodsigoa*. *Chodsigoa* appears later in the record of China than *Notiosorex* in North America, but the Old World origin of the lineage is unquestionable.

2 The Hemphillian is remarkably poor in varieties of cricetine as well as microtine rodents. About the only cricetine rodents represented are varieties of *Peromyscus*-like species. The three genera listed here are believed to be Blancan immigrants from Eurasia, largely because no possible ancestor is known in North America and all resemble earlier Old-World cricetine rodents.

3 The origin of the lemmings is unknown; the late Blancan record of *Synaptomys* is the oldest record of this distinctive group. Considering their modern distribution and the great number of pre-Pleistocene Old-World microtines, one is tempted to postulate a boreal Old-World origin for the lemmings, which, as yet, is undocumented by fossil evidence.

4 The early and late Blancan *Trigonictis kansasensis* appears to be at least generically identical with *Pannonictis pliocaenica* of the European Villafranchian faunas.

5 The late Blancan *Canimartes(?) idahoensis* and *C.(?) cooki* appear to be derived from Old-World mustelids, but it is not certain which; the late Miocene to early Pliocene *Trochichtis* appears most similar to *Canimartes(?)*.

6 The origin of *Enhydra*, the sea otter, is unknown. The earliest record is from North America in beds tentatively considered to be of Blancan age (Leffler, 1964, p. 63). This record is of a femur that is very similar to the femur of living *Enhydra*. If the reported English occurrence of *Enhydra* in the Norwich Crag of post-Villafranchian age is correct, it probably represents introduction of this North Pacific otter into the North Atlantic by means of the Bering Strait. However, this specimen, an $M_1$, suggests a morphologic stage about intermediate between living *Enhydra* and more conventional otters, particularly the African clawless otter (Pohle, 1919, p. 196). It would appear to be less similar to living *Enhydra* than the Blancan femur, strongly suggesting that it may be a different lineage than *Enhydra*.

7 An undescribed mustelid from the San Joaquin fauna (late Blancan) is a species of *Enhy-*

14 Old World genera appear to have been introduced into the North American fauna and only four or possibly five genera seem to have moved westward into Asia.

Simpson (1947, p. 644–645) has theorized that the competitive environment was much more demanding in temperate Eurasia than in North America, both because of the larger area with resulting greater population differences and because of continued introduction of the more aggressive elements evolving in the Asian and African tropics. He concludes, "It would thus be expected that ability to migrate and to survive under changing conditions would be selected far more rigorously in Eurasia than in North America." Unquestionably, the increasing rate of development of climatic zonality in temperate regions during Blancan time accentuated the advantages of the adaptable members of the Eurasian fauna. This becomes increasingly apparent in consideration of later Pleistocene intercontinental mammalian exchange discussed in following sections of this report.

Those mammals involved in dispersal across the Bering Land Bridge during Blancan time do not differ greatly in environmental significance from those migrants of Hemphillian time. Nearly all appear to be forest or aquatic forms. Only the zebra *Plesippus* suggests the need for appreciable grassland. The camel *Paracamelus* is ancestral to modern Old World camels and could have, by Blancan time, become adapted to open-steppe environments, but was derived from a browsing forest form of the Hemphillian (S. D. Webb, written communication, 1964).

OLD WORLD CORRELATES OF THE BLANCAN

As has been stated, there appears to be little question that the Blancan faunas of North America are correlative to both the Csarnotan faunas and the Villafranchian faunas of Europe. As can be seen from Tables 5 and 6, there are five genera in the Blancan faunas of North America that cannot be older than the Csarnotan faunas of the Old World. Two of these, *Chasmaporthetes* (or *Euryboas*) and *Pannonictis* (or *Trigonictis*) are known only from the Villafranchian in Europe. (They are, however, known from early Blancan localities in North America.) Also, there are three genera in the Old World Villafranchian or Csarnotan faunas that cannot be

---

*driodon.* It is less advanced than the Villafranchian *Enhydriodon sivalensis* of India, however, and is much closer to *Enhydriodon lluecai* from the late Pontian of Los Algezares, Spain, or to *Enhydriodon campani* from the Pontian of Monte Bamboli, Italy.

[8] Largely unpublished discoveries of "narrow-toothed" hyaenid remains from early and late Blancan localities (Repenning, 1962, p. 555) may all belong to the genus *Chasmaporthetes*, which appears to be the senior synonym of the Villafranchian genus *Euryboas*.

[9] Some doubt about the origin of *Odocoileus* has been raised (see Flerow, 1952, p. 10, 13, 14); it is here considered a derivative of *Cervus*.

TABLE 6. *Holarctic Genera of Blancan Age Not Representing Dispersal to Nearctica in Blancan Time*

| Genus | Zoogeographic Significance[1] | China | India | Europe |
|---|---|---|---|---|
| *Sorex* | 1 | Post-"Villafran-chian" to living | Essentially absent | Pannonian to living |
| *Hypolagus* | 1 | "Villafranchian" | Absent | Pannonian to late Villafranchian |
| *Dipoides* | 1 | "Pontian" | Absent | Pannonian to Csarnotan |
| *Vulpes* | 1 or 2 | "Villafranchian" to living | "Villafranchian" to living | Pannonian (?) to living |
| *Lutra* | 1 | "Pontian" to living | "Pontian"(?) to living | Pannonian to living |
| *Felis* | 1 | "Pontian" to living | "Pontian"(?) to living | Pannonian to living |
| *Tapirus* | 1 | "Pontian" to Pleistocene | Absent | Pannonian to early Villafranchian |
| *Marmota* | 2 | "Villafranchian" to living (Csarnotan to living, Siberia) | Pleistocene (?) to living | Elster I to living |
| *Canis* | 2 | "Villafranchian" to living | "Villafranchian" to living | Csarnotan to living |
| *Plesippus* | 2 | "Villafranchian" to middle Pleistocene | Absent | Late Villafranchian |
| *Paracamelus*[2] | 2 | "Villafranchian" to middle Pleistocene | "Villafranchian" to middle (?) Pleistocene | Villafranchian (USSR) |
| *Martes* | 3 | "Pontian" to living | Pleistocene (?) to living | Pannonian to living |
| *Mammut* | 3 | "Pontian" to late (?) Pleistocene | "Pontian" to late (?) Pleistocene | Pannonian to Elster I |

[1] (1) Holarctic in pre-Blancan faunas. (2) Emigrant from Nearctica. (3) Uncertain.

[2] Probably derived from North American Pliocene *Procamelus* (S. D. Webb, written communication, 1964) but not recognized in Blancan faunas.

older than Blancan. All three are present in what appear to be Csarnotan equivalent deposits in Asia, but one, the zebra *Plesippus*, did not reach Europe until late Villafranchian time; another, the marmot *Marmota*, did not reach Europe until middle Pleistocene time; and the third, the camel *Paracamelus*, did not reach Europe at all.

From this it appears that the North American Blancan can be no older than the Csarnotan of Europe, but, disregarding Csarnotan correlations to Asian faunas, it could also be no older than late Villafranchian of Europe (the time of arrival of *Plesippus* into European faunas). The correlation of the Blancan mammalian age to both the Csarnotan and the Villafranchian mammalian ages of Europe thus depends on the rapid dispersal of the North American wolf, *Canis*.

*Canis*, which by itself could be as old as Hemphillian, arrived in China essentially simultaneously with *Marmota*, *Plesippus*, and *Paracamelus*; these occurrences are in faunas that Teilhard and Leroy call Villafranchian. In India, *Canis* and *Paracamelus* also appear together in faunas considered to be Villafranchian. In Siberia, *Marmota* is reported from a Csarnotan fauna (Tologoi) near Lake Baikal (N. K. Vereshchagin, oral communication, 1965). And, finally, *Canis* is first reported in the Csarnotan faunas of Europe, long before the arrival of *Plesippus* and *Marmota*. It thus appears that the immigration of *Canis*, *Plesippus*, *Marmota*, and *Paracamelus* into Eurasia was a Csarnotan event, but that in varying degrees *Plesippus*, *Marmota*, and *Paracamelus* experienced a lag in dispersal across this continent to Europe. It also appears, therefore, that faunas called Villafranchian in China and India contain parts that must be as old as the occurrence of *Canis* in the European Csarnotan faunas. In addition, these faunas can be no older than the North American Blancan mammalian age, as determined by the evolutionary significance of *Plesippus*, *Marmota*, and *Paracamelus*. This conclusion is in agreement with the K/Ar age determinations cited earlier, 3.5 million years B.P. being the approximate age of a mid-Blancan fauna, and 3.3 million years B.P. being the approximate age of an earliest Villafranchian fauna.

Hibbard has suggested that the hare *Lepus* also indicates that part of the Blancan mammalian age is older than Villafranchian faunas of Europe. The suggestion based on this mammal seems less secure to me, however. He (Hibbard, 1963) has observed *Lepus*-like and *Oryctolagus*-like mutations in a population of early Blancan *Nekrolagus*, and postulates the derivation of these to genera from *Nekrolagus*. Thus, *Oryctolagus* and *Lepus*, first known from the Villafranchian of Europe, must be younger than early Blancan. However, the lack of North American *Oryctolagus*, the failure to discover any North American *Lepus* before the Irvingtonian mammalian age, and the lack of any record of *Nekrolagus* in the Old World all suggest greater complexity to the story than is now known: possibly, *Lepus* and *Oryctolagus* evolved in the Old World from either an Old World *Alilepus* or from an unknown Old World *Nekrolagus*. In this case there would be no demonstrable temporal significance in *Lepus*-like and *Oryctolagus*-like mutations from the early Blancan.

*Cosomys* has often been placed in the Old World genus *Mimomys*, and has been used both as a basis of correlation with the Villafranchian and as evidence of intercontinental dispersal. It has also been pointed out that *Cosomys* is more primitive than the most primitive Villanfranchian species

of *Mimomys*, *M. pliocaenicus*, the implication being that *Cosomys* is probably pre-Villafranchian in age. It is, however, no more primitive than several Csarnotan vole species variously referred to as either *Mimomys* or
*Cseria*. *Cosomys* need not so much reflect, as parallel, the evolution of *Mimomys*. Largely through the works of Hibbard it has become evident that
there is a complete morphologic gradation between the Hemphillian *Promimomys*, the early Blancan *Ogmodontomys*, and the late Blancan *Cosomys*.
In view of this gradation and in view of the fact that *Cosomys* represents a
stage of evolution comparable to a pre-Villafranchian stage of *Mimomys*
evolution, strong evidence will be required to state that late Blancan *Cosomys* is not a native North American vole derived from Hemphillian ancestors.

The end of the Blancan mammalian age of North America is marked by
the appearance, in Irvingtonian faunas, of the genera *Mammuthus*, *Equus*,
*Lepus*, *Microtus*, *Euceratherium*, and *Dinobastis*, as well as of several other
genera that have less bearing on correlation with Old World faunas. In
China, a comparable event marks the end of the "Villafranchian" of Teilhard and Leroy (1942): the arrival of *Equus*, *Lepus*, *Microtus*, and *Dinobastis* (*Machairodus* [*Epimachairodus?*] *ultimus* Teilhard *fide* Kurtén
[1963b, p. 98]). However, in China, the earliest *Mammuthus* is pre-"Villafranchian" of Teilhard and Leroy, and is a more primitive species than is
known in North America; but this is consistent with the known Asiatic
origin of the genus. Clearly, the late Villafranchian arrival of *Mammuthus*
in Europe is not relevant to North American correlations.

There appears to be little problem in correlating the end of "Villafranchian" time in China with the end of Blancan time in North America. Correlation of this time to Europe is, however, much more obscure. Of those
mammals that mark the beginning of the middle Pleistocene in China and
of the Irvingtonian age in North America, only *Dinobastis* and *Microtus*
appear to be significant. *Lepus* arrives in Europe in the late Villafranchian
(upper Val d'Arno fauna), and the horse *Equus* is first known earlier in the
Villafranchian than all other genera considered to indicate North American
Irvingtonian time. *Equus*, in fact, is found earlier than the North American
Blancan zebra *Plesippus*, according to Kurtén's tabulation of Villafranchian
faunas of Europe (1963a).

Kurtén (1963b) shows that *Dinobastis* is a characteristic Holarctic
middle Pleistocene saber-toothed cat, and that (Kurtén, 1960a, p. 31) its
earliest occurrence in Europe is in the latest Villafranchian Tiglian fauna.
The first European occurrence of the meadow mouse *Microtus* appears to
be from Cromerian (Günz-Mindel) faunas (for example, Fejfar, 1961, p.

193) and is used by Kretzoi (see Kretzoi and Vértes, 1965, p. 79) as one of the genera that characterize his Biharian stage in Europe. The continent of origin of *Dinobastis* may be questionable, since older homotheriine predecessors were present in both hemispheres, but the Old World origin of *Microtus* seems unquestionable in view of available information on the fossil history of these rodents. At present, therefore, it would appear that the earliest occurrence of *Microtus* in North America must be considered no older than the earliest record in Eurasia, and that, therefore, the end of Blancan time in North America can be no older, within the margin of error imposed by our fossil record, than the end of "Villafranchian" time in China and of the end of Villafranchian time in Europe.

This correlation, however, carries certain implications that are not accepted by all authorities. If *Lepus* and *Equus* are first known in Europe in the Villafranchian, and are unknown in both China and North America until middle Pleistocene time, they can hardly be of North American origin.

Because of the essential lack of mammalian dispersal from North America to Eurasia at the beginning of Irvingtonian time, there is no faunal control to suggest that the Blancan-Irvingtonian faunal boundary cannot be younger than the Villafranchian-Biharian faunal boundary. Some help in establishing the correlation is found in recent K/Ar age determinations. Evernden and others (1964, p. 164) list an Irvingtonian age determination of 1.36 million years B.P.; they also (p. 170) give a Villafranchian age determination from France of 1.61 million years B.P. This would suggest that the Blancan-Irvingtonian boundary could have been no more than a quarter of a million years later than the Villafranchian-Biharian boundary. For the present, I am considering the two faunal boundaries as synchronous.

IRVINGTONIAN FAUNA

As defined by Savage (1951, p. 289), the Irvingtonian North American mammalian age is characterized by species less advanced than Ranchola-brean and Recent species and by the absence of *Bison*. However, as with the Blancan, the beginning of Irvingtonian time is strongly marked by the appearance in North America of a variety of Old World immigrants. Significant additions to the North American fauna are the first elephantids (*Mammuthus*), bovids (*Euceratherium* but not *Bison*), new hypsodont microtines (*Microtus* and *Pitymys*), hares (*Lepus*), horses (*Equus*), and saber-toothed cats (*Dinobastis* and *Smilodon*). Genera of Old World origin first recorded in the North American fauna during the Irvingtonian age are shown in Table 7, along with their Old World records.

TABLE 7. *Irvingtonian Genera Representing the First Nearctic Record of Their Lineage*

| Irvingtonian Genus | Earliest Record in China | Earliest Record in Europe |
|---|---|---|
| *Lepus*, hare | Middle Pleistocene | Late Villafranchian, Europe (Csarnotan(?), USSR) |
| Possibly *Tamiasciurus*, chickaree or red squirrel | Villafranchian (*Sciurotamias*) | No record |
| *Pitymys*, pine vole | Middle Pleistocene | Cromerian |
| *Microtus* (two subgenera), meadow mice | Middle Pleistocene | Günz |
| *Gulo*, wolverine | Middle Pleistocene | Cromerian |
| *Dinobastis*, saber-toothed cat | Middle Pleistocene | Tiglian |
| *Smilodon* (from Old-World *Meganteron*), saber-toothed cat | Middle Pleistocene | Villafranchian |
| *Mammuthus* (*M. haroldcooki*), mammoth | Pontian | Late Villafranchian |
| *Euceratherium*, shrub ox | Villafranchian bovids | Villafranchian bovids |
| *Equus*, horse | Middle Pleistocene | Late Villafranchian |

The uncertainty of the place of origin of *Lepus* has been discussed. According to the correlation of Blancan and Villafranchian here considered most reasonable, *Lepus* must have originated in the Old World from native *Alilepus* stock. Several records of *Lepus* from faunas presumably of Csarnotan age in the southern part of European USSR (Alekseieva, 1961) would seem most significant. The similar problem of *Equus* has also been mentioned.

Although microtines are reasonably common in the Blancan, they remain conspicuously primitive and undiversified in comparison with the variety seen in the Villafranchian of Europe. Only *Pliolemmus* and *Synaptomys* from the late Blancan have rootless cheek teeth. In Irvingtonian time, the North American spectrum of microtines matured in its native stock and enlarged through immigration of Old World forms. For the first time the North American microtine fauna resembled that of Europe. However, not until Rancholabrean time can the Old World microtine fauna be considered to have become Holarctic in distribution. Significantly, although Nearctic cricetine rodents continue to diversify, none are introduced from the Old World in Irvingtonian time; quite likely, this is the effect of Pleistocene climatic changes on these predominantly warm-temperate to tropical rodents. Kurtén (1963b) has discussed the immigration of *Gulo*, *Dinobastis*, and *Smilodon* into North America in Irvingtonian time.

The imbalance in exchange of faunal elements between Old and New Worlds during Blancan time was accentuated during Irvingtonian time.

Although about 11 genera of Old World origin appear for the first time in North America in Irvingtonian faunas, only two genera of New World origin appear to have dispersed into Palearctica. The North American ground squirrel *Spermophilus* is first reported in the Old World in the middle Pleistocene of China and the Günz of Europe. Although not represented in the Irvingtonian, the shrew *Shikamainosorex* (Hasegawa, 1957) from the middle Pleistocene of Japan is so similar to the Blancan *Paracryptotis* that it must be considered a North American immigrant presumably of middle Pleistocene time. The possibility of a marine dispersal of *Enhydra* to the North Atlantic, based on the specimen from the middle Pleistocene Norwich Crag, has been mentioned (Table 5).

As with Blancan faunas, correlation of Irvingtonian faunas to those of the middle Pleistocene in China is much more obvious than their correlation to the middle Pleistocene faunas of Europe (here arbitrarily including faunas dated from post-Tiglian Günz to Mindel glacial ages, or essentially equivalent to Kretzoi's [1941] Biharian Stage). Seven of the 11 Eurasian genera introduced to North America in the Irvingtonian faunas also first appear in the middle Pleistocene of China. *Tamiasciurus* (questionably as *Sciurotamias*), *Mammuthus*, and the bovids appear earlier in China. Four of these genera appear first in the middle Pleistocene of Europe and five are Villafranchian or older in their earliest records. The microtine rodents appear to be the definitive element in this correlation, for it is reasonable to postulate that these are derived from Old World sources; the Irvingtonian microtines are quite clearly at a stage of evolution comparable to the European middle Pleistocene microtines, though not as diverse.

Insofar as published records show, the youngest radiometric age determinations on rocks associated with Blancan faunas are about 2 million years B.P. (Evernden *et al.*, 1964, p. 164, and appendix). The same publication lists a radiometric date of about 1.36 million years B.P. for an Irvingtonian fauna. The Blancan-Irvingtonian faunal transition is thus between these ages.

The Irvingtonian-Rancholabrean faunal transition is even less certain. Between the above-cited Irvingtonian date of 1.36 million years B.P. and the limit of carbon-14 dating, there are no radiometric data available that can be related to North American mammalian ages.

The hare *Lepus*, the mammoth *Mammuthus*, the shrub ox *Euceratherium*, the horse *Equus*, and the ground squirrel *Spermophilus*, which crossed the Bering Land Bridge in Irvingtonian times, would seem to have preferred open grasslands although probably none would have been out of place in open forests with scattered glades. The other migrants would appear to have been better adapted to life within forests. There is, however, a much stronger suggestion of grassland habitats for the fauna crossing the Bering Land Bridge during Irvingtonian time than for those crossing during Blancan time.

### RANCHOLABREAN

As defined by Savage (1951, p. 289), the Rancholabrean North American mammalian age is characterized by the presence of *Bison* and by many species inseparable from Recent species. As with older mammalian ages, the Rancholabrean is also characterized by a good number of Old World immigrants, about 23. Eight of these are known only from the Rancholabrean fauna of Alaska, however, and it is questionable whether these should be considered Rancholabrean immigrants to North America, because it is obvious from the distribution of continental ice that Alaska was a part of the Northeastern Siberian faunal province during periods of Pleistocene glaciation. Five of these eight have obviously become a part of the North American fauna in Recent times, however.

From the modern distribution of those species still living, and from the distribution of boreal forms not recorded as fossil, some further inferences might be drawn about times of immigration. One genus (*Alopex*) and one species (*Lepus timidus*) have not been recognized in the North American fossil record, but appear to have immigrated into the New World during Rancholabrean time.

Those Rancholabrean taxa that represent new immigration into North America are shown in Table 8. Some 21 forms are represented in the fossil record; nine of these new immigrants are reported in early Rancholabrean faunas (pre-Wisconsin), and only 12 are reported from the Pleistocene of North America outside of Alaska. There appears to have been no dispersal at all from North America to Eurasia at this time. The adaptive superiority of the Palearctic fauna seems to have reached a maximum in Rancholabrean time.

There is great holarctic faunal similarity in the late Pleistocene. The greatest similarity of the Rancholabrean fauna is, expectably, to the Northeastern Siberian late Pleistocene fauna. This applies particularly to Alaska. However, most North American immigrants have a middle Pleistocene or earlier record in Eurasia and offer little basis for correlation of the Rancholabrean to the late Pleistocene of the Old World. Although five genera in Table 8 are known in the Old World only from the late Pleistocene, all but one—*Dicrostonyx*—are found in North America exclusively in Rancholabrean faunas of Alaska. *Dicrostonyx* itself is known from only two North American fossil localities outside of Alaska.[c]

Eleven of the 23 mammals considered to have immigrated from Eurasia to the New World in Rancholabrean time have been reported only from late Rancholabrean localities correlated with the Wisconsin Glaciation. This suggests the possibility of a twofold immigration from Asia during Rancholabrean time. In addition, the modern distribution of several species of boreal mammals (see notes in Table 8) suggests that they were introduced only late in the Rancholabrean, during the last glacial advance. Although the observed data are not conclusive and could be interpreted in other ways, it seems most reasonable at present to infer two periods of immigration during this mammalian age.

A glance at Table 8 will show that the nature of Rancholabrean immigrants is highly boreal. Nearly all forms are mammals of either tundra or taiga, and the list of immigrants differs distinctly from those of Irvingtonian and earlier immigrations in having a pronounced climatic as well as other ecologic significance. For the first time in the Pleistocene history of mammalian dispersal across Beringia, distinctly arctic and chiefly opensteppe and tundra conditions are indicated.

## Glacial Chronology

Although glacial ages have been used in the preceding discussions of correlations of North American faunas to Eurasian faunas, neither the degree of detail of this review nor the actual fossil record with its correlation to glacial chronology is adequate to equate particular European glaciations with those of North America. Current assignments to glacial ages of Pleistocene mammalian faunas of Europe (Kurtén, 1960a,b, 1963a) and North America (Hibbard *et al.*, 1965) have been used in this report. In China, the "Villafranchian" and Middle and Late Pleistocene assignments of faunas follow the usage of Teilhard and Leroy (1942), although Kurtén (1960b) has proposed a correlation of these faunas with European glaciations. Because of the uncertainties of correlation with European glacial ages, the Russians have prudently used subdivisions of the Anthropogene in Siberia without reference to glacial ages. However, because the Russian terminology

[c] According to Peter Robinson (oral communication, 1965), *Dicrostonyx* has recently been found in a Rancholabrean fauna in Colorado. It has also been found in deposits of this age in Pennsylvania (Guilday and Doutt, 1961).

TABLE 8. *Rancholabrean Taxa Representing Immigration from Palearctica*

| Taxon | Rancholabrean Record | | Earliest Record in East Asia | Earliest Record in Europe |
| | Early-Late | Alaska Only or Northern America Generally | | |
| --- | --- | --- | --- | --- |
| *Homo* | Late | N. America | Middle Pleistocene, China | Mindel |
| *Lepus timidus*[1] | No fossil record | | Late Pleistocene | Würm |
| *Spermophilus undulatus* | Late | Alaska | "Riss," N.E. Siberia (An$_2$1) | No record |
| *Lemmus*[2] | Late | Alaska | Post-"Würm," N.E. Siberia | Cromerian |
| *Dicrostonyx*[3] | Early | N. America | "Riss," N.E. Siberia (An$_2$1) | Eemian |
| *Clethrionomys*[1] | Late | N. America | Middle Pleistocene, China; "Riss," N.E. Siberia (An$_2$1) | Günz |
| *Microtus* (*Stenocranius*)[1] | Late | Alaska | "Riss," Siberia (An$_2$1) | Cromerian |
| *Alopex*[4] | No fossil record | | "Cromerian," Siberia (An$_2$1) | Würm |
| *Lynx canadensis*[5] | Late | Alaska | Late Pleistocene, China | Würm |
| *Xenocyon*[6] | Early | Alaska | Middle Pleistocene, China[6] | Cromerian |
| *Mammuthus primigenius* | Early | N. America | Late Pleistocene, China "Riss," N.E. Siberia (An$_2$1) | Mindel |
| *Sangamona* | Late | N. America | Eurasian Pleistocene Cervidae | |
| *Cervalces* | Early[7] | N. America | Eurasian Pleistocene Cervidae | |
| *Alces* | Early[7] | Alaska | "Günz," N.E. Siberia (An$_1$2) | Villafranchian |
| *Rangifer* | Early[7] | Alaska | "Riss," N.E. Siberia (An$_2$1) | Riss |
| *Bison* | Early | N. America | "Günz," N.E. Siberia (An$_1$2) Villafranchian, India and China | Mindel |
| *Bos* | Late | Alaska | "Würm," N.E. Siberia (An$_2$2) ; Late Pleistocene, China | Holstein |
| *Saiga* | Late | Alaska | "Riss," N.E. Siberia (An$_2$1) | Würm |
| *Oreamnos* *Ovis* | Late and Early[7] | N. America | *Ovis*—Villafranchian, China; "Eemian," N.E. Siberia (An$_2$2) | *Rubicapra*— sub-Recent |
| *Bootherium* *Symbos* *Ovibos* | Early[7] | N. America | *Ovibos*—"Riss," N.E. Siberia; *Boopsis*—middle Pleistocene, China | *Ovibos*—Würm *Praeovibos*— Cromerian |

1 Rausch (1963) cites reasons for inferring a late Rancholabrean (Wisconsin) introduction of *Lepus timidus, Clethrionomys rutilus, Microtus oeconomus, Microtus (Stenocranius) gregalis,* and *Ovis nivicola* into North America.

2 *Lemmus* and *Spermophilus undulatus* are known as fossil in North America only from the Rancholabrean of Alaska, where they are associated with *Dicrostonyx torquatus* (not *D. hudsonius*) (Repenning and others, 1964).

3 *Dicrostonyx* appears to have entered North America in earlier Rancholabrean time, to have been forced southward with the tundra during the Wisconsin Glaciation, and to have returned northward with the retreat of the ice, creating an isolated population east of Hudson Bay (Guilday, 1963). This population is believed by Guilday to have remained isolated as *D. hudsonius*, while the population west of Hudson Bay and in the Old World was replaced by the modern *D. torquatus*

may not be familiar to some, glacial terminology has been applied to the Anthropogene subdivision in Table 8, following the estimation of the correlations by Vangengeim (1961, Table 1).

If these correlations with the glaciations are correct, the correlation of North American and European glaciations shown in Table 9 seems most reasonable in light of the present review. Certain difficulties are apparent—the Old and New World glaciations do not match in succession. Resolution of the problem will require more data than has here been reviewed, and probably will emerge from further radiometric dating rather than from further study of mammalian faunas.

## Summary and Significance for the History of the Bering Land Bridge

During the late Cenozoic, the number of genera involved in Palearctic-Nearctic mammalian faunal exchanges remains roughly equal for each North American mammalian age: about 14 in Hemphillian, about 19 in Blancan, about 13 in Irvingtonian, and about 23 in Rancholabrean time.

The duration of each mammalian age is progressively and rapidly shortened: about six million years for the Hemphillian, about two million years for the Blancan, and perhaps one million years each for the Irvingtonian and Rancholabrean mammalian ages.

The potential for adaptability, as reflected in direction of intercontinental dispersals, is progressively biased in favor of Palearctic genera. The proportion of Palearctic immigrants to the New World, relative to Nearctic

in late Wisconsin time. The discovery of *D. hudsonius* in Rancholabrean deposits of Pennsylvania strongly supports this hypothesis (Guilday and Doutt, 1961). A different view held by Rausch (1963, p. 36) is based upon his opinion that the morphologic differences between the two species are of no taxonomic significance.

[4] There is no published Nearctic fossil record of *Alopex*, but the modern distribution of the genus in North America is comparable to *Dicrostonyx*, except that the Arctic fox has included the forests south of Hudson Bay in its range. In the Old World, the genus is known from the middle and late Pleistocene of Northeastern Siberia. Presumably it is at least a late Rancholabrean immigrant.

[5] Kurtén and Rausch (1959, p. 44) have suggested that the modern distribution and the degree of similarity of the North American *Lynx rufus*, *L. canadensis*, and the Old World *L. lynx* are such that *L. rufus* may represent an earlier immigration to North America, perhaps earliest Rancholabrean, and that *L. canadensis* may represent a later introduction, perhaps in early Wisconsin time. There is one unpublished record of *L. canadensis* from the Rancholabrean of the Arctic Slope of Alaska.

[6] Undescribed specimen, No. F:A.M. 67180, in the Frick Laboratories, American Museum of Natural History, from Cripple Creek sump, Fairbanks area, Alaska (see Péwé and Hopkins, this volume). *Cuon* cf. *alpinus*, Pei (1934, p. 39–40, specimen K) from Choukoutien Locality 1, belongs in this genus, as indicated by its size, by the *Canis*-like prominence of the metaconid, and by the greater (than *Cuon*) width of the trigonid relative to talonid on $M_1$. I am not certain whether the genus should be placed in the Caninae or in the Simocyoninae.

[7] Péwé and Hopkins, this volume.

TABLE 9. *Conventional Assignment of Glacial Ages to Mammalian Faunas, and Suggested Nearctic-Palearctic Faunal Correlations*[1]

| European Mammalian Correlation with: | | Correlation with Exchange of Nearctic-Palearctic Faunas | | | North American Mammalian Correlation with: | |
|---|---|---|---|---|---|---|
| Interglaciation | Glaciation | Europe | East Asia | North America | Glaciation | Interglaciation |
| Eemian | Würm | LATE PLEISTOCENE | LATE PLEISTOCENE | RANCHOLABREAN 23 immigrants | Wisconsin | Sangamon |
| Holsteinian | Riss | | | | Illinoian | Yarmouth |
| Cromerian | Mindel | BIHARIAN 1 immigrant | MIDDLE PLEISTOCENE 2 immigrants 3 immigrants | IRVINGTONIAN 11 immigrants | Kansan | Aftonian |
| Tiglian | Günz | VILLAFRANCHIAN 3 immigrants | "VILLAFRANCHIAN" 1 immigrant | LATE 3 immigrants BLANCAN ? | Nebraskan | |
| (Pliocene) | Donau | CSARNOTAN 1 immigrant 1 immigrant | 4 immigrants | EARLY 11 immigrants | (Pliocene) | |
| | | PANNONIAN undifferentiated (3 immigrants) | "PONTIAN" AND MIDDLE PLIOCENE (5 immigrants) | HEMPHILLIAN undifferentiated (9 immigrants) | | |

1 Migrants shown between Europe and Asia are only those related to North American dispersals.

immigrants to the Old World, is about 2 : 1 in the Hemphillian, 3 : 1 in the Blancan, 5 : 1 in the Irvingtonian, and 23 : 0 in the Rancholabrean.

Since Hemphillian time, there appear to have been at least three and possibly four surges of immigration of Old World genera into the North American fauna: early Blancan, early Irvingtonian, early Rancholabrean, and quite likely late Rancholabrean. It is tempting to correlate these surges of immigration with times of glacial advances, lowering sea levels, and emergence of the Bering Land Bridge. However, the timing of the surges may also have been affected, at least in part, by times of glacial retreats and openings of ice-free passages between Alaska and conterminous United States, across western Canada.

Time lag is demonstrable in the intercontinental dispersal of some mammalian genera. However, the possibility of significant dispersal time lag is eliminated in some other cases. Mutual intercontinental exchange of chronometrically significant taxa that occur jointly with the earliest records of the native form establish contemporaneity within the limits of the sensitivity of the fossil record. Hence, the greater the number of genera involved in mutual intercontinental exchange, the more secure are correlations based on fossil mammals. Because of the increasing west-east bias in direction of intercontinental dispersal, intercontinental correlations based on mammalian faunas become increasingly speculative from Hemphillian time on. This uncertainty is accentuated by the increasing brevity of the mammalian ages.

The adaptive superiority in Palearctic mammals resulting from the existence of larger areas of temperate environments in the Old World, as Simpson postulated, would be accentuated during glaciations because of the great extent of the ice sheets in North America, as opposed to the limited and spotty glaciation that was characteristic of Pleistocene Eurasia. Except for Alaska, which during these times had to be a part of the Siberian faunal province, there was essentially no area in North America, during glacial maxima, in which boreal faunas could evolve.

Therefore, both the increasing west-east bias in direction of intercontinental dispersal and the increase in number of boreal forms that crossed the Bering Land Bridge agree with other lines of evidence indicating progressive increase in climatic severity during the Pleistocene.

Mammals are noted for their environmental adaptability, and it is difficult to interpret environment from their records. However, the records of those mammals involved in Hemphillian intercontinental dispersal suggest that the Bering Land Bridge at that time was warm-temperate, humid, and forested; in the Blancan, dispersing mammals suggest temperate, humid, and forested conditions, and a few forms suggest some grasslands; in the

Irvingtonian, dispersing mammals suggest an even division between forest and open grasslands, and temperate climate; only in the Rancholabrean does the record of dispersing mammals strongly suggest arctic conditions for the Bering Land Bridge, with tundra, steppe, and only broken arctic forest (taiga) suggested.

## REFERENCES

Alekseieva, L. I. 1961. O rannei faze razvitiia chetvertichnoi fauny mlekopitaiushchikh na territorii iuga ievropeiskoi chasti SSSR (An early phase of the development of Quaternary mammalian fauna in south-European USSR): Izvestia Akad. Nauk SSSR, Geol. Ser., no. 12, p. 87–96. In Russian; English translation p. 71–79 in running transl. pub. by Am. Geol. Inst.

Curtis, G. H. 1965. Potassium-argon date for early Villafranchian of France (abs.): Am. Geophys. Union Trans., v. 46, p. 178.

Evernden, J. F., D. E. Savage, G. H. Curtis, and G. T. James. 1964. Potassium-argon dates and the Cenozoic mammalian chronology of North America: Am. Jour. Sci., v. 262, p. 145–198.

Fejfar, O. 1961. Review of Quaternary Vertebrata in Czechoslovakia, in Quaternary of Central and Eastern Europe: Poland, Instytut Geologiczny, Prace, t. XXXIV, Warsaw, p. 109–118.

Flerow, C. C. 1952. Fauna SSSR, Mlekopitaiushchie (Fauna of USSR, mammals); v. 1, no. 2, Kabargi i oleni (Musk deer and deer): Akad. Nauk SSSR, Inst. Zool., n.s. no. 55, 247 p. In Russian; English transl. pub. for U.S. Nat. Sci. Found. by Israel Program for Scientific Transl. 1960.

Guilday, J. E. 1963. Pleistocene zoogeography of the lemming, *Dicrostonyx*: Evolution, v. 17, p. 194–197.

Guilday, J. E., and J. K. Doutt. 1961. The collared lemming (*Dicrostonyx*) from the Pennsylvania Pleistocene: Proc. Biol. Soc. Washington, v. 74, p. 249–250.

Hasegawa, Y. 1957. On a new insectivore from the upper Kusuii formation in Japan: Yokohama Nat. Univ. Sci. Repts., sec. 2, no. 6, p. 65–69.

Hibbard, C. W. 1963. The origin of the $P_3$ pattern of *Sylvilagus, Caprolagus, Oryctolagus*, and *Lepus*: Jour. Mammalogy, v. 44, p. 1–15.

———— 1964. A contribution to the Saw Rock Canyon local fauna of Kansas: Papers, Michigan Acad. Sci., Arts, and Letters, v. 49, p. 115–127.

Hibbard, C. W., C. E. Ray, D. E. Savage, D. W. Taylor, and J. E. Guilday. 1965. Quaternary mammals of North America, p. 509–525, in H. E. Wright, Jr., and D. G. Frey, eds., The Quaternary of the United States: Princeton Univ. Press, Princeton, N.J., 922 p.

International Geological Congress Commission. 1950. Recommendations of Commission appointed to advise on the definition of the Pliocene-Pleistocene boundary, in The Pliocene-Pleistocene boundary: Proc. 18th Internat. Geol. Cong. (London), 1948, pt. 9, sec. H, p. 6.

Kretzoi, M. 1941. Betrachtungen über das Problem der Eiszeiten; ein Beitrag zur Gliederung des Jungtertärs und Quartärs: Ann. Hist.–Nat. Mus. Nation. Hungarici, v. 34, p. 56–82.

———— 1955. *Promimomys cor* n.g. n. sp., ein altertümlicher Arvicolide aus dem ungarischen Unterpleistozän: Acta Geologica Acad. Sci. Hungaricae, v. 3, p. 89–94.

Kretzoi, M., and L. Vértes. 1965. Upper Biharian (Intermindel) pebble-industry occupation site in western Hungary: Current Anthro., v. 6, p. 74–87.

Kurtén, B. 1960a. Chronology and faunal evolution of the earlier European glaciations: Soc. Sci. Fennica, Comm. Biol., v. 21, 62 p.

———— 1960b. An attempted parallelization of the Quaternary mammalian faunas of China and Europe: Soc. Sci. Fennica, Comm. Biol., v. 23, 12 p.

———— 1963a. Villafranchian faunal evolution: Soc. Sci. Fennica, Comm. Biol., v. 26, p. 1–18.

———— 1963b. Notes on some Pleistocene mammal migrations from the Palearctic to the Nearctic: Eiszeitalter u. Gegenwart, v. 14, p. 96–103.

Kurtén, B., and R. Rausch. 1959. Biometric comparisons between North American and European mammals: Acta Arctica, v. 11, 44 p.

Leffler, S. R. 1964. Fossil mammals from the Elk River Formation, Cape Blanco, Oregon: Jour. Mammalogy, v. 45, p. 53–61.

Matthew, W. D. 1932. New fossil mammals from the Snake Creek quarries: Am. Mus. Nat. Hist. Novitates 540, p. 1–8.

Moore, J. C. 1959. Relationships among living squirrels of the Sciurinae: Am. Mus. Nat. Hist. Bull., v. 118, art. 4, p. 153–206.

Pei, Wen-Chung. 1934. On the Carnivora from Locality 1 of Choukoutien: Palaeo. Sinica, Series C, v. VIII, fasc. 1, 216 p., 24 pl., 47 figs.

Pohle, H. 1919. Die Unterfamilie der Lutrinae: Arch. f. Naturgeschichte, Abt. A, Heft 9, p. 1–220.

Rausch, R. L. 1963. A review of the distribution of Holarctic Recent mammals: 10th Pacific Sci. Cong. (Hawaii), 1961, Symposium "Pacific Basin Biogeography," p. 29–43.

Repenning, C. A. 1962. The giant ground squirrel *Paenemarmota*: Jour. Paleontology, v. 36, p. 540–556.

Repenning, C. A., D. M. Hopkins, and M. Rubin. 1964. Tundra rodents in a Late Pleistocene fauna from the Tofty placer district, central Alaska: Arctic, v. 17, p. 176–197.

Savage, D. E. 1951. Late Cenozoic vertebrates of the San Francisco Bay region: Univ. Calif. Pub. Geol. Sci., v. 28, no. 10, p. 215–314.

Shotwell, J. A. 1956. Hemphillian mammalian assemblage from northeastern Oregon: Geol. Soc. America Bull., v. 67, p. 717–738.

———— 1958. Evolution and biogeography of the aplodontid and mylagaulid rodents: Evolution, v. 12, p. 451–484.

Simpson, G. G. 1947. Holarctic mammalian faunas and continental relationships during the Cenozoic: Geol. Soc. America Bull., v. 58, p. 613–687.

Sulminski, A. 1964. Pliocene Lagomorpha and Rodentia from Weze 1 (Poland): Acta Palaeontol. Polonica, v. 9, p. 149–262.

Teilhard de Chardin, P., and P. Leroy. 1942. Chinese fossil mammals: Inst. de Géo-Biologie, Pékin, no. 8, 142 p.

Vangengeim, E. A. 1961. Paleontologicheskoe obosnovanie stratigrafii Antropogenovykh otlozhenii severa vostochnoi Sibiri (Paleontological basis of the stratigraphy of the Anthropogene deposits of Northeast Siberia): Akad. Nauk SSSR, Tr. Geol. Inst., Vyp. 48, 182 p. In Russian; English translation of parts available from Am. Geol. Inst.

Wood, H. E., II, *et al.* 1941. Nomenclature and correlation of the North American continental Tertiary: Geol. Soc. America Bull., v. 52, p. 1–48.

# 17. The Stratigraphy of Tjörnes, Northern Iceland, and the History of the Bering Land Bridge

THORLEIFUR EINARSSON
*University Research Institute, Reykjavik*

DAVID M. HOPKINS and RICHARD R. DOELL
*U.S. Geological Survey, Menlo Park, California*

Recent studies of the upper Pliocene and Pleistocene rock sequences of Iceland contribute to knowledge of the history of land bridges and sea passages in the Bering Sea area, and also cast light on several other aspects of the history of the Pleistocene Epoch. On the peninsula Tjörnes[a] in northern Iceland is a well-exposed sequence of Pliocene and Pleistocene marine and nonmarine sedimentary rocks intercalated with basaltic lava flows and, in its higher part, with tillite layers recording at least ten glacial episodes. The principal time of opening of Bering Strait is recorded in the lowest sequence of sedimentary rocks, the Tjörnes beds, by the sudden appearance of a host of boreal molluscan taxa of Pacific ancestry. The ten tillite strata higher in the sequence are separated by strata or erosion surfaces that record long nonglacial intervals, suggesting that the conventional sequence, established in North America and Europe, of only four or five Pleistocene glaciations, may be incomplete and that Pleistocene climatic history is much more complex than has been previously recognized.

The intercalated lava flows provide an opportunity to relate geologic and climatic events on Tjörnes to the sequence of Quaternary geomagnetic polarity epochs and events established by A. V. Cox, R. R. Doell, and G. B. Dalrymple (1964; Doell *et al.*, 1966; Doell and Dalrymple, 1966). Lava flows retain a "memory" of the geomagnetic field at the time of eruption, because their ferromagnetic minerals acquire a weak but stable magnetism that is ordinarily parallel to the local magnetic field at the time the lavas

Einarsson's participation in our joint studies in Iceland and Alaska was supported by the Arctic Institute of North America, under contractual arrangement with the U.S. Office of Naval Research, and by the Icelandic National Science Foundation.
[a] The suffix "-nes" in Icelandic means peninsula, "-vík" means bay, and "-á" means stream.

cool past their curie points. Field and laboratory studies of ancient lava flows have shown that the geomagnetic field has undergone a series of reversals in orientation during the Cenozoic Era. The reversals of orientation have been rather sudden, and they have been separated by protracted or brief intervals during which the field has retained the same general orientation. The radiometric ages in years of the field reversals have been established by potassium-argon age determinations on lavas from many parts of the world. The protracted intervals of similar geomagnetic orientation are termed "geomagnetic polarity epochs"; some of these have been punctuated by brief intervals, termed "polarity events," during which the field was oriented 180° to the orientation that prevailed during the remainder of the polarity epoch (Fig. 2, central column). Epochs and events during which the geomagnetic field was oriented as at present are termed "normal," and epochs and events during which the field was oriented opposite to its present orientation are termed "reversed."

The approximate orientation of the thermo-remanent magnetism of a lava flow can be easily determined at the outcrop, using a small portable magnetometer and an oriented hand specimen (Doell and Cox, 1962). Thus, stratigraphic units of normally and reversely magnetized lavas can be traced laterally or related to intercalated sedimentary rocks in the course of ordinary geologic mapping. Field and laboratory studies of the remanent magnetism of the lavas intercalated in the stratigraphic sequence on Tjörnes provide an unparalleled opportunity to relate a paleomagnetic sequence to significant Quaternary events.

Tjörnes lies at the northwestern edge of the Icelandic central graben, the most active locus of Quaternary volcanic activity in Iceland (Fig. 1, inset). Tjörnes has been a restless, tectonically active area during the development of the graben. Repeated and still-active faulting has created a basin-and-range topography. Several small, short-lived basins have been formed and then filled with marine, fluviatile, lacustrine, and sometimes glacial deposits. Except for the table-mountain Búrfell, which was built up during a subglacial eruption in late Pleistocene time, there have been no active volcanoes on Tjörnes itself during the Pliocene and Pleistocene Epochs (Búrfell lies just south of the southeast corner of the area shown in Fig. 1). However, lavas that were erupted from vents or fissures in the nearby central graben have invaded Tjörnes from time to time and have become incorporated into the upper Pliocene and Pleistocene sequence. Finally, the whole sequence has been uplifted and tilted. In the southeastern part of Tjörnes the uplift amounts to 500–600 meters.

The marine sediments and lignite beds of Tjörnes and their significance for the geology of Iceland were first mentioned by Ólafsson in 1749 and by Ólafsson and Pálsson in 1772. Since that time, numerous geologists, of whom only a few will be mentioned, have studied the Tjörnes sequence. At the beginning of this century the Icelandic geologist Pjeturss (1910) studied and first recognized the tillite beds in the higher part of the sequence. In the 1920's, Bárdarson (1925) made a thorough study of the Tjörnes sediments, and his work is now classical for the sequence. The latest work on the sequence was done by Áskelsson (1960a,b) and the German geologist Strauch (1963). Trausti Einarsson (1957a) made the first paleomagnetic study of the interbedded lavas.

In 1964 we reexamined this relatively thick and continuous upper Pliocene and Pleistocene sequence and made a thorough study of the paleomagnetic records contained in the interbedded lava flows. Our conclusions on the age of various parts of the Tjörnes sequence are based mainly upon paleontological determinations by F. S. MacNeil of the United States Geological Survey and upon the paleomagnetic stratigraphy. The age estimates may be modified in some details when potassium-argon determinations have been made.

## Stratigraphy

The sedimentary sequence on Tjörnes comprises, from older to younger, the Tjörnes beds, the Furuvík beds, and the Breidavík beds (Fig. 2). These are covered and stratigraphically separated from one another by moderately thick sequences of lava flows.

*The Tjörnes beds.* The Tjörnes beds are a sequence of dominantly marine deposits approximately 500 meters thick that lie discordantly on a tectonically strongly disturbed and hydrothermally altered pile of Tertiary plateau basalts. The Tjörnes beds dip 5–10° to the northwest. They crop out in sea cliffs and river canyons for about 6 km along the coast of the bay Barmur on the west side of the peninsula (Fig. 1). Bárdarson (1925) divided the Tjörnes beds into three zones. The lowest and oldest is the "Tapes" zone, named after *Tapes aurea* (synonym *Paphia aurea* and *Venerupis aurea*), which is typical for the lowest part of the Tjörnes beds; this is overlain by the "Mactra" zone, named after *Mactra* sp. (synonym *Spisula* sp.), and then the "Cardium" zone, named after *Cardium grönlandicum* (synonym *Serripes grönlandicum*).

The Tapes and Mactra zones consist mainly of shallow water and littoral sediments intercalated with numerous thin lignite seams and lacustrine beds. The molluscan faunas of the Tapes and Mactra zones have an Atlantic char-

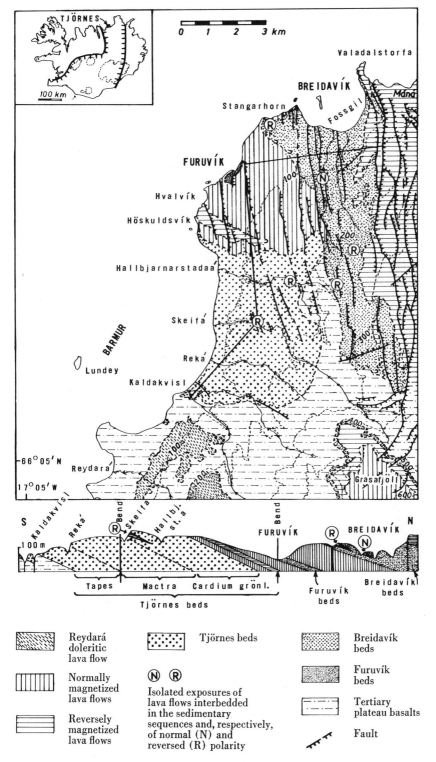

Fig. 1. Geological map of Tjörnes area, Northern Iceland (location shown in inset).

acter, but they include a few warm-water mollusks of Pacific ancestry that were presumably participants in a late Miocene migration from the Pacific to the Atlantic. Thus, the mollusk fauna of the two lowest groups is similar to that of the Coralline Crag (Astian or late Pliocene) of East Anglia, England (Baden-Powell, 1955; Áskelsson, 1960a,b; Strauch, 1963; Durham and MacNeil, this volume).

The pollen flora of the lignite beds in the Mactra zone was studied by Pflug (Schwarzbach and Pflug, 1957), and indicates a mixed coniferous and hardwood forest vegetation and a temperate climate. Pflug found pollen of *Pinus* of the *haploxylon* group, *Pinus* of the *silvestris* group, *Picea*, *Abies*, *Larix*, *Alnus*, *Corylus*, *Myrica gale* type, *Betula*, *Salix*, *Fagus silvatica* type, *Quercus*, *Platanus*, *Cyrilla* or *Castanea*, *Rhamnus*, *Ilex aquifolium* type, *Phellodendron*, and *Hedera*. The molluscan faunas of the Tapes and Mactra zones also contain species that indicate a milder climate than the present climate of Iceland. Among these are *Corbulomya*, which is now confined to the Mediterranean Sea; *Cardium tuberculatum*, which now lives south of England; *Tapes aurea*, which now lives as far north as southern Norway; and *Glycimeris* (the *Pectunculus* of Bárdarson), found in the highest part of the Mactra zone, which now lives around England and may range to southern Norway. The *Mactra* species are extinct.

The Cardium zone consists mostly of littoral marine sediments, but near the top it contains sandstone and claystone of probable nonmarine origin and thin seams of lignite. The molluscan fauna of the Cardium zone contrasts with those of the Tapes and Mactra zones in containing abundant boreal mollusks of Pacific ancestry. The Pacific mollusks first appear about 150 meters below the top of the Tjörnes beds. Within the next 20 meters upward, the assemblage is transformed, and the Pacific boreal species come to dominate the fauna. Among the newly arrived Pacific migrants are a species of *Neptunea* (a dextral form close to Late Pliocene or Early Pleistocene forms in the Gulf of Alaska, according to MacNeil), *Musculus nigra*, *Clinocardium ciliatum*, *Panomya arctica*, *Mya pseudoarenaria*, *Macoma* cf. *incongrua* (or *M. obliqua*), *Macoma calcarea*, and finally, *Serripes grönlandicum* (or *Cardium grönlandicum*), which is so abundant that Bárdarson (1925) chose this species to characterize the uppermost zone of the Tjörnes beds.

An exposure in the canyon of Skeifá about 2.2 km upstream from the coast displays two thin layers of pillow lava of reversed remanent magnetism interbedded in the Cardium zone a few meters above the level at which Pacific boreal mollusks first appear (Figs. 1 and 2). Similar reversed pillow basalts interbedded in nonfossiliferous sedimentary rocks in the

Hallbjarnarstadaá valley probably represent the same horizon. These seem to be the only volcanic rocks incorporated in the Tjörnes beds.[b]

The Cardium zone appears to be of the same age as the Red Crag of England, where many Pacific-derived mollusks also make their first appearance (Baden-Powell, 1955; Áskelsson, 1960a). An earliest Pleistocene age is indicated for the Red Crag by molluscan faunas that are similar to those in the basal Pleistocene Calabrian beds of Italy and by the presence of Villafranchian mammals in a basal bone breccia. Thus the Cardium zone of the Tjörnes beds is largely and perhaps entirely of Early Pleistocene age.

*Lava flows and sedimentary rocks between the Tjörnes beds and the Breidavík beds.* The Tjörnes beds are overlain by a thick lava flow of normal remanent magnetism exposed at Höskuldsvík (Figs. 1 and 2), then by a pair of lava flows of rather obliquely reversed magnetism followed by a clearly reversed flow, and in turn by a thicker sequence of normally magnetized flows exposed at Hvalvík. A thin sequence of nonfossiliferous sedimentary rocks, the Furuvík beds, is interbedded in the normally magnetized lava sequence. The Furuvík beds consist, from bottom to top, of iceberg tillite, conglomerate, sandstone of probable marine origin, then a thin lava flow, and another tillite layer. The tillite layers at Furuvík are the oldest glacial deposits exposed in the Tjörnes sequence.

*The Breidavík beds.* Sea cliffs of the small bay Breidavík on the north coast of Tjörnes expose a sequence of sediments about 125 meters thick containing four or five tillite beds separated from one another by interglacial beds. The Breidavík beds can be traced far southward and inland to Grasafjöll and Búrfell. They rest upon a surface of unconformity planed across all of the older stratigraphic units in the Tjörnes sequence. In the sea cliffs at Breidavík, the lowermost tillite rests on the striated surface of the uppermost of the normally magnetized lava flows. Farther south, the erosion surface is planed across the Furuvík beds and the Tjörnes beds, and on Grasafjöll and Búrfell the Breidavík beds rest on the Tertiary plateau basalts (Fig. 1).

Most of the interglacial beds at Breidavík are of marine origin, and fossiliferous marine beds with a cold-water fauna also intertongue with some of the tillite beds. The molluscan faunas consist mostly of modern

[b] We agree with earlier authors that the basalt that Strauch (1963) considered to represent lava flows interbedded in the basal part of the Tapes zone actually consists of Tertiary plateau basalt lying unconformably beneath the Tjörnes beds. Stratigraphic relationships in the area of the Tapes zone are obscured by landsliding. Jumbled topography and anomalous paleomagnetic orientations suggest that some of the exposures interpreted by Strauch as representing interbedded lava flows actually consist of landslide blocks.

species, but the *Chlamys* differ slightly from the typical *C. islandica*. Faunas in the interglacial beds are more arctic in character than those in any part of the Tjörnes beds.

Interglacial beds separating the two lowest tillite layers are nonmarine and probably of lacustrine origin. They contain a pollen flora that includes *Alnus, Myrica,* and *Salix,* as well as small quantities of *Picea, Abies, Larix,* and *Pinus* of the *silvestris* type (Schwarzbach and Pflug, 1957). Some of the pollen in these beds may have been redeposited from older sedimentary rocks on Tjörnes.

A reversely magnetized lava flow is interbedded in the lower part of the Breidavík beds, a normally magnetized lava flow occupies a slightly higher stratigraphic level, and another reversely magnetized lava flow is interbedded in the middle part of the sequence (Fig. 2).

*Volcanic and sedimentary rocks of middle and late Pleistocene age.* The Breidavík beds are overlain at Valadalstorfa by a thick sequence of reversely magnetized lavas, and these, in turn, are overlain by normally magnetized lava flows and pillow breccias that are exposed on Grasafjöll, on Búrfell, and in the cliffs on the east coast of Tjörnes, east of the area shown in Fig. 1. The lowest normally magnetized lavas on Búrfell and Grasafjöll are separated from underlying reversely magnetized lavas by a frost breccia; a thick tillite bed occupies the same position in the coastal cliffs of northeastern Tjörnes. Búrfell itself is a table mountain—a volcano built up in a subglacial eruption. Its normally magnetized pillow breccias rest on another, higher, tillite bed (Fig. 2), and the top of the mountain bears weathered glacial grooves and striae.

The youngest volcanic rocks on northwestern Tjörnes consist of remnants of a lava flow of normal magnetism that occupy much of the areas southwest of Kaldakvisl stream (Fig. 1). This lava flowed into the area from the south after Búrfell had been deeply dissected and the surrounding area had been drastically lowered by glacial erosion. In the canyons of Bakkaá (south of the map area) and Reydará, the young lava flow rests on a tillite bed. The flow has in turn been deeply eroded by renewed glaciation.

## Age and Correlation

We do not yet have potassium-argon age determinations for the volcanic rocks of Tjörnes. However, some conclusions on probable ages in years can be based on a comparison of the succession of lava flows of normal and reversed magnetism on Tjörnes with the standard sequence of geomagnetic polarity epochs and events (Cox *et al.*, 1964; Doell *et al.*, 1966; Doell and Dalrymple, 1966). Two alternative correlations will be discussed: one assumes that the paleomagnetic record is complete (Fig. 2, left column); the other assumes that the newly discovered and evidently brief Jaramillo normal event is not represented (Fig. 2, right column).

The young interglacial lava flow at Reydará is obviously of Late Pleistocene age and was poured out during the Brunhes or present normal geomagnetic polarity epoch. The much more dissected but normally magnetized lava flows and breccias of Búrfell and Grasafjöll must also have been erupted during the Brunhes epoch. Thus, these lava flows and interbedded tillite layers are less than 0.7 million years old.

If we assume that the paleomagnetic record is complete (Fig. 2, left column), then the reversely magnetized lava flows covering the Breidavík beds were erupted during the last segment of the Matuyama reversed-geomagnetic-polarity epoch, as was the highest reversely magnetized lava flow within the Breidavík beds. The single normally magnetized flow interbedded in the Breidavík beds would then represent the Jaramillo normal event, which took place 0.9 million years ago. The reversely magnetized flow near the base of the Breidavík beds would represent the middle part of the Matuyama reversed epoch; the thick sequence of normally magnetized lavas enclosing the Furuvík beds would represent the Olduvai normal event, which took place about 1.9 million years ago; and the three flows of reversed magnetism intercalated with normal lavas between the Tjörnes and the Furuvík beds would represent the early part of the Matuyama reversed epoch, which began about 2.4 million years ago. The basal normal flow in this sequence would record the last part of the Gauss normal epoch, and the two reversely magnetized pillow lavas within the Cardium zone in the Tjörnes beds would represent the Mammoth reversed event, which took place about 3.0 million years ago.

According to this correlation, the Breidavík beds range in age from slightly more than 0.7 to slightly more than 1.0 million years old; the tillite in the Furuvík beds is about 1.9 million years old; and the invasion of Pacific boreal mollusks recorded at the base of the Cardium zone took place slightly more than 3.0 million years ago.

An alternative correlation assumes that the Jaramillo normal event is not recorded on Tjörnes (Fig. 2, right column). The period of nondeposition recorded by frost breccia on Búrfell and Grasafjöll and the record destroyed by local glacial erosion recorded by tillite on the northeast coast are assumed to encompass the time of the Jaramillo normal event and the brief late segment of the Matuyama reversed epoch. According to this correla-

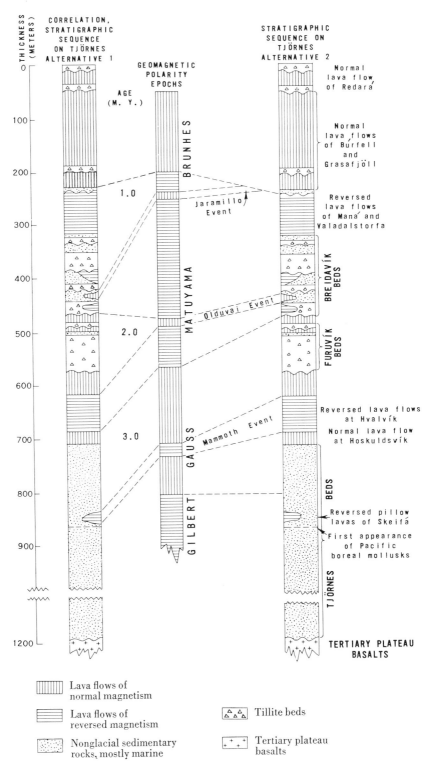

Fig. 2. Alternative correlations of the Tjörnes sequence with the standard geomagnetic polarity sequence.

tion, the reversely magnetized lavas that cover the Breidavík beds and the single reversely magnetized flow interbedded in the middle part of the Breidavík beds represent the middle segment of the Matuyama reversed epoch; the single normally magnetized flow low in the Breidavík sequence was erupted during the Olduvai event; and the reversely magnetized flow near the base of the sequence represents the early part of the Matuyama reversed epoch. The thick sequence of normally magnetized lava flows between the Breidavík and Tjörnes beds was erupted during the Gauss normal polarity epoch, and the three reversed flows in the lower part of this lava pile represent the Mammoth reversed event. The reversely magnetized pillow lavas interbedded in the upper part of the Tjörnes beds, just above the level at which the Pacific boreal mollusks first appear, would have been erupted during the Gilbert reversed-geomagnetic-polarity epoch. According to this correlation, the deposition of the Breidavík beds began between 2.4 and 1.9 million years ago and ended between 1.9 and 0.9 million years ago; the first glaciation on Tjörnes recorded by tillite in the Furuvík beds took place between 2.4 and 3.0 million years ago; and the first invasion of Pacific boreal mollusks took place more than 3.35 million years ago.

The correlation that assumes the absence of the Jaramillo event (Fig. 2, right column) is attractive, because then most of the longer segments of the standard paleomagnetic sequence are represented on Tjörnes by thick sequences of lava flows recording many eruptions or by sedimentary sequences recording several glacial events, and the shorter segments of the standard sequence encompass correspondingly less complex stratigraphic intervals on Tjörnes. But either correlation is reasonably consistent with G. H. Curtis's potassium-argon age determination, quoted by Obradovich (1965), of 3.3 million years for the early Villafranchian mammal fauna at Perrier, France, and with Obradovich's potassium-argon age determination of 3.04 million years for glauconite from the Lomita Marl Member of the San Pedro Formation of early Pleistocene age in California (Obradovich, 1965).

SIGNIFICANCE OF THE PALEOMAGNETIC CORRELATIONS

Both of the alternative paleomagnetic correlations point to the conclusion that the Pacific boreal mollusks invaded the waters around Tjörnes more than 3.0 million years ago, at or shortly before the beginning of the Pleistocene Epoch. Because most of the newly arrived species have a boreal distribution, their sudden appearance in the Cardium zone of Iceland and in the Red Crag of England has customarily been interpreted as recording a southward expansion of cold-water species in response to sudden climatic cooling. But this abrupt invasion of Icelandic waters and of the North Sea Basin by a flood of mollusks of Pacific origin requires more than a simple climatic explanation. As MacNeil (1957) and Hopkins (1959) have shown, the Bering Land Bridge was an effective barrier to transarctic migrations throughout most of Tertiary time. Complex events, not yet well understood, permitted a relatively few thermophile mollusk genera to escape by a northern route from one ocean to the other as early as Pontian (or Late Miocene) time (MacNeil, 1965; Durham and MacNeil, this volume); some of these mollusks are present in the Tapes and Mactra zones of the Tjörnes beds. However, the more complete and more dramatic opening of a sea passage through Bering Strait evidently took place later, and is well recorded by the sudden influx of Pacific boreal mollusks in the Tjörnes beds and in the English Crag sequence. The Pacific boreal mollusks arrived in Tjörnes more than 3 million years ago, at least half a million and perhaps more than a million years before the first Pleistocene glaciation was recorded there. The migration must have taken place at a time when the Arctic Ocean was warmer than at present, for some of the migrating taxa no longer range as far north as the Arctic Ocean.

Whether the Furuvík beds are 1.9 or between 2.4 and 3.0 million years old, the glacial record that they contain is remarkably early[c] and marks the beginning of a remarkably large number of glacial cycles. A total of nine or ten glaciations are recorded in the Tjörnes sequence. Of these, at least three took place during the Brunhes normal-polarity epoch, but the number of glaciations recognized within this interval on Tjörnes must be regarded as minimal. The Tjörnes sequence appears to record more glacial events than any single area elsewhere in the world.

There are several reasons to think that the Icelandic glacial record, especially the Tjörnes record, is not anomalous but merely exceptional in its completeness. The persistent volcanic activity in Iceland has resulted in the covering of glacial sediments by lava flows and therefore in their protection against erosion during subsequent glacial advances. One might assume that these tillite beds record local mountain glaciations fed by nearby volcanic or faulted highlands during intervals that were not especially cold in other

---

[c] Tillite beds of probably similar age have been found by Wensink (1964a,b) in eastern Iceland. However, it is likely that neither these nor the Furuvík beds represent the oldest record of glaciation in Iceland. There are one or two tillite horizons in southwestern Iceland whose positions in the paleomagnetic sequence (Einarsson, 1957b; Sigurgeirsson, 1957) suggest that they are older than Furuvík beds, and a tillite horizon in southeastern Iceland, described by Jónsson (1955), that underlies lignite containing a pollen flora similar to that in the Mactra zone on Tjörnes (Schwarzbach and Pflug, 1957). These may have been deposited inland in relatively mountainous regions and thus may record local mountain glaciations.

parts of the world. However, the fault-block mountains on Tjörnes are higher now than they have ever been before, as is evident from the uplifted marine sediments extending far inland. Moreover, there have been no high local volcanoes in the area, except the young, subglacially formed volcano Búrfell. Thus we must assume that the tillite beds of Tjörnes were deposited near sea level by large ice sheets whose centers of nourishment lay far away, in central Iceland. Ice sheets of these dimensions could have existed in Iceland only during worldwide cold intervals.

If we are correct in concluding that the large number of glacial and interglacial episodes recorded on Tjörnes were worldwide in their effects, then they must have been associated with an equal number of major sea-level fluctuations. The effects of these sea-level fluctuations upon land bridges and seaways in the Bering Sea area would be affected by tectonic events there (Sainsbury, this volume). If the topography of the Bering shelf has remained much the same as at present throughout Quaternary time, then the ten glaciations on Tjörnes have probably been represented by as many as ten intervals when sea level was low enough to bring the Bering Land Bridge into existence again, and the nine interglacial intervals have probably been associated with nine episodes when sea level was high enough to reopen a sea passage from the Pacific Ocean through Bering Strait to the Arctic and Atlantic Oceans.

## Conclusions

Bering Strait opened in its present form during late Pliocene or early Pleistocene time, allowing a flood of mollusks of Pacific boreal ancestry to reach Iceland more than three million years ago. The Arctic Ocean was still relatively warm and probably ice-free; the migrations took place at least half a million and perhaps more than a million years before the first widespread Pleistocene glaciation in Iceland.

The first glaciation on Tjörnes took place at least 1.9 and perhaps between 2.4 and 3.0 million years ago; it was followed by at least nine later glaciations separated by lengthy nonglacial intervals. This record is probably unique only in its completeness. A reconstruction of the former topography indicates that the local relief on Tjörnes was even less conducive to local glaciation during most of Pleistocene time than it is at present; the tillite beds on Tjörnes evidently were deposited by large ice sheets originating in central Iceland. The ten glacial cycles recorded on Tjörnes probably represent a minimum estimate of the total number of important worldwide Pleistocene cold fluctuations and glacial cycles. This conclusion implies that Pleistocene climatic history is much more complex than is suggested by the classical concepts evolved for the Alps and for the mid-continent of North America, which involve only four or five glaciations, and it casts serious doubts on intercontinental correlations of the older Quaternary glacial deposits. It also implies that the Bering Land Bridge may have been renewed as many as ten times by glacio-eustatically lowered sea level during the Pleistocene Epoch.

## REFERENCES

Áskelsson, J. 1960a. Fossiliferous zenoliths in the Móberg Formation of South Iceland: Acta. Nat. Islandica, v. 2, no. 3, 30 p.

———— 1960b. Pliocene and Pleistocene fossiliferous deposits, in On the Geology and geophysics of Iceland: 21st Internat. Geol. Cong. (Copenhagen), 1960, Guide to Excursion A2, p. 28–32.

Baden-Powell, D. F. W. 1955. The correlation of the Pliocene and Pleistocene marine beds of Britain and the Mediterranean (with discussion): Geol. Assoc. London Proc., v. 66, pt. 4, p. 271–292.

Bárdarson, G. 1925. A stratigraphical survey of the Pliocene deposits at Tjörnes in Northern Iceland: Kong. Danske Vidensk. Selskab, Skr. Medd. Biol., v. 4, no. 5, 118 p.

Cox, A., R. R. Doell, and G. B. Dalrymple. 1964. Reversals of the earth's magnetic field: Science, v. 144, p. 1537–1543.

Doell, R. R., and A. Cox. 1962. Determination of the magnetic polarity of rock samples in the field: U.S. Geol. Survey Prof. Paper 450-D, p. 105–108.

Doell, R. R., and G. B. Dalrymple. 1966. Geomagnetic polarity epochs: a new polarity event and the age of the Brunhes-Matuyama boundary: Science, v. 152, p. 1062–1064.

Doell, R. R., G. B. Dalrymple, and A. Cox. 1966. Geomagnetic polarity epochs— Sierra Nevada data, 3: Jour. Geophys. Research, v. 71, p. 531–541.

Einarsson, Trausti. 1957a. A survey of the geology of the area Tjörnes-Bárdardalur in Northern Iceland: Soc. Sci. Islandica, v. 32, 79 p.

———— 1957b. Magneto-geological mapping in Iceland with the use of a compass: Advances in Physics, v. 6, p. 232–239.

Hopkins, D. M. 1959. Cenozoic history of the Bering Sea land bridge: Science, v. 129, p. 1519–1528.

Jónsson, J. 1955. Tillite in the basalt formation in eastern Iceland. The Hoffellssandur, Part II: Geografiska Annaler, v. 37, p. 170–175.

MacNeil, F. S. 1957. Cenozoic megafossils of northern Alaska: U.S. Geol. Survey Prof. Paper 294-C, p. 99–126.

———— 1965. Evolution and distribution of the genus Mya with a discussion of Tertiary faunal migrations: U.S. Geol. Survey Prof. Paper 483-G, 51 p.

Obradovich, J. D. 1965. Isotopic ages related to Pleistocene events (abs.): 7th Internat. Assoc. Quaternary Research Cong. (Boulder and Denver), 1965, Gen. sess., p. 364.

Ólafsson, E. 1749. Enarrationes historicae de Islandiae natura et constitutions, etc.: Copenhagen.

Ólafsson, E., and B. Pálsson. 1772. Reise igiennem Island: Sorö (Danish).

Pjeturss, H. 1910. Island—Handbuch der regionalen Geologie, v. 4, no. 1, p. 1–22, Heidelberg.

Schwarzbach, M., and H. D. Pflug. 1957. Das Klima des Jüngeren Tertiärs in Island: Beiträge zur Klimageschichte Islands Neues Jahrb. Geol. und Paläontol., Abh. 104, p. 279–298. Feb. 1957.

Sigurgeirsson, Th. 1957. Direction of magnetization in Icelandic basalts: Advances in Physics, v. 6, p. 240–246.

Strauch, F. 1963. Zur Geologie von Tjörnes (Nordisland): Sonderveröffentlichungen Geol. Inst. Univ. Köln, no. 8, 129 p. In German; English translation available from Am. Geol. Inst.

Wensink, H. 1964a. Paleomagnetic stratigraphy of younger basalts and intercalated Plio-Pleistocene tillites in Iceland: Geol. Rundschau, v. 54, no. 1, p. 364–384.

———— 1964b. Secular variation of earth magnetism in Plio-Pleistocene basalts of eastern Iceland: Geologie en Mijnbouw, v. 43, no. 9, p. 403–413.

## 18. Cenozoic Migrations of Marine Invertebrates Through the Bering Strait Region

J. WYATT DURHAM
*University of California, Berkeley*

F. STEARNS MACNEIL
*U.S. Geological Survey, Menlo Park, California*

Biogeographers concerned with terrestrial biotas have given much atten-tion to the Bering Strait region because of its obvious importance as a bridge between Eurasia and North America. Much less attention has been given to the region by marine biogeographers, although its significance has been considered by some, notably Davies (1930, 1934), Soot-Ryen (1932), and Ekman (1953), in the last four decades. Nevertheless, it is obvious that any feature serving as a pathway for terrestrially limited or-ganisms must, at the same time, serve as a barrier to the migrations of those restricted to the marine environment. Any biogeographic evaluation of a key region such as the Bering Strait area should include a synthesis of the data from inhabitants of both realms. As a contribution toward this goal, this study provides a biogeographic analysis of the larger marine Ceno-zoic invertebrates of the North Pacific, Arctic, and North Atlantic regions.

## Biogeographical Analyses

### METHODS AND LIMITATIONS

Because of variations in local geologic histories and in the extent of ex-ploration from one area to the next, the data here presented have been ob-tained from a relatively few areas: England and the adjacent Low Countries (Belgium and the Netherlands), Iceland, the arctic coast of Alaska, the Bering Strait area and the vicinity of Nome, the Chukotka Peninsula of Northeastern Siberia, the Aleutian Islands, and the Gulf of Alaska region. In order to evaluate the histories of various taxa, the Tertiary faunas of Western Europe, eastern and western North America, Japan, Sakhalin, and

Kamchatka have also been considered. Both the fossil record and our knowledge of the record are highly biased reflections of past biotas, and thus analyses of the biotas are nearly always based on minimal materials. Because gastropods, pelecypods, brachiopods, and echinoids have well-developed and substantial hard parts, usually of easily observable size, they are not only commonly preserved as fossils, but are also more conspicuous in the rocks and have therefore been collected and recorded far more often than other marine invertebrates. For this reason, the present study has been restricted to these groups.

Although substantial reference collections are available, a considerable portion of the data presented here has been derived from an analysis of the literature, and as such is variable in its immediate level of significance. Specific and higher taxonomic-level concepts vary greatly from one systematist to another. Likewise, age assignments are subject to controversy and change. Two notable instances of changed age assignments affecting this study and its comparison with earlier analyses lead to apparent contradictions unless they are clearly understood. The earlier of these is concerned with the age assignments of the Empire Formation and Coos Conglomerate of Oregon, the Jacalitos and Etchegoin Formations of central California, and unnamed beds in the Pribilof Islands. In the reports of Dall, and of other paleontologists working with the U.S. Geological Survey until about 1920, these formations and their faunas were referred to the Miocene, although present data clearly indicate that the Empire, Coos, Jacalitos, and Etchegoin Formations should all be assigned to the Pliocene (Weaver *et al.*, 1944), and that the marine beds on the Pribilof Islands are entirely of Pleistocene age (Cox *et al.*, 1966). In consequence, many of the "first occurrences" recorded from these formations in previous analyses have been erroneously assigned to the Miocene rather than to the Pliocene or early Pleistocene.

Similarly, the decision made at the Eighteenth International Geological Congress in London (1948) to place the Pliocene-Pleistocene boundary at the beginning of Calabrian-Villafranchian time in northern Italy has resulted in an early Pleistocene age assignment for many biotas previously referred to the late Pliocene. Thus, one of us (Durham), as a result of studies currently in progress, now places the Pliocene-Pleistocene boundary near the middle (about at the base of the "Pecten zone") of the San Joaquin "stage," which was in its entirety assigned to the Pliocene in the correlation chart by Weaver *et al.* (1944). It is worth noting that this altered position of the Pliocene-Pleistocene boundary has probably increased the length of the Pleistocene from an interval of about one million years to one of around three million years, according to current "absolute" age determinations (Evernden *et al.*, 1964; Obradovich, 1965).

A somewhat similar confusion exists with respect to the position of the Miocene-Pliocene boundary in both marine and terrestrial sequences (see Durham *et al.*, 1954). Many marine paleontologists and stratigraphers, including the present authors, accept the type Pontian and Sarmatian of Europe as of late Miocene age. Many vertebrate paleontologists, concerned with the terrestrial mammal record of the late Cenozoic, assign these same stages to the early Pliocene. As a result, apparently different age assignments may in reality signify identical ages, and considerable effort may be necessary to determine whether this is so, inasmuch as many investigators are unaware of these differing standards and so do not indicate which they are using.

Another uncertainty affecting biogeographic analyses and inferences is the degree of resolution with which age assignments may be made. Ideally, some of the radiometric dating methods such as potassium-argon will yield nearly perfect results (in the perspective of geologic time), but unfortunately material suitable for these techniques is rare and often not in juxtaposition to the fossiliferous sequences. As a result, most age assignments have to be made by paleontologic methods, at a much poorer level of resolution. Although in local areas a high degree of precision of correlation may be attained with paleontologic methods, interregional correlations are much more difficult because of limited geographic distribution of critical organisms, differing ecologies, biogeographic barriers, and, finally, lack of data from critical areas. In the marine realm at the present time, it appears highly improbable that any Cenozoic paleontologic correlations can be made between the North Pacific and Western Europe within a probable error of one million years. Inasmuch as numerous climatic oscillations occurred within the later Cenozoic and many of them within the late Pleistocene of current usage (an interval approximating a million years in magnitude), it is apparent that we have not yet reached a degree of resolution in correlation by paleontologic methods that will permit us to outline more than the broad pattern of events.

DISTRIBUTION OF GENERA

In preparing the accompanying distributional lists, we have encountered several difficulties. Some genera, such as the gastropods *Neptunea* and *Nucella*, with well-documented histories in one region, have, with the opening of Bering Strait, migrated into the adjacent region and diversified there;

for these genera, a complete distributional documentation is attempted. In other genera, presumably of more ancient origin or of more cosmopolitan dispersion, distribution and history are complex and poorly understood at the genus level; within these genera, only a few species appear to be significant to the present study. In these instances, distributions are given only for the significant species.

Because of the uncertainties in interregional correlation discussed above, age occurrences in the lists are given in terms of commonly used local sequences. On the Pacific Coast, local correlations are largely as given in the chart by Weaver *et al.* (1944). Except in special instances, authorities for occurrences in Recent faunas have not been cited. For the eastern Pacific north to Point Barrow, the checklist of Keen (1937) has been the basic authority, but MacGinitie's report (1959) on the Point Barrow mollusca extends some previously reported ranges. Soot-Ryen's (1932) inclusion of *Saxidomus gigantea* in the Arctic fauna is not documented, and an extensive search of the literature has not disclosed the basis for the claim. Consequently, it is not accepted here. Since the present study is concerned primarily with migrations through Bering Strait, the Arctic and North Atlantic ocean areas (except Iceland and Spitzbergen) have been treated as a unit for the purposes of Table 1, and the strait is considered the boundary with the North Pacific. One of us (MacNeil) has studied the new collections obtained in the course of the restudy by Einarsson, Hopkins, and Doell (this volume) of the late Cenozoic marine sequence in Iceland; these data, in conjunction with Bárdarson's (1925) list, have been combined to furnish the basis for Table 2. The distribution and ancestry of the taxa which cannot be adequately considered in the checklists, are discussed further in the following sections.

BRACHIOPODA

*Hemithyris psittacea* (Gmelin), a living species, is widespread throughout the polar regions, the North Atlantic, and the North Pacific (Elliot, 1956; Hertlein and Grant, 1944, p. 46–54). It is reported from the Miocene of the Yakataga Formation (Miocene and Pliocene) of southern Alaska, and from younger beds along the northern Pacific Coast of America, as well as from possible late Oligocene strata in Japan (Hertlein and Grant, 1944). Other species referred to this genus occur in California in the early Eocene (termed middle or late Eocene, originally) Sierra Blanca Limestone of Nelson (1925) (Hertlein and Grant, 1944, p. 46). *H. psittacea* occurs in the Red Crag at Sutton, England, and in the "Crag" near Antwerp, Belgium, as well as in other, younger Pleistocene localities in Europe.

TABLE 1. *Genera and Species of North Pacific Origin and Their Times of Appearance in Arctic and West European Areas*

Column groups: **PACIFIC** (Recent; Later Pleistocene; Pliocene: San Joaquin*, Etchegoin, Jacalitos; Miocene; Oligocene) — **ATLANTIC and ARCTIC** (Recent; Later Pleistocene) — **WESTERN EUROPE**: England (Early Pleistocene: Butleyan, Newbornian; Pliocene: Waltonian, Gedgravian); Netherlands/Belgium (Later Pleistocene; Early Pleistocene: Late Amstelian, Poederlian; Scaldisian; Pliocene: Late Diestian).

| Genera and Species | Recent | Later Pleist. | San Joaquin* | Etchegoin | Jacalitos | Miocene | Oligocene | Recent | Later Pleist. | Butleyan | Newbornian | Waltonian | Gedgravian | Later Pleist. | Late Amstelian | Poederlian | Scaldisian | Late Diestian |
|---|---|---|---|---|---|---|---|---|---|---|---|---|---|---|---|---|---|---|
| **BRACHIOPODA** | | | | | | | | | | | | | | | | | | |
| Hemithyris (see discussion) | X | X | X | X | X | X | X | X | X | X | X | X | X | | | X | | |
| H. psittacea (Gmelin) | X | X | X | X | X | X | | X | X | X | X | X | X | | | X | | |
| **PELECYPODA** | | | | | | | | | | | | | | | | | | |
| Acila | X | 6 | 6 | 6 | 6 | 1 | 1 | | 17 | 17 | | | | 4 | 4 | | | |
| A. cobboldiae (Sowerby) | | 13 | 13 | 6 | 6 | 1 | | | 17 | 17 | | | | 4 | 4 | | | |
| Axinopsis viridis Dall | X | 13 | 13 | 6 | | 6 | | X | | | | | | | | | | |
| Clinocardium | X | 6 | 6 | 6 | | | | X | | | | | | | | | | |
| C. californiense (Deshayes) | X | 10 | 6 | | | | | X | | | | | | | | | | |
| C. ciliatum (Fabricius) | X | 10 | 6 | | | | | X | | | | | | | | | | |
| Crenella decussata (Montagu) | X | 6 | 6 | | | | | X | | | | | | | | | | |
| Diplodonta aleutica Dall | X | 10 | 10 | | | | | X | | | | | | | | | | |
| Fortipecten | | | | 16 | 16 | | | | | See discussion | | | | | | | | |
| F. hallae (Dall) | X | 6 | 6 | 15 | | | | | 17 | 17 | 17 | | | 4 | 4 | | | |
| Kellia suborbicularis and laperousii | X | 10 | 6 | | | | | X | | | | | | | | | | |
| Liocyma fluctuosa (Gould) | X | 6 | 6 | | | | | X | | | | | | | | | | |
| Lyonsia | X | 6 | | | | | | X | | | | | | | | | | |
| L. arenosa (Möller) | X | 19 | | | | | | 19 | | | | | | | | | | |
| L. norvegica (Gmelin) | X | 6 | 6 | | | 1 | 1 | X | 4 | 4 | X | X | X | 4 | X | X | 4 | 4 |
| Macoma | X | 6 | 6 | | | 1 | | X | 4 | 4 | X | X | X | 4 | X | X | 4 | 4 |
| M. balthica (Linnaeus) | X | 6 | | | | | | X | | | | | | | | | | |

330

| Species | | | | | | | | | | | | | |
|---|---|---|---|---|---|---|---|---|---|---|---|---|---|
| M. incongrua von Martens | | | | | | | X | | | | | | |
| M. moesta (Deshayes) | | | | | | | X | | | | | | |
| Megayoldia thraciaeformis (Storer) | 10 | | | | | | X | | | | | | |
| Modiolus | 6 | X | 6 | 1 | 6 | 1 | X | | 17 | X | 17 | X | 17 |
| M. modiolus (Linnaeus) | 6 | | | | | | X | | 17 | | 17 | | |
| Montacuta planata (Dall) | 10? | | | | | | X | | | | | | |
| Musculus | 6 | | | | | | X | | | | | | |
| M. corrugatus (Stimpson) | 19 | | | | | | 19 | | | | | | |
| M. discors laevigatus (Gray) | 19 | | | | | | 19 | | | | | | |
| M. niger (Gray) | 6 | | | | | | X | | | | | | |
| Mya (except M. producta Conrad) | 14 | 14 | 14 | 14 | 14 | | X | | 14 | 14 | 14 | | 14 |
| M. arenaria Linnaeus, subsp. | 14 | 14 | 14 | | | | X | | 14 | 14 | 14 | | 14 |
| M. pseudoarenaria Schlesch | 14 | | | | | | X | | | | 14 | | 14 |
| M. pullus Sowerby | | | | | | | | 14 | | | | | |
| M. truncata Linnaeus | 14 | 14 | 14 | 14 | | | X | 14 | 14 | | 17 | | 14 |
| Mytilus | 14 | 6 | 6 | 6 | 1 | | X | 4 | 4 | | 4 | 4 | 4 |
| M. edulis Linnaeus | 6 | 6 | 6 | | | | X | 4 | 4 | | 4 | 4 | 4 |
| Pandora | 6 | 7 | 1 | | | | X | | | | | X | X |
| P. glacialis Leach | 6 | | | | | | X | | | | | | 4 |
| Panomya | 6 | 11 | | | | | X | | | | 4 | | 4 |
| P. ampla Dall | 10 | | | | | | X | | | | | | |
| P. arctica (Lamarck) | | | | | | | X | | | | | 4 | 4 |
| Placopecten | | | | 16 | | | See discussion | | | | | | |
| Protothaca staminea (Conrad) | 6 | 6 | | | | | 21 | | | | | | |
| Serripes | 6 | 6 | | | | | X | | 4 | 17 | 4 | | 4 |
| S. groenlandicus (Bruguière) | 8 | | | | | | X | | 4 | 17 | 4 | | 4 |
| Spisula polynyma voyi (Gabb) | 6 | | | | | | X | | | | | | |
| Tellina lutea (Gray) | 6 | | | | | | X | | | | | | |
| Thracia ? | 6 | × | 1 | 1 | | | X | | | | | | |

331

* One of us (Durham) currently regards the San Joaquin "stage" as partly Pliocene and partly Pleistocene.

References:

1 Weaver, 1943.
2 Beets, 1946.
3 Durham, 1944.
4 Heering, 1950.
5 Harmer, 1914-25.
6 Grant and Gale, 1931.
7 Keen and Bentson, 1944.
8 MacNeil, 1957a.
9 Woodring and Bramlette, 1950.
10 MacNeil, unpublished data, 1965.
11 Faustman, 1964.
12 Durham, 1957.
13 Woodring et al., 1946.
14 MacNeil, 1965.
15 Hopkins and MacNeil, 1960
16 Masuda, 1962.
17 S. V. Wood, 1874.
18 MacNeil et al., 1943.
19 MacGinitie, 1959.
20 Merklin et al., 1962.
21 Soot-Ryen, 1932.

TABLE 1. *Continued*

| Genera and Species | PACIFIC — Recent | PACIFIC — Later Pleistocene | Pliocene — San Joaquin* | Pliocene — Etchegoin | Pliocene — Jacalitos | Miocene | Oligocene | ATLANTIC and ARCTIC — Recent | England — Later Pleistocene | England — Early Pleistocene Butleyan | England — Early Pleistocene Newbornian | England — Early Pleistocene Waltonian | England — Pliocene Gedgravian | Neth./Belg. — Later Pleistocene | Early Pleistocene Late Amstelian | Early Pleistocene Poederlian | Early Pleistocene Scaldisian | Plio-cene Late Diestian |
|---|---|---|---|---|---|---|---|---|---|---|---|---|---|---|---|---|---|---|
| *T. adamsi* MacGinitie | 19 | | | | | | | 19 | | | | | | | | | | |
| *T. curta* Conrad | × | 10? | | | | | | × | | | | | | | | | | |
| *T. myopsis* (Möller) | 19 | | | | | | | 19 | | | | | | | | | | |
| *Thyasira bisecta* Conrad | × | | 6 | 6 | 6 | 6 | 1 | × | | See discussion | | | | | | | | |
| *Yoldia hyperborea* Torrell | × | 10 | 6 | | 1 | | | × | | | | | | | | | | |
| *Y. scissurata* Dall | × | 6 | 6 | | | | | 19 | | | | | | | | | | |
| *Zirfaea* | × | 6 | 6 | 6 | 6 | 6 | | × | 4 | 4 | 4 | 4 | 4 | 4 | | | 4 | |
| *Z. crispata* (Linnaeus) | × | 6 | 6 | 6 | 6 | 6 | | × | 4 | 4 | 4 | 4 | 4 | 4 | | | 4 | |
| **GASTROPODA** | | | | | | | | | | | | | | | | | | |
| *Acirsa* | × | | | | | 7 | | × | 5 | | | | | | | | | |
| *A. estrichti* (Holböll) | | | | | | | | × | 5 | 5 | 5 | 5 | 5 | | | | | |
| *Acmaea* | × | | | | 6 | 1 | 1 | × | 5 | 5 | 5 | 5 | 5 | | | | | |
| *A. rubella* (Fabricius) | | | | | 6 | 1 | 1 | × | 5 | 5 | 5 | 5 | 5 | | | | | |
| *A. virginea* (Müller) | | | | | | 1 | | × | 5 | 5 | 5 | 5 | | | | | | |
| *Admete* | × | 6 | 9 | | | 1 | | × | 5 | 5 | 5 | 5 | 5 | | | 2 | 2 | |
| *A. couthouyi* (Jay) | × | 6 | | | | | | × | 5 | 5 | 5 | 5 | 2 | | | | | |
| *A. middendorfiana* Dall | × | 10? | | | | | | × | | | | | | | | | | |
| *A. regina* Dall | 19 | | | | | | | 19 | | | | | | | | | | |
| *A. viridula* Fabricius | × | | | | | | | × | 5 | 5 | 5 | 5 | 5 | | | 2 | 2 | |
| *Amauropsis* | × | | | | 6 | 1 | 1 | × | 5 | 5 | 5 | 5 | 5 | | | 2 | 2 | |
| *A. islandica* (Gmelin) | × | | | | | 1 | 1 | × | 5 | 5 | 5 | 5 | 5 | | | | | |
| *Boreoscala* | × | 6 | | | | 1 | 1 | × | 5 | 5 | 5 | 5 | 5 | | | | | |

332

| | | | | | | | | | | | | | | | | | | |
|---|---|---|---|---|---|---|---|---|---|---|---|---|---|---|---|---|---|---|
| *B. sinuus* (Sowerby) | | | | | | | | | | | | | | | | | | |
| *Boreotrophon* | × | 6 | 6 | 11 | | | 5 | 5 | 5 | 5 | 5 | | | 5 | 5 | | |
| *B. beringi* Dall | 19 | | | | | | 19 | 5 | 5 | 5 | 5 | | | 5 | | | |
| *B. clathratus* (Linnaeus) | × | | | | | | × | 5 | 5 | 5 | 5 | | | 5 | | | |
| *B. pacificus* Dall | 19 | | | | | | 19 | | | | | | | | | | |
| *B. truncatus* (Ström) | 19 | | | | | | 19 | | | | | | | | | | |
| *Buccinum* | × | 6 | 6 | 6 | 8 | | × | 5 | 5 | 5 | 5 | | | 5 | | | 5 |
| *B. angulosum* Gray | 19 | | | | | | 19 | | | | | | | | | | |
| *B. ciliatum* Fabricius | 19 | | | | | | 19 | | | | | | | | | | |
| *B. fringillum* Dall | 19 | | | | | | 19 | | | | | | | | | | |
| *B. glaciale pardellum* Dall | × | 10? | | | | | × | | | | | | | | | | |
| *B. plectrum* Simpson | × | 10? | | 8? | | | × | | | | | | | | | | |
| *B. polare* Gray | 19 | | | | | | 19 | | | | | | | | | | |
| *B. tenue* Gray | × | 10? | | | | | × | 5 | 5 | 5 | 5 | | | 5 | | | |
| *B. undatum* (Linnaeus) | 19 | | | | | | 19 | | | | | | | | | | |
| *B.* spp. (numerous) | × | 6 | 6 | 8 | 8 | | × | 5 | 5 | 5 | 5 | | 5 | 5 | 5 | | 5 |
| *Colus* [Sipho] | × | | | | | | × | 5 | 5 | 5 | 5 | | 5 | 5 | 5 | | 5 |
| *C. herendeeni* (Dall) | | | | | | | | | | | 5 | | | | | | |
| *C. martensi* (Krause) | 19 | | | | | | 19 | | | | | | | | | | |
| *C. spitzbergensis* (Reeve) | × | 10 | | | | | × | 5 | 5 | 5 | 5 | | | 5 | | | 5 |
| *C.* spp. (numerous) | × | | | | | | × | | | | | | | | | | |
| *Crepidula grandis* Middendorff | 19 | 20 | | | | | 19 | 5 | 5 | 5 | 5 | | | 5 | | | 5 |
| *Cryptonatica clausa* (Broderip & Sowerby) | × | 6 | 6 | 6 | | | × | 5 | 5 | 5 | 5 | | 5 | | | | |
| *Cylichnella* | × | 6 | 6 | 6 | | | × | 5 | 5 | 5 | 5 | | 5 | | | | |
| *C. alba* (Brown) | × | | | | | | × | 5 | | | | | | | | | |
| *Lacuna* | × | | | 7 | | | × | 5 | | | | | | | | | |
| *L. divaricata* (Fabricius) | × | 6 | 6 | 6 | | | × | 5 | | | | | | | | | |
| *Lora* | × | 10? | | | | | × | 5 | 5 | 5 | 5 | | | 5 | | | 2 |
| *L. chiachiana* Dall | × | | | | | | × | 5 | 5 | 5 | 5 | | | 5 | | | 2 |
| *L. harpularia* '(Couthouy) | × | | | | | | × | 5 | 5 | 5 | 5 | | | | | | |
| *L. nobilis* (Möller) | × | | | | | | × | 5 | 5 | 5 | 5 | | | | | | |

\* One of us (Durham) currently regards the San Joaquin "stage" as partly Pliocene and partly Pleistocene.

References:

1 Weaver, 1943.
2 Beets, 1946.
3 Durham, 1944.
4 Heering, 1950.
5 Harmer, 1914–25.
6 Grant and Gale, 1931.
7 Keen and Bentson, 1944.
8 MacNeil, 1957a.
9 Woodring and Bramlette, 1950.
10 MacNeil, unpublished data, 1965.
11 Faustman, 1964.
12 Durham, 1957.
13 Woodring *et al.*, 1946.
14 MacNeil, 1965.
15 Hopkins and MacNeil, 1960.
16 Masuda, 1962.
17 S. V. Wood, 1874.
18 MacNeil *et al.*, 1943.
19 MacGinitie, 1959.
20 Merklin *et al.*, 1962.
21 Soot-Ryen, 1932.

333

TABLE 1. Concluded

| | PACIFIC | | | | | | | ATLANTIC and ARCTIC | WESTERN EUROPE | | | | | | | | | |
| | | | Pliocene | | | | | | England | | | | | Netherlands/Belgium | | | | |
| Genera and Species | Recent | Later Pleistocene | San Joaquin* | Eichegoin | Jacalitos | Miocene | Oligocene | Recent | Later Pleistocene | Early Pleistocene | | | Plio-cene | Later Pleistocene | Early Pleistocene | | | Plio-cene |
| | | | | | | | | | | Butleyian | Newbornian | Waltonian | Gedgravian | | Late Amstelian | Poederlian | Scaldisian | Late Diestian |
| L. reticulata (Brown) [trevelliana] | X | | | | | | | X | | | | | | | | | | |
| L. similis (Nyst) | | | | | | | | | 5 | 5 | | 5 | 5 | | | 2 | 2 | |
| L. simplex (Middendorff) | 19 | | | | | | | 19 | 5 | 5 | 5 | 5 | | | | | | |
| L. tenuilirata Dall | 19 | | | | | | | 19 | 5 | 5 | 5 | 5 | | | | | | |
| L. turricula (Montagu) | X | 6 | | | | | | X | 5 | 5 | 5 | 5 | | 5 | | | | |
| L. viridula (Fabricius) [exarata] | X | 6 | | | | | | X | 5 | 5 | 5 | 5 | | 5 | | | | |
| Lunatia | X | 10 | 15 | 6 | 6 | | | X | | | | | | | | | | |
| L. groenlandica (Möller) | X | 10? | | 6 | | | | 19 | 5 | 5 | 5 | 5 | | | | | | |
| L. monterona (Dall) | 19 | | | | | | | 19 | 5 | 5 | 5 | 5 | | | | | | |
| L. pallida (Broderip & Sowerby) | X | 10 | 15 | | | | | X | 5 | 5 | | | | | | | | |
| Margarites | X | 13 | 13 | | 7 | | | X | | | | | | | | | | |
| M. avensooki MacGinitie | 19 | | | 8 | 8 | 8 | | 19 | | | | | | | | | | |
| M. cinereus (Couthouy) | X | | | | | | | X | 5 | 5 | 5 | 5 | | | | | | |
| M. frigidus Dall | 19 | | | | | | | 19 | 5 | 5 | 5 | 5 | | | | | | |
| M. helicinus (Fabricius) | X | | | | | | | X | 5 | 5 | 5 | 5 | | | | | | |
| M. vahli (Möller) | 19 | | | | | | | 19 | 5 | | | | | | | | | |
| Neptunea | X | 6 | 6 | 8 | 8 | 8 | | X | 5 | 5 | 5 | 5 | | | | | | |
| N. antiqua (Linnaeus) | X | | | | | | | X | 5 | 5 | 5 | 5 | 5 | 5 | | 2 | 2 | |
| N. contraria (Linnaeus) | X | | | | | | | X | 5 | 5 | 5 | 5 | 5 | 5 | | 2 | 2 | |
| N. despecta (Linnaeus) | X | | | | | | | X | 5 | 5 | 5 | 5 | | 5 | | 2 | | |
| N. heros Gray | 19 | 6 | 6 | 6 | 6 | 1 | | 19 | 5 | | | | | | | | | |
| N. middendorffiana MacGinitie | 19 | | 6 | 6 | 6 | | | 19 | | | | | | | | | | |
| N. ventricosa (Gmelin) | 19 | 6 | 6 | 6 | 6 | 1 | | X | | | | | | | | | | |

334

| | | | | | | | | | | | |
|---|---|---|---|---|---|---|---|---|---|---|---|
| N. tetragona (Sowerby) | × | | | | 1 | 3 | × | 5 | 5 | 5 | 5 | | 2 | 2 |
| Puncturella | 19 | 6 | 13 | | | | × | 5 | 5 | 5 | 5 | |
| P. noachina (Linnaeus) | × | 10? | | | | | × | 5 | 5 | 5 | 5 | 5 |
| Pyrulofusus deformis (Reeve) | × | 6 | 6 | 6 | | | × | 5 | |
| Searlesia | × | 6 | 6 | | 1 | 1 | × | 5 | 5 | 5 | 5 | |
| S. costifer (Wood) | × | | | | | | × | 5 | 5 | |
| Tachyrhynous | × | 6 | | | | | × | 5 | 5 | |
| T. erosus (Couthouy) | × | | | | | | × | |
| T. reticulatum (Mighels) | 19 | 6 | | | 1 | 1 | 19 | 5 | 5 | |
| Trichotropis | 19 | | | | | | × | 5 | 5 | |
| T. bicarinata (Sowerby) | 19 | 6 | | | | | 19 | 5 | |
| T. borealis Sowerby | × | | | | | | × | 5 | 5 | 5 | |
| T. insignis Middendorff | × | 6 | | | | | × | 5 | 5 | |
| T. kroyeri Philippi | × | 6 | | | | | × | |
| Velutina | × | 6 | 13 | | | | × | 5 | 5 | |
| V. laevigata (Linnaeus) | × | 6 | 13 | | | | × | 5 | 5 | |
| V. lanigera Möller | 19 | | | | | | 19 | |
| V. plicatilis (Müller) | 19 | | | | | | 19 | |
| V. undata Brown | × | 10 | | | | | × | 5 | 5 | 5 | |
| Volutopsius | × | 6 | | | | 6 | × | 5 | 5 | 5 | 5 | 5 |
| V. norvegica (Chemnitz) | × | | | | | | × | 5 | 5 | 5 | 5 | 5 |
| V. stefanssoni Dall | × | 10? | | | | | × | |

ECHINOIDEA

| | | | | | | | | | | |
|---|---|---|---|---|---|---|---|---|---|---|
| Echinarachnius (see discussion) | × | × | × | × | | | × | |
| E. parma (Lamarck) | 12 | × | × | × | | | × | | × | × | × |
| Strongylocentrotus (see discussion) | × | × | × | ? | | | × | × | × | × | × |
| S. droebachiensis (Müller) | × | × | × | | | | × | × | × | × | × |
| S. pallidus (Şars) | × | | | | | | × | × |

* One of us (Durham) currently regards the San Joaquin "stage" as partly Pliocene and partly Pleistocene.

References:

1 Weaver, 1943.
2 Beets, 1946.
3 Durham, 1944.
4 Heering, 1950.
5 Harmer, 1914–25.
6 Grant and Gale, 1931.
7 Keen and Bentson, 1944.
8 MacNeil, 1957a.
9 Woodring and Bramlette, 1950.
10 MacNeil, unpublished data, 1965.
11 Faustman, 1964.
12 Durham, 1957.
13 Woodring et al., 1946.
14 MacNeil, 1965.
15 Hopkins and MacNeil, 1960.
16 Masuda, 1962.
17 S. V. Wood, 1874.
18 MacNeil et al., 1943.
19 MacGinitie, 1959.
20 Merklin et al., 1962.
21 Soot-Ryen, 1932.

TABLE 2. *Icelandic Fossil Mollusca with Pacific Affinities*

| Genera and Species | Later Pleistocene | Serripes zone | Mactra zone | Tapes zone |
|---|---|---|---|---|
| **PELECYPODA** | | | | |
| *Clinocardium ciliatum* (Fabricius) | | X | | |
| *Macoma calcarea* Gmelin | | X | | |
| *M. incongrua* von Martens | | X | | |
| *Modiolus modiolus* (Linnaeus) | | | O | |
| *M.* (?) sp. | | | | X |
| *Musculus niger* (Gray) | | X | | |
| *Mya pseudoarenaria* Schlesch | | X | | |
| *M. truncata* Linnaeus | X | | | |
| *Mytilus edulis* (Linnaeus) | | O | O | X |
| *Nucula tenuis* Montagu | X | | | |
| *Panomya arctica* (Lamarck) | | X | | |
| *Serripes groenlandicus* (Bruguière) | | X | | |
| *Zirfaea crispata* (Linnaeus) | | X | | |
| *Z.* sp. | | | O | |
| **GASTROPODA** | | | | |
| *Admete couthouyi* (Jay) | | X | | |
| *Beringius turtoni* (Bean) | | X | | |
| *Boreotrophon* sp. cf. *truncatus* (Ström) | X | | | |
| *Buccinum* sp. cf. *groenlandicum patula* Sars | | | X | |
| *B. inexhaustum* Verkreuzen of Harmer | | X | | |
| *B.* sp. aff. *plectrum* Stimpson | | X | | |
| *B.* sp. cf. *totteni islandicum* Harmer | | X | | |
| *B.* sp. cf. *undatum* Linnaeus | | X | | |
| *Colus sabinii* (Gray) | | X | | |
| *C.* sp. cf. *togatus* (Mörch) | | | | cf. |
| *Cryptonatica clausa* (Broderip & Sowerby) | | X | | |
| *Lora* sp. cf. *borealis* (Reeve) | | X | | |
| *L. decussata tjörnesensis* Schlesch | | X | | |
| *Neptunea* sp. cf. *decemcostata* (Say) | | X | | |
| *N.* sp. cf. *despecta* (Linnaeus) | | X | | |
| *Nucella tetragona* (Sowerby) | | | X | |
| *Searlesia costifer* (Wood) | | X | | |
| *Tachyrhyncus erosus* (Couthouy) | X | | | |

X from MacNeil, unpublished data 1965.    O from Bárðarson 1925 (selected species).

TABLE 3. *Genera and Species of North Atlantic Origin and Their Times of Appearance in the North Pacific*

| Genera and Species | NORTH PACIFIC | | | | | | | NORTH ATLANTIC | | | | |
| --- | --- | --- | --- | --- | --- | --- | --- | --- | --- | --- | --- | --- |
| | Recent | Later Pleistocene | Late Pliocene and Early Pleistocene | Late Pliocene | Middle Pliocene | Early Pliocene | Late Miocene | Recent | Pleistocene | Pliocene | Miocene | Oligocene |
| **BRACHIOPODA** | | | | | | | | | | | | |
| Glaciarcula (see discussion) | 17 | | | | | | | 17 | 17 | | | |
| **PELECYPODA** | | | | | | | | | | | | |
| Arca boucardi-tetragona stock | X | 15 | | | | | | X | | | | |
| Astarte | X | 13 | 13 | 4 | 14 | | 3 | X | 6 | 6 | 6 | 6 |
| A. alaskensis Dall (elliptica Brown) | X | 13 | 13 | | | | 3 | X | 6 | 6 | 6 | 6 |
| A. arctica (Gray) | X | 13 | 13 | | | | | | 6 | | | |
| A. borealis (Schumacher) | X | 13 | | | | | | X | 6 | | | |
| A. fabula Reeve | X | 7? | | | | | | | | | | |
| A. soror Dall | | 7? | | | | | | X | | | | |
| Bathyarca | | 13 | | | | | | X | 16 | 16 | 16 | 16 |
| B. glacialis | | 13 | | | | | | X | | | | |
| Chlamys islandica group | X | 11 | 11 | | | | | X | X | | X | |
| Cyrtodaria | X | | | | | | | X | 6 | 6 | 6 | |
| C. kurriana Dunker | X | 13 | | | | | | X | 7 | | | |
| Hiatella arctica lineage | X | 7 | 7 | | 5 | | 3 | X | 3 | 3 | 3 | 3 |
| Portlandia arctica (Gray) | X | 7 | | | | | | X | 6 | | | |
| Siliqua patulamedia stock | X | 3 | | 3 | | 3 | | | 3 | 3 | | |
| Thyasira flexuosus-gouldii stock | X | 11 | 11 | 3 | 5 | 3 | 3 | X | 10 | 10 | 8? | 3 |
| **GASTROPODA** | | | | | | | | | | | | |
| Liomesus | X | | 7 | 4 | 5 | | | X | 1 | 1 | 1 | 1 |
| Molleria costulata (Möller) | | | | | | | | X | 10 | | | |
| **ECHINOIDEA** | | | | | | | | | | | | |
| Echinocyamus pusillus stock | | 5 | | | | | | X | 9 | 9 | 9 | 9 |

References:
X Present.
1 Beets, 1946.
2 Weaver, 1943.
3 MacNeil, 1965.
4 Woodring et al., 1940.
5 Durham, unpublished data, 1965.
6 Heering, 1950.
7 MacNeil, unpublished data, 1965.
8 MacNeil, 1957b.
9 Engel, 1958.
10 Harmer, 1914–25.
11 Grant and Gale, 1931.
12 MacNeil et al., 1943.
13 Merklin et al., 1962.
14 Durham, 1950.
15 Vedder and Norris, 1963.
16 Reinhart, 1935.
17 Elliot, 1956.

TABLE 4. *Species Whose Direction of Migration Is Uncertain*

| Genera and Species | PACIFIC Recent | Later Pleistocene | Pliocene San Joaquin* | Pliocene Etchegoin | Pliocene Jacalitos | Miocene | Oligocene | ATLANTIC and ARCTIC Recent | W. EUROPE Later Pleistocene | Butleyan | Newbornian | Waltonian | Gedgravian |
|---|---|---|---|---|---|---|---|---|---|---|---|---|---|
| **PELECYPODA** | | | | | | | | | | | | | |
| Axinopsida orbiculata (Sars) | ? | 8 | | | | | | ? | | | | | |
| Cardita crassidens (Broderip & Sowerby) | × | | | | | | | × | | | | | |
| C. crebricostata (Krause) | × | 6 | 6 | | | | | × | | | | | |
| Nucula tenuis Montagu | × | 3 | 8 | | | | | × | 1 | 5 | | 1 | 1 |
| Nuculana minuta (Fabricius) | ? | | | | | | | ? | | | | | |
| N. radiata (Krause) | ? | | | | | | | ? | | | | | |
| Pseudopythina compressa Dall | ? | | | | | | | ? | | | | | |
| Yoldia myalis Couthouy | ? | | | | | | | ? | | | | | |
| **GASTROPODA** | | | | | | | | | | | | | |
| Aquilonaria turneri Dall | ? | | | | | | | ? | | | | | |
| Bela simplex (Middendorff) | × | 4? | | | | | | × | | | | | |
| Beringius beringi (Middendorff) | ? | | | | | | | ? | | | | | |
| B. stimpsoni (Gould) | ? | | | | | | | ? | | | | | |

| Species | | |
|---|---|---|
| Cingula castanea alaskana Dall | | 7 |
| Cylichna occulta (Mighels) | | 7 |
| Littorina palliata (Say) | 6 | × 2 |
| L. rudis (Maton) | × | × 2 |
| Margaritopsis grosvenori (Dall) | 2 | 7 |
| M. pribiloffensis (Dall) | | 7 |
| Nodotoma impressa (Mörch) | | 7 |
| Odostomia cassandra (Dall & Bartsch) | | 7 |
| Oenopota tenuicostata (Sars) | | 7 |
| O. harpa Dall | | 7 |
| "Oenopota" elegans (Möller) | | 7 |
| O. nazanensis (Dall) | | 7 |
| O. pyramidalis (Strøm) | | 7 |
| Onchidiopsis glacialis (Sars) | | 7 |
| Piliscus commodus (Middendorff) | | 7 |
| Plicifusus kroyeri Möller | | 7 |
| Ptychatractus occidentalis Stearns | | 7 |
| Solariella obscura (Couthouy) | 4 | × 2 |
| S. varicosa (Mighels & Adams) | × | × 2 |
| **AMPHINEURA** | | |
| Symmetrogephyrus vestitus (Broderip & Sowerby) | | 7 |
| Trachydermon albus (Linnaeus) | | 7 |

* One of us (Durham) currently regards the San Joaquin "stage" as partly Pliocene and partly Pleistocene.

× Present.    2 Harmer, 1914–1925.    4 MacNeil, unpublished data, 1965.    6 MacNeil et al., 1943.    8 Merklin et al., 1962.
1 Heering, 1950.    3 Grant and Gale, 1931.    5 S. V. Wood, 1874.    7 MacGinitie, 1959.

339

It is also present in various Pleistocene localities in New England and in the Gulf of St. Lawrence area. The genus is not known in the pre-Pleistocene of the Atlantic or Arctic areas.

Elliot (1956) proposed the new genus *Glaciarcula*, with *Terebratella spitzbergensis* Davidson as type, and noted that it is widely distributed in the recent fauna of the North Atlantic and Arctic and is also known from a single occurrence off Japan. It also occurs in the Pleistocene of Scandinavia. On this scanty basis, he suggests that it is of Atlantic origin and has been a late immigrant into the Pacific through Bering Strait.

Elliot (1956) considered that the three closely related genera *Dallina*, *Campages* (including *Japanithyris*), and *Macandrevia* had probably migrated from the Pacific to the Atlantic in the early Pliocene, either by way of the Arctic or through the Panamic portal. He noted that *Dallina* and *Japanithyris* occur in the so-called Pliocene of Sicily. At that time, *Japanithyris* was recorded only from the recent fauna off Japan, although the other two genera are known in the Miocene of Japan. Since then, Cooper (1957, p. 2) has concluded that *Japanithyris* is a synonym of *Campages* and occurs in the Pliocene of Okinawa, further supporting the conclusion that these genera are of Pacific origin.

### ECHINOIDEA

The genus *Strongylocentrotus* seems to be clearly of North Pacific origin, although the Miocene records (both Atlantic and Pacific) cited by Grant and Hertlein (1938, p. 32–33) have been negated. Mortensen (1943, p. 197) pointed out that the Miocene records from France are better referred to *Paracentrotus*. The "Miocene of Oregon" occurrence of Arnold, cited by Grant and Hertlein (p. 33, 38), is based on a listing of *Strongylocentrotus cf. purpuratus* by Arnold (1906, p. 120) from beds near Yachats, Oregon. The original specimen has not been seen, but echinoids subsequently collected from this locality are certainly not referable to this genus, and the associated fauna suggests a late Eocene or early Oligocene age. Arnold's listing of the genus appears extremely doubtful. Dall (1909, p. 18, 139) doubtfully recorded *Strongylocentrotus* from the Astoria Formation (Miocene) at Astoria, Oregon, and from the Empire Formation (then assigned to Miocene) at Coos Bay, Oregon. The Empire Formation is now referred to the early Pliocene, and the spines from there are probably referable to *Glyptocidaris*, a genus that has recently been identified from this locality. However, despite the failure to confirm the published pre-Pliocene records of *Strongylocentrotus*, other evidence indicates that the genus has considerable antiquity and that it is of North Pacific origin. Only two species

(*S. droebachiensis* and *S. pallidus* [or *S. echinoides*], see Swan, 1962) occur in the Atlantic. Both also occur in the North Pacific, where nine species
are recognized by Mortensen (1943). Only three species (*S. droebachiensis,
S. purpuratus,* and *S. franciscanus*) have been reported as fossils in the Pacific, but each is recognizable in the middle Pliocene Falor Formation of
coastal northern California (Manning and Ogle, 1950). The fact that these
three divergent species have their distinctive characters in the middle Pliocene indicates that they must have separated even earlier and that the genus
had a considerably earlier origin. Isolated spines in the so-called Santa
Margarita Formation of late Miocene age at the south end of the San Joaquin
Valley, California, appear to be referable to *Strongylocentrotus* and thus
seem to substantiate its occurrence in the Miocene of the North Pacific. The
large number of species present in the North Pacific also supports the conclusion of long residence of the genus in the area.

*Strongylocentrotus droebachiensis* has been reported by Clark and
Twitchell (1915) from the Caloosahatchee Formation (Pliocene or Pleistocene) of Florida, although Cooke (1959, p. 25) notes that the two specimens
on which the record is based are poorly preserved, and doubts the validity
of the identification. Kier (1963), in his monograph of the Caloosahatchee
echinoids, strangely omits any notice of Clark and Twitchell's record, but
comparison of his illustrations of *Echinometra lucunter* suggests that Clark
and Twitchell's specimens may be referable to the latter species.

In Europe, *S. droebachiensis* is widespread as a fossil in the Pleistocene
Red Crag of England (Reid, 1890) and in equivalent beds in the Low Countries. It is recorded from the Icenian, Scaldisian, and Diestian of Netherlands (Engel, 1941).

An incomplete adult and a very small immature specimen of *Echinocyamus* have been collected by personnel of the U.S. Geological Survey from
Pleistocene beds at Einahnuhto Bluff on St. Paul Island in Bering Sea. The
adult is unlike any species known from the North Pacific but is closely similar to *E. pusillus* (Müller) from the eastern North Atlantic areas (where this
species has a well-documented fossil record; see Engel, 1958), and in consequence it is tentatively referred to that species. It thus appears to be one
of the few organisms of Atlantic origin that succeeded in passing through
the Bering Strait portal, but it probably failed to colonize permanently in the
Pacific basin. To date, no living species of *Echinocyamus* has been recorded
north of the Hawaiian Islands, and the species found in that area are quite
different from *E. pusillus*.

The genus *Echinarachnius* is clearly of Pacific origin; *E. subtumidus*
Nisiyama & Hashimoto, from the Miocene of Hokkaido, Japan, is the earliest

species validly referred to the genus (Durham, 1955, p. 165). *E. parma* (Lamarck) is now known to occur in the Recent fauna at Point Barrow (Durham, 1957, p. 629), as well as in Cuba and Jamaica (Yale Oceanographic Expedition, *Atlantis*, 1933, unpublished data), which would seem to give this species a Recent distribution from the North Pacific (Puget Sound to northern Japan), through the Arctic, and southward along the American coast to Cuba and Jamaica. These southern occurrences are quite unexpected in a genus of temperate and boreal preferences. Cooke noted in 1959 (p. 44) that *E. parma* was not known as a fossil at that time. Since then, it has been found in the Pleistocene of the Pribilof Islands, including the locality at which *Echinocyamus pusillus*(?) occurs. Seemingly, *E. parma* is a very late migrant through Bering Strait and the Arctic, and has rapidly extended its range southward along the Atlantic Coast of North America.

*Echinarachnius? woodii* Forbes (1872) from the British Red Crag does not appear to be an *Echinarachnius*, and in any case is certainly not closely related to *E. parma*.

PELECYPODA

Most of the mollusks that have achieved transarctic migration are listed in Tables 1–4. The extinct pelecypod genus *Fortipecten* should be discussed separately, however; although it has never penetrated to the Atlantic Ocean, the distribution of its fossil occurrences in the Pacific Ocean, Bering Sea, and Chukchi Sea provides evidence of an early opening of Bering Strait.

*Fortipecten hallae* (Dall) occurs in the Kivalina area north of Bering Strait (Hopkins and MacNeil, 1960). Although *Fortipecten* was proposed originally as a subgenus of *Patinopecten*, Masuda (1962, p. 222) ranks it as a genus coordinate with *Patinopecten*, which occurs in the Oligocene to Recent of the North Pacific Basin. *Fortipecten* is known in Japan only from beds assigned to the early and middle Pliocene in the Japanese chronology.

## Migrations

Early Cenozoic marine faunas have a subtropical to tropical aspect in much higher latitudes than those living today (see Durham, 1950, 1952, 1959). At that time the Tethyan seaway and the Panamic portal were open, and there was free communication between the warmer parts of the Atlantic and Pacific Oceans. However, we are unable to recognize any evidence of migration through Bering Strait and the Arctic during this interval. This is in accord with the observation of Hopkins (1959) that there is no geologic evidence of submergence of the Bering Land Bridge in the early Cenozoic. Following the Eocene, climates in northern latitudes gradually became cooler, heralding the approach of Pleistocene glaciations, and recognizably frigophilic organisms appear in Alaskan faunas. During the late Cenozoic, taxa of foreign aspect appear in the biotas of the northern areas of both oceans. It is with these "invaders" that this study is primarily concerned.

Our analyses show that a surprisingly large number of species have migrated through Bering Strait from one ocean to the other, largely in one direction. Over 125 species, mostly gastropods and pelecypods, of Pacific origin have entered the Arctic-Atlantic region. The direction of migration of another 33 species is uncertain, but many of them may be of Pacific origin. In marked contrast, we have been able to recognize in North Pacific Late Cenozoic faunas no more than about 16 species or species groups that are of Atlantic origin. This marked dominance of Pacific emigrants may be merely a reflection of the prevailing eastward current pattern in the Arctic (see MacNeil, 1965, p. 9–11), or it may in part be a function of the richer biotas of the Pacific (Ekman, 1953), particularly in the tropics (also see Ladd, 1960). As shown by Ekman, there is a greater amount of endemism in Pacific faunas than in those of the Atlantic; this in turn appears to be a reflection of the essentially unidirectional migration through Bering Strait.

Table 1 lists all species and species groups that we recognize as of Pacific origin, and that have reached areas other than Iceland and Spitzbergen. This group includes over 69 gastropods, more than 41 pelecypods, three echinoids, and one brachiopod.

The Icelandic material is presented separately in Table 2, because better comparative material was available to us during the preparation of this report. Thirty-two species of Pacific origin that have reached Iceland are presented in Table 2. This group includes 14 pelecypods and 18 gastropods.

The 16 species and species groups that are considered to have entered the Pacific from the Atlantic region are given in Table 3. These include 12 pelecypods, two gastropods, one echinoid, and one brachiopod. The available data seem to indicate that the echinoid and three of the mollusks were not able to maintain themselves after reaching the Pacific and are now extinct.

Thirty-three species (Table 4), in large part known only in the Recent Arctic and North Pacific faunas, are of uncertain origin. They include 23 gastropods, eight pelecypods, and two chitons.

These northern interocean migrations should be considered from two aspects: first, what data do they provide on the time or times that the Bering Strait was open; and second, what do they indicate about the correctness of present late Cenozoic correlations from one ocean to the other?

The earliest suggestion of interoceanic migration via a northern route is furnished by the genera *Siliqua*, *Astarte*, *Hiatella*, *Mya*, and *Placopecten* in the Late Miocene or earliest Pliocene. The pelecypod *Siliqua* occurs in the Eocene and Miocene (Helvetian) of France (Cossmann and Peyrot, 1913–14, p. 428), the Eocene of the Mississippi embayment (Harris, 1919, p. 196, pl. 59, fig. 4), and the Miocene of the eastern United States, as well as in the Recent North Atlantic, Arctic, and North Pacific faunas. The earliest appearance of *Siliqua* in the North Pacific is in the late Miocene Briones Formation of California (Trask, 1922, p. 142), and it occurs in appropriate lithologies of younger sedimentary deposits from central California northward. A *Siliqua* of undoubted northern type occurs in the Empire Formation (Grant and Gale, 1931) of earliest Pliocene age. Although other species of the genus occur in Recent tropical faunas, they are characterized by a high narrow internal buttress, in contrast to the low broad buttress of the northern species, and thus seem to be separable from the temperate and boreal species.

The cold-water pelecypod genus *Mya* has a well-documented Pacific history (MacNeil, 1965), and the earliest records are in the late Eocene or early Oligocene of Japan. The first member of the Pacific group to appear in the Atlantic is *M. arenaria* in the late Miocene Yorktown Formation of Virginia (MacNeil, 1965, p. 13), and the genus is now widespread in higher latitudes. *M. producta*, from the middle Miocene Choptank and Kirkwood Formations of Maryland and New Jersey, does not appear to be closely related to other members of the genus (MacNeil, 1965, p. 8).

The scallop *Placopecten* is represented by four species in the early Miocene of Japan (Masuda, 1962, p. 192–194), but is not otherwise known from the Pacific basin. In the Atlantic, *P. clintonianus* and *P. virginiana* appear in the late Miocene of the Atlantic States (Gardner, 1943, p. 37–39); the stock is represented by *P. magellanicus* (Gmelin) in the Recent Western Atlantic fauna. The Atlantic Miocene occurrences are in the same beds (Gardner, 1943, p. 139) as the first Atlantic occurrences of the Pacific group of *Mya*, which suggests that representatives of the two genera migrated together from the North Pacific.

*Hiatella arctica* (Linnaeus) has a well-documented history in the Atlantic, tracing its ancestry back at least to the Oligocene (MacNeil, 1965, p. 9). This species and another emigrant from the North Atlantic–Arctic, *Astarte alaskensis* (or *elliptica*) Dall, first appear in the Pacific in a part of the Yakataga Formation of Alaska that is of either latest Miocene or very early Pliocene (MacNeil, 1965, p. 8). It is now widespread from Japan eastward along the Alaskan and Pacific Coasts.

On the basis of the well-documented eastward migration of the cold-water *Mya*, the apparently similar movement of *Placopecten*, and the apparent westward migration through the Arctic of *Siliqua* at about the same time in the late Miocene, it appears that the ancestral Bering Strait marine passageway first opened up at this time. Whether the first Pacific occurrences of *Hiatella arctica* and *Astarte alaskensis* represent a migration synchronous with that of *Mya* and *Placopecten* is uncertain, but if not, the movement was only slightly later.

The pelecypod *Thyasira bisecta* (Conrad) has a well-documented history in the Pacific Basin, from at least the late Oligocene (Grant and Gale, 1931, p. 281–282; Weaver, 1943, p. 142) to Recent. It also occurs in Spitzbergen (see MacNeil, 1965, p. 7) in beds once referred to the Paleocene but which must be younger than the earliest opening of the Bering Strait portal —possibly late Miocene. It is possible that this species may have accompanied *Mya* and *Placopecten* on their eastward journey but did not get as far south. This stock of *Thyasira* is not known in the Atlantic province, except for the Spitzbergen occurrence.

By undoubted early Pliocene time, *Mytilus edulis* and *Strongylocentrotus droebachiensis* (see discussion above) reached Western Europe (Diestian of the Low Countries), and by late Pliocene (Scaldisian of Low Countries, Coralline Crag or Gedgravian of Britain), more than 25 species (17 genera) of Pacific ancestry reached Western Europe. These include the gastropod genera *Acmaea, Admete, Boreoscala, Boreotrophon, Buccinum, Colus* [*Sipho*], *Lora, Neptunea, Nucella, Trichotropis,* and *Volutopsius,* and the pelecypods *Macoma, Modiolus, Mya, Serripes,* and *Zirfaea.* In the late Pliocene (see discussion of the Pliocene-Pleistocene boundary on p. 327, above) of the Pacific States, the gastropod *Liomesus* and pelecypods of the *Arca boucardi–A. tetragona–A. sisquocensis* stock appear, both of well-established European ancestry.[a] Migration through Bering Strait has continued since then, mostly in an eastward direction, but a few emigrants from the Atlantic continue to appear in the Pacific. Some of these, such as *Bathyarca glacialis, Molleria costulata,* and *Echinocyamus pusillus*(?), did not succeed in establishing permanent colonies and soon died out.

The Icelandic sequence (Table 2; Bárdarson, 1925; MacNeil, unpublished data, 1965) is apparently largely of Pleistocene age in terms of current placement of the Pliocene-Pleistocene boundary, although the "Tapes" zone and the "Mactra" zone may be of latest Pliocene age. The fossiliferous

---

[a] *Note added in proof:* Noda (1966, p. 55–56, Table 3) reports *Arca boucardi* from strata as old as early Miocene. If his identifications are correct, the evaluation of *Arca* presented here is incorrect.

strata are farther north (Tjörnes at lat. 66° N.) than the critical West European faunas, and thus presumably were deposited under somewhat more Arctic conditions than prevailed then in England and the Low Countries. In this respect, the fauna is interesting in that it contains 12 species of boreal mollusks of Pacific ancestry that did not reach the lower latitudes of Western Europe.

The data presented here (within the precision of correlation noted above) seem to indicate that the late Miocene and younger correlations used in this study are approximately correct. With the limited data available from the Bering Sea–Arctic area, we are unable to recognize any intervals, subsequent to its first opening, in which Bering Strait was closed, but neither is there evidence that it was continuously open. Recognition of oscillations in that vicinity requires either more data from the marine paleontologic record or other bases for establishing criteria.

## Conclusions

Marine molluscan genera that have simple biogeographic histories and that originated and developed in the middle and higher latitudes of the North Pacific and North Atlantic Oceans are most significant in evaluating marine migrations through Bering Strait and the Arctic Ocean. The available evidence suggests that no transarctic Cenozoic marine migrations occurred earlier than about late Miocene (Pontian) time; during late Pliocene and Pleistocene time, migrations are well documented. These data do not require the connecting seaways to have remained open continuously after their first establishment. It seems more probable that the seaways were intermittently open and closed, but the molluscan paleontologic data on hand do not permit accurate dating of these fluctuations.

### REFERENCES

Arnold, R. 1906. The Tertiary and Quaternary pectens of California: U.S. Geol. Survey Prof. Paper 47, 264 p.

Bárdarson, Gudmundur G. 1925. A stratigraphical survey of the Pliocene deposits at Tjörnes, in Northern Iceland: Kong. Danske Vidensk. Selskab, Skr. Medd. Biol., v. 4, no. 5, 118 p.

Beets, C. 1946. The Pliocene and lower Pleistocene gastropods in the collections of the Geological Foundation in the Netherlands (with some remarks on other Dutch collections): Netherlands, Geol. Stichting, Med., ser. C-IV-1, no. 6, 166 p.

Clark, W. B., and M. W. Twitchell. 1915. The Mesozoic and Cenozoic Echinodermata of the United States: U.S. Geol. Survey Monogr. 54, 341 p.

Cooke, C. W. 1959. Cenozoic echinoids of eastern United States: U.S. Geol. Survey Prof. Paper 321, 106 p.

Cooper, G. A. 1957. Tertiary and Pleistocene brachiopods of Okinawa, Ryukyu Islands: U.S. Geol. Survey Prof. Paper 314-A, p. A1–A20.

Cossmann, M., and A. Peyrot. 1913–14. Conchologie Néogénique de L'Aquitaine: Imprimerie A. Saugnac et Cie, Bordeaux, v. 2, 496 p.

Cox, A., D. M. Hopkins, and G. B. Dalrymple. 1966. Geomagnetic polarity epochs: Pribilof Islands, Alaska: Geol. Soc. America Bull., v. 77, p. 883–910.

Dall, W. H. 1909. Contributions to the Tertiary paleontology of the Pacific Coast. I. The Miocene of Astoria and Coos Bay, Oregon: U.S. Geol. Survey Prof. Paper 59, 278 p.

Davidson, T. 1874–82. A monograph of the British fossil *Brachiopoda*. v. 4, Tertiary, Cretaceous, Jurassic, Permian, and Carboniferous supplements; and Devonian and Silurian *Brachiopoda* that occur in the Triassic pebble bed of Budleigh Salterton in Devonshire: London, Palaeontographical Soc., 16 p.

Davies, A. M. 1930. Faunal migrations since the Cretaceous Period: Geol. Assoc. London Proc., v. 40, p. 307–327.

———— 1934. Tertiary faunas. Vol. II. The sequence of Tertiary faunas: Thomas Murby and Co., London, 252 p.

Durham, J. W. 1944. Megafaunal zones of the Oligocene of northwestern Washington: Univ. Calif. Pub. Geol. Sci., v. 27, no. 5, p. 101–211.

———— 1950. Cenozoic marine climates of the Pacific Coast: Geol. Soc. America Bull., v. 61, no. 11, p. 1243–1264.

———— 1952. Early Tertiary marine faunas and continental drift: Am. Jour. Sci., v. 250, no. 5, p. 321–343.

———— 1955. Classification of clypeasteroid echinoids: Univ. Calif. Pub. Geol. Sci., v. 31, no. 4, p. 73–198.

———— 1957. Notes on echinoids: Jour. Paleontology, v. 31, no. 3, p. 625–631.

———— 1959. Palaeoclimates, p. 1–16 in L. H. Ahrens et al., eds., Physics and chemistry of the earth, V. III: Pergamon Press, London.

Durham, J. W., R. H. Jahns, and D. E. Savage. 1954. Marine-nonmarine relationships in the Cenozoic section of California, p. 59–71 in R. H. Jahns, ed., Geology of southern California: California Div. Mines Bull. 170, pt. III.

Ekman, Sven. 1953. Zoogeography of the Sea: Sidgwick and Jackson, Ltd., London, 417 p. Translated from the Swedish by Elizabeth Palmer.

Elliot, G. F. 1956. On Tertiary transarctic brachiopod migration: Ann. & Mag. Nat. Hist., ser. 12, v. 9, no. 100, p. 280–286.

Engel, H. 1941. Tertiaire en quartaire echinodermen uit boringen in Nederland: Geol. en Mijnbouw, 3e Jaarg., no. 1, p. 5–17.

———— 1958. *Echinocyamus pusillus* (O. F. Müller): Netherlands, Geol. Stichting, Med., n.s., no. 11, p. 41–42, pl. 27.

Evernden, J. F., D. E. Savage, G. H. Curtis, and G. T. James. 1964. Potassium-argon dates and the Cenozoic mammalian chronology of North America: Am. Jour. Sci., v. 262, p. 145–198.

Faustman, W. 1964. Paleontology of the Wildcat Group at Scotia and Centerville Beach, California: Univ. Calif. Pub. Geol. Sci., v. 41, no. 2, p. 97–160, 3 pls.

Forbes, E. 1852. Monograph of the *Echinodermata* of the British tertiaries: London, Palaeontographical Soc., 36 p.

Gardner, Julia. 1943. Pelecypods, pt. 1; Scaphopoda and Gastropoda, *in* Mollusca from the Miocene and lower Pliocene of Virginia and North Carolina, with a summary of the stratigraphy by Wendell Clay Mansfield (1874–1939): U.S. Geol. Survey Prof. Paper 199-A, 178 p.

Grant, U. S., IV, and H. R. Gale. 1931. Catalogue of the marine Pliocene and Pleistocene mollusca of California and adjacent regions: San Diego Soc. Nat. Hist. Mem., v. 1, 1036 p.

Grant, U. S., IV, and L. G. Hertlein. 1938. The West American Cenozoic Echinoidea: Univ. Calif. Pub. Math. Phys. Sci., v. 2, 225 p.

Harmer, F. W. 1914–25. The Pliocene *Mollusca* of Great Britain, being supplementary to S. V. Wood's Monograph of the Crag *Mollusca*: London, Palaeontographical Soc., 2 vols.

Harris, G. D. 1919. Pelecypoda of the St. Maurice and Claiborne stages: Bulls. Am. Paleontology, v. 6, no. 31, 260 p.

Hatai, K., and S. Nisiyama. 1952. Checklist of Japanese Tertiary marine Mollusca: Tohoku Univ. Sci. Repts., 2d ser. (Geology), spec. v., no. 3, 464 p.

Heering, J. 1950. Pelecypoda (and Scaphopoda) of the Pliocene and older-Pleistocene deposits of the Netherlands: Netherlands, Geol. Stichting, Med., ser. C-IV-1, no. 9, 225 p.

Hertlein, L. G., and U. S. Grant, IV. 1944. The Cenozoic Brachiopoda of western North America: Univ. Calif. Pub. Math. Phys. Sci., v. 3, 236 p.

Hopkins, D. M. 1959. Cenozoic history of the Bering Land Bridge (Alaska): Science, v. 129, no. 3362, p. 1519–1528.

Hopkins, D. M., and F. S. MacNeil. 1960. A marine fauna probably of late Pliocene age near Kivalina, Alaska: U.S. Geol. Survey Prof. Paper 400-B, art. 157, p. B339–B342.

Hopkins, D. M., F. S. MacNeil, and E. B. Leopold. 1960. The coastal plain at Nome, Alaska—A late Cenozoic type section for the Bering Strait region: Rept., 21st Internat. Geol. Cong. (Copenhagen), 1960, pt. 4, p. 46–57.

Hopkins, D. M., F. S. MacNeil, R. L. Merklin, and O. M. Petrov. 1965. Quaternary correlations across Bering Strait: Science, v. 147, no. 3662, p. 1107–1114.

Keen, A. Myra. 1937. Abridged check list and bibliography of west North American marine mollusca: Stanford Univ. Press, Stanford, Calif., 84 p.

Keen, A. Myra, and H. Bentson. 1944. Check list of California Tertiary marine Mollusca: Geol. Soc. America Spec. Paper 56, 280 p.

Kier, P. M. 1963. Tertiary echinoids from the Caloosahatchee and Tamiami formations of Florida: Smithsonian Misc. Coll., v. 145, no. 5, 63 p.

Ladd, H. S. 1960. Origin of the Pacific island molluscan fauna: Am. Jour. Sci., v. 258-A (Bradley Volume), p. 137–150.

Lagaaij, R. 1952. The Pliocene Bryozoa of the Low Countries and their bearing on the marine stratigraphy of the North Sea region: Netherlands, Geol. Stichting, Med., ser. C-V, no. 5, 233 p.

Laursen, D. 1950. The stratigraphy of the marine Quaternary deposits in West Greenland: Meddl. Grønland, v. 151, no. 1, 142 p.

MacGinitie, Nettie. 1959. Marine mollusca of Point Barrow, Alaska: U.S. Nat. Mus. Proc., v. 109, no. 3412, p. 59–208.

MacNeil, F. S. 1957a. Selected mollusks from the Poul Creek and Yakataga formations, Yakataga and and Malaspina districts, Alaska, showing tentative identifications and stratigraphic range: U.S. Geol. Survey Oil & Gas Inv. Map OM-187, sheet 2, table 1.

———— 1957b. Cenozoic megafossils of northern Alaska: U.S. Geol. Survey Prof. Paper 294-C, p. 99–126.

MacNeil, F. S. 1965. Evolution and distribution of the genus *Mya*, and Tertiary migrations of Mollusca: U.S. Geol. Survey Prof. Paper 483-G, p. G1–G51.

MacNeil, F. S., J. B. Mertie, Jr., and H. A. Pilsbry. 1943. Marine invertebrate faunas of the buried beaches near Nome, Alaska: Jour. Paleontology, v. 17, no. 1, p. 69–96.

Manning, G. A., and B. A. Ogle. 1950. Geology of the Blue Lake quadrangle, California: California Dept. Nat. Res., Div. Mines, Bull. 148, 36 p.

Masuda, K. 1962. Tertiary Pectinidae of Japan: Tohoku Univ. Sci. Repts., ser. 2 (Geology), v. 33, no. 2, p. 117–238.

Merklin, R. M., O. M. Petrov, and O. V. Amitrov. 1962. Atlas-Guide of mollusks of the Quaternary deposits of the Chukotka Peninsula: Acad. Sci. USSR Comm. Study Quaternary Period, Moscow, 56 p. In Russian; translation available from Office of Tech. Services, U.S. Dept. Commerce, and from American Geol. Inst.

Mortensen, T. 1943. A monograph of the Echinoidea: v. 3, pt. 3, p. 1–446, atlas of 66 pls.

Nelson, R. N. 1925. Geology of the hydrographic basin of the upper Santa Ynez River, California: Univ. Calif. Pub. Geol. Sci., v. 15, no. 10, p. 327–396.

Noda, H. 1966. The Cenozoic Arcidae of Japan: Sci. Repts., Tohoku Univ., Sendai, Second Ser., Geol., v. 38, p. 1–161.

Obradovich, J. D. 1965. Isotopic ages related to Pleistocene events: Intern. Assoc. Quaternary Research (INQUA), 7th Intern. Congress, Abstracts, p. 364.

Petrov, O. M. 1963. The stratigraphy of the Quaternary deposits of the southern parts of the Chukotka Peninsula: Bull. Comm. Study Quaternary Period, no. 28, p. 135–152. In Russian; translation available from Office of Tech. Services, U.S. Dept. Commerce, and from Am. Geol. Inst.

Reid, Clement. 1890. The Pliocene deposits of Great Britain: Geol. Survey Mem. Great Britain, 326 p.

Reinhart, P. W. 1935. Classification of the pelecypod family Arcidae: Bull. Mus. Roy. His. Nat. Belgique, v. 11, no. 13, 68 p.

Soot-Ryen, T. 1932. Pelecypoda, with a discussion of possible migrations of Arctic pelecypods in Tertiary times: Norwegian North Polar Expedition "Maud," 1918–1925, Sci. Results (pub. by Geofysisk Inst., Bergen), v. 5, no. 12, 35 p.

Swan, E. F. 1962. Evidence suggesting the existence of two species of *Strongylocentrotus* (Echinoidea) in the northwest Atlantic: Canadian Jour. Zoology, v. 40, p. 1211–1222.

Trask, P. D. 1922. The Briones Formation of middle California: Univ. Calif. Pub. Geol. Sci., v. 13, no. 5, p. 133–174.

Vedder, J. G., and R. M. Norris. 1963. Geology of San Nicolas Island, California: U.S. Geol. Survey Prof. Paper 369, 65 p.

Weaver, C. E. 1943. Paleontology of the marine Tertiary formations of Oregon and Washington: Univ. Washington Pub. Geol., v. 5, pts. 1–3, 789 p.

Weaver, C. E., *et al.* Correlation of the marine Cenozoic formations of western North America: Geol. Soc. America Bull., v. 55, no. 5, p. 569–598.

Wood, S. V. 1874. Supplement to the Crag Mollusca, Bivalvia: London, Palaeontographical Soc. Monogr., v. 27, p. 99–231.

Woodring, W. P., and M. N. Bramlette. 1950. Geology and paleontology of the Santa Maria district, California: U.S. Geol. Survey Prof. Paper 222, 185 p.

Woodring, W. P., M. N. Bramlette, and W. S. W. Kew. 1946. Geology and paleontology of Palos Verdes Hills, California: U.S. Geol. Survey Prof. Paper 207, 145 p.

Woodring, W. P., R. B. Stewart, and R. W. Richards. 1940. Geology of the Kettleman Hills oilfield, California; stratigraphy, paleontology, and structure: U.S. Geol. Survey Prof. Paper 195, 170 p.

# 19. Marine Mammals and the History of Bering Strait

VICTOR B. SCHEFFER
*Bureau of Commercial Fisheries,*
*Marine Mammal Biological Laboratory, Seattle*

We deal here with the question: "Do differences between the marine mammal faunas of Pacific and Atlantic sectors of the Arctic-Subarctic provide evidence of interchange via Bering Strait during the late Cenozoic?" The differences to be discussed are geographical and morphological. It is well to caution that taxonomic arrangements currently expressing these differences are still provisional. For example, of the 27 genera mentioned, 19 are monospecific. To a land-mammal taxonomist, this ratio of genera to species will seem high. How truly the ratio represents the phylogeny of the marine mammal groups is by no means clear.

The "Arctic" is defined broadly as the region where marine mammals are regularly associated with permanent ice, the "Subarctic" with temporary ice. The words "race" and "subspecies" are used interchangeably.

Writers who have speculated on population spread and speciation among northern marine mammals include Davies, King, McLaren, Simpson, and Udvardy (see references). I have been helped by studying their conclusions, and especially by discussing them with V. Standish Mallory.

## General Features of Marine Mammals

Recent marine mammals include most of the Cetacea, the Sirenia, and the Pinnipedia (all three orders also have freshwater representatives), and one of the Carnivora, the sea otter. All have flippers—flattened digits—on at least one pair of limbs. They gather submarine food and cache no part of it for future use. They give birth in the open to single, large, precocious young, and they do not use nest materials. No other mammal has all of the preceding characteristics.

Scheffer and Rice (1963) recognized 116 species of recent marine mammals. We shall review 16 of these in some detail, concentrating especially on those that regularly enter the Bering-Chukchi region or could have entered it during late Cenozoic time. Other species are mentioned briefly, including those that live in temperate waters south of, and adjacent to, the Subarctic.

At least four marine-mammal traits have implications for the spread of populations and the splitting of races.

First, many cetaceans and some pinnipeds are migratory, feeding in one place and breeding in another (Clarke, 1957). The migratory route of the northern fur seal is about 5,000 km each way, and of the California gray whale about 5,500 km. Migration is a curious trait; some mammals have adopted it and others have not: the fur seal moves north in summer to breed; the gray whale moves south in winter to breed; the sea otter breeds in any month without leaving its home waters. In the evolution of the migratory habit, the seasonal exodus to food-rich waters has probably been a more important factor than seasonal return to suitable breeding grounds. When temperatures dropped and seasons became more pronounced in the late Miocene and early Pliocene (Wolfe and Leopold, this volume), food organisms such as plankton, fish, and squid in the marine pastures of the north must have changed both quantitatively and qualitatively. Coincident with these changes, the seasonal movements of certain marine mammals may have begun. Today, migration routes are deeply engraved in the behavior patterns of many marine mammals.

Second, individuals may stray far from home. An Alaskan(?) ribbon seal was captured in Morro Bay, California, as were a Mexican(?) elephant seal in southeastern Alaska, an Atlantic harp seal and a Pacific fur seal near the mouth of the Mackenzie River, and ice-dependent hooded seals in Portugal and Florida. As a result of wandering, new colonies have become established.

Third, all pinnipeds are able to travel on land and ice. In this respect they differ from cetaceans; they are free to cross a barrier into new waters. Mummified crabeater seals have been found 30 miles from Ross Sea and up to 3,000 feet above sea level in Antarctica. Walruses may travel 15 to 20 miles overland in an emergency.

Fourth, many cetaceans and pinnipeds are "antitropical" (Davies, 1963). They shun warm water. Though the ancestors of certain species are presumed to have crossed the equator in cooler epochs (post-Miocene), the modern species now hold to higher latitudes. According to Davies (1960), antitropical representatives of the pilot whale *Globicephala melaena* are approaching subspecific distinction after separation near the end of the last ice age, about 11,000 years ago.

## Arctic and Subarctic Marine Mammal Faunas

Sixteen species of northern marine mammals are listed in Table 1. The list is based largely on distributional records from the following sources: marine mammals, Scheffer and Rice (1963); cetaceans, Nasu (1963), Sergeant (1961), Slijper (1962), and Tomilin (1962); pinnipeds, King (1964) and Scheffer (1958). Fossil records are largely from King (1964), Lepiksaar (1964), Matthes (1962), and Simpson (1945).

Each number in the right-hand column of Table 1 refers to a species discussed below.

CETACEA; ODONTOCETI

1. The killer whale *Orcinus orca* is monotypic, cosmopolitan, and temperature-tolerant. It ranges north to Novaya Zemlya, Svalbard, Greenland, Baffin Bay, Chukotka Peninsula, and Alaska. It is believed to range widely through the Arctic Ocean; certainly in Antarctica it penetrates to permanent ice.

2. The harbor porpoise *Phocoena phocoena* inhabits coastal waters of the North Atlantic and North Pacific, north to the Arctic Ocean in summer.

TABLE 1. *Comparison of Recent Marine Mammal Faunas of Pacific and Atlantic Sectors of the Arctic and Subarctic*[1]

| Species | Earliest Known Fossil Occurrence of Genus | Levels at Which Pacific and Atlantic Forms Differ | Rank |
|---------|----------|----------|------|
| ARCTIC REGION | | | |
| *Delphinapterus leucas* | Pleistocene | Subspecific | 3 |
| *Monodon monoceros* | Pleistocene | | 4 |
| *Balaena mysticetus* | Pliocene | | 8 |
| *Pusa hispida* | Pliocene | Subspecific | 10 |
| *Histriophoca fasciata* | Pleistocene | Generic (i.e., in Pacific only) | 12 |
| *Pagophilus groenlandicus* | Pleistocene | Generic (i.e., in Atlantic only) | 13 |
| *Erignathus barbatus* | Pleistocene | Subspecific | 14 |
| *Cystophora cristata* | Pleistocene | Generic (i.e., in Atlantic only) | 15 |
| *Odobenus rosmarus* | Pleistocene | Subspecific | 16 |
| SUBARCTIC REGION | | | |
| *Orcinus orca* | Pliocene | | 1 |
| *Phocoena phocoena* | Pleistocene | | 2 |
| *Balaenoptera physalus* | Pliocene | | 5 |
| *Balaenoptera acutorostrata* | Pliocene | | 6 |
| *Eschrichtius gibbosus* | Pleistocene | Subspecific (?) | 7 |
| *Phoca vitulina* | Miocene | Subspecific | 9 |
| *Halichoerus grypus* | Pleistocene | Generic (i.e., in Atlantic only) | 11 |

[1] The species are ranked in the right-hand column roughly in order of phyletic age and specialization for marine life; those first are thought to be the most remote from terrestrial ancestors.

Circumboreal communication since the last glaciation has presumably allowed some intermingling of Pacific and Atlantic stocks, for members of the two are indistinguishable.

3. The beluga *Delphinapterus leucas* ranges around the rim of the Arctic Ocean and regularly into cold waters of adjacent seas (Kleinenberg *et al.*, 1964, p. 257). At least three subspecies are recognized—from the White Sea, Arctic, and Far-Eastern Seas of Siberia—though Sergeant (1962, p. 2) stated that there are "probably many local stocks or races." In the waters between Alaska and Siberia, the beluga moves regularly as far south as northern Bering and Okhotsk Seas, and occasionally to Sakhalin. With the coming of each glaciation, the beluga was presumably able to find refuge as isolated populations in marine, brackish periglacial, or estuarine waters along a shoreline that then lay at the outer edge of the continental shelf.

4. The narwhal *Monodon monoceros* is nearly confined to the Arctic Ocean; it is the most polar of all cetaceans and is monotypic. It is more abundant on the Atlantic than on the Pacific side. In Bering Sea, tusks have been found as far south as St. Lawrence Island. Narwhals move about through channels in ice along the shore, sometimes in large groups. During glacial stages they would have taken refuge in broken ice of the North Atlantic. They would have been pushed south not only by fast ice but also by lowering of sea levels and seaward migration of shorelines.

CETACEA; MYSTICETI

5. The finback whale *Balaenoptera physalus* is cosmopolitan. It moves north in spring in the western Bering Sea and penetrates far into the summer ice of Chukchi Sea. It is the most abundant Pacific subarctic whale. Finbacks from the North Pacific and North Atlantic are indistinguishable.

6. The little piked whale *Balaenoptera acutorostrata* is cosmopolitan. It penetrates to Arctic ice fronts in summer and is believed to go farther into polar ice than any other rorqual. It tends to be coastal, strongly migratory, often solitary. Little is known of its movements in the North Pacific and adjacent seas (Norris and Prescott, 1961; Fiscus and Niggol, 1965), though its habits have been studied near Norway. Little piked whales from the North Pacific and North Atlantic are indistinguishable.

7. The gray whale *Eschrichtius gibbosus* is represented by two populations—Californian and Korean. A third (North Atlantic) population, now extinct, may have persisted into historic time; it is known from subfossil remains from Sweden, England, and Holland. It "must have occurred still along the European coasts in the first centuries A.D. Along the American coast of the Atlantic this species still must have been present in the begin-

ning of the 18th century, for at that time it was still caught by the whalers" (Van Deinse and Junge, 1935, p. 184). The California gray whale winters and breeds in Mexico and summers in Bering, Chukchi, and Beaufort Seas. The Korean gray whale winters and breeds near Korea and summers in Okhotsk Sea (Andrews, 1914; Mizue, 1951). Resemblance of Atlantic fossils to the bones of living Pacific gray whales suggests either that Atlantic and Pacific stocks were in cómmunication via Central America near the time of the Miocene-Pliocene boundary (Whitmore and Stewart, 1965) or via the Arctic during the Quaternary, or both. Since the gray whale will feed among ice floes, individuals must often have strayed from one ocean to the other, north of the Canadian Arctic archipelago. (For a delightful drawing of "California grays among the ice," see Scammon, 1874, plate 5.)

8. The bowhead whale *Balaena (Balaena) mysticetus* is distinctly Arctic. It summers at the edge of ice in the Arctic Ocean, and winters at the edge of ice in adjacent southern waters. Three populations are recognized. The first winters between eastern Greenland and the Svalbard Archipelago; its members were hunted almost to extermination by 1887. The second population winters west of Greenland in Davis Strait and the Gulf of St. Lawrence; a few individuals remain. The third and largest group winters in Bering and Okhotsk Seas; its members move northward and eastward through Bering Strait in April and May, and return south in autumn. (Dale W. Rice has kindly called my attention to Townsend's [1935] Map D, which indicates that Bering and Okhotsk stocks may be distinct.)

The beluga, the narwhal, and the bowhead are the only cetaceans that have solved the problem of reproducing in polar waters. The bowhead seems to be the result of an early split from right whale stock (p. 358) that has now widened to subgeneric distinction. The present tripartite North Pacific population of the right whale suggests that the split more likely took place in the Pacific Arctic than in the Atlantic Arctic. A series of Bering land bridges may have forced the issue in early Pleistocene time. During glacial maxima, the early bowheads could have taken refuge in either Pacific (Okhotsk?) or Atlantic Basins; they were subsequently free to intermingle.

PINNIPEDIA; PHOCIDAE

Udvardy (1963) developed the theme that the distribution patterns of northern pinnipeds and alcids (auks, murres, and puffins) are similar. The animals in both groups are linked to linear coastlines rather than to two-dimensional territories. The large climatic fluctuations of the Pleistocene dispersed seals and sea birds in similar fashion.

9. The harbor seal *Phoca vitulina* is circumboreal, temperature-tolerant, and nonmigratory, except in the northern part of its range, where it moves seasonally with the ice. It ranges north to Point Barrow, Ellesmere Island, and northern Norway. It has four indistinct marine subspecies; their postulated origin is outlined in Table 2. The zoogeography of the harbor seal is stressed here because it seems to exemplify the effect of alternate contraction and expansion of the polar ice sheet, a divisive effect seen also in the present distributional patterns of certain other seals.

A fifth subspecies of *P. vitulina* was proposed by Doutt (1942) to identify a landlocked form in lakes of Quebec. He wrote (p. 118): "The seals may have been isolated there for approximately 4,000 years, and it is assumed that this is the length of time which has been required to make a new subspecies."

10. The ringed seal *Pusa hispida* is circumboreal at the edge of ice, ranging to the North Pole. It is the only northern seal that keeps open holes in the ice through the entire winter. Four indistinct marine subspecies are recognized in the Arctic Ocean and in Okhotsk, Bering, and Baltic Seas, respectively. Two freshwater subspecies are recognized in Finnish lakes (Scheffer, 1958). In Bering Sea, the ringed seal drifts on seasonal ice as far south as the Pribilof Islands.

The common ancestor of the ringed seals may have been *Pusa pontica*, lectotype from the Sarmatian of Crimea, dated as late Miocene or early Pliocene (McLaren, 1960a). The taxonomy of modern *P. hispida* races is confused, mainly because of variation introduced by age, sex, and environment. Davies (1958a) made a scholarly attempt to reconstruct the Pleistocene history of the subspecies. Perhaps it is enough to state that *Pusa* is adaptable to subzero weather and to both fresh and salt water, and that its

TABLE 2. *Postulated Origin of Subspecies of the Harbor Seal* (Phoca vitulina) [1]

| Years Ago | Glacial Stage | History |
|---|---|---|
| 0–11,000 | Postglacial | The subspecies *richardi, largha, vitulina,* and *concolor* as now recognized. |
| > 10,000 | 4th glacial | Four groups substantially as known today are separated: E. and W. Pacific, E. and W. Atlantic. |
| > 120,000 | 3d interglacial | The groups reunite, though distinction between Pacific and Atlantic populations persists. |
| > 340,000 | 3d glacial | Arctic ice forces seals south, separating them into Pacific and Atlantic groups (or smaller groups?). |
| > 420,000 | 2d interglacial | Easy movement around the Arctic Ocean and adjacent seas; ancestral harbor seal undifferentiated. |

[1] Largely after Davies (1958a) and Ericson *et al.* (1964).

principal food (macroplankton) is widely available. Thus, Pleistocene climatic changes may have pushed ringed seals back and forth without appreciably changing their morphology.

11. The gray seal *Halichoerus grypus* is confined to the North Atlantic; it is nonmigratory and monotypic. It is represented by at least three breeding stocks: Canadian, northeastern Atlantic, and Baltic. Gray seals penetrate the Arctic to about 70° N. above Norway. Though fossil evidence is lacking, the three gray seal stocks may have separated during the late Pleistocene, always remaining within the North Atlantic basin (Davies, 1957, p. 302).

Gray seals of the British Isles, composing more than half of the world population, breed on land; others breed on ice or land. Because the gray seal everywhere gives birth to a white-coated pup, it is presumed to have originated in an icy environment. It was exposed to ice during the last glaciation and has retained a white birthcoat which can surely have no survival value today on dark-colored land nurseries.

12. The ribbon seal *Histriophoca fasciata* is monotypic and has a small range at the edge of ice in the Okhotsk, Bering, and Chukchi Seas. Its center of abundance is northwestern Bering Sea. It is regarded as one member of an allopatric pair; *Pagophilus groenlandicus* is the other. The two are the only striped pinnipeds. Because of clear taxonomic distinctions, they are thought to have split apart in early Pleistocene time. Both are strongly migratory and pelagic, and thus the confined environment of the Arctic Ocean has apparently not been attractive to them for a long time. At times in glacial cycles when the Bering and Chukchi shelves were dry land, ribbon seals may have moved into Okhotsk Sea and the deeper part of Bering Sea (Davies, 1958a, p. 107; D. M. Hopkins, written communication, 1966).

13. The monotypic harp seal *Pagophilus groenlandicus* is confined to the North Atlantic. Like the ribbon seal it breeds on ice floes, penetrating to 80° N. It has three breeding populations: White Sea, Jan Mayen, and Newfoundland. The land barrier of Greenland was responsible for the creation of slight morphological differences between the White Sea–Jan Mayen group and the Newfoundland group (Yablokov and Sergeant, 1963).

14. The bearded seal *Erignathus barbatus* is circumboreal at the edge of ice. It feeds in shallows and does not venture into the deep North Polar Basin. It has two indistinct subspecies: Pacific Arctic and Atlantic Arctic. Davies (1958a) suggested that the bearded seal was pushed south by ice and has not yet fully reclaimed areas suited to it, though King (1964) concluded that it now ranges completely around the Arctic.

15. The hooded seal *Cystophora cristata* is confined to the western

North Atlantic, at the edge of ice from Bear Island and Svalbard to eastern Canada. Though monotypic, it has two breeding centers—Jan Mayen and Newfoundland. It is a poor indicator of Pleistocene events in Bering Sea, for its geographic range is small and its ancestral affinities are with the elephant seal (*Mirounga*) of the Southern Hemisphere.

PINNIPEDIA; ODOBENIDAE

16. The monospecific walrus *Odobenus rosmarus* is nearly circumboreal. Distinct Pacific Arctic and Atlantic Arctic subspecies have long been recognized, though where they intergrade (if they do) is not clear. A recent map (King, 1964, p. 37) shows gaps in the distribution of walrus in the high latitudes of the Canadian Arctic Archipelago and Severnaya Zemlya. The walrus favors water depths of 20 to 30 meters.

Pliocene walruses are represented by *Alachtherium* from Belgium and *Valenictus* from California; also by an "odobenid, possibly new genus and species" from Santa Cruz, California, described by Mitchell (1962, p. 4). "The Santa Cruz odobenid is probably related to the ancestral stock which gave rise to the living species, *Odobenus rosmarus*" (p. 18). By implication, *Odobenus* is most likely of Pacific origin.

Pleistocene *Trichechodon*, which resembled the modern walrus but lived in warm water, is known from Florida, South Carolina, and England. *Trichechodon* and *Odobenus* coexisted in the late Pleistocene. "During the glacial period... both *Odobenus* and *Trichechodon* would have moved farther south, the two populations both dividing into American and European parts, and retreating again to the north during the interglacial periods" (King, 1964, p. 42).

## North Temperate Marine Mammal Faunas

Species that live in temperate waters south of, and adjacent to, the Subarctic are treated lightly because they do not seem to offer good evidence concerning the Pleistocene history of Bering Strait. To judge from their present habits and temperature tolerances, some of them have never crossed the Arctic Ocean and others have never crossed in numbers sufficient to leave a trace. Where fossil records are known, they are given.

CETACEA

The dolphin *Lagenorhynchus* has distinct North Pacific and North Atlantic species. There are two of the latter, one of which is the only common dolphin north of Norway.

The Dall porpoise *Phocoenoides dalli* is a North Pacific autochthone with two forms that seem to be at least subspecifically distinct. Their ranges overlap.

The beaked whale *Mesoplodon* has distinct North Pacific and North Atlantic species, as well as species in nearly all temperate and tropical waters of the world. It is known from the upper Miocene.

The beaked whale *Berardius* ranges through most of the temperate oceans of the world except the North Atlantic. The North Pacific stock has presumably never sent an offshoot through the cold-water barrier of the Arctic Ocean; the Southern Hemisphere stock (known to be cold-adapted) has presumably never sent an offshoot northward through the warm-water barrier of the Caribbean. Of the beaked whale *Hyperoodon*, the reverse is true; it ranges through most of the temperate oceans of the world except the North Pacific.

The blue whale *Balaenoptera musculus* and sei whale *B. borealis* are cosmopolitan. They are seldom seen during summer migration as far north as Bering Strait.

The black right whale *Balaena* (*Eubalaena*) *glacialis* is represented by three subspecies in temperate waters of the North Atlantic, the North Pacific, and the Southern Hemisphere, respectively. Furthermore, Klumov (1962) concluded that the North Pacific population contains three independently migrating groups: an Okhotsk, an Asian Pacific, and an American Pacific. The northern feeding range of the right whale overlaps the southern breeding range of its relative, the bowhead, in Bering and Okhotsk Seas.

The humpback whale *Megaptera novaeangliae* is cosmopolitan. In the north it ranges to Chukchi Sea, Svalbard, and Novaya Zemlya. Specimens from the North Pacific and North Atlantic are indistinguishable.

SIRENIA, PINNIPEDIA, AND CARNIVORA

Also included in the temperate marine-mammal faunas are the Steller sea cow, two or three otariid seals, and the sea otter. Though tolerant of ice, all need ice-free shallows or shores for feeding or breeding. All belong to genera known exclusively from the North Pacific basin; here they presumably originated and here they have remained.

The Steller sea cow *Hydrodamalis gigas* was discovered on Bering Island in 1741 and was exterminated by 1768. There are no distributional records, fossil or recent, outside the Commander Islands, though a rib was picked up on Attu Island. The sea cow is the only known member of the subfamily Hydrodamalinae (Simpson, 1932, p. 424). It is the only recent sirenian living outside the tropics and the only one lacking functional teeth. It was said to feed upon seaweeds (quite certainly the kelps *Thallasiophyllum, Nereocystis, Laminaria,* and *Alaria* [Stejneger, 1936, p. 355; Scheffer in Murie, 1959, p. 367]). Reinhart (1959, p. 1) proposed that *"Hydrodamalis* is . . . on a direct line of descent from the very large Miocene *Halianassa* from California." *Halianassa* is known from both North Atlantic and North Pacific beds. Mitchell and Repenning (1963, p. 14) stated that "sirenians are found in the Eocene of the Caribbean, and apparently have continuously occupied the Caribbean since then." Only two, or perhaps three, species of *Halianassa* are known from North Pacific beds, from Mexico and California. They date from late Miocene to early Pliocene (Mitchell and Repenning, p. 15). Until evidence to the contrary is presented, I presume that ancestral sea cows entered the Pacific from the Caribbean well before late Miocene time; they came increasingly to depend upon the kelp pastures of the North Pacific; they eventually extended their range northward to the southern edge of winter drift ice; they increased in body size in response to a cold environment; they lost their teeth in response to a soft diet; and, finally, along one coast after another they succumbed to man, the hunter.

The Steller sea lion *Eumetopias jubatus* is monotypic; it breeds as far north as the Pribilof Islands and wanders to the Diomedes. Fossils are known from Pliocene (or Miocene?) beds within the geographic range of living sea lions.

Another sea lion, *Zalophus californianus,* was never tolerant of cold water and perhaps should not be listed among the north-temperate faunas. It now breeds in the Galapagos Islands, southern Japan (if not recently exterminated), and Mexico–California. It may have disappeared from the Aleutian arc when the waters chilled during late Pleistocene time.

The northern fur seal *Callorhinus ursinus* breeds only on treeless islands near the southern limit of temporary ice in Bering and Okhotsk Seas. There are no good biological reasons why it does not breed farther south. Perhaps it formerly did, and its colonies were exterminated by prehistoric man. Its main breeding grounds (Commander and Pribilof Islands) were uninhabited by man at the time of their discovery in the mid-1700's.

The range of the sea otter *Enhydra lutris* has shrunk in historic time to southern California, the Alaska Peninsula, the Aleutian, Commander, and Kurile Islands, and Kamchatka. There are no authentic fossil or recent distribution records of *Enhydra* outside the North Pacific Ocean and adjacent seas. The northern limit in historic time was the Pribilof Islands (57° N.), where life for otters must have been marginal, for drift ice frequently surrounds the islands. The oldest North American fossil is a right femur from a "presumed early Pleistocene" (*Elphidiella*) bed of coastal Oregon (Leffler, 1964, p. 53). An unerupted molar tooth from a lower Pleistocene bed of Norwich Crag, England, was assigned to *Latax* (or *Enhydra*) by Pohle, who called it "ein Mittelding zwischen *Lutra* und *Latax*" (1920, p. 167). I agree with Repenning (this volume and written communication, 1965) that *Enhydra* stock apparently originated in eastern Asia in Pliocene time. At least one branch reached the North Pacific and was successful.

## Conclusion and Summary

We may assume that during each glacial stage of the Pleistocene, when a Bering land bridge appeared as a barrier to the passage of marine mammals, permanent ice over the Arctic Ocean was thick and extensive, perhaps like the fast ice of Antarctica now. V. Standish Mallory states (in ms.) that even as far south as the Strait of Juan de Fuca, shelf ice was extensive during the Wisconsin maximum. (See also Easterbrook, 1963.)

The Pleistocene series of land-and-ice barriers was effective in shaping the evolution of subspecies of beluga, ringed seal, bearded seal, walrus, harbor seal, and (?) gray whale. It may have split the common ancestor of the ribbon seal and harp seal, now subgenerically or generically distinct.

Racial differences could conceivably have arisen temporarily between North Pacific and North Atlantic stocks of the narwhal, killer whale, harbor porpoise, finback whale, and little piked whale while trans-Arctic communication was closed during the last glaciation. The time available for the evolution of differences was many tens of thousands of years. The stocks are now able to communicate more or less freely, however, and are not visibly different. We may assume that, if they deviated during the Wisconsin, they did not deviate far. (The narwhal is a special case; it presumably was not forced by ice into both Pacific and Atlantic Subarctic basins but only into the Atlantic.)

Subarctic hooded seals and gray seals are known in the North Atlantic only; no evidence exists that they have crossed the Arctic.

Known in northern temperate waters of the Pacific are six genera, represented by the Dall porpoise, Baird beaked whale, Steller sea cow, Steller sea lion, northern fur seal, and sea otter. All are unknown in the North Atlantic. Conversely, the North Atlantic bottle-nosed whale is unknown in the North Pacific. Evidence is lacking that any have crossed the Arctic during late Cenozoic time.

The inference to be drawn from the above two paragraphs is that cold water, not always coincident with ice, has acted as a barrier. It is difficult otherwise to explain the failure of certain cetaceans, sirenians, and seals to cross the Arctic during interglacial periods, at least one of which may have lasted half a million years (Ericson *et al.*, 1964).

## REFERENCES

*See Oppenheimer, 1960, for a more complete list of sources.*

Andrews, R. C. 1914. Monographs of the Pacific Cetacea. I. The California gray whale (*Rhachianectes glaucus Cope*): Mem. Am. Mus. Nat. Hist., n.s., v. 1, p. 227–287.

Clarke, R. 1957. Migration of marine mammals: Norwegian Whaling Gaz., v. 46, no. 11, p. 609–630.

Davies, J. L. 1957. The geography of the gray seal: Jour. Mammalogy, v. 38, p. 297–310.

———— 1958a. Pleistocene geography and the distribution of northern pinnipeds: Ecology, v. 39, p. 97–113.

———— 1958b. The Pinnipedia: An essay in zoogeography: Geogr. Rev., v. 48, p. 474–493.

———— 1960. The southern form of the pilot whale: Jour. Mammalogy, v. 4, p. 29–34.

———— 1963. The antitropical factor in cetacean speciation: Evolution, v. 17, p. 107–116.

Doutt, J. K. 1942. A review of the genus *Phoca*: Ann. Carnegie Mus., v. 29, p. 61–125.

Easterbrook, D. J. 1963. Late Pleistocene glacial events and relative sea-level changes in the northern Puget lowland, Washington: Geol. Soc. America Bull., v. 74, p. 1465–1483.

Ericson, D. B., M. Ewing, and G. Wollin. 1964. The Pleistocene epoch in deep-sea sediments: Science, v. 146, p. 723–732.

Fiscus, C. H., and K. Niggol. 1965. Observations of cetaceans off California, Oregon, and Washington: U.S. Fish and Wildlife Serv., Spec. Sci. Rep.—Fish. 498, 27 p.

King, J. E. 1964. Seals of the world: British Mus. (Nat. Hist.), 154 p.

Kleĭnenberg, S. E., A. V. Iablokov, V. M. Bel'kovich, and M. N. Tarasevich. 1964. Belukha [Beluga]: Moskva, Izd-vo. Nauka, 456 p. In Russian; English translation of p. 104–234, 370–394 (Chaps. 4–7, 14) available, Clearinghouse Fed. Sci. Tech. Info., Springfield, Va., 22151, as JPRS 27607, TT:64-51831, 1964.

Klumov, S. K. 1962. Gladkie (Yaponskie) kity Tikhogo okeana (The right whales of the Pacific Ocean): Trudy Instituta Okeanologii (Akad. Nauk SSSR), v. 58, p. 202–297.

Leffler, S. R. 1964. Fossil mammals from the Elk River formation, Cape Blanco, Oregon: Jour. Mammalogy, v. 45, p. 53–61.

Lepiksaar, J. 1964. Subfossile Robbenfunde von der schwedischen Westküste: Zeitsch. Säugetierkunde, v. 29, p. 257–266.

McLaren, I. A. 1960a. On the origin of the Caspian and Baikal seals and the paleoclimatological implication: Am. Jour. Sci., v. 258, p. 47–65.

———— 1960b. Are the Pinnipedia biphyletic? Syst. Zool., v. 9, p. 18–28.

Matthes, H. W. 1962. Verbreitung der Säugetiere in der Vorzeit: Walter de Gruyter & Co., Berlin, Handbuch der Zoologie, Band 8, Lieferung 28, Teil 11, Beitrag 1, p. 1–198.

Mitchell, E. D., Jr. 1962. A walrus and a sea lion from the Pliocene Purisma formation at Santa Cruz, California, with remarks on the type locality and geologic age of the sea lion *Dusignathus santacruzensis* Kellogg: Los Angeles County Mus. Contrib. Sci. 56, 24 p.

Mitchell, E. D., Jr., and C. A. Repenning. 1963. The chronological and geographic range of desmostylians: Los Angeles County Mus. Contrib. Sci. 78, 20 p.

Mizue, K. 1951. Grey whales in the East Sea area of Korea: Sci. Repts. Whales Res. Inst. 5, p. 71–79.

Murie, O. J. 1959. Fauna of the Aleutian Islands and Alaska Peninsula, with notes on invertebrates and fishes collected in the Aleutians, 1936–38: U.S. Fish and Wildlife Serv. North American Fauna 61, 406 p.

Nasu, K. 1963. Oceanography and whaling ground in the subarctic region of the Pacific Ocean: Sci. Rpts. Whales Res. Ins. 17, p. 105–155.

Norris, K. S., and J. H. Prescott. 1961. Observations on Pacific cetaceans of Californian and Mexican waters: Univ. Calif. Pub. Zool., v. 63, p. 291–402.

Oppenheimer, G. J. 1960. Reference sources for marine mammalogy: U.S. Fish and Wildlife Serv. Spec. Sci. Rept.—Fish. 361, 9 p.

Pohle, H. 1920. Die Unterfamilie der Lutrinae (eine systematisch-tiergeographische Studie an dem Material der Berliner Museum): Arch. f. Naturgeschichte, v. 85, Abt. A, pt. 9, 247 p.

Reinhart, R. H. 1959. A review of the Sirenia and Desmostylia: Univ. Calif. Pub. Geol. Sci., v. 36, 146 p.

Scammon, C. M. 1874. The marine mammals of the north-western coast of North America, described and illustrated, together with an account of the whale fishery: John H. Carmany, San Francisco, 325 p.

Scheffer, V. B. 1958. Seals, sea lions, and walruses: a review of the Pinnipedia: Stanford Univ. Press, Stanford, Calif., 179 p.

Scheffer, V. B., and D. W. Rice. 1963. A list of the marine mammals of the world: U.S. Fish and Wildlife Serv., Spec. Sci. Rep.—Fish. no. 431, 12 p.

Sergeant, D. E. 1961. Whales and dolphins of the Canadian east coast: Fish. Res. Bd. Can., Arctic Unit, Circ. 7, 17 p.

———— 1962. The biology and hunting of beluga or white whales in the Canadian Arctic: Fish. Res. Bd. Can., Arctic Unit, Circ. 8, 14 p.

Simpson, G. G. 1932. Fossil Sirenia of Florida and the evolution of the Sirenia: Amer. Mus. Nat. Hist. Bull., v. 59, p. 419–503.

———— 1945. The principles of classification and a classification of mammals: Amer. Mus. Nat. Hist. Bull., v. 85, 350 p.

Slijper, E. J. 1962. Whales: Hutchinson, London, 475 p.

Stejneger, L. 1936. Georg Wilhelm Steller, the pioneer of Alaskan natural history: Harvard Univ. Press, Cambridge, 623 p.

Tomilin, A. G. 1962. Kitoobraznye fauny morei SSSR (Cetacean fauna of Soviet seas): Opredeliteli po faune SSSR (Zool. Inst., Akad. Nauk SSSR) 79, 212 p. Moskva, Izd-vo. Nauka.

Townsend, C. H. 1935. The distribution of certain whales as shown by logbook records of American whaleships: Zoologica, v. 19, p. 1–50.

Udvardy, M. D. F. 1963. Zoogeographical study of the Pacific Alcidae, p. 85–111 *in* Pacific Basin Biogeography: Bishop Museum Press, Honolulu.

Van Deinse, A. B., and G. C. A. Junge. 1935. Recent and older finds of the California gray whale in the Atlantic: Temminckia, v. 2, p. 161–188.

Whitmore, F. C., Jr., and R. H. Stewart. 1965. Miocene mammals and Central American seaways: Science, v. 148, p. 180–185.

Yablokov, A. V., and D. E. Sergeant. 1963. Izmenchivost' kraniologicheskikh priznakov grenlandskogo tiulenia (*Pagophilus groenlandicus* Erxleben, 1777) (Cranial variation in the harp seal [*Pagophilus groenlandicus* Erxleben, 1777]): Zoologicheskiĭ Zhurnal, vol. 42, p. 1857–1865. In Russian; English translation available, Fish. Res. Bd. Can., Transl. Ser. 485, 15 p., 1964.

## 20. Depth Changes in Bering Sea During the Upper Quaternary, as Indicated by Benthonic Foraminifera

H. M. SAIDOVA

*Institute of Oceanology, Academy of Sciences of the USSR*

The present distribution of benthonic foraminifera in the northwestern part of Bering Sea has been established on the basis of bottom samples from 170 stations ranging in depth from 5 to 4,400 meters. A total of 80 species of calcareous and arenaceous foraminifera have been found. Calcareous benthonic species have been found thus far in littoral areas on the shelf and continental slope of Bering Sea and on the Shirshov Submarine Ridge; they do not inhabit the abyssal floors of the Aleutian and Kamchatka Basins. The maximum depth at which calcareous foraminifera have been found in Bering Sea is the same as in the boreal area of the Pacific Ocean and does not exceed 3,000–3,500 meters; they are found in greatest abundance at depths of 0–200 meters, 700–1,500 meters, and 2,500–3,250 meters. Each abundance maximum is represented by a different foraminiferal species-complex (Saidova, 1961). Arenaceous species are found mainly in two areas in Bering Sea: in the littoral zone in depths of less than 200–250 meters, and in open-sea areas where depths exceed 2,500 meters. The abyssal plains of the Aleutian and Kamchatka Basins are inhabited only by arenaceous foraminifera. This deep-water complex penetrated into Bering Sea through the deep straits in the Aleutian-Commander arc from the Pacific Ocean.

Thus, the variations in the species composition of foraminiferal faunas in Bering Sea depend mostly upon depth (Saidova, 1958, 1961, 1965). A study of the vertical distribution of individual species has shown large differences in the species composition of foraminiferal faunas on the continental shelf, on the continental slope, and in the deep-sea basins; the number of species in common is very small. Species found within two adjacent geomorphic zones have a quantitative maximum in only one. Thus, the hypsometric zonation is very distinctly expressed by changes in the species-composition of benthonic foraminifera, and is quite distinct at all latitudes.

This zonation is also expressed in an alternation of minimum and maximum quantities of foraminifera.

In boreal and arctic areas, the vertical variation in the quantitative and species composition of foraminiferal faunas is determined mainly by pressure. Water temperature apparently plays a smaller role, especially in the polar area, where fluctuations with depth are insignificant and do not exceed one or two degrees. In tropical areas, the temperature of the water can change with depth from 1.5° to 25° C.; there, both temperature and depth are of decisive importance for foraminiferal faunas.

Four complexes of benthonic foraminifera can be distinguished in the modern foraminiferal fauna of Bering Sea: the complex of the continental shelf, the complex of the upper continental slope, the complex of the lower part of the continental slope, and the complex of the abyssal basins.

The past distribution of foraminifera in Bering Sea has been studied in 16 cores penetrating the bottom sediments. Cores that penetrate Upper Pleistocene deposits encounter species complexes that differ from the modern foraminiferal faunas at the localities where the cores were taken. In every case, the Upper Pleistocene sediments contain species complexes indicating that the water was shallower than at present. A description of three typical cores (Fig. 1) follows.

Four horizons can be distinguished in core 619, taken at a depth of 3,700 meters in the Kamchatka Basin, where deep-sea arenaceous foraminifera now live. Horizon I, the uppermost, is 70 cm thick and contains the same foraminifera that live there at the present time—*Adercotryma glomerata* (Brady), various species of *Rhabdammina*. Horizon II, extending from 70 to 820 cm, differs by the presence of calcareous foraminifera that are characteristic of modern faunas found at depths less than 2,000 meters. The most abundant tests are those of *Cassidulinoides ochoticus* Saidova, *Uvigerina ochotica* Saidova, *U. peregrina* Cushman, *Globobulimina auriculata* (Bailey), *Nonionellina scapha* (Fichtel & Moll), *Heterolepa vehemens* Saidova, and *Pseudoparella pacifica* (Cushman). Horizon III, extending from 820 to 1,200 cm, contains a great deal of tuffaceous material and no foraminifera (Saidova and Lisitsyn, 1961). Horizon IV, extending from 1,200 cm to the base of the core at 1,400 cm, contains calcareous species with a predominance of *Globobulimina auriculata* (Bailey), *Nonionellina scapha* (Fitchel & Moll), and *Valvulineria ochotica* (Stschedrina).

Two horizons can be distinguished within core 1039, taken at·a depth of 876 meters on the east slope of the Shirshov Ridge, an area presently inhabited by foraminifera of the upper part of the continental slope. Horizon I, 300 cm thick, contains the local foraminiferal complex, but foramini-

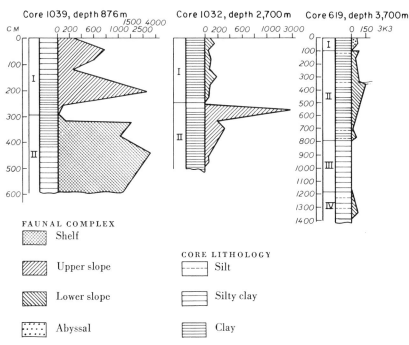

Fig. 1. Distribution of foraminiferal species complexes in typical cores from northwestern Bering Sea.

fera are much more abundant in the middle part of this horizon than at the surface. Horizon II, extending from 300 cm to the bottom of the core at 600 cm, is characterized by the appearance of foraminifera that now inhabit the continental shelf at depths less than 300 meters. Predominant among these are *Cassidulina californica* Cushman & Hughes, *C. smechovi* Voloshinova, *Elphidiella argutum* Saidova, and *Nonionellina labradorica* (Dawson).

Two horizons can also be distingiushed in core 1032, taken from a depth of 2,700 meters on the lower continental slope east of Cape Olyutorskiy. Horizon I, 260 cm thick, again contains the species living in the vicinity at the present time. Horizon II, extending from 260 cm to the base of the core at 500 cm, is distinguished by the appearance of species that are now characteristic of the upper continental slope, where depths are less than 2,000 meters. Predominant are *Globobulimina auriculata* (Bailey), *Cassidulinoides ochoticus* Saidova, *Bulimina exilis* (Heron-Allen & Earland), *Chilostomellina fimbriata* Cushman, and *Nonionellina scapha* (Fichtel & Moll). Foraminifera are more abundant than in the first horizon.

A similar picture can be observed in all of the other cores studied from Bering Sea. Horizon I corresponds to the Holocene, Horizon II to the Wisconsin Glaciation, Horizon III to the Sangamon Interglaciation, and Horizon IV to the Illinoian Glaciation (Saidova, 1960, 1961). Thus, during the Wisconsin Glaciation we find relatively shallow-water calcareous species now characteristic of the lower part of the continental slope rather than the modern deep-sea arenaceous foraminifera, in both the Aleutian and Kamchatka Basins. In cores from the lower part of the continental slope, species are found in Wisconsin-aged beds that are characteristic of the middle part of the continental slope at the present time, and Wisconsin-aged beds in cores from the uppermost part of the continental slope contain species now belonging to the continental shelf.

This indicates that there has been a subsidence of the sea floor and consequently an increase in water depths during late Wisconsin time or at the beginning of the Holocene. The greatest subsidence was experienced by the basin (Fig. 2), which subsided about 1,000 meters. The extent of sub-

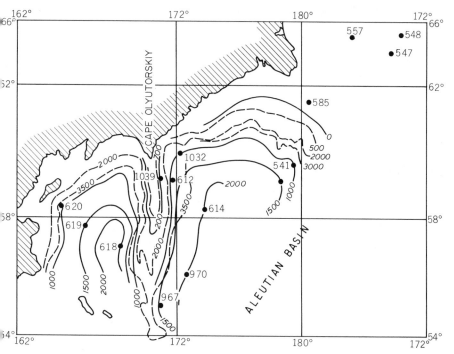

Fig. 2. A comparison of the depth of northwestern Bering Sea at present (*dashed lines*) and during the Wisconsin Glaciation (*solid lines*). Assumed depth contours for Wisconsin time are based upon study of foraminifera in the cores indicated by the numbered dots.

mergence diminishes toward the coast; submergence did not exceed 500 meters on the upper part of the continental slope, and it was less than 250 meters on the shelf.

Even shallower water is indicated by Illinoian foraminiferal faunas. These decreases in depth during the Wisconsin and Illinoian Glaciations must have resulted in lesser water depths in Bering Strait, but, in any case, in depths sufficient to lead to the formation of the Bering Land Bridge.

## REFERENCES

Saidova, H. M. 1958. Novye dannye po ekologii foraminifer (New data on the ecology of foraminifera) : Priroda, no. 10, p. 107–110.

——— 1960. Stratigrafiia osadkov i paleogeografiia severo-zapadnoi chasti Tikhogo Okeana i dal'nevostochnykh morei SSSR po donnym foraminiferam (Stratigraphy of sediments and paleogeography of the northwestern part of the Pacific Ocean and the Far East Seas of the USSR according to benthonic foraminifera) : Mezd. Geol. Kongr. 21st, Dokl. Sov. Geol., Probl. 10, p. 59–68, Izd-vo. Akad. Nauk SSSR, Moskva. In Russian with English abstract.

——— 1961. Ekologiia foraminifer i paleogeografiia dal'nevostochnykh morei SSSR i severo-zapadnoi chasti Tikhogo Okeana (Ecology of foraminifera and paleogeography of the far-eastern seas of the USSR and the northwestern part of the Pacific Ocean) : Akad. Nauk SSSR, Inst. Okeanologii, 182 p.

——— 1965. Raspredelenie donnykh foraminifer v Tikhom Okeane (Distribution of benthonic foraminifera in the Pacific Ocean) : Okeanologiya, v. 5, p. 99–110.

Saidova, H. M., and A. P. Lisitsyn. 1961. Stratigrafiia osadkov i paleogeografiia Beringova moria v chetvertichnoe vremia (Sedimentary stratigraphy and paleogeography of the Bering Sea during the Quaternary Period) : Akad. Nauk SSSR, Dokl., v. 139, p. 1221–1224. In Russian; English translation available, p. 59–63 in Dokl. Acad. Sci. USSR: Oceanology, p. 136–141, Am. Geol. Inst., 1961.

# 21. Diatom Floras and the History of Okhotsk and Bering Seas

A. P. JOUSÉ

*Institute of Oceanology, Academy of Sciences of the USSR*

The data given in this paper are summarized from a monograph describing recent investigations on diatoms found in the bottom sediments of the northwestern part of the Pacific Ocean and the far-east seas of the USSR (Jousé, 1962).

Three types of diatom floras can be distinguished in the surficial sediments of Bering Sea, the Sea of Okhotsk, and the northwestern Pacific Ocean: a neritic flora in which neritic diatoms constitute 60–100 per cent of the total flora; an oceanic type in which oceanic species constitute 70–80 per cent; and a mixed zone in which both plankton complexes are present in equal quantities. The recent neritic floras are found principally on the continental shelf, the mixed floras on the continental slope, and the oceanic floras in the central and southern bathyal regions.

A study of 64 cores from this region shows that five stratigraphic units can be distinguished in the bottom sediments on the basis of diatom floras. None of the cores penetrated through Quaternary beds into Tertiary beds, and most of them show only the two upper stratigraphic units. The thickness of individual stratigraphic units in Bering Sea and the Sea of Okhotsk far exceeds the thickness of corresponding horizons in the northwestern Pacific.

The five stratigraphic units correspond to postglacial, glacial, and interglacial intervals (Fig. 1). Horizon I represents postglacial time, Horizons II and IV represent glacial episodes of lowered temperature, and Horizons III and V are synchronous with interglacial episodes of increased temperature.

During the cold-climate episodes represented by Horizons II and IV, the arctic and arctoboreal diatom floras expanded throughout most of the area of the Sea of Okhotsk and Bering Sea and forced out the oceanic flora.

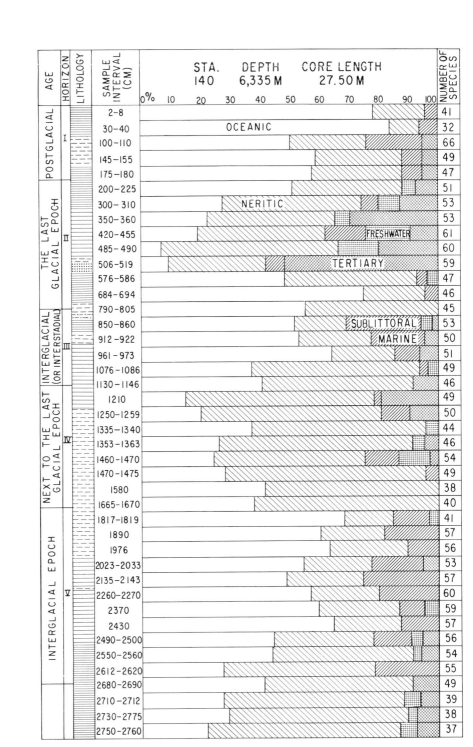

Fig. 1. Composition of diatom floras in a long core from the Sea of Okhotsk.

Single representatives of arctic and arctoboreal species are also found in the northwestern Pacific in these stratigraphic units. Diatoms are less abundant in the sediments of the glacial epochs than in the interglacial ones. This is partly because the annual duration of diatom development was decreased and partly because the diatoms were diluted to a greater degree by terrigenous material. The ice cover that must have been present during much of the year shortened the vegetation season, and the lack of biogenous elements in the upper water mass reduced the productivity of the waters. Conditions were probably much like present-day conditions in the Polar Basin.

During the epochs of warm climate represented by Horizons III and V, the neritic and oceanic floras extended northward to their present limits. Thermophile diatoms not characteristic of the modern floras were present.

The postglacial thermal maximum is recorded within Horizon I by an expansion of south-boreal diatoms to the northern limits of the boreal zone. At this time, the southern border of the boreal zone was displaced approximately 1,000 km northward.

The distribution of diatom floras was influenced by changes in sea level as well as by climatic factors. In general, diatom analysis does not give direct evidence of variations in sea level, because diatoms are planktonic organisms inhabiting the surface water mass. However, the diatom floras of Horizons I, III, and V in Bering Sea and the Sea of Okhotsk indicate improved connections with the Pacific Ocean during the interglacial and postglacial episodes, thus confirming that sea level was relatively high. On the other hand, masses of redeposited Tertiary diatoms are present in terrigenous sediments of glacial age (Horizons II and IV, see Fig. 2), suggesting lowered sea level, accentuated erosion, and an increase in the washing-out of Tertiary rocks on the shores. Sediments with redeposited ancient diatoms are, in fact, good horizon markers. Redeposited organic remains are especially abundant in the horizons of glacial age in the northwestern Pacific.

Direct evidence on the lowered position of sea level during the last glaciation is found in the sediments of Horizon II in Bering Sea. Cores taken at depths of 150–180 meters at the outer edge of the Anadyr Gulf in lat. 62° N., long. 175–179° E. contain a recent neritic diatom flora in Horizon I, changing at a depth of 40–80 cm into a typical sublittoral marine flora. This sublittoral flora contains benthonic and planktonic shallow-water species (*Melosira sulcata, Rhabdomena arcuatum*) that now live in the 0-to-25-meter depth zone. Horizon II is characterized by a scarcity of diatom remains; arctic and arctoboreal species predominate, and a significant admixture of freshwater diatoms is present. Diatom analysis indicates that the second horizon accumulated in the nearshore, shallow-water zone of

Bering Sea in an arctic climate. The seacoast at that time lay not far from the position in which the cores were taken. Similar results were obtained in cores from depths of about 200 meters near the northwestern and northeastern shores of the Sea of Okhotsk.

The change in the diatom floras during the time that Horizons I and II accumulated provides evidence concerning the magnitude of the reduction in temperature and the extent of regression of the shore during the last glacial interval. Judging from the material studied, the sublittoral zone of Bering Sea lay near lat. 62° N., 400–500 km south of its present position. At that time, Bering Strait was not in existence, and a broad land mass served as a bridge between Asia and North America. This conclusion is confirmed by investigations of the diatoms in the sediments of Providence Bay and of the Anadyr Gulf. Beds analogous to Horizon II and synchronous with the last glaciation were not found in either area, and cores 288–355 cm long consist entirely of sediments of postglacial age.

## REFERENCE

Jousé, A. P. 1962. Stratigraficheskie i paleogeograficheskie issledovaniia v severo-zapadnoi chasti Tikhogo Okeana (Stratigraphic and paleogeographic researches in the northwestern Pacific) : Akad. Nauk SSSR, Inst. Okeanologii, 258 p. In Russian with English summary.

## 22. On Migrations of Hunters Across the Bering Land Bridge in the Upper Pleistocene

HANSJÜRGEN MÜLLER-BECK
*University of Freiburg, Germany*

In attempting to deal with the problem of early migrations of hunters across the Bering Land Bridge, one must first state that so far there is no indisputable local evidence that man was present in the region earlier than about 8,000 years ago (Black and Laughlin, 1964). Nevertheless, enough data have been collected in northeastern Eurasia as well as in America during the last ten years to permit us to reconstruct some features of earlier migrations. The evidence with respect to the Bering region is indirect, but it is strong enough to allow us to discuss the possibility of migrations or "cultural contacts" in the Bering region as early as 28,000 years ago.

So far, there are no facts available that would rule out even earlier migration over the Bering Land Bridge. My discussion, however, must be restricted to events leading to migrations within the last 25,000 to 30,000 years, since these are the only events that can be reconstructed from stratigraphic and archaeological evidence. Furthermore, argument must be confined largely to two particular archaeological traditions: an early Mousteroid tradition whose evolution can be traced in the steppes of Eastern Europe and Western Siberia prior to the critical time level that is significant for the Bering Land Bridge migration, and a later Aurignacoid tradition that probably reached America only at the end of the Pleistocene. This restriction to only one or two main archaeological traditions means, of course, that one cannot exclude the possibility that there might have been additional contemporary migrations of different technologies, and perhaps even populations, coming from parts of Asia other than the Siberian steppes (Laughlin, this volume). The plains of Siberia and Europe are the areas in the Old

Of the figures in this report, all except *1*, *2*, and *9* appeared in a closely related article in *Science*, v. 152, May 27, 1966, p. 1191–1210, copyright 1966 by the American Association for the Advancement of Science.

World to which this discussion must be restricted, because it is there, during
the late Pleistocene, that the specialized technological traits evolved which
form the basis for our reconstruction of diffusion events.

It must also be made clear that our evidence in the Old World relates to
the diffusion of technological traits rather than to population migrations in
themselves; population movements are difficult to trace (Grigor'ev, 1965)
and have little relevance to the present problem. The situation is different,
of course, in the Bering region itself; there, because a series of climatically
controlled events acted upon unique topographic features, an actual migration of human groups around 28,000 years ago is the only plausible solution
for some of the archaeological problems we shall encounter. The problem
becomes more complex there in later times, as I shall show in the course of
this discussion.

I am indebted to the Deutsche Forschungsgemeinschaft, Bad Godesberg,
German Federal Republic, for support during the studies and journeys from
1960 to 1965 that provided the basis for this paper, and to Mr. and Mrs.
R. E. Morlan of the University of Wisconsin, who assisted me in putting
this paper into readable English.

## Pleistocene Chronology

A brief but critical review of Pleistocene geologic and climatic chronology is given below and in Fig. 1 to provide an understanding of the basis
for archaeological correlations and of the changing landscapes in which
Upper Palaeolithic cultures evolved prior to and during man's first invasion
of North America. The discussion is based primarily upon European stratigraphic sequences, because it is there that Late Pleistocene history has been
most thoroughly studied and is most completely known. For purposes of this
discussion, the Late Pleistocene Subepoch will be divided into early, middle,
and late parts, defined below.

The Late Pleistocene is customarily restricted to the last full interglaciation and the following glaciation (Woldstedt, 1962, Lüttig, 1964), and includes the transitional interval (the "advance period" of Stehlin, 1933) between these two. Thus, the Late Pleistocene begins with the beginning of
the European Eemian. In this paper the early Late Pleistocene subepoch will
be considered to correspond to the European Eemian.

The Eemian is well known paleobotanically from more than 100 sections
in different areas of Europe (Frenzel, 1959; Andersen, 1961; Behre, 1962;
Selle, 1962); it is generally contemporary with the Sangamon interglacial
sequence of North America (Flint, 1957), whose palynology is described

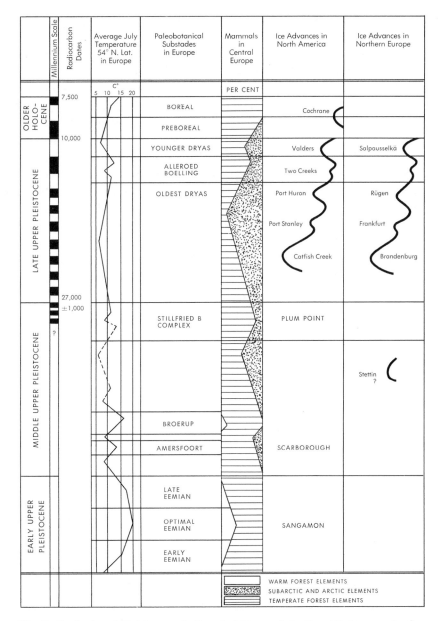

Fig. 1. Geologic subdivisions and climatic history of the Late Pleistocene in the Northern Hemisphere. Time scale is divided into millennia back to 32,000 years ago. Average July temperatures are based on Andersen (1961), Zagwijn (1961), and Müller-Beck (1961) and are given as straight lines connecting known points; data are insufficient for construction of a smooth curve.

by Terasmae (1960). Pollen profiles show clearly that the Eemian was a single climatic cycle, with only one main optimum but with several other minor oscillations. Consequently, only one large marine transgression can be expected during this interval. A large and steadily growing body of evidence indicates that the Eemian Interglaciation is represented by the classical Monastirian I transgression, which reached an altitude of about 15 meters (Zeuner, 1958). The maximum of this transgression probably was synchronous with the Eemian pollen zones III and IV of Selle (1962).

The boundary between early and middle Late Pleistocene is most conveniently defined on the basis of the pollen sequence in northwestern Europe (Netherlands, Denmark, and northern Germany). I would place the boundary somewhere in the "Kiefernzeit" (pollen zone VI of Selle, 1962, and pollen zone VII of Behre, 1962), or, more precisely, between pollen zones E 6b (Eemian) and EW 1a (early Weichselian) as defined by Zagwijn (1961).

The middle Late Pleistocene, as defined here, includes a series of warm and cold oscillations (see Fig. 1). Well-controlled paleobotanical records from a number of well-preserved and uninterrupted sections (Andersen and others, 1960; Andersen, 1961; Zagwijn, 1961) indicate that the Eemian Interglaciation was followed by a quite cold interval of relatively short duration, which was interrupted briefly by a rather minor warm oscillation, the so-called Amersfoort interval. These events, which could be regarded as a small "ice age" in themselves, were followed by another and larger warm stage, the rather complex Broerup interval, which might legitimately be regarded as an independent and smaller interglaciation. During the Broerup warming, boreal forests, which had retreated during the previous colder times, reconquered the larger part of Denmark and other parts of the European coast.

At the end of the Broerup, the record weakens. It is known only that a relatively cold period, of unknown absolute length, follows. Whether or not there were minor oscillations before the climatic minimum in the middle Late Pleistocene is thus far open to question. It is clear only that within this interval, for the first time in the Late Pleistocene, loess was deposited over wide areas of Europe, and cold faunal elements such as *Ovibos moschatus* penetrated to the western part of the Alps (Koby, 1955).

It is well established, however, that this poorly understood middle Late Pleistocene cold period is followed by a number of warmer oscillations. These are especially well represented by buried soils in Late Pleistocene loess sections in parts of eastern Central Europe, including the "Stillfried B" (Fink, 1954) and PK I (Klíma and others, 1961; Ložek, 1965). The climatic sequence within this series of oscillations is not yet well understood; one can say only that at least two main phases must be distinguished (Chmielewski, 1961; Müller-Beck, 1965). The second or last of these is the classical "Paudorf," if this term is to be used at all (it may easily be misunderstood, having often been used for the whole range of the "Stillfried B"). These oscillations took place mostly during a time range within which radiocarbon dates are not very reliable unless large groups of stratigraphically significant samples are available. There are not yet enough radiocarbon dates from stratigraphically well-placed samples to permit one to say when this series of oscillations began, except that they were initiated more than 35,000 years ago. It is not even clear whether the cooler oscillation between the two main warmer phases took place between 32,000 and 33,000 years ago or more recently. However, there are enough stratigraphically secure dates within a more reliable range of radiocarbon dating to indicate that the warm oscillations ended about $27,000 \pm 1,000$ years ago. The dated specimens are covered by solifluction layers formed in a colder climate (Kukla and Klíma, 1961; Fink, 1962). This stratigraphically well-placed and radiocarbon-dated horizon in the loess sequence of the region between Brno, Czechoslovakia, and Vienna, Austria, is taken here to define the boundary between the middle and late parts of the Late Pleistocene.

Correlations can hardly be drawn between minor climatic events in the Old and New Worlds, especially in view of the low reliability of radiocarbon dates in this time range. Thus, for example, the Scarborough beds near Toronto (Dreimanis, 1959, 1960) can be definitely placed in the lower part of the middle Upper Pleistocene, but whether they are entirely contemporary with the Amersfoort warming or, in their higher parts, even with the Broerup remains open to question. It is possible, or even likely, that the St. Pierre interval (Terasmae, 1958, 1960) is really identical with the Broerup, but so far there is no definite proof of this. One can state more certainly that the Plum Point interval (Dreimanis, 1959, 1960) must lie somewhere near the end of the oscillations during the later part of the middle Late Pleistocene. The relatively late dates connected with this interval are understandable when it is realized that sedimentation was ended by a new ice advance, probably well after the end of the truly warm interval defined in upper Middle Pleistocene deposits in the Brno region.

The late part of the Late Pleistocene, as defined here, begins with the onset of another cold period, during which the classical and in most regions maximal Late Pleistocene ice advance took place, represented by the original Würm (Penck and Brückner, 1901–1909), the late Weichsel (Woldstedt, 1950), the Valdai (Moskvitin, 1960), the Sartan (Kind, this volume) and

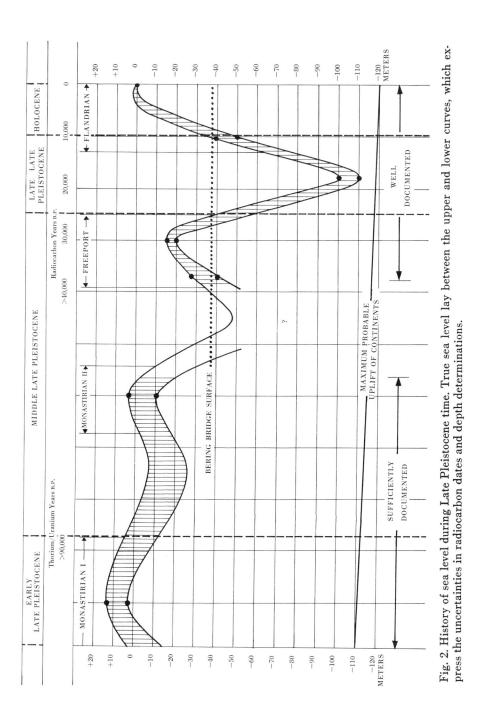

Fig. 2. History of sea level during Late Pleistocene time. True sea level lay between the upper and lower curves, which express the uncertainties in radiocarbon dates and depth determinations.

the "classical" Wisconsin (Flint, 1963). In most areas, the furthest advance seems to have taken place around 20,000 to 18,000 years ago (Flint, 1963, Cepek, 1965). Relatively warm oscillations are represented by the well-dated European Boelling and the following Alleroed (Fig. 1), which are both to be placed between about 12,500 and 10,800 years ago and which are separated by a colder event, the Older Dryas, dating between 12,000 and about 11,700 years ago (Tauber, 1962; Lang, 1962). The Two Creeks interval in North America seems to fit roughly in the same time range (Flint, 1963). Another colder stage follows, represented by a well-documented glacial readvance during the time of the Younger Dryas in Europe and the Valders glacial stade in the United States.

In the European paleobotanical sequence, the Younger Dryas (pollen zone III of Firbas, 1949) ends about 10,000 years ago (Müller, 1962). This date and the transition between Firbas' pollen zones III and IV are used here as the stratigraphic definition for the boundary between the Upper Pleistocene and the Holocene. From this time on, the interference of man in paleontological and sedimentological events can be clearly seen, lending the Holocene a different aspect from all previous geological epochs.

Eustatic movements of sea level must also be considered in discussing early human migrations in the Bering area. However, owing to the lack of detailed information, especially for the older parts of the Late Pleistocene, sea-level movements can be reconstructed here only as a pair of curves showing the total possible ranges of levels at different moments in time (Fig. 2).

As we have seen, the early part of the Late Pleistocene includes only one complete climatic cycle, the Eemian; the main transgression at this time was represented by the Monastirian I, which reached a maximum of about 15 meters above present sea level. The ensuing cold period, which opened the middle part of the Late Pleistocene and which itself was interrupted by the brief Amersfoort warm episode, evidently coincided with a pronounced sea-level regression to at least −8 meters and perhaps to as much as −25 meters (Wetzel and Haller, 1945; Pfannenstiel, 1952). The Broerup interval was warm enough and long enough to cause a new transgression, and was probably represented by the Monastirian II, which reached about 6 meters (Pfannenstiel, 1952; Zeuner, 1958). The following cold stages must have coincided with a new regression whose extent is not yet known but which may have reached −50 to −70 meters or even lower at its lowest point. The warmer oscillations at the end of the middle Late Pleistocene must be correlated with another transgression, here named the Freeport transgression (Fig. 2), which seems to have reached a level of −15 to −18 meters about 30,000 years ago (Curray, 1961).

A renewed regression must be placed in the late Late Pleistocene. The minimum Late Pleistocene sea level at about −110 meters occurred in the same time range as the maximum Late Pleistocene glacial advance, about 20,000 to 18,000 years ago (Curray, 1961). Then, as glaciers retreated, sea level began to rise again, and a new transgression, the Flandrian, was reached during the Holocene.

In spite of the uncertainties, one can state with the highest probability that a land bridge emerged during two distinct intervals in the Late Pleistocene. A land bridge would emerge if sea level were to fall about 40 meters below its present level (Creager and McManus, this volume), making allowance for possible submarine erosion in narrow channels. This means that the land bridge was open for some time during the middle Late Pleistocene (Fig. 2). There are not enough reliable dates available to provide any absolute time range for this event, but it must have been later than the Broerup warming and earlier than the group of warm oscillations at the end of the middle Late Pleistocene. Thus, the middle Late Pleistocene land bridge must have existed earlier than 35,000 and perhaps earlier than 40,000 years ago. The bridge was submerged again during the climatic oscillations at the end of the middle Late Pleistocene, but it reopened with the beginning of the late part of the Late Pleistocene, when sea level dropped below the −40-meter mark about 28,000 to 25,000 years ago (Fig. 2). The bridge must have widened as the regression progressed until about 20,000 years ago, when a new transgression began, coinciding with the initial retreat of the glaciers. The bridge was below water again during the Two Creeks interval or possibly a bit earlier (Creager and McManus, this volume) (Fig. 9), but sea level probably fell enough during the final cold oscillation of the late Late Pleistocene (the Younger Dryas of Europe and the Valders readvance of North America) to reopen the bridge, though only in a relatively narrow form. The final submergence during the Flandrian transgression probably occurred about 10,000 years ago.

It is also significant for archaeological problems that glaciers generally reached their Late Pleistocene maxima about 20,000 years ago. Bayrock (1965) has established conclusively that glaciers flowing from the Canadian shield coalesced with those originating in the Rocky Mountains during some part of the Wisconsin Glaciation; this coalescence most likely occurred in the late part of the Late Pleistocene. The general climatic evidence discussed above suggests that coalescence may have lasted from as early as 23,000 until as late as 13,000 years ago. During most of this interval, when Alaska was connected with Siberia by a wide Bering land bridge, an ice barrier would have separated Alaska from central North America (Figs. 9 and 11a), and contact between Alaska and central North America would have been extremely difficult for land animals and man. Contact would have been difficult, even during early and late parts of this period when a narrow but barren corridor may have been opened between the two ice fronts.

Thus, there were only three periods within the Late Pleistocene during which direct overland contact was possible between Siberia and central North America. The first took place during an undatable interval that coincided with the colder part of the middle Late Pleistocene, the second during an interval between about 28,000 and 23,000 years ago, and the third during a shorter and possibly interrupted interval from about 13,000 to 10,000 years ago. These last two "contact periods" are especially relevant to our discussion; however, I must emphasize that Alaska northwest of the ice barrier was uninterruptedly connected with Siberia between these last two intervals.

## Archaeological Events in Europe

It is, of course, impossible to discuss here in any detail the complex history of technological improvements in the Palaeolithic of northern Eurasia during the Late Pleistocene. But the evolution of certain general traits, at least, must be reviewed.

By early Late Pleistocene time, there appeared in Central Europe, besides more or less refined flake-tool industries and rather unspecialized industries with a biface technology, an industry already characterized by specialized projectile points. This industry is well represented in the Weimar-Ehringsdorf site (Fig. 3), which is well dated into the later part of the Eemian optimum (Kahlke, 1958; Behm-Blancke, 1960). Its assemblage includes refined, partially bifacially retouched projectile points, end scrapers, numerous other bifacial forms, and various cutting implements. Similar complexes of middle Late Pleistocene age are found throughout the plains of central and eastern Europe (Fig. 4). Some have similar stone points, as at Pradnik Czerwony in southern Poland (Jura, 1939; Kozlowski, 1960) and Volgograd (Zamiatnin, 1961), and some even have early specialized bone points, as at Salzgitter-Lebenstedt (Tode *et al.*, 1953). Preliminary analysis of faunal remains at Salzgitter-Lebenstedt shows that these hunters had already adapted very well to cold conditions: *Rangifer* (reindeer) represent about 72 per cent; *Mammonteus* (mammoth) 14 per cent; *Bison* 5.4 per cent; and *Coelodonta* (woolly rhinoceros) 2 per cent of the total identifiable number of animals (Tode *et al.*, 1953). Not a single warm or temperate species is represented in this assemblage or in those of other similarly

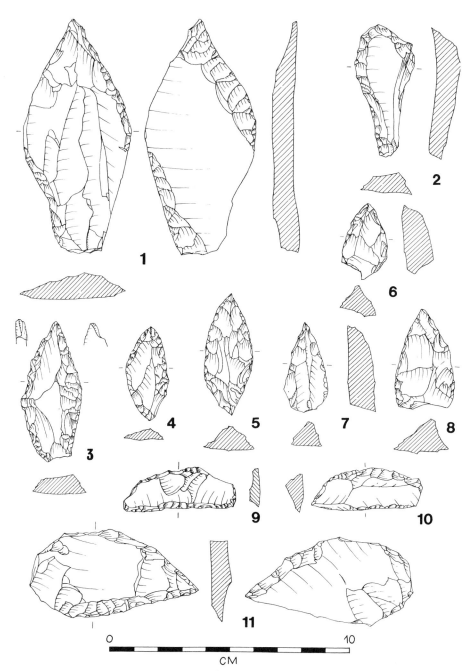

Fig. 3. Stone implements of the Weimar complex from Ehringsdorf, Germany. (*1*) Weimarian projectile points; (*2*) end scraper; (*3*) point showing burin blows; (*4–8*) double-ended and single-ended thick points; (*9–10*) small backed knives; (*11*) bifacially retouched knife.

dated sites representing the same or a similar technological tradition. All of these middle Late Pleistocene industries with specialized points in the European plains show a strong bifacial retouching tradition, as had been typical for the Acheuloid industries of the Middle Pleistocene (the "Holstein" and "Riss" complexes). They evidently represented the most advanced hunting technologies of their time. They share some features with the Mousterian cultures of Western Europe, and therefore may be called, in a more basic and more general sense, Mousteroid.

Flake-tool complexes that are contemporary with the industries characterized by specialized points are occasionally found in the plains, and in the low mountainous regions of Central and Eastern Europe (the Mittelgebirge) other aspects of the bifacial retouch tradition are found, mostly in caves, along with the more simple-looking flake-tool industries, but these are almost without a trace of specialized points (Fig. 4). It is safe to assume that the specialized points of the plains are genuine projectile points representing a technical improvement in the weapons of hunting, an improvement that was critical for survival in the environment represented by the mammalian fauna of Salzgitter-Lebenstedt.

These projectile-point industries of the European plains were contemporary with the classical sequence of so-called Micoquian and Mousterian industries in France (Bordes and Bourgon, 1951), but resemble them in only a few aspects and only in certain sites. For this reason, it is unwise to expand the regional French terminology into Central and Eastern Europe except in a generalized way.

The industries mentioned thus far were replaced in most areas at the end of the middle Late Pleistocene by still more specialized, but in a general way still Mousteroid, complexes (Fig. 5). These are probably related, for the most part, to earlier local traditions and hence to older already regionally restricted technologies. The relatively wide and short Weimarian stone projectile points were increasingly replaced by narrower types such as the Jerzmanovice points (Fig. 6, 2–4) found, for example, in the upper levels of the Nietoperzowa cave in southern Poland (Chmielewski, 1961). Some of the contemporary projectile-point sites, such as Mauern, Bavaria (Bohmers, 1951; Zotz and others, 1955), or Ranis, Eastern Germany (Hülle, 1936; Otto, 1951), have yielded a higher number of completely bifacially retouched leaf-shaped points that are basically identical with the only partially bifacially retouched Jerzmanovice types. The percentage of well-made blades is relatively high at all these sites. The sites are distributed throughout Central and Eastern Europe (Fig. 5) and form a well-defined and stratigraphically secure time horizon within the early part of the warm oscillations at

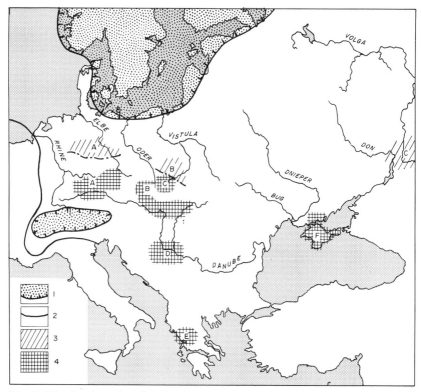

Fig. 4. Distribution of archaeological groups (industry areas) in Central and Eastern Europe during the earlier part of the middle Late Pleistocene. (*1*) Probable maximum extent of glaciers; (*2*) northeastern limit of area of Micoquoid industries of the Mousterian of Acheulian tradition in Western Europe; (*3*) Mousteroid projectile-point industries of the open plains; (*4*) Mousteroid bifacial industries of the mountainous regions.

Regional aspects: (*3A*) northern German (Salzgitter-Lebenstedt); (*3B*) southern Polish; (*3C*) Volgograd (Stalingrad); (*4A*) southern German; (*4B*) Czechoslovakian-Hungarian; (*4C*) Polish Karst; (*4D*) Croatian-Bosnian; (*4E*) Greece; (*4F*) Crimea.

the end of the middle Late Pleistocene. Bifacial knife types (Fig. 6, *1*), which can be distinguished from true projectile points by the lack of a tip and their asymmetric shape, are still common in this horizon, as they had been earlier.

A relatively strong local flake-tool tradition of basic Mousteroid character can be recognized in a number of sites in the mountainous regions of Europe—in the Mittelgebirge—notably in southern Moravia (Valoch, 1959) and Hungary (Vértes, 1958, 1959) (the Szeletoid industries of Fig. 5). The number of classical Mousteroid implement types north of the moun-

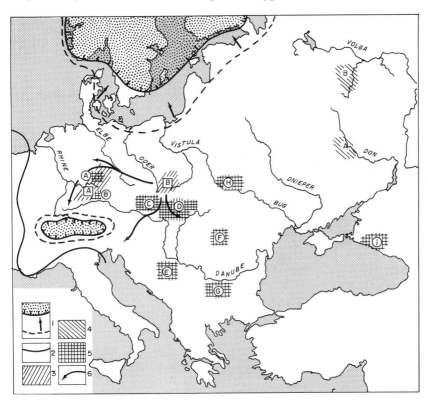

Fig. 5. Distribution of archaeological groups in Central and Eastern Europe at the end of the middle Late Pleistocene. (*1*) Probable extent of glacial retreat; (*2*) northeastern limit of the late Mousterian and early Perigordian industries; (*3*) Jerzmanovice and similar industries of the northwestern plains; (*4*) Kostyenki and related industries of the Russian plains; (*5*) Szeletoid groups of the mountainous areas; (*6*) direction of diffusion of early bone points of the Mladeč or Lautsch types.

Regional aspects: (*3A*) central-southern German; (*3B*) southern Polish; (*4A*) Kostyenki; (*4B*) Sungir; (*5A*) Franconian; (*5B*) Bavarian; (*5C*) Moravian; (*5D*) Hungarian; (*5E*) Bosnian; (*5F*) Bulgarian; (*5G*) northern Rumanian; (*5H*) western Ukrainian; (*5i*) Caucasian.

tains is rather restricted, however, and implements of more typical Late Paleolithic character, especially refined blade tools, are more abundant; these are already within the range of aspects found in the Aurignacian industries of Europe (Müller-Beck, 1965). Also contemporary are the first well-made large bone points of the so-called Mladeč type, whose origin very likely centered in the eastern part of Central Europe and the western part of Eastern Europe. Very-well-made specimens manufactured of mammoth ivory and possessing sharp cutting edges have been found in the Mammu-

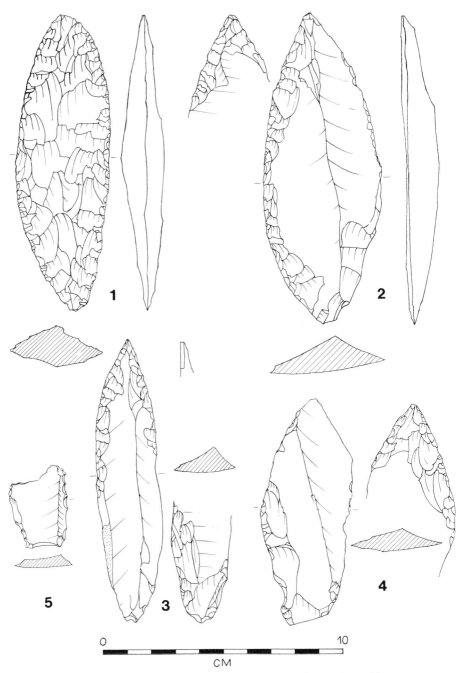

Fig. 6. Stone implements of the Jerzmanovician industry from Nietoperzowa Cave, Poland. (*1*) bifacial knife; (*2–4*) projectile points; (*5*) multiedge scraper.

tova Cave of southern Poland (Kozlowski, 1922, 1924). These, too, are among the earliest evidence of the development of a more characteristic Late Paleolithic technology. Industries possessing stronger Mousteroid and thus more conservative traits existed at the same time in the form of the Lower Perigordian in climatically more favorable southern France (Movius, 1960; Sonneville-Bordes, 1960), and even more notably in the type spectrum of the stone-projectile industries of southern Moravia, Hungary, and other areas in the Balkans.

Industries with stone projectile points are also documented in the same time range in Eastern Europe. The inventory of level 5 in Kostyenki site I near Voronesh on the Don River (Fig. 5, *4A*) is very typical (Okladnikov, 1954; Rogachev, 1957). Included again are bifacially retouched knives (Fig. 7, *10*), but there are also points which, because of their concave bases and well-developed basal thinning (Fig. 7, *1–5*), are quite distinct from projectile forms found to the west. There are even some specialized end scrapers (Fig. 7, *7–8*) which are unknown in contemporary Western industries. The occurrence in the archaeological horizon of *Mammonteus primigenius*, *Equus* sp., *Bison* sp., and *Saiga tartarica* (steppe antelope) shows that the hunters of this culture were well adapted to cold conditions. This is still better proved in the related but somewhat younger site of Sungir, east of Moscow, where the presence of *Rangifer tarandus* (reindeer) and *Alopex lagopus* (arctic fox) indicates a subarctic environment (Bader, 1961; Bader *et al.*, 1964).

Toward the end of the series of slightly warmer oscillations at the end of the middle Late Pleistocene, the industries with stone projectile points were replaced in many areas of Europe by locally differing aspects of industries close to the classical Aurignacian. One relatively early and well-dated example is the Aurignacian industry of level 5 of the Vogelherd site in southwestern Germany (Riek, 1934; Müller-Beck, 1965) (Fig. 8). Typical of all of these Aurignacian industries—in reality the earliest aspect of a long and locally distinct technological tradition that might be called Aurignacoid to set it apart from Mousteroid—are bone points with split bases (Fig. 8, *10–12*). The percentage of blade tools is relatively high, the bone-working technique is markedly improved, and the first well-worked figurines appear (Riek, 1934). It seems that the early Aurignacoid technology spread rapidly toward the west, at least partially as a true migration (Müller-Beck, 1961), and formed in southern France the classical Aurignacian (Sonneville-Bordes, 1960), which differs in some respects from the similar complex at the Vogelherd site. The first diffusion and possible migration of true Aurignacoid components is very likely related to the cold

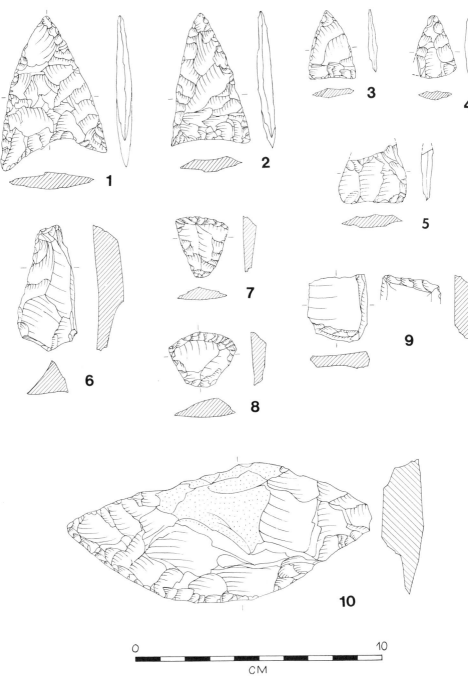

Fig. 7. Stone implements from Kostyenki, site I, level 5, USSR. (*1–3*) Triangular projectile points; (*4–5*) fluted triangular projectile points; (*6*) atypical end scraper; (*7–8*) sharp-angled end scraper (Eckkratzer); (*9*) burin; (*10*) bifacial knife.

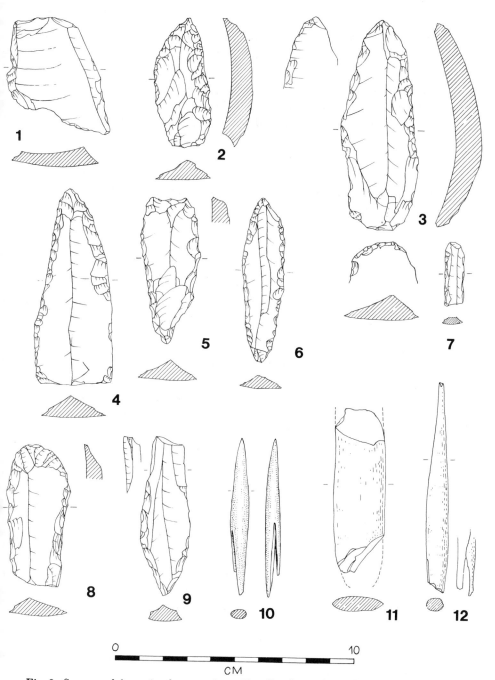

Fig. 8. Stone and bone implements from Vogelherd, southern Germany; *1* is from the still-Mousteroid level 6 with Mladeč points; *2–12* from level 5. (*1*) Retouched flake; (*2–3*) pointed blades resembling projectile points; (*4*) pointed blade; (*5*) atypical end scraper; (*6*) retouched blade; (*7*) slightly retouched microblade; (*8*) end scraper; (*9*) burin; (*10, 12*) bone projectile points with split base; (*11*) fragment of a Mladeč (Lautsch) point.

interval that separates the two warm periods of our series of climatic oscillations at the end of the middle Late Pleistocene.

With the beginning of the late part of the Late Pleistocene, more-developed Aurignacoid industries were established in most of the open plain areas in the European lowlands as well as on the loess steppes. The Pavlovian industry, which belongs to this complex, is well dated and stratified in southern Moravia and neighboring areas (Klíma, 1961). The two best-known sites are the Pavlov (Pollau) site and the nearby Dolní Věstonice site, both situated in the solifluction layer that postdates the climatic oscillations mentioned above. Pavlov has yielded a radiocarbon date of 25,020 ± 150 years (GrN-1325) B.P. and Dolní Věstonice a date of 25,820 ± 170 years (GrN-1286) B.P. (Klíma, 1963). These and other age determinations, including some from soils formed during the preceding climatic oscillations (Klíma et al., 1961) confirm our estimate of the age of the boundary between the middle and late parts of the Late Pleistocene as defined in the type area between Brno and Vienna. They show clearly that the warmer episode represented by the soil was succeeded about 26,000 years ago by solifluction during a new colder period.

I must emphasize, however, that bifacially retouched stone projectile points were produced and used even during the late part of the Late Pleistocene, especially in areas of less severe climate (Müller-Beck, 1966), such as some parts of southern France, some areas in southern Russia, and locally in Hungary and probably even in the southern part of Central Siberia. But the influence of generally Aurignacoid technologies is evident in all of these regions, even though the local industries differ from one another in many aspects and evidently were no longer in contact with one another. This is true, for example, of the French Solutrean, which forms one of these local complexes (Smith, 1964). The strong Aurignacoid influence clearly distinguishes these late Late Pleistocene industries from the industries of the end of the middle Late Pleistocene that also produced bifacially and partially bifacially retouched leaf-shaped points.

One may state, therefore, that the Aurignacoid technology was widespread in Europe during the late Late Pleistocene. It is found in numerous local industrial traditions in what were then the subarctic parts of the continent, including southern France, parts of Spain, and other areas in the European Mediterranean that now have a much warmer climate. Even the French Magdalenian and its Central European counterparts might be considered a special development within the Aurignacoid technological tradition. It is easily understandable that the often highly developed bone technologies must differ significantly from one area to another, especially with respect to ornaments. On the other hand, advanced inventions, especially those pertaining to weapons for hunting, very likely diffused quickly throughout the whole area occupied by the subarctic hunters. This is most probably the case with the harpoon, which represented a new projectile point, and the bow and arrow, which represented an improved weapon. And it is also very likely that all of these different and surely not very populous human groups possessed basically similar religious beliefs, judging from the art work that has been found. One wonders whether, despite the local technological differences, we should not consider these groups to have formed a single "culture," especially when we realize that they had no competition in the environment to which they were adapted.

## Archaeological Events in Siberia

The older part of the record that I have reviewed above is largely restricted to Central and Eastern Europe. In Siberia, we do not yet possess any stratigraphically well-controlled evidence for archaeological events earlier than 15,000 or 20,000 years ago, but there is a good archaeological record spanning the last 15,000 years. Material found in the famous Mal'ta site near Lake Baikal is clearly Aurignacoid (Gerasimov, 1935, 1941, 1958), even though it has a rather special local aspect. This is also true of the industries of the Lena River (Okladnikov, 1953), which are generally younger, and of the late Late Pleistocene industries of Hokkaido (Yoshizaki, 1959, 1961), found in an area that remained a part of Siberia as long as sea level was low enough to provide a dry bridge to the continent.

During the part of the late Late Pleistocene that can be documented by dated archaeological remains, there is no doubt that Siberia was technologically an extension of the subarctic areas of Europe (Fig. 9). The technological differences in Siberia are no larger than local differences elsewhere. One may assume that this was also generally true during a preceding interval that is not yet well documented by actual finds. It is likely, nevertheless, that technical improvements reached Central and Eastern Siberia rather belatedly; harpoons, for example, seem to have reached Eastern Siberia only at the end of the late Late Pleistocene. Taking this expected lag into account, it is highly probable that the horizon of stone projectile points documented everywhere in Europe, from at least the Rhine River in the west to the Don and Vyazma Rivers in the east during the middle Late Pleistocene, spread across Siberia at the end of the middle Late Pleistocene, as the

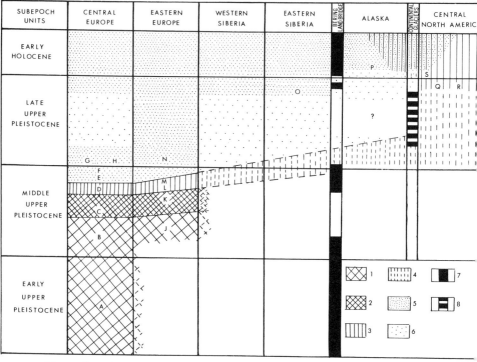

Fig. 9. Late Pleistocene and Early Holocene archaeological chronology in the Northern Hemisphere. Patterns indicate technological stages: (1) early advanced industries with bifacial technology (hand ax–scraper I, etc.); (2) later advanced industries with bifacial technology (hand ax–scraper II, etc.); (3) documented industries with stone projectile points (Jerzmanovician, etc.); (4) industries with stone projectile points assumed to have been present; (5) documented Aurignacoid industries; (6) Aurignacoid industries assumed to have been present. Narrow columns indicate history of natural barriers; (7) Bering Land Bridge submerged; (8) continental and Cordilleran ice sheets probably merged in Canada.

Letters indicate position of particular sites and components: (A) Weimar-Ehringsdorf; (B) Tata; (C) Salzgitter-Lebenstedt; (D) Nietoperzowa; (E) Vogelherd 5; (F) Vogelherd 4; (G) Pavlov; (H) Dolní Věstonice; (J) Volgograd; (K) Il'skaia; (L) Kostyenki, site I, level 5; (M) Sungir; (N) Kostyenki, site I, level 4; (O) Mal'ta; (P) Anangula; (Q) Lehner Ranch; (R) Blackwater Draw, Llano level; (S) Levy Rockshelter.

Aurignacoid industries did later. Stone projectile-point industries in Siberia at the end of the middle as well as at the beginning of the late parts of the Late Pleistocene would be expected to be more similar to those of Kostyenki I/5 and Sungir than to the more western complexes. In fact, we may suppose that the projectile-point industries of the easternmost European subarctic sites were the westernmost representatives of an archaeological aspect in Eastern Europe and Siberia that was just earlier than the Aurignacoid industries.

The older industries of the Volgograd type may even have spread into Siberia during the early part of the middle Late Pleistocene, but it is difficult, thus far, to determine how large an area they might have covered. However, the climatic adaptation shown in various sites of this technological level leaves little doubt that these complexes could also already exist under severe subarctic conditions.

## Archaeological Events in America

The well-documented part of the archaeological record in America is even shorter than in Siberia. One of the presumably earliest aspects, the industries with a high percentage of crude artifacts and without stone projectile points (the "pre-projectile-point complexes" of Krieger, 1964), seems to have no contact of any sort with the technological evolution in the northern steppes of the Old World that we have discussed thus far. But the stratigraphically and radiometrically documented record, which begins 12,000 or 13,000 years ago with the so-called "Llano Complex" (Krieger, 1964), *can* be related to the archaeological traditions discussed above.

The Llano is an early plains hunting complex known mainly from sites in the southwestern United States and Mexico (Fig. 11a), of which Lehner Ranch (Haury and others, 1959) and Blackwater Draw (Sellards, 1952) are typical. The earliest well-dated sites in this complex undoubtedly belong in the late part of the Late Pleistocene, but thus far none can be shown to be more than about 12,000 (Haynes, 1964) or 13,000 (Müller-Beck, 1966) years old. Typical of all Llano industries are well-worked fluted stone projectile points and other artifacts of general Mousteroid character (Fig. 10). There is practically nothing that would make one of these Llano inventories Aurignacoid in the sense that I have used this term in the discussion of Old World events. And the typical "advanced Aurignacoid" traits that are so common in all the known late Late Pleistocene industries in Siberia are completely lacking in Llano sites.

Before the Pleistocene ended, the Older Llano (Müller-Beck, 1966) underwent certain modifications and, in the form of a Younger Llano, spread into areas of northeastern United States that had previously been covered by the ice (Fig. 11a). In the plains and nearby areas, it was transformed into the Folsom-Llano (Müller-Beck, 1966), distinguished from the older

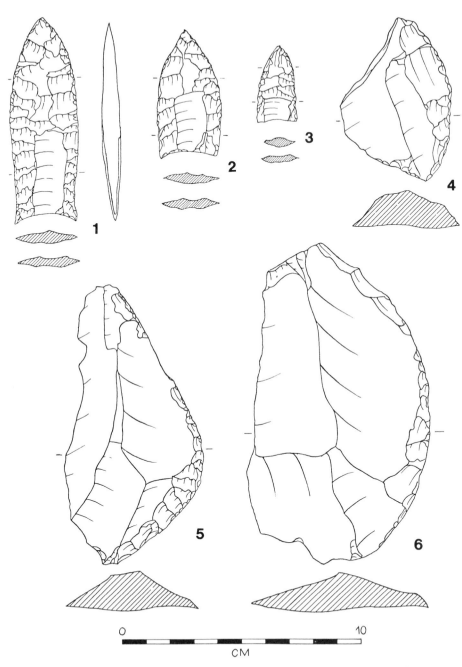

Fig. 10. Stone implements of the "Older Llano" from the Lehner Ranch site, Arizona. (1–3) Clovis points; (4–6) Mousteroid side scrapers.

horizon only by minor typological changes. Another complex rooted in the Older Llano crossed the topographically difficult Mesoamerican isthmus; the isthmus was probably easiest to pass at the beginning of a brief regression of sea level that would have opened a less densely forested strand zone in at least some areas (Müller-Beck, 1966). The evidence for this movement, which probably was a true and spectacularly rapid migration, is the Toldense complex (Menghin, 1957) of South America, represented in Patagonia by the Fell's Cave site (Bird, 1938; Rubin and Berthold, 1961) (Fig. 11a).

At the beginning of Early Holocene time, industries with numerous local variants of stone projectile points covered large areas of North and South America (Fig. 11b). These can be subdivided into many differing local traditions (Krieger, 1964) which, however, differ primarily in the highly and easily variable point forms. The local differences in projectile points are important for detailed discussions of local traditions, but the other tools are more or less uniform over much wider areas (Müller-Beck, 1966). This uniformity allows the grouping of the different local industries into larger complexes, as I have done in Fig. 11b, which distinguishes in North America the areas of a Northern and a Southern Plano, the Old Cordilleran complex (Butler, 1961), and the Desert Culture (Krieger, 1964). The Desert Culture shows some features that are alien to the Llano and that seem to be rooted in early aspects of the American industries with relatively crude tools and no stone projectile points.

During this interval in Early Holocene time, stone industries are known in Alaska and the Aleutians that show clear and advanced Aurignacoid traits. The oldest Aurignacoid site reported thus far is Anangula in the Aleutian Islands (Laughlin and Marsh, 1954; Black and Laughlin, 1964; Laughlin, this volume), dated to at least 8,000 years ago. Aurignacoid traits are also characteristic of later stone and bone industries in this area, mainly from proto-Eskimo contexts. Evidently, true Aurignacoid traits diffused from this region into the more southern part of continental North America (Müller-Beck, 1966), because typical and well-made burins that would fit into any Aurignacoid complex have been found in the Levi Rockshelter in Texas (Alexander, 1963). The lowest level at which burins are found in the Levi Rockshelter is dated as $10,000 \pm 175$ years old (0–1106). This diffusion initially may have been restricted to the western part of the continent (Fig. 11b). An Aurignacoid diffusion might easily have been anticipated within this time range, even though we do not yet possess direct evidence that Aurignacoid industries were present in northwestern North America. It is also significant that the highly specialized burins in Texas

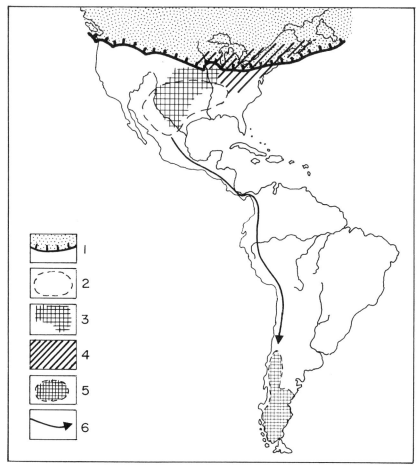

Fig. 11a. Distribution of stone projectile-point complexes in America at the end of the Late Pleistocene: (1) maximum extent of Wisconsin ice; (2) core area of the Older Llano; (3) Folsom-Llano; (4) Younger Llano; (5) Toldense; (6) probable route of migration into South America.

are found isolated in a Southern Plano industry and are not accompanied by other typical Aurignacoid traits.

Conversely, a Holocene movement of the Northern Plano into northwestern Canada and southern Alaska is quite evident (MacNeish, 1964; Müller-Beck, 1966) (Fig. 11b).

These facts demonstrate beyond doubt that two totally different archaeological aspects or traditions were present in North America at least as early as the beginning of the Holocene (a possible third tradition, the "pre-projectile complex," is, as I have stated, not discussed here). One tradition, found in the north, is totally Aurignacoid. The other, found in the south,

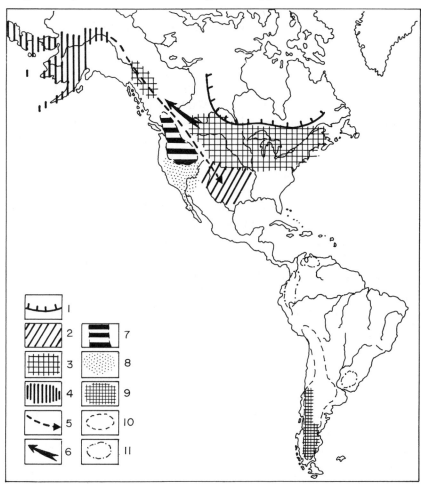

Fig. 11b. Distribution of selected stone industries in America during the Early Holocene: (*1*) mean position of ice boundary; (*2*) main area of the Southern Plano; (*3*) generalized area of the Northern Plano; (*4*) Aurignacoid industries; (*5*) probable route of burin diffusion; (*6*) route of northwestward diffusion or migration of the Northern Plano; (*7*) Old Cordilleran; (*8*) Desert Culture; (*9*) Toldense; (*10*) Ayampitin of Müller-Beck (1966); (*11*) other stone projectile-point complexes in South America. The Eastern Plano of Müller-Beck (1966) is not shown because of the difficulty of separating it from the Northern Plano and the Younger Llano.

and usually termed "paleo-Indian," is characterized by well-made and specialized stone projectile points, but the basic typology of all those tools that are not projectile points shows a rather retarded Mousteroid character that prevails even in the rare later Southern Plano sites in which some Aurignacoid influence becomes visible.

## Archaeological Events in the Bering Region

I have shown above that the Bering Land Bridge must have been open at some time during middle Late Pleistocene time and that it was submerged again much of the time during the following warmer oscillations. It existed once more from about 28,000 until 10,000 years ago, with a brief submergence during the Two Creeks interval. Between about 23,000 and 13,000 years ago, however, contact between Alaska and the interior of continental North America was hampered by the ice advances from the Canadian Shield and the Rocky Mountains.

During the first Late Pleistocene opening of the Bering Land Bridge, more than 40,000 years ago, technological developments on the northern plains of Europe and probably in Western Siberia had reached a stage represented by sites such as Salzgitter-Lebenstedt and Volgograd. No counterpart of this technological stage has been found in America thus far, and therefore the question whether or not man used the Bering Land Bridge during this early period must remain open. This must be stressed, even though man was already sufficiently adapted to cold conditions in northern Eurasia to have been able to use the land bridge. The assumption by Hopkins (1962) that man had by this time developed the capacity to live in a tundra environment can be proved by several clear and direct lines of evidence for the period constituting the coldest part of the middle Late Pleistocene, as I have noted in more detail above. It is known with certainty that hunting cultures existed during that period on the Eurasian plains, and that these peoples not only survived in subarctic environments but were also adapted to life in the open tundra and cold steppes—much better places to hunt than the taiga. The Salzgitter-Lebenstedt site, for example, with its clear documentation of the presence of a tundra-like environment, demonstrates the presence of such cultures beyond any doubt.

The following stage of technological improvement, represented by sites in the northern Eurasian plains dating from the end of the middle Late Pleistocene, is more interesting for our problem. Numerous and quite variable projectile points are found in this horizon, and the inventory at one site—Kostyenki I, level 5—resembles the American Llano complex more closely than any other known thus far in the Old World, not only with respect to projectile points but also in the form of end scrapers, bifacial knives, and other persisting Mousteroid features. The projectile points differ to some degree, as of course they must in view of the time lapse involved, but the differences represent merely a logical and effective technical improvement from the simple but skillful basal thinning seen in points from Kostyenki I/5 to the more refined fluting seen in the American points. It is significant that the somewhat younger material from Sungir, which shows a larger proportion of Aurignacoid features even though it is still close to Kostyenki I/5, is already more distant from the Llano in a number of aspects. That the projectile-point horizon of the northern Eurasian plains is definitely older than the classical Aurignacians as represented in the Vogelherd site is proved by stratigraphic superposition in several sites.

The complex of industries of which the inventory of Kostyenki I/5 is, of course, only one regionally restricted part would have reached Eastern Siberia and the Bering region just at the end of the middle Late Pleistocene or at the beginning of the following subepoch (Fig. 12a), regardless of whether it spread by migration or diffusion. The well-documented adaptation of the makers of these industries to a hunting life in the tundra means that they would have been easily able to follow the reindeer, mammoth, musk-ox, arctic fox, and bison, that they were accustomed to hunt, onto the emerging land bridge; consequently, they could have crossed over together with these animals in a true migration to the new continent (Fig. 12b), perhaps without even being aware of it. There is every reason to believe that this actually happened about 28,000 years ago. The way south into the heart of the continent would have been open for the following several millennia and would not have been at all difficult (Fig. 9).

The question of what happened next is more complex. In many areas in the western part of the Eurasian plains, the first truly Aurignacoid industries clearly began to replace the older traditions. At this time, the climate was growing colder and the environment of the northern tundras was becoming more severe. This provided strong pressure to improve the already well-developed bone technology, as hard wood for weapons and tools became more and more difficult to find. But at this same time the tundras began to spread into more southern latitudes. Hunters adapted to the tundra environment would automatically have moved southward, sometimes encountering there people of other traditions who were trying anything to adapt themselves to the new situation or, failing, were themselves moving to more southerly latitudes. Some parts of the "Frostschutt tundra" of Frenzel (1959, 1960, 1964) were very likely abandoned during this southward retreat, for, unlike the normal tundra and the tundra-like loess-steppe regions, this zone was unable to support enough animal life to fill the needs of a human population employing even the most advanced hunting methods of the time. If, as is likely, a similar retreat took place in the Bering region at the beginning of the coldest phase of the late part of the Late Pleistocene, the logical routes for human movements would have been southwestward, back into Siberia, and southeastward and forward into continental North America (Fig. 12c).

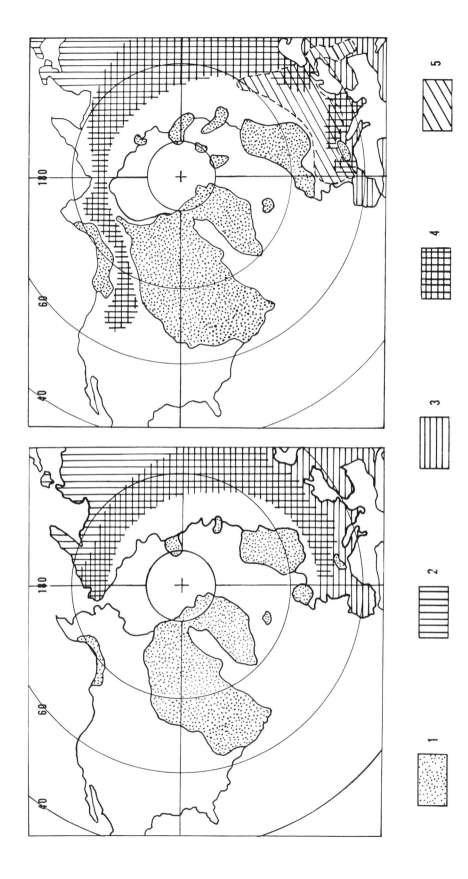

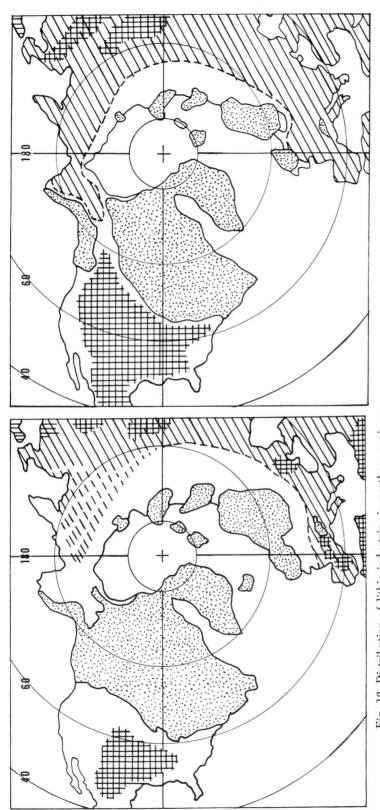

Fig. 12. Distribution of lithic industries in northern regions.

12a (*above, left*): At end of middle Late Pleistocene, 40,000 to 35,000 years ago. 12b (*above, right*): At transition from middle to late part of Late Pleistocene, 28,000 to 26,000 years ago. 12c (*below, left*): At time of maximum glaciation in late part of Late Pleistocene, 20,000 years ago. 12d (*below, right*): At end of Late Pleistocene.

Symbols: (*1*) glaciers; (*2*) advanced pebble-tool industries; (*3*) more conservative Mousteroid industries; (*4*) advanced Mousteroid industries with bifacial traditions and stone projectile points; and (*5*) Aurignacoid industries.

The available European evidence suggests that a southward retreat of populations did take place at this time. But a small population may, of course, have been able to survive in such parts of the Bering region as central Alaska, southern Chukotka, and Kamchatka. This remaining population would necessarily have remained in contact with the Siberian technological developments into the Aurignacoid industries; only this could have allowed the switch to the well-developed bone technology that was essential for survival under the increasingly severe climatic conditions.

The fact that the Llano complex is not influenced by Aurignacoid traits, on the other hand, indicates that an earlier and as yet unknown paleo-Indian stage that preceded it was somehow isolated from the Aurignacoid diffusion. The most likely explanation is isolation in central North America south of the ice barrier, where it would have been easy to go on with an improved but essentially unchanged technology. That the isolation might have taken place in Siberia is out of the question. All industries there show strong Aurignacoid features beginning with the oldest sites known, occupied at least 15,000 years ago. Isolation in the Bering region proper is extremely unlikely, for the reasons given above.

The situation becomes clearer at the end of the Late Pleistocene. Siberia, as we have seen, was then full of industries showing more or less pronounced Aurignacoid traits. Most typical is the site of Mal'ta, which has yielded a splendid bone technology and figurines, in addition to well-made huts or tentlike structures. The Mal'ta industry is especially similar to the later but also Aurignacoid Anangula site in the Aleutian Islands; in a number of aspects it constitutes a very good ancestor for the subsequent proto-Eskimo industries in northern arctic America. Precisely when the first truly Aurignacoid industries appeared in the Bering region is not yet known. Possibly they had a long history there, comparable to the rest of the Old World, but as I have said, it is rather more likely that the whole area was virtually abandoned by man for some time. However, the climatic situation probably improved sufficiently, shortly after 15,000 years ago, that, from that time on, one might expect the presence of Aurignacoid industries both in Alaska and on the southern part of the land bridge itself. These industries would have spread both northward and eastward into areas abandoned by the retreating glaciers (Fig. 12d). There, as early as 10,000 years ago and perhaps earlier, they would have come into contact—and from then on in lasting contact—with paleo-Indian groups spreading northward after the retreat of the Valders Ice. Evidence that this contact took place—the burins found in a Llano site in Texas—has already been mentioned.

## Conclusions

A review of climatic history, of the geographic history of the Bering
Land Bridge, and of Paleolithic technological developments in the Old and
New Worlds during the Late Pleistocene allows a partial reconstruction of
diffusion and migration events in the Bering region.

The Llano complex of North America differs from Aurignacoid industries in numerous aspects and cannot be derived from either an early or a
late Aurignacoid technological level. Instead, the ancestry of the Llano
complex is to be placed in a complex of Mousteroid industries with stone
projectile points, older than the local Aurignacoid industries, that was widespread on the plains of Eurasia at the end of the middle Late Pleistocene.
The inventory from level 5 of Kostyenki site I is the aspect most closely resembling the American Llano among Eurasian industries known thus far.
Industries of this type, recording human groups that were well adapted to
subarctic conditions, arrived by diffusion or migration in the Bering region
at the end of the middle Late Pleistocene and must have crossed over the
emerging land bridge in the course of hunting activities about 28,000 years
ago.

Aurignacoid industries were present on the Siberian plains at least
15,000 years ago. It can be assumed that they would also have been present
in Alaska beginning at about this same time. The uninfluenced continuation
of the more Mousteroid projectile-point tradition represented by the earliest-
known Llano industries requires isolation from later Late Pleistocene technological developments in Eurasia. Isolation from Aurignacoid influence
was clearly impossible in Siberia; isolation in Alaska is highly unlikely, although it cannot be completely ruled out. More probable, however, is an
isolation of the ancestors of the Llano complex in interior North America
south of the coalescing glaciers of the Canadian Shield and the northern
Rocky Mountains.

The Anangula site of the Aleutian Islands and slightly later proto-Eskimo and proto-Aleut industries are to be connected with the later, advanced
Aurignacoid traditions of Eurasia. There must have been a direct genetic
contact between these. The most closely related ancestral site for this tradition in Siberia would be Mal'ta, where a highly advanced bone technology
is provided.[a]

---

[a] Laughlin (this volume) emphasizes the affinities between the stone industry at Anangula and pre-pottery industries of Aurignacoid cast in northern Japan—a view that
amplifies and is not necessarily inconsistent with that espoused here by Müller-Beck. ED.

Physical anthropological observations are in agreement with these conclusions. The American Indians are clearly distinct from even the earliest skeletons from Alaska and the Aleutian Islands (Laughlin, this volume). A temporary isolation of the ancestors of the American Indians, such as is proposed here for strictly archaeological reasons, would make this difference easier to understand. The human remains found at Sungir (Bader and others, 1964; Bader 1965), although perhaps not in the direct ancestral line of the Llano people and later Indians, nevertheless may represent a closely related side branch. They show very typical Cro-Magnon features, which in similar combination are common among the earlier and more recent American Indians.

Our present knowledge of Pleistocene archaeology makes it extremely difficult, if not truly impossible, to propose any alternative to the hypothesis that the ancestral traditions of the Llano complex originated in the horizon of stone projectile-point industries of the Eurasian plains.

## REFERENCES

Alexander, H. L. 1963. The Levi site: A Paleo-Indian campsite in central Texas: Am. Antiquity, v. 28, p. 510–528.

Andersen, S. T. 1961. Vegetation and environment in Denmark in the Early Weichselian glacial (last glacial): Danm. Geol. Unders. II.R., no. 75, 175 p.

Andersen, S. T., H. de Vries, and W. H. Zagwijn. 1960. Climatic change and radiocarbon dating in the Weichselian glacial of Denmark and the Netherlands: Geol. en Mijnbow, v. 39, p. 38–42.

Bader, O. N. 1961. Stoianka Sungir', ieie vozrast i mesto v paleolite Vostochnoi Evropy (The Sungir site, its importance and its position in the Palaeolithic of Eastern Europe): Akad. Nauk SSSR, Trudy Komm. po Izucheniiu Chetvertichnogo Perioda, v. 18, p. 122–131.

———— 1965. Drevneishiie verkhnepaleoliticheskiie pogre beniia bliz Vladimira (Oldest Upper Palaeolithic burials near Vladimir): Vestnik AN SSSR, v. 35, no. 5, p. 77–80.

Bader, O. N., V. I. Gromov, and V. N. Soukatchev. 1964. Soungir, une station de Paléolithique superieur de la Russie Centrale: Report 6th Int. Cong. on Quaternary, Warsaw, 1961; Lodz, 1964, v. 4, p. 229–240.

Bayrock, L. A. 1965. Incomplete continental glacial record in Alberta: 7th Cong. INQUA (Boulder), 1965, Abstracts, p. 18.

Behm-Blancke, G. 1960. Altsteinzeitliche Rastplätze im Travertingebiet von Taubach, Weimar, Ehringsdorf: Alt-Thüringen, v. 4, 246 p.

Behre, K. E. 1962. Pollen- und diatomeenanalytische Untersuchungen an letztinterglazialen Kieselgurlagern der Lüneburger Heide: Flora, v. 152, p. 325–370.

Bird, J. 1938. Antiquity and migrations of the early inhabitants of Patagonia: Geogr. Rev., v. 28, p. 250–275.

Black, R. F., and W. S. Laughlin. 1964. Anangula: a geologic interpretation of the oldest archeologic site in the Aleutians: Science, v. 143, p. 1321–1322.

Bohmers, A. 1951. Die Höhlen von Mauern, Teil 1: Kulturgeschichte der altsteinzeitlichen Besiedlung: Palaeohistoria, v. 1, 107 p.

Bordes, F., and M. Bourgon. 1951. Le complex moustérien: Moustérien, Levalloisien et Tayacien: L'Anthropologie, v. 55, p. 1–23.

Butler, B. R. 1961. The Old Cordilleran Culture in the Pacific Northwest: Occas. Papers of the Museum, Idaho State College, no. 5, 111 p.

Cepek, A. G. 1965. Geologische Ergebnisse der ersten Radiokarbondatierungen von Interstadialen im Lausitzer Urstromtal: Geologie, v. 14, p. 625–657.

Chmielewski, W. 1961. Civilisation de Jerzmanovice: Inst. Hist. Kult. Mat. Polskiej Akad. Nauk, Warszawa-Kraków, 92 p.

Curray, J. R. 1961. Late Quaternary sea level: a discussion: Geol. Soc. America Bull., v. 72, p. 1707–1712.

Dreimanis, A. 1959. Proposed local stratigraphy of the Wisconsin glacial stage in the area south of London, southwestern Ontario: Univ. West. Ontario, Contrib. Dept. Geol., no. 25, p. 24–30.

——— 1960. Pre-classical Wisconsin in the eastern portion of the Great Lakes region, North America: Rept. 21st Internat. Geol. Congr. (Copenhagen), 1960, Pt. 4, p. 108–119.

Fink, J. 1954. Die fossilen Böden im österreichischen Löss: Quartär, v. 6, p. 85–108.

——— 1962. Studien zur absoluten und relativen Chronologie der fossilen Böden in Österreich: II: Wetzleinsdorf und Stillfried: Archaeol. Austriaca, v. 31, p. 1–18.

Firbas, F. 1949. Spät- und nacheiszeitliche Waldgeschichte Mitteleuropas nördlich der Alpen. I. Allgemeine Waldgeschichte: Gustav Fischer, Jena, 480 p.

Flint, R. F. 1957. Glacial and Pleistocene geology: Wiley, New York, 553 p.

——— 1963. Status of the Pleistocene Wisconsin stage in central North America: Science, v. 139, p. 402–404.

Frenzel, B. 1959. Die Vegetations- und Landschaftszonen Nord-Eurasiens während der letzten Eiszeit und während der postglazialen Wärmezeit. I. Allgemeine Grundlagen: Akad. Wiss. u. Lit. Mainz, Abh. math. naturwiss. Kl. 1959, Pt. 3, p. 937–1099.

——— 1960. Die Vegetations- und Landschaftszonen Nord-Eurasiens während der letzten Eiszeit und während der postglazialen Wärmezeit. II. Rekonstruktionsversuch der letzteiszeitlichen und wärmezeitlichen Vegetation Nord-Eurasiens: Akad. Wiss. u. Lit. Mainz, Abh. math. naturwiss. Kl. 1960, Pt. 6, p. 287–453.

——— 1964. Zur Pollenanalyse von Lössen: Eiszeitalter u. Gegenwart, v. 15, p. 5–39.

Gerasimov, M. M. 1935. Raskopki paleoliticheskoi stoianki v sele Mal'ta (Excavations in the Paleolithic site of the village of Mal'ta): Izv. Gosudarstvennoi Akad. Inst. Mat. Kultury, no. 118, p. 78–124.

——— 1941. Obrabotka kosti na paleoliticheskoi stoianke Mal'ta (Worked bone from the Paleolithic site of Mal'ta): Mat. i. Issl. po Arkheol. SSSR, no. 2, p. 65–85.

——— 1958. Paleoliticheskaia stoianka Mal'ta (raskopki 1956–1957 gg) (The Paleolithic site Mal'ta [excavations of 1956–1957]): Sov. Etnografiia, no. 3, p. 28–52. English translation p. 4–32 *in* H. N. Michael, ed., The archaeology and geomorphology of northern Asia: Arctic Inst. North America, Anthropology of the North: Translations from Russian Sources, no. 5 (1964).

Grigor'ev, G. P. 1965. Migrations, indigenous development and diffusion in the Upper Palaeolithic: Arctic Anthro., v. 3, p. 116–121.

Haury, E. W., E. B. Sayles, and W. W. Wasley. 1959. The Lehner Mammoth Site, Southeastern Arizona: Am. Antiquity, v. 25, p. 2–30.

Haynes, C. V. 1964. Fluted projectile points; their age and dispersion: Science, v. 145, p. 1408–1413.

Hopkins, D. M. 1962. Comment to R. J. Mason, The Paleo-Indian tradition in eastern North America: Current Anthro., v. 3, p. 254.

Hülle, W. 1936. Die Bedeutung der Funde aus der Ilsenhöhle unter Burg Ranis für die Altsteinzeit Mitteldeutschlands: Jahresschr. Mitteld. Vorgesch., v. 24.

Jura, A. 1939. Le paléolithique des Cracovie et des environs: Bull. Int. de l'Acad. Polonaise des Sc. et Lettr. 1939.

Kahlke, H. D. 1958. Die Jungpleistozänen Säugetierfaunen aus dem Travertingebiet von Taubach-Weimar-Ehringsdorf: Alt-Thüringen, v. 3, p. 97–130.

Klíma, B. 1961. Paleolitické osídlení Pavlovskych Vrchů: Anthropos, v. 14, p. 173–178.

———— 1963. Dolní Věstonice. Výzkum tábořiště lovců mamutů v letech 1947–1952: Náklad. Českoslov. Akad. Věd, Praha, 427 p.

Klíma, B., J. Kukla, V. Ložek, and H. de Vries. 1961. Stratigraphie des Pleistozäns und Alter des paläolithischen Rastplatzes in der Ziegelei von Dolní Věstonice (Unter-Wisternitz): Anthropozoikum, v. 11, p. 93–146.

Koby, F. E. 1955. Découverte d'un ossement d'avibos dans la couche à ours du Schnurenloch: Actes Soc. Jur. d'Émulation 1954–55, p. 117–131.

Kozlowski, J. K. 1960. Paleolit: Prace Archeologiczne, Zeszyt 1: Pradzieje Powiatu Krakowskiego, tom 1, p. 18–65.

Kozlowski, L. 1922. Starsza Epoka kamiena w Polsce (Paleolit): Poznanskie Towarz. Przy. Nauk Prace Kom. Archeol., v. 1, 53 p.

———— 1924. Die Ältere Steinzeit in Polen (translation of L. Kozlowski, 1922): Die Eiszeit, v. 1, p. 112–160.

Krieger, A. D. 1964. Early man in the New World, in J. D. Jennings and E. Norbeck, eds., Prehistoric man in the New World: Univ. Chicago Press, p. 23–81.

Kukla, J., and B. Klíma. 1961. Comment in: More on Upper Palaeolithic archaeology: Current Anthro., v. 2, p. 437–439.

Lang, G. 1962. Die spät- und frühpostglaziale Vegetationsentwicklung im Umkreis der Alpen: Eiszeitalter u. Gegenwart, v. 12, p. 9–17.

Laughlin, W. S., and G. H. Marsh. 1954. The lamellar flake manufacture site on Anangula Island in the Aleutians: Am. Antiquity, v. 20, p. 27–39.

Ložek, V. 1965. The relationship between the development of soils and faunas in the warm Quaternary phases: Anthropozoikum, v. A-3, p. 7–33.

Lüttig, G. 1964. Prinzipielles zur Quartär-Stratigraphie: Geol. Jb., v. 82, p. 177–202.

MacNeish, R. S. 1964. Investigations in southwest Yukon: Papers Peabody Found. Archaeology, v. 6, no. 2, p. 201–488.

Menghin, O. F. A. 1957. Vorgeschichte Amerikas, in W. D. Barloewen, ed., Abriss der Vorgeschichte: Oldenbourg, München, p. 162–211.

Moskvitin, A. I. 1960. Über wärmere und kühlere Interglaziale in der UdSSR. ————. Berichte, Geol. Ges. DDR, 1960, Heft ½, p. 5–20.

Movius, H. L. 1960. Radiocarbon dates and Upper Palaeolithic archaeology in Central and Western Europe: Current Anthro., v. 1, 355–391.

Müller, W. 1962. Der Ablauf der holozänen Meerestransgression an der südlichen Nordseeküste und Folgerungen in bezug auf eine geochronologische Holozängliederung: Eiszeitalter u. Gegenwart, v. 13, p. 197–226.

Müller-Beck, H. 1961. Comment in: More on Upper Palaeolithic archaeology. Current Anthro., v. 2, p. 439–444.

———— 1965. Eine "Wurzel-Industrie" des Vogelherd-Aurignaciens: Fundberichte a. Schwaben, v. N. F. 17, p. 43–51.

———— 1966. Paleohunters in America: Origins and diffusion: Science, v. 152, p. 1191–1210.

Okladnikov, A. P. 1953. Sledy paleolita v doline r. Leny (Palaeolithic remains in the Lena River basin): Mat. i. Issl. po Arkheol. SSSR, no. 39, p. 227–265. English translation, p. 33–79 in H. N. Michael, ed., The archaeology and geomorphology of Northern Asia: Arctic Inst. North America, Anthropology of the North: Translation from Russian sources, no. 5 (1964).

———— 1954. Nekotorye voprosy izucheniia verkhnego paleolita SSSR v svete noveishikh issledovanii (Some studies on the Upper Paleolithic in the USSR in the light of recent research): Sov. Arkheol., v. 21, p. 5–29.

Otto, K. H. 1951. Zur Chronologie der Ilsenhöhle in Ranis, Kr. Ziegenrück: Jahresschr. Mitteld. Vorgesch., v. 35, p. 8–15.

Penck, A., and E. Brückner. 1901–1909. Die Alpen im Eiszeitalter: Tauchnitz, Leipzig, 1199 p.

Pfannenstiel, M. 1952. Das Quartär der Levante. I. Die Küste Palästina Syriens: Mainz. Akad. Wiss. Lit., Abh. math. naturwiss. Kl. 1952, pt. 7, p. 373–475.

Riek, G. 1934. Die Eiszeitjägerstation am Vogelherd: Heine, Tübingen, 338 p.

Rogachev, A. N. 1957. Mnogosloinye stoianki Kostyenkovsko-Borshevskogo raiona na Donu i problema razvitiia kultury v epochu verkhnego paleolita na Russkoi ravnine (The multi-level sites from the area of Kostyenki-Borshevo near the Don River and the problem of the cultural evolution during the Upper Paleolithic epoch in the Russian Plains): Mat. i. Issl. po Arkheol. SSSR, no. 59, p. 9–134.

Rubin, M., and S. M. Berthold. 1961. U.S. Geological Survey radiocarbon dates, VI: Radiocarbon, v. 3, p. 86–98.

Sellards, E. H. 1952. Early man in America: Univ. Texas Press, Austin, 211 p.

Selle, W. 1962. Geologische und vegetationskundliche Untersuchungen an einigen wichtigen Vorkommen des letzten Interglazials in Nordwestdeutschland: Geol. Jb., v. 79, p. 295–352.

Smith, P. E. L. 1964. The Solutrean culture. Sci. American, v. 211, p. 86–94.

Sonneville-Bordes, D. de. 1960. Le Paléolithique supérieur en Périgord, I and II: Delmas, Bordeaux, 558 p.

Stehlin, H. G. 1933. Epilogue, in A. Dubois and H. G. Stehlin. La grotte de Cotencher, station moustérienne: Mém. Soc. Paléontol. Suisse, v. 52–53, 292 p.

Tauber, H. 1962. Copenhagen radiocarbon dates V: Radiocarbon, v. 4, p. 27–34.

Terasmae, J. 1958. Contributions to Canadian palynology No. 1. Pt. 2, Nonglacial deposits in the St. Lawrence lowlands, Quebec; pt. 3, Nonglacial deposits along Missinaibi River, Ontario: Geol. Survey Canada Bull. 46, p. 13–34.

———— 1960. Contributions to Canadian palynology No. 2. Pt. 2, A palynological study of Pleistocene interglacial beds at Toronto, Ontario: Geol. Survey Canada Bull. 56, p. 23–40.

Tode, A., F. Preul, K. Richter, W. Selle, K. Pfaffenberg, A. Kleinschmidt, E. Guenther, A. Müller, and W. Schwartz. 1953. Die Untersuchung der paläolithischen Freilandstation von Salzgitter-Lebenstedt: Eiszeitalter u. Gegenwart, v. 3, p. 144–220.

Valoch, K. 1959. Lösse und paläolithische Kulturen in der Tschechoslowakei: Quartär, v. 10/11, p. 115–149.

Vértes, L. 1958. Die archäologischen Funde der Szelim-Höhle: Acta Archaeol. Acad. Sci. Hungaricae, v. 9, p. 5–17.

———— 1959. Das Moustérien in Ungarn: Eiszeitalter u. Gegenwart, v. 10, p. 21–40.

Wetzel, R., and J. Haller. 1945. Le Quaternaire de la région de Tripoli: Notes et Mém. Sect. géol. Délég. gén. de France au Levant, v. 4, p. 1–48.

Woldstedt, P. 1950. Norddeutschland und angrenzende Gebiete im Eiszeitalter: Enke, Stuttgart, 464 p.

———— 1962. Über die Gliederung des Quartärs und Pleistozäns: Eiszeitalter u. Gegenwart, v. 13, p. 115–124.

Yoshizaki, M. 1959. Tachikawa, preceramic industries in Southern Hokkaido: Hakodate Munic. Mus. Research Bull., v. 6, 64 p.

———— 1961. The Shirataki site and the preceramic culture in Hokkaido: Minzkugaku Kenkyu, v. 26, p. 13–23.

Zagwijn, W. H. 1961. Vegetation, climate and radiocarbon datings in the Late Pleistocene of the Netherlands. Pt. 1. Eemian and Early Weichselian: Med. Geol. Stichting Netherlands, n.s., v. 14, p. 15–45.

Zamiatnin, S. N. 1961. Stalingradskaia paleoliticheskaia stoianka (The paleolithic site of Stalingrad): Kratk. soob. Inst. Arkheol. Akad. Nauk CCCR, v. 82, p. 5–36.

Zeuner, F. E. 1958. Dating the past: an introduction to geochronology, 4th ed.: Methuen, London.

Zotz, L., G. Freund, F. Heller, E. Hofmann, and G. C. Vojkffy. 1955. Das Paläolithikum in den Weinberghöhlen bei Mauern: Quartär. Bibl., v. 2, 330 p.

## 23. Human Migration and Permanent Occupation in the Bering Sea Area

W. S. LAUGHLIN
*University of Wisconsin*

Vitus Bering's discovery in 1741 of the Aleutian Islands and of their flourishing population of over 16,000 maritime hunters opened an area crucial to the solution of two generic questions that had been posed by Lief Ericsson's original discovery of America in A.D. 1000. Where did the aboriginal populations of the New World come from, and how did they get here? These two simple queries involve several basic questions in human evolution and population genetics, and they require for their solution the combined use of evidence from many different disciplines. Recognition of the fact that the eastern Aleutian Islands were formerly part of the Bering Land Bridge, and of the probability that the living Aleuts are descendants of earlier populations extending continuously back to some of the original inhabitants of the land bridge provides our most valuable insight into an understanding of the origins of the Bering Sea Mongoloids and of the American Indians.

Bering's earlier discovery of the strait that separates America and Asia —the strait that now bears his name—proved in 1728 that the New World was surrounded by water and thereby encouraged the erroneous idea that the American aborigines, like their discoverers, must have arrived by boat. During the first half of the present century, the proportion of opinion favoring a sea route versus a land route was strangely reminiscent of Paul Revere's signaling system for warning of still another immigration, of redcoats rather than redskins—two if by sea, one if by land. More recently, the acquisition of geological perspective has made it possible and necessary to think more seriously about land routes and to correlate their availability with our knowledge of the biology of the New World inhabitants and their history (Fewkes, 1912; Laughlin, 1963).

Recognition that the early invaders probably traveled a land route has been acknowledged mostly as a concession to the growing body of evidence compiled by geologists, while the implications of a land bridge in terms of human occupancy have scarcely been visualized. The Bering Strait area is still commonly visualized as a narrow path or trail over which people hustled, in one direction, on their way to take up positions in which they would presently be discovered (Collins, 1964; Laughlin, 1965a). The view presented here is that, in fact, the Bering Land Bridge was an enormous continental area extending nearly 1,500 km from its southern extremity, now the eastern Aleutians, to its northern margin in the Arctic Ocean. It was an area that could accommodate many permanent residents, human and animal, and it endured for a longer time than that documented for the entire period of human occupancy in America. The southern coastal area was ecologically quite different from the interior regions and provided the basis for the differentiation of the sea-oriented Mongoloids, ancestors of the Aleuts and Eskimos, and the land-oriented big-game hunters of the interior, the ancestors of the American Indians. It is obvious that several generations of occupants, some of them moving slowly eastward and southward, were unaware that most of their territory would eventually be submerged. In its last phases, the postglacial rise in sea level was visually apparent to the coastal dwellers, whose economy was improved by the creation of more coastline and who therefore remained at the migrating edge of the sea.

The long-recognized external similarities of the American aborigines to Asiatic Mongoloids can now be converted to biological affinities attributable to a common ancestry, on the basis of new genetic information. The distribution of various groups within the New World and the differences among them provide basic evolutionary information as well as some indication of the length of time that these groups have been in the New World. The origin of these people thus becomes more clear, and a vast laboratory for the study of recent human evolution is created. The greatest difference between groups is the substantial biological difference between American Indians on the one hand, and the Aleuts and Eskimos on the other. The Aleuts and Eskimos are more similar to each other than either is to the American Indians, and they are more similar to Asiatic Mongoloids than are American Indians. The Aleuts and Eskimos, along with the Chukchi, the Koryak, and probably the Kamchadal as well, encircle the Bering Sea and compose a biologically related group, the Bering Sea Mongoloids. When we add to the biology of these peoples, archaeological evidence, the dating of old village sites, and some cultural evidence from the living peoples, including their languages and patterns of ecological adaptation, we can sketch the outline for a prehistoric permanent occupation as well as migration through the Bering Land

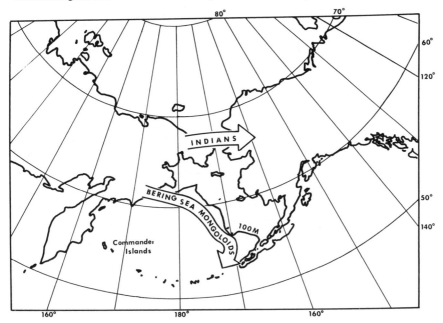

Fig. 1. Migration routes of ancestral American Indians and ancestral Bering Sea Mongoloids.

Bridge. The ancestral Aleuts and Eskimos, possibly the Chukchi as well, were the permanent residents of the coastal margin. The ancestral American Indians were the interior inhabitants, some of whom moved on into the New World. Our picture of the occupation of the coastal part of the Bering Land Bridge is considerably enhanced by the living Aleuts, for whose ancestors there is evidence of over 8,000 years of continuous occupation in the eastern Aleutian Islands. This area itself survives the Bering Land Bridge (Fig. 1).

For several reasons, we shall focus on the Aleuts. Most simply, only two major groups are involved in the occupation of the New World: Bering Sea Mongoloids and Indians. Of the Bering Sea Mongoloids who presently inhabit the shores of the Bering Sea, only the Aleuts occupy a position that is directly relevant to the Bering Land Bridge. The Eskimos occupy areas that were formerly in the interior. The prehistory of this southern group, the contemporary Aleuts, is better known for a longer period of time than that of any other ancient or contemporary group of Bering Sea Mongoloids. The western tip of Umnak Island was the terminus of the Bering Land Bridge, and it is fortunate that the people now living there have patent connections with the earliest inhabitants.

## Diversity and Origins of American Indians

The unity or plurality of the American aborigines has been debated since at least the eighteenth century, and the possible correlation between different kinds of Indians and different periods of entry from Asia has concerned many investigators (Stewart and Newman, 1951). An extreme point of view was expressed by Ulloa in 1772, who said that he who has seen one Indian may be said to have seen them all. In view of the great diversity in weight and stature among the Indians, such a statement suggests that the author indeed saw only one, or perhaps none at all. At one end of the size range are the tall groups, such as the Ona of the tip of South America, the Yuma of the American southwest, the Cheyenne, and many tribes of the eastern seaboard, especially the Seneca, who appear to have been the largest of the American Indians (Newman, 1957). These groups were generally in excess of 172 cm tall and provided a marked contrast with the much shorter Pueblo Indians, Mayas, Alacaluf, and many others whose mean heights were well below 160 cm. Some slight geographical patterning in the distribution of statures is apparent (Newman, 1953): short peoples tend to be concentrated in the lower latitudes about the equator, and stature shows irregular increases toward north and south, with an important discontinuity provided by the Eskimos.

Head form, as reflected in cephalic index (vault breadth divided by vault length), shows less geographical patterning over large areas, but important distinctions between more local groups. The scanty data on early skeletal remains reveal a tendency toward narrow heads in relation to breadth, whereas the increasingly larger series of skeletons derived from later strata clearly indicate a change in head shape favoring increased frequencies of broad-headedness. There is also some evidence that shorter-statured peoples in the warmer climates had taller ancestors, and that the taller Indians in the areas of colder climate evolved from shorter-statured ancestors (Newman, 1962). Whether or not these are adaptive changes, they are changes that have, in all likelihood, taken place since the peopling of the New World. The change in cephalic index does appear to characterize large areas; it is seen not only among the Indians but among the Aleuts and Eskimos as well.

Similar change is found in many other areas of the world, and does not require long periods of time (Laughlin, 1956, 1958; Suzuki, 1958; Levin, 1963). There is, of course, considerable diversity between various groups of American Indians in several other traits, and this diversity includes some extreme values for the human species. The large, prominent nose and projecting chin of the Plains Indians provides an extreme contrast with Bering Sea Mongoloids. Other variations include high frequencies of dislocated hip among the Apache and Navaho, of beard hair among various Paiute and Coahuilla, of albinism in the Hopi and San Blas, of obesity among Papago and others, and of very large chest circumferences among Andean Indians. Turning to serological variations, the world's highest frequencies of the Diego factor are found in Venezuela, and many groups of South American Indians entirely lack the A and B blood-group antigens. The Blood, Piegan, and Blackfoot Indians display the world's highest frequencies of blood type $A_1$. In generalizing on genetic features of American Indians, it is important to note that the frequency of blood type N is very low, that $R_2$ is high, and that very few or no abnormal-hemoglobin types occur. At the same time it must be noted that variation between groups is marked, a feature of the distributions suggestive of founder effect and subsequent genetic drift: new communities are often founded by a handful of migrants who are closely related, and who therefore carry only a fragment of the genetic variation of the total population of their origin (see Mayr, 1963, p. 211).

The ultimate Asian ancestry of the American aborigines can be demonstrated with certainty, and in fact, as the geneticist Robert Kirk has observed, the relation of the Indians and Asiatic Mongoloids provides the best example of affinity over distance (Spuhler, 1951; Kirk, personal communication, 1964; Neel and Salzano, 1964). Considering that the American aborigines originated in Asia, that there is considerable diversity between different groups, and that they display extreme values for several traits, can we use this diversity to estimate the time that they have been in America or the number of waves of migration? Are there relict populations who have survived in isolation and who show unusual resemblance to earlier types of man in Asia? Are there elements of morphology or genetic traits that reflect an earlier stage of evolution, or whose absence indicates migration before the appearance of these traits in Asia? Is there any distributional evidence contained in clines and discontinuities that reflects different moments of migration? Is there evidence that any of the differing groups were separated in time by a glacial advance or interstadial? Is there evidence of ecological separation, maintained over time, that correlates with morphological and genetic discontinuities?

The answers to these questions appear mostly in the negative or must be placed in the suspense account (Dahlberg, 1963), with the important exception of the last, which pertains to the dichotomy between Indians on one hand and the Aleuts and Eskimos on the other. There does not appear to be any division between any groups of American Indians that corresponds in magnitude to that between the Indians and their northern neighbors. For forensic purposes it is often possible to distinguish the skulls of many tribes, and no trained investigator is likely to mistake a Seneca skull for that of a Hopi, or an Ona skull for that of an Arikara, and, of course, Indian skulls can be distinguished in most cases from those of all other continental and insular divisions of mankind. However, the ability to sort out skulls and skeletons of many different tribes does not in itself indicate any great time depth. Noting the kinds of evolutionary changes in stature and head form described by Newman, and the adaptations—physiological as well as morphological—to high altitude seen in Andean Indians, T. D. Stewart has remarked, "Indeed, it is safe to say that no population of comparable size has remained so uniform after expanding, in whatever time has been involved, over such a large land area" (Stewart, 1960, p. 262). No evidence has been found of Neanderthal man, either as fully developed individuals or as individual traits attributable to Neanderthal man or to an Asiatic counterpart. There are no early skeletal materials that do not appear to fall within the range of American Indians, and this can be demonstrated for series of skulls, such as those of Lagoa Santa, and Tehuacan, as well as for individual finds. The Lagoa Santa series is of special interest because it includes data on 17 skulls, all narrow-headed and all exhibiting Inca bones, dehiscences, and shovel-shaped incisors, all of which remove any doubt about their status as American Indians. The series was removed from the cave of Sumidouro in the Lagoa Santa region of the State of Minas Geraes, Brazil. They were apparently associated with a fauna containing both extinct and contemporary species. It appears likely that they may be dated somewhat in excess of 10,000 years B.P. A smaller series of skeletal materials, more firmly dated at between 8,800 and 7,000 B.P., are represented in six burials from the El Riego phase, Tehuacan, Mexico. These are within the range of American Indians, and are somewhat broader-headed than those of Lagoa Santa (Anderson, 1965, p. 496–497). Comparisons with a small series of skeletons from Kawumkan Springs, Oregon, suggest that similarities between early American Indians in South, Middle, and North America were more prominent than differences. Though the skeletons of Kawumkan Springs are not certainly dated, they may be older than 7,500 years B.P. (Laughlin, 1956, p. 475–480).

It is tempting and plausible to envisage more than one group of immigrants. For example, it would be plausible to postulate a group carrying only blood type O and the Diego factor into South America, and another group that brought blood type $A_1$ into North America; both might well have preceded the ancestral Aleuts and Eskimos on the Bering Land Bridge, or might have passed through the interior while the ancestors of the Aleuts and Eskimos were evolving on the south coast of the land bridge. It is important to study such possibilities, but they cannot yet be accepted as likely.

The most plausible argument for non-Mongoloid traits in the American Indians has been made by Birdsell (1951), who cites various series of crania, such as those of Santa Catalina Island, California, for their similarities with crania of Australian aborigines; he also notes such various traits in the living as beard and body hair and large ears with pendulous lobes as being of the sort found among the living Australian aborigines and the Ainu of northern Japan. In reviewing the evidence he has called attention to an important fact: hybrids of the mixing of two races are not necessarily, and in fact often are not, intermediate between the two parental types. There is much masking and buffering, and a great deal of cryptic variation, which is difficult to investigate. In support of his dihybrid thesis for the origin of American Indians, Birdsell suggests that there was an archaic Caucasoid population, termed Amurian, living in northern Asia, and that, though mixed with Mongoloids, they are reflected in such groups as the Cahuilla tribes of inland southern California, and to a certain extent in the Pomo and Yuki of northern California. But earlier antecedents of contemporary Indians obviously are not identical with their contemporary representatives. It would appear possible to contain all the variability demonstrated in New World populations within the overall division of Mongoloids without invoking hybridization between Mongoloids and Caucasoids, especially when the variations over time are recognized.

Another consideration in the interpretation of the variations in American Indians and of their affinities with Asiatic Mongoloids lies in the character of skeletal remains in China and Japan. Three undated crania from the Upper Cave of Choukoutein, likely more than 15,000 years old, have been characterized by W. W. Howells as "unmigrated American Indians" in appearance (Howells, 1959, p. 300). Early American Indian crania have been described by J. L. Angel as "proto-Mongoloid," and many of the Asiatic finds, such as those of Choukoutein, may fit the same general category. It is important to recognize, then, that the peoples of these two continental areas have been evolving separately since their geographic separation at least 15,000 years ago.[a] Contemporary Asiatic Mongoloids differ from

[a] The estimate of 15,000 years for the time since the peoples in the old and new worlds were separated is based upon the assumption that some of the North American "early-man sites" known thus far may be as much as 12,000 or 13,000 years old, although most are clearly in the 9,000–11,000-year age range, and that man's time range in North America will be extended back another 2,000 or 3,000 years when more sites have been discovered. The 15,000-year figure is consistent with the assumption that the overall diversity between groups of American Indians is less than that between indigenous groups in comparable areas in Europe and Africa. It is best to consider all of these assumptions only as working hypotheses to be tested with all possible forms of evidence and independent methods of analysis.

American Indians in many traits that have probably been elaborated since their separation. It is unlikely that any relict groups of Indians can be found to match specific groups in Asia.

When more early skeletons are recovered in Asia, it will be possible to secure some estimates of rates of evolution in various traits. For genetic traits not preserved in the skeleton, the estimates of original frequencies at the time of separation and of rates of change must depend upon studies of contemporary populations. Whether the earliest immigrants had no genes for blood groups A and B, or whether they did have the genes but lost them, is not yet amenable to a dogmatic statement. If they were contained in the earliest immigrants, they must have been in very low frequencies and must have been dropped in many of the earliest groups in order to compose the picture seen today. The high frequency of blood type O in South America, the high frequencies of $A_1$ and O in North America, and the presence of B as well as $A_1$ and O in the Aleuts and Eskimos is in itself strongly suggestive of early differences that have been retained in general outline; this is similar, to a limited extent, to the picture in Australia, where B is absent except in a small area of northern Australia. The presence of the Diego factor in American Indians and in many Asiatic Mongoloids but not in Eskimos can be easily explained by the assumption—plausible but by no means proved— that the Bering Sea Mongoloids living on the coasts of the land bridge were bypassed by the ancestors of the American Indians moving through the interior.

A useful parameter in speculations about the early immigrants has been provided by J. V. Neel and F. M. Salzano, who show that an original group of only 400 individuals could have increased to 10,000,000 in a period of 15,000 years. Assuming an average generation length of 20 years and a modest increase of 1.4 per cent per generation, the 750 elapsed generations are sufficient to provide more Indians than were actually present in A.D. 1000 (Neel and Salzano, 1964, p. 86). The picture of only a few small groups actually completing the crossing of the bitterly inhospitable interior of the land bridge is compatible both with the landscape conditions between 24,000 and 11,000 years ago and with the observed degree of diversity among the Indians.

An interesting speculative system for correlating time depth with diversity between living groups can be attempted by noting that the Indians of North and South America appear to be generally similar to each other in polygenic traits (traits such as hair form, face form, and stature, which have several genes involved in their inheritance), whereas there are obvious dissimilarities in the single-gene traits such as the Diego factor, and blood group $A_1$. One would assume that a longer time is involved in the development of substantial differences in polygenic traits than in single-gene traits. The distances involved are so great that more genetic differences should be present, if the Indians had occupied these areas for a long period of time. In addition to the distance barrier to gene flow, there is the important and unnoticed fact that the partitioned population density of Middle America was an insurmountable barrier to gene flow for at least the last 6,000 years. This point has been mentioned by M. T. Newman: "Without actual migration it is inconceivable that the gene for blood group A could spread several thousands of miles through an unknown number of breeding populations in a historically reasonable period of time" (Newman, 1958, p. 34).

There is still another speculative system for estimating time depth from biological characteristics, this one employing a general characterization rather than the diversity between groups. Assuming that skin pigmentation is adaptive and responds to sunlight, then the aborigines in the equatorial regions of America should be as heavily pigmented as those in Africa and other regions of heavy pigmentation. That this is not the case has been used as an indication that the Indians have been in the New World for a period not much longer than 10,000 years (Haldane, 1956).

The Asiatic origin of the American aborigines is beyond doubt. However, dating of migration times from biological evidence is not similarly well founded. The most useful observation that can be made is that there is no evidence incompatible with a time lapse of some 15,000 years since the earliest migration into North America. This does not exclude older population arrivals; it remarks only that there is no evidence for a longer population history.

## Dichotomy of Bering Sea Mongoloids and American Indians

In considering American aborigines, most authors have separated the Eskimos and Aleuts from the Indians for the reason that these more northern groups were quite different and had more obvious Asiatic affinities (Laughlin, 1962). Viewed from North America, the Aleuts and Eskimos look like displaced Asiatics. Viewed from Asia, they appear most similar to other peoples inhabiting the shores of the present Bering Sea—the Kamchadal, Koryak, and Chukchi (Fig. 2). Though the time of divergence is in doubt, the similarities between the presently mutually unintelligible Aleut and Eskimo languages indicate that they are derived from a common stem (Swadesh, 1962). Eskimo has two major divisions in North America, Inyupik and Yupik. Inyupik is spoken from Norton Sound around all of north-

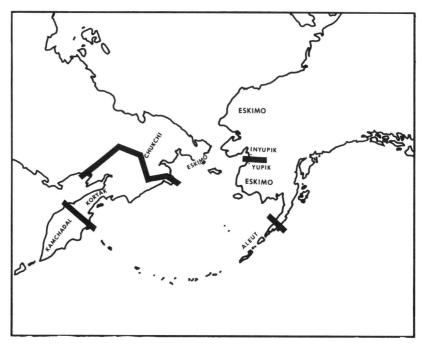

Fig. 2. Location of Bering Sea Mongoloids. Separation of Aleut from Eskimo lies at Port Moller on the Alaska Peninsula.

ern North America to Greenland; it extends westward to the Diomede Islands in Bering Strait. Yupik is spoken in western Alaska beginning at Port Moller and extending up the coast to Unalakleet on Norton Sound, as well as on Nunivak and St. Lawrence Islands and in a portion of Chukotka,[b] and on Kodiak Island, Cook Inlet, and Prince William Sound. Though various estimates of the time of divergence have been calculated employing lexico-statistical methods, their principal value lies in confirming the observation that Aleut and Eskimo must have differentiated much earlier (4,500 years ago or more) than the division between Yupik and Inyupik (probably 1,500 years ago). There is fair evidence that the Eskimo language has moved from south to north, and there is very good evidence that southwestern Alaska, the area of greatest linguistic diversity, is the area of greatest time depth as well (Sapir, 1916). Linguists agree that Aleut and Eskimo have an "Asiatic flavor," and have suggested fairly close connections with Chukchi, but formal demonstrations have not yet been made.

[b] Menovshchikov, the distinguished Russian linguist specializing in the languages of Siberia, has recently suggested that the Serenik dialect of Siberian Eskimos is a separate language.

There is clearly no "Asiatic flavor" shared between Eskimo and American Indian languages. The congruence between the genetic characteristics of the Eskimos and Aleuts and their languages is especially clear and illustrates the way in which language, as well as ecological, physiographic, and cultural differences, serve as isolating mechanisms between populations.

In their physical characteristics, the Aleuts (Laughlin, 1951) and Eskimos (Laughlin, 1966a) display many common elements that establish an especially close affinity with the Chukchi and with Asiatic Mongoloids in general, rather than with American Indians. Common elements are large heads and faces (cranial capacities exceeding those of Indians); large mandibles, with the world's largest minimum breadth of the ascending ramus; high frequencies of mandibular torii in excess of palatine torii (Indians have lower frequencies of mandibular torii and their frequencies of palatine torii, like those of the Norse, exceed the mandibular torii frequencies); distinctive temporal areas of the skull in which exostoses of the ear are absent (these are present in high frequencies in some Indians); thickenening of the tympanic plate often pronounced, especially in north Alaskan, Canadian, and Greenlandic Eskimos; and narrow nasal bones, again achieving a world extreme in Eastern Eskimos. There are many dental traits common to Eskimos, Aleuts, and Asiatic Mongoloids, among which are the frequent absence of the third molars, the large lateral as well as median incisors, and the three-rooted first lower molar (Moorrees, 1957). The latter may be singled out for the way in which it epitomizes the dichotomy with Indians: the recent studies of C. G. Turner (1965) reveal a frequency of 43 per cent in male Aleut crania, 25 per cent in male Eskimos, and none in male Indians. Physiologically, the Eskimos display differences from Indians in their cold adaptations, especially in their elevated basal metabolism (Milan, 1963). In the very high frequency of separate neural arches and other anomalies of the spinal column (Merbs, 1963) the Eskimos again display their greater affinity with Asiatic Mongoloids than with Indians. The serological differences are expressed primarily in the presence of blood type B in Eskimos and Aleuts, apparently representing the terminal distribution for this antigen, which continues through Asia to reach its maximum value among central Asiatics such as the Buryats, the Kalmucks, and the peoples of northern India (Zolatoreva, 1965; Laughlin, 1966a). The fact that blood type B has not been found even in contiguous groups of Tlingit, Athabascan, or Algonquin Indians is a reflection of the mutual lack of esteem in which these people have held one another, as well as of the patent Asiatic affiliations of the Eskimos.

There are of course important regional variations within the Eskimo area, the most outstanding being the serological characteristics (low A and B) of the central Arctic Eskimos, including the Polar Eskimos of Greenland, and the increase in cranial breadth and decrease in vault height in western Alaska, from the Eskimos of Norton Sound south to the Koniag Eskimos of Kodiak Island and the Alaska Peninsula (Laughlin, 1950; Jørgensen, 1953; Gessain, 1960). Thus the central Arctic Eskimos in many ways form a slump block, showing a discontinuity with both the Greenlandic Eskimos to the east and the Alaskan Eskimos to the west. Whether their low $A_1$ and B frequencies are due to random genetic drift, to low frequencies carried by the founders, to bottlenecks in the population in the course of their history, or to some other combination of agencies is not known. It is important that this area is characterized by low population density, great distances between isolates, and a small total population for the central Arctic as a whole.

The low and broad cranial vaults of the recent southwestern Alaskan Eskimos and Aleuts have most likely resulted from internal evolutionary processes in those areas. In each local area—in the Aleutian Islands, on Kodiak Island, in the Kuskokwim area, and elsewhere—the earlier populations are more narrow-headed (Fig. 3). A slight trend toward an increase

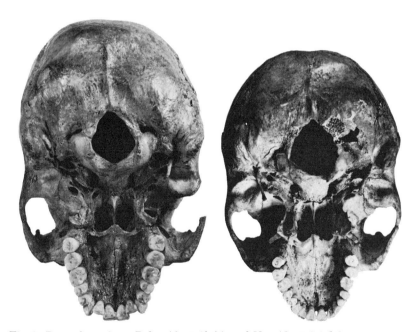

Fig. 3. Base of cranium: Paleo-Aleut (*left*) and Neo-Aleut (*right*).

in head breadth is evident among the Angmagssalik Eskimos of the south-eastern coast of Greenland, the terminal isolate in the Eskimo movement across the North American Arctic (Laughlin, 1958).

Thus, in morphology, dentition, growth, pathology, physiology, and serology, the Aleuts and Eskimos are quite distinct from American Indians and more similar to Asiatic Mongoloids, such as the Chukchi, Koryak, and Kamchadals of the western shores of the Bering Sea.

## Living Conditions along the Interior Route

The Bering Land Bridge provided two distinct kinds of routes for migration into North America, and at the same time provided sites for permanent residence based on two distinctly different kinds of adaptation to the ecological resources of the bridge. It is in fact likely that movements across the land bridge were so slow that the people themselves were scarcely, if at all, aware they were migrating. The idea that these people moved onto the bridge in order to accomplish a crossing—or perhaps even that they knew there *was* something to cross—is not tenable. The long-term result was, of course, migrations, both along the coast and through the interior (see Fig. 1). From the point of view of those living on the bridge it was permanent occupation, and from the point of view of one who has investigated 8,000 or more years of Aleut habitations, it continues to be permanent occupation.

At its widest, the Bering Land Bridge extended nearly 1,500 km from its northern shore to a southern margin on the older and larger Alaska Peninsula, which then incorporated land subsequently partitioned into islands (see also Hopkins, 1959, Figs. 1 and 3). There was ample space in the interior for small groups of hunters in bands of 50 or 100 and constituting a total population at any one time of only a few hundred to move about in pursuit of game without intruding upon each other. The interior landscape was evidently a low rolling plain, for the most part devoid of relief, studded with bogs and swamps, frozen much of the time, and lacking in trees or even many bushes. Grass-eating herbivores may have been present in fair numbers. The human adaptation to this region must surely have been that of big-game hunters, living by means of scavenging dead mammoths and such bovids as caribou, bison, and musk-ox, and by intentionally hunting live animals. The big-game hunting tradition was well developed at an early date in the New World, as is evidenced by the Folsom and Clovis cultures of ten to twelve thousand years ago.

Unless their physiological adaptation to cold was considerably greater than that of the cold-adapted contemporary Eskimos, these early hunters would have required shelter, fire, and clothing, including boots and mittens of some sort. Shelter could well have consisted of tents, and the use of double-walled tents, such as those presently in use by the Chukchi and by Asiatic Eskimos, would have been practicable. Lacking wood for fuel, they would have had to resort to fat, oil, and marrow for heat; whether they rendered fat to produce oil for lamps or used fat lamps, they required fire.

Life for the interior hunters was without doubt difficult (Colinvaux, 1964). The screening effect of survival in this extremely harsh environment has undoubtedly had some effect on various aspects of disease resistance in the descendants of the early big-game hunters—the American Indians.

The settlement patterns and the patterns of annual movements of the interior herbivore hunters of historical time are different from those of contemporary coastal marine hunters, and comparable differences must have characterized the coastal and interior land-bridge peoples of 15,000 years ago. Contemporary interior hunters, such as the Nunamiut Eskimos of the Brooks Range in Alaska, the Caribou Eskimo of Canada, and various Athabascan Indian groups of Alaska, are characteristically partitioned into small bands often numbering fewer than 50 persons, and there are great distances between bands. The camps must be moved annually or sometimes frequently in the course of a single year in order to intercept game, and every few years the people must concentrate in a totally different area or at least one that overlaps the previously exploited area only partially; thus the total exploitational area is very large. An extremely important consequence is that when such hunting bands migrate, they pick up the entire camp—tents, utensils, and people—and move, leaving no one behind and little evidence of their occupation. Prospects for finding sites that yield artifacts, human skeletons, and faunal remains are considerably less than obtains for coastal areas with permanent villages. Furthermore, the more frequent kill sites or hunting camps characterized by projectile points do not yield the full suite of artifacts to indicate methods of manufacture of tools and living shelters, or other contextual materials indicating the way of life. Occasional complete depopulation of an area, which occurred recently on the Arctic slope of Alaska, and occasional extinction of entire settlements or bands, recorded historically in the Canadian Arctic, illustrate the marginal existence of the Arctic-interior big-game hunters. The northern coast of the land bridge probably was unoccupied, because year-round sea ice prevented the development of sea-mammal populations that establish rookeries on the shore; the north coast must have been even more inhospitable than the interior.

## Living Conditions along the South Coast of the
## Bering Land Bridge

People who inhabit a coastal area can command a large portion of the rich marine resources of the ocean in addition to those of the land, without leaving the shore. Each step in increasing adaptation to marine life proceeds logically with a system of increasing rewards, beginning with an initial economy based on gathering, scavenging, hunting, and fishing along the shoreline, on through the use of various kinds of boats that permit the invasion of additional ecological habitats, and culminating in the development of the skin-covered kayak, which makes possible complete and expert exploitation of the sea. The population densities of various peoples inhabiting the coast of Bering Sea reflect both the varying richness of the natural resources and the varying number of habitats actually invaded by the populations. The Kamchadal focused on the shoreline and made only limited use of the sea by remaining close to shore in open wood boats. More northerly inhabitants did not have the rich shoreline habitats of the Aleutian Islands; they were forced to hunt on the ice and to make maximum use of the skin-covered umiak, with much less use of kayaks. The Aleutian Islanders are employed as the primary model for this discussion because they had the full suite of ecological habitats available to them and they also had the necessary technology to achieve full exploitation of all of them.

A coastal life derives its cluster of economic benefits at the most simple or primitive level from the ease of exploitation of a rich intertidal zone, making use of the rich animal and plant communities along the beach, on the reefs, and in the bays and lagoons. A great variety of marine-intertidal algae and edible invertebrates includes such plentiful forms as sea urchins, limpets, whelks, mussels, chitons, clams, and octopi. Sea-urchin spines and tests, for example, compose a large percentage of the midden remains of Aleutian village sites but not of Eskimo coastal sites to the north.

The intertidal zone is amenable to exploitation by simple, though clever, methods that can be used by otherwise disadvantaged segments of the community. Sea urchins and the like can be gathered in baskets, and edible algae are also easily collected. Octopus and various fish can be secured by means of gaff hooks at low tide, when large portions of the reef system are exposed and when the crevices in which the octopi reside can be reached by a man in watertight boots or no boots at all. Dip nets, set nets, and fish lines and hooks, as well as fish spears, comprise a major part of the simple technology for exploiting the intertidal zone. It is maximally important that in the Aleutian area these food resources are accessible for a large part of the year. It is also maximally important that they are available to women, children, the elderly, and the disabled, so that these kinds of people are able to make a significant contribution to the economy, a sustaining contribution that they could not make in an interior economy or in areas where winter sea ice blankets the shores. These simply obtained resources provide a buffer against starvation; they remove communities from the extremes of the feast-or-famine cycle so characteristic of interior peoples who depend upon migrations of caribou or the uncertain movements of musk-ox or, in earlier times, of bison, mammoth, and horse.

With the addition of anadromous fish—those that run up streams from the sea—the coastal economy receives another major food source. Though annual runs of salmon are restricted to streams and last through only four months of the year, they are enormously important because of the large amounts of food made available, the simple methods needed to exploit them, and therefore the utility of women, children, and old people in exploiting them. It is not possible to certify a date for the beginning exploitation of salmon runs, but it is likely that some stream fish and some anadromous fishes were taken by the earliest coastal inhabitants of the Bering Land Bridge and of the Siberian coast south of the Anadyr River. The use of fresh-water and anadromous fish favored settlement around stream mouths and contributed to the development of relatively permanent villages.

The use of boats permits the exploitation of additional ecological niches. Open skin boats—coracles or bull boats—are widely distributed and undoubtedly appeared early in the history of the human species, and were surely available to the coastal land-bridge people. The marine hunter or fisher has only to move a relatively short horizontal distance in order to sample much deeper zones in which many additional fish, such as halibut and cod, may be found in large numbers. The land hunter cannot change ecological zones as quickly or as easily. He must walk up a mountainside to enter new zones in person, rather than simply lowering a 200-meter fishing line. Furthermore, the use of boats, even simple open skin boats such as small coracles or umiaks, opened the way to exploitation of offshore islands and skerries. Offshore islands, even small ones, are important as nesting sites for cliff birds, such as cormorants, puffins, and other *Alcidae* (Udvardy, 1963). Several other birds may also be accommodated at these sites, including eiders, kittiwakes, and gulls, depending upon the configuration and position of the islands. The eggs are a renewable income, for once the hunter removes the eggs, the cormorant, puffin, or duck will often replace them with others. Eggs were certainly an important food resource especially available to the coastal hunters, and the skins of cormorants and puffins in particular provided warm and light parkas. The isolation of offshore islands protects eggs and young from such terrestrial predators as foxes. Islands also provide hauling grounds in which seals and sea lions may breed. All pinnipeds (seals, sea lions, and walrus) must come to a land or ice front for breeding; this behavior congregates them into rookeries for at least a part of the year and thereby facilitates their capture. Offshore islands also add significant amounts of coastline for the stranding of dead animals and driftwood.

The addition of kayaks—decked-over skin boats propelled by double-bladed paddles—added still another dimension to the exploitation of the marine environment. It cannot automatically be assumed that the earliest coastal inhabitants of the land bridge had developed kayaks, though the possibility does exist. Those resident sea mammals that haul up on shores (sea lions, seals, and sea otters) can be hunted on foot. However, open boats are an obvious advantage in reaching offshore islands and permanent rookeries, and in addition the transient sea mammals (fur seals and many of the whales), with few exceptions, can be hunted only on the high seas.

The kayak is an engineering triumph, as is seen in the complex construction of its skeleton, sometimes made with a three-piece keelson in certain kinds of Aleutian kayaks. It has a circumscribed distribution restricted to Aleuts, Eskimos, Chukchi, and Koryak. This distribution, limited to Bering Sea Mongoloids (including those who migrated to Greenland) but not known to all Bering Sea Mongoloids, indicates its Bering Sea origin. The high state of technical development of the kayak and its most extensive use in the Aleutians suggests that it originated in this area. No Indians or Europeans have ever mastered the complexities of kayak hunting. Although it has been possible for such people to learn to hunt sea mammals, or to paddle kayaks, the combination of hunting sea mammals from kayaks never diffused from the Bering Sea Mongoloids to other peoples. Extensive childhood training as well as technical knowledge is required for efficient use of the kayak.

The ecological functions of kayak hunting provide insight into the nature of the human coastal ecology and therefore into the differentiation and history of the Bering Sea occupants.

Kayak hunting makes possible more efficient hunting of mammals at sea, providing the speed and maneuverability essential to harpooning humpback whales, fur seals, sea lions, porpoises, and sea otters on the open sea. The kayak can be operated in stormy seas and in fact can be launched into surf that ordinarily prohibits the use of the umiak. Though umiaks can be decked over, and were formerly covered for long inter-island voyages, their maneuverability is small compared with the one-man or two-man kayak. Portability is a prominent feature—one man can carry the boat across an island to a favorable launching site, proceed with his hunting or travel, and haul up the boat without help when he is finished. A unique function of the kayak derives from the fact that marine hunters are, in large part, horizon hunters. This means that they spend much of their hunting time scanning the horizon for the interrupting outline of a sea mammal. The sea mammals, in turn, must frequently return to the surface for air, and thus there is a common point of intersection between mammal and man. The kayak hunter uses the kayak as an efficient moving platform for scanning the sea and also the coastline. He cannot use the tracks or spoor that are available to the land hunter, but he enjoys much faster scanning of the open-sea horizon and of the coastline where the sea mammals must appear.

The ability to navigate out of sight of land enlarged the effective exploitational area of the marine hunters. Not all kayak-hunting groups could navigate in fog or far from land, but the Aleuts did and also the Kodiak Eskimos. At the same time, kayaks and also open boats were highly efficient for retrieval of game and its transport to the home base, owing to the fact that mammals could be floated and towed, or butchered and stowed inside boats, for the return trip. Thus, one man could tow a few hundred kilos of meat or carry one to two hundred kilos inside his kayak. The open boats could, of course, carry much more. The contrast with interior hunters, in energy and time expenditure, is illustrated as an important difference. Land hunters had only their back for transport; dogs were probably not in use by the earliest interior hunters. Pack dogs came into use, but not until comparatively recent times did dog-drawn sleds come into use. Water transport has another obvious inherent advantage in the fact that boats do not eat the food they are transporting. They do not need to be fed and at most will take only a little water.

Tidal phenomena, especially critical in passes between islands, enhance the usefulness of the kayak. Certain places characterized by high-velocity tidal bores can be successfully crossed by kayak hunters, whereas slower boats are inadequate for the task.

In summary, the coastal biosystem is inherently richer than that of the interior, and this contrast has obtained throughout the existence of the land bridge. A natural progression of greater rewards for improved technology exists in the coastal area. Each additional step, from strand hunting and collecting, through use of a simple coracle, through use of a boat with keelson (umiak), and finally to the kayak—and with it the culmination of open-sea hunting—adds another series of exploitational areas without detracting from the others. All segments of the population are able to make contributions to the economy, at all times of the year. This is effectively reflected in the historical demographic picture.

By 1741, the Aleutian Islanders had achieved a large total population; there were over 16,000 Aleut speakers, including those on the mainland. Approximately 10,000 lived in the eastern area, 5,000 in the central Aleutians, and some 1,000 or more in the western Aleutians. This large population was characterized by a comparatively low rate of infant and early-childhood mortality, and a relatively high proportion of elderly people. This population profile contrasts remarkably with that of the extinct Sadlermiut Eskimo of Southampton Island, Hudson Bay, whose infant and early-childhood mortality was approximately twice that of the eastern Aleuts, and who seem to have died, generally, before age 55 (Laughlin, 1963, p. 6). The low population numbers and densities of the barren-ground Eskimos and of the interior Eskimos of Alaska are probably matched with high infant mortality and reduced longevity. The advantage to people inhabiting the coast of the Bering Land Bridge lay not only in larger numbers but also in age composition and in the reduction of genetic wastage reflected in reduced infant mortality. Increased longevity has many cultural and genetic rewards. An increase in the average length of life and in the number of persons continuing into old age means that information is preserved and transmitted that would otherwise be lost. One specific example of benefit to the population lies in the development of anatomical and medical knowledge contributing to the reduction of infant mortality and the prolonging of life generally (Marsh and Laughlin, 1956).

Seaweeds, sea urchins, seals, and sea otters, and the role they played in the coastal ecosystem, are key points of focus in understanding how people lived on the coast and why they prospered. Clubs, baskets, fishhooks, and fish spears were adequate for appropriation of a rich diet and a sustained, year-around food supply.

## Prehistory of the Umnak Corner of the Bering Land Bridge

The dual relevance of the prehistory of the eastern Aleutian Islands to the occupation of the Bering Land Bridge and to living populations lies in the geologic fact that the eastern Aleutians were formerly a part of the land bridge. Umnak Island was the terminus of the then-enlarged Alaska Peninsula. The artifacts and skeletons provide a sample of information from the bridge itself. Other Alaskan sites are for the most part located on present coastlines that were several hundred miles inland during the history of the land bridge, or are located in interior areas even farther removed from either the coast or the interior of the land bridge. With the exception of St. Lawrence Island and the two Pribilof Islands, the former coastal areas of the land bridge have been completely submerged.

The first people or peoples who encroached upon the land bridge were of course unaware of the enormous faunal resources at the Umnak gateway to Bering Sea. The fact that the coastal area has been everywhere more reward-ing and congenial to human occupancy than the interior should not obscure two related facts: the eastern side of Bering Sea is richer than the western side, and the more southern portions of the eastern coast were richer and more congenial to occupation than the more northern shores of the same coast. I have stressed the fact that simple hunting methods and the collec-tion of intertidal foods contribute a significant portion to the diet of the Aleuts. There was a system of increasing rewards with each step in increased adaptation to marine life, culminating in open-sea hunting with the use of the kayak. However, the first step was adequate to maintain thriving com-munities. The numbers and densities of people living around the Bering Sea, the directions of their movements, and the length of time they have remained in particular areas are all tangibly related to the economic bases.

The terminus of the Alaska Peninsula passed through at least three dis-cernible stages. It was originally a constituent part of the bridge (Fig. 4). As sea level rose during Late Wisconsin time, a large island comprising present-day Unalaska, Anangula (Ananiuliak on Coast and Geodetic Survey charts), and Umnak was separated in the region of Akutan Pass. Subse-quently, Umnak and Anangula were separated from Unalaska; our first actual occupational remains represent this period of the Umnak-Anangula juncture, some 8,000 years ago. Later, the shallow pass separating Anan-gula from Umnak was inundated, and these two became separated islands sometime prior to 5,000 years ago. A profile of eastern Aleutian passes re-veals no serious obstacle to the movement of people (Fig. 5). Owing to its strategic position, commanding the first entrance to Bering Sea, the Umnak corner functioned as an ecological magnet. This magnet attracted humans as well as whales, and it explains why people have remained continuously in this area over the eight or more millennia since their arrival.

The Alaskan shores of the Bering Sea are richer and warmer than the Siberian shores. Relevant to temperature distribution and the life forms de-pendent upon temperature is the observation that "The present contrast in the temperatures of waters along the coast of western Alaska and along the coast of Chukotka seems to have existed throughout much of Quaternary time" (Hopkins et al., 1965, p. 7). Focusing more directly on the eastern

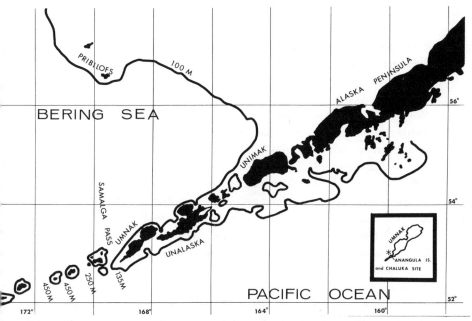

Fig. 4. Former Bering Land Bridge as defined by the 100-meter bathymetric contour. Umnak Island was then the terminus of the Alaska Peninsula.

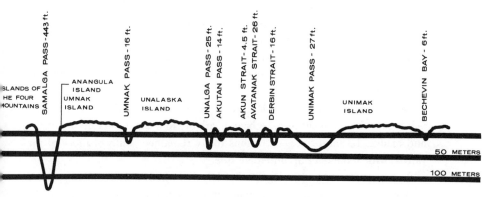

Fig. 5. Profile of the eastern Aleutians. Source: Coast and Geodetic Survey Charts 8802 and 8861. Not drawn to scale.

Aleutian portion of Bering Sea, R. F. Black has remarked, "Gross climatic zonation, including major atmospheric circulation, oceanic currents, and upwelling along the south flank of the Aleutian Islands, should not have changed significantly from the Pleistocene to the present. Local intensification of the circulation and upwelling would likely have occurred in the vicinity of Umnak Island when it was formerly the western extension of the continent and the southern edge of the Bering Sea platform and partly covered with continental ice" (Black, oral communication, October 15, 1965; for an overall perspective of the geology of this area I have drawn upon Black, 1966). The population densities of the aboriginal groups inhabiting the shores of Bering Sea at the time of European discovery provide a useful index of food distribution and thus of the abundance of marine animal life. The historical population was relatively sparse along much of the Siberian coast, and with the exception of the uninhabited Commander Islands, the number of fur seal, sea lion, sea otter, and walrus was less on the Siberian side generally than on the Alaskan side.

The temperature differential has existed a long time (Hopkins *et al.*, 1965). Present faunal differentials, though not actual numbers, can be reasonably attributed to a much longer period than that beginning 15,000 years ago. The major pinnipeds of economic importance to native populations have inhabited Bering Sea for longer periods of time than simply since the last appearance of the land bridge. Their habits, when coupled with the temperature-related aspects of their distributions, permit the speculation that some were available to the earliest human occupants. An important behavior of the pinnipeds, and one that places them within range of land-based hunters, is the annual return to a land front or ice front during the reproductive season.

Walrus have clearly been present at all times since the modern species first evolved, because the Pacific form, *Odobenus rosmarus divergens*, is concentrated in the northern portion of Bering Sea, primarily from the Gulf of Anadyr around to Bristol Bay. During the last century, they were found in large numbers as far south and west as Walrus Island on the north side of the Alaska Peninsula.

Steller's sea lion is a north Pacific form, present on both sides of Bering Sea. It must be assumed to have been available to coastal inhabitants of the land bridge. Its major concentrations likely remained south of those of the walrus, with some overlap on either side.

Fur seals still return to the Pribilof Islands, which were formerly part of the land bridge, and to the Commander Islands, for delivery of their offspring and for breeding. Assuming that for the past 15,000 years the fur seal maintained generally similar habits of dispersal into Pacific waters on both the American and Asiatic sides during the greater part of the year, and returned to northern breeding grounds in summer, they would have then been accessible to the inhabitants of the Umnak corner as they entered the first major pass from the Pacific. That they may have hauled up in the Aleutians, as well as in ice-free portions of the land bridge, is a possibility, though they have rarely been seen on shore in the Aleutians.

Two seals, the harbor seal and the ringed seal, may have been of importance to the earliest land-bridge inhabitants. The ringed seal is rarely seen in Aleutian waters, but is an Arctic mainstay. The bones of harbor seal, sea lion, and fur seal are well represented in Aleutian archaeological sites.

Sea otters, like sea lions, are year-round residents and are obviously old inhabitants of Bering Sea. Their temperature preferences suggest that they have been represented in their present areas for a long time, though their southern forms, not distinct species, on the North Pacific and California coasts, may have had much larger numbers. Scheffer has suggested that the sea otter may have evolved in the North Pacific: "Did the protected, food-rich, kelp reefs of the North Pacific serve as a launching platform for both otariid and sea otter stocks?" (1958, p. 33). Many of the foods eaten by the sea otter, fish, octopus, and sea urchin are also eaten by the Aleuts and southern Eskimos.

The special richness of the Umnak corner of the old Alaska Peninsula depended in part on the upwelling due to major currents striking the edge of the platform. An ice sheet that centered south of the Peninsula and blanketed the Peninsular-Kodiak area, preventing the inland movement of people inhabiting the coastal fringe, would not have altered the direction of currents or their basic behavior. Inside Bering Sea, the major current system was counterclockwise (Zenkevich, 1963). Thus the area south and east of the present-day Pribilof Islands (Fig. 4) was warmer than areas to the north of that coastline inflection. The waters of Samalga Pass, the 16-mile pass separating Umnak Island from the Islands of the Four Mountains to the west (Fig. 5), were the first entry into Bering Sea on the American side, and were a natural concentration area for migrating species, such as fur seals and whales.

In brief, the transient mammals passing around the Umnak terminus of the bridge (fur seals and whales), the resident sea mammals (sea otters, sea lions, harbor seals), the fish (greenling, cod, salmon, and halibut), and the birds (especially cormorants, puffins, and ducks), as well as the intertidal invertebrates (especially sea urchins), made this area one of the richest in the world, and the promontory thus provided a uniquely productive hunting station.

THE ANANGULA UNIFACIAL CORE AND BLADE INDUSTRY

The Anangula site (Figs. 5 and 6), the oldest dated archaeological site found thus far in Alaska, displays Asiatic similarities and thus, in its position on a remnant of the land bridge, substantially enhances the probability that a coastal migration from Asia took place. Anangula is a small island within sight of the present village of Nikolski, which lies at the south end of Umnak Island (Laughlin and Marsh, 1954). The occupation site, border-

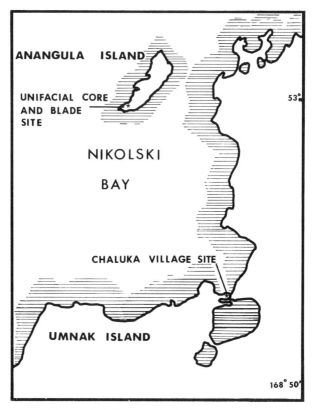

Fig. 6. Location of the unifacial core and blade site on Anangula Island and of the Chaluka village site on Umnak Island.

ing on a small, shallow pond that is usually dry, faces Bering Sea at a point where the water is 70 meters deep today. The site is 17 to 20 meters above present sea level. Owing to this unique topographic position, it has survived the inundation that must have submerged many other sites on lower portions of the land bridge.

Three radiocarbon age determinations on charcoal from the cultural layer establish a minimum age for the occupation. (Black and Laughlin, 1964, suggest that the samples may be slightly contaminated and that the ages may be slightly too young.[c])

[c] A series of new radiocarbon determinations confirms that the Anangula occupation layer is about 8,400 years old. Four new determinations (P-1102 to P-1105) range in age from 7,933 ± 96 to 8,419 ± 90 years, using a half-life of 5,730 years, probably the most accurate half-life in this range. The most reliable specimens are the oldest, P-1103 and P-1104, because these were large enough to permit additional NaOH pretreatment to remove possible humic contaminants. These average 8,396 ± 67 years (R. Stuckenrath, Applied Science Center for Archaeology, Univ. Pennsylvania, written communication, May 3, 1966).

| Laboratory No. | Years Ago |
|---|---|
| I-715 ................................. | 8,425 ± 275 |
| W-1180 ................................ | 7,660 ± 300 |
| I-1046 ................................ | 7,990 ± 230 |

The dates establish the fact that people lived on this coast at a time when Anangula and Umnak were joined and when access from Unalaska Island involved little more than wet feet. Viewed critically, the dating does not permit the conclusion that the inhabitants walked to this place from less mountainous parts of the land bridge, though that is a good possibility. The dating does indicate that the people who occupied this site are most likely descendants of people who did walk into this area, and owing to the former presence of an ice sheet to the east, they could only have come from the Bering Sea coast of the land bridge.

The stone tools from this large living area belong to a unifacial core and blade industry. Bifacial tools are absent, and this promptly distinguishes the stone industry of Anangula from all other reported Alaskan sites (Laughlin, 1965b). The essential inventory of stone tools and utensils includes cores, platform tablets, lamellar (prismatic) blades, and the blade tools such as knives, pointed blades (gravers, thrusting points), angle burins made on blades, scrapers, rubbing stones (with red-ochre stain), abraders of scoriaceous lava and pumice, hammer stones, and two fragments of a single stone vessel or dish.[d] The large numbers of tools recovered show all stages of manufacture and thus provide new information on the nature of such industries. This facilitates comparisons with the related Asiatic industries. Depressions in the iron-stained living floor may represent house or tent pits; and charcoal hearths indicate the use of driftwood for fuel. The house

[d] Of the more than 6,500 specimens from Anangula, fully one-third are blades. About 25 per cent of the blades are unifacially edge-retouched, in contrast to the flake tools, of which less than 10 per cent show unifacial edge retouching. The blades range from less than two centimeters up to 12 centimeters in length; their lengths plot as a normal curve with the modal class between 4.5 and 5.0 cm and the average and median classes between 5.0 and 5.5 cm. The 39 cores from which blades and flakes were struck also represent a complete size range and are in full accord with the evidence from the blades in indicating that emphasis was *not* on the production of so-called microblades. No "type core" can be recognized at Anangula—several are prismatic, but the spectrum includes cores from which only blades were struck and others from which only flakes were removed, judging from the scars remaining on the cores. The "burins" reported from Anangula are actually scraping tools; they consist of flakes and blade fragments from which smaller flakes have been removed transversely and slightly ventrally; this treatment has generally produced a uniformly smooth scraping edge across the flake or blade. Aside from the "burins," which evidently were used as scraping tools, the Anagula chipped-stone material cannot be "typed" according to function in the traditional sense. Clearly, the blade tools served as cutting *and* piercing instruments and are best characterized as multifunctional. Except for removal of transverse flakes, there has been no extensive retouching or reshaping of the blades and flakes.

pits and the rich variety of tools identify this site as a coastal village and manufacturing site rather than a hunter's camp or a kill site.

A key aspect of this industry lies in the fact that the tools are flaked on only one surface, and usually on only one or two edges. Consequently, the steps in the manufacturing process can be seen, and comparisons with similar industries in Japan and Siberia (Müller-Beck, this volume) are facilitated. The method and techniques of manufacture, not just the end-products, indicate the greater similarities with Asia and the dissimilarities with the New World. The manufacture of prismatic blades began with raw material, such as a cobble or pebble. The blades were detached from one end of the core, the striking platform, which was first prepared by the removal of large, broad, irregular cortex flakes. The remaining cortex had ordinarily to be peeled in order to prepare the surface for the production of prismatic blades. The first course or round of long flakes, detached by striking on the platform, yielded flakes that were usually longer than broad, but were not prismatic in cross section owing to the weathered cortex surface. Successive courses or rounds of blades had little or no cortex and had two or more lamellar blade surfaces on their dorsal or external (noncore) surface. Thus, they are prismatic in cross section, and are relatively long, narrow, and straight. With the removal of each blade, a single concave surface (facet or blade scar) was left on the side of the core. Two contiguous facets formed one ridge. Consequently, blades with two or more facets on their dorsal surfaces could be detached at the discretion of the knapper. The detached blades are characterized by one core surface and two, three, or four dorsal surfaces (Fig. 7).

Many of the relatively early Alaskan industries (4,000 years or less in age) have been called microblade industries, owing to the fact that only

Fig. 7. Variation in widths and number of facets of prismatic blades. The right blade is 5.0 cm long.

Fig. 8. Blade cores. 8a (*Left*): Chert blade core, rotated. Prismatic blades have been removed from two directions. The surface of one striking platform and the concave scars of the detached blades may be seen in the lower right corner. The remainder of the surface consists of the scars of prismatic blades detached from a striking platform on the left. 8b (*Center*): Obsidian blade core. 8c (*Right*): Chert blade core with core tablet showing method of platform rejuvenation by method of removing a tablet. Note the contrast in size between this core, which would yield larger blades, and the core to the left (8b), which would yield "microblades." The center core (8b) is 2.3 cm long.

small blades were made. The continuous range of blades at Anangula can be only artificially partitioned into three stereotyped categories, micro-, meso-, and macroblades, and the Anangula industry is very important in showing how both large and small blades could be struck from the same core. As the striking platform became battered or otherwise unsuitable, it was rejuvenated by the removal of the platform (Fig. 8c). This involved striking from one side and detaching relatively broad and irregularly shaped flakes, or sometimes removing the entire surface of the striking platform in one large thick flake. Such rejuvenation might reduce the core as much as 2 cm in length. The next round of prismatic blades, or of flakes, removed was necessarily shorter than those rounds preceding the shortening of the core. With successive redressing of the striking platform, progressively shorter blades were removed (compare Fig. 8, b and c). The important consequence is that a large range of blade lengths could be struck from the same core. Though it is correct to say that microblades and microcores are present at Anangula, it is an incomplete statement of fact and omits some two-thirds or more of the core and blade industry represented here. This size range is not duplicated in any other Alaskan site.

Still another method of altering the shape and size of the core appears to have resulted when the blades became too curved, rather than being relatively straight in longitudinal section. The core was then rotated or inverted.

Fig. 9. Reassembled core indicating that first prismatic blades were struck, then nonprismatic flakes were struck from the same core. (*Left*) Side view. (*Right*) Superior view, enlarged.

A side of the core was used for a new striking platform and the blades removed accordingly (Fig. 8a). Reassembly of cores and blades or flakes removed from them indicates that blades were struck from a core and then nonprismatic flakes were struck from the same core (Figs. 8 and 9).

Reworking of these blades consisted of retouching by light flaking of one or two edges but only one surface, to produce knives and pointed blades (Fig. 10). It is likely that many of the pointed blades were projectile or thrusting points. An important category of tool made on blades, which shows affinity with traits found in Japan and Siberia, is the so-called angle burin (Fig. 11). This is quite dissimilar to the typical burins of Iyatayet at Cape Denbigh and of other Arctic cultures (Giddings, 1964, p. 209–219). At first sight it resembles a blade that has been snapped off, but upon closer inspection a transverse flake is seen to have been removed by a blow delivered from one edge. The negative bulb of percussion distinguishes this surface from that produced by a hinge fracture. The uses of such a tool remain to be determined. The scrapers are most frequently made of obsidian, apparently obtainable only at Cape Chagak at the northwest end of Umnak Island. Pointed tools were manufactured either on prismatic blades or on ridged flakes that were similar to true prismatic blades. A fuller discussion of the stone working methods is found in Laughlin and Aigner, 1966.

That the Anangula site represents a village of coastal and marine hunters is inferred from its position overlooking the sea, and from the absence of occupation material around a larger lake a little more inland, overlooking

Fig. 10. Three kinds of pointed tools. (*Left*) Naturally pointed prismatic blade. (*Center*) Prismatic blade with natural point at end of dorsal median edge. (*Right*) Retouched, symmetrical point on a ridge flake.

the now submerged connection with Umnak Island. A primary focus on the sea does not of course exclude hunting caribou or any other land animals, as is regularly done by coastal Eskimos and Aleuts whenever the opportunity presents itself. No osteological remains are present, human or nonhuman, most likely a consequence of the high acidity of the soil and volcanic ash. This limitation also obtains for the younger sites of the Denbigh microblade complex and for the Arctic Small Tool Tradition as a whole.

The affinities of the Anangula industry clearly lie more with Asia than with Alaska, with the tools of the Sakkotsu microblade industry of Hokkaido, the Araya site on Honshu, and the Budun site in Siberia, rather than with other presently known sites of Alaska. The Japanese pre-ceramic scholar M. Yoshizaki has worked at the Anangula site and has noted that the material from the site could fit easily on Hokkaido in the general period 9,000 to 12,000 years ago. Substantial differences from the early Alaskan mainland sites are evident, even when allowances are made for distance and for the fact that they are later in time. Thus, the Denbigh flint complex, as seen at Iyatayet in the authoritative description of the original finds (Giddings, 1964), and in its extensions and derivatives constituting the Arctic Small Tool Tradition (Irving, 1962; Collins, 1964), is quite unlike Anangula, though an ultimate derivation of Denbigh from Anangula or from a common base is possible.

Fig. 11. Dorsal views (*above scale*) and ventral views (*below scale*) of burinized flakes and blade fragments. Read from left to right, top to bottom. (1) Chert flake with secondary retouch and a single ventral burin blow. (2) Chert blade fragment with a single ventral burin blow. (3) Chert flake with retouch and two ventral burin blows, proximal and distal. (4) Chert blade fragment with a single ventral burin blow. (5) Siliceous chert blade fragment with secondary retouch and a single ventral burin blow. (6) Chert blade with secondary retouch and a single ventral burin blow. (7) Chert flake or blade fragment with retouch and two ventral burin blows, proximal and distal. (8) Chert flake with two ventral burin blows.

CHALUKA: FOUR THOUSAND YEARS OF CONTINUITY
WITH CHANGE

The deep, stratified, village site of Chaluka now forms the southern margin of Nikolski, a village of some 55 Aleuts at the southern end of Umnak Island (Fig. 6; Laughlin and Reeder, 1962). The mound rests on an elevation related to the 3 meter beach of the postglacial thermal maximum, and thus cannot be older than 5,000 years. The oldest radiocarbon date obtained thus far is 3,750 ± 180 years, for a sample approximately 50 cm above the sterile floor, and an extrapolated age of 4,000 years has been proposed for the beginning of occupation. The reasons for the long-term occupation of this site are apparent in the regional resources, and are documented in the varied faunal remains contained within the site. Freshwater lakes, an enclosed bay, extensive reef systems, offshore rocks and islands such as Anangula, a complex coastline, and a varied submarine topography are prominent physical features contributing to the exploitational area of the people who lived at Chaluka and in nearby sites.

The characterizing artifacts of the Chaluka midden consist of a preponderance of chipped stone tools, such as scrapers, knives, thrusting points, end points for slotted harpoon heads, and spear points. Lamellar blade tools (unifacial) and their derivatives occur mostly in the lower levels and diminish in frequency in the younger levels. Ground slate occurs sparsely, and only in the youngest levels. Whalebone and ivory harpoon heads of several kinds occur in substantial numbers. The predominant forms are detachable (nontoggle) harpoon heads. Toggle heads, in contrast, are more characteristic of Kodiak and other Eskimo cultures. Chipped stone end points in harpoon and spear heads, and in arrow points, indicate another distinctive combination. No side blades have been found, in contrast to some Eskimo cultures. Stone dishes (shallow stone vessels), stone pots (deep, with carbon on the outside), and lamps (shallow basins with carbon on the inside) are found throughout, but with variation in form of the lamps especially marked. Compound fishhooks and sinkers show temporal changes. The importance of woodworking throughout the last 4,000 years is demonstrated by whalebone splitting wedges, chipped-stone adz blades, and whalebone adz heads. Labrets (ivory lip decorations) and red ochre (on grinders and pallette stones, and on human skeletons) occur throughout. Whalebone root-diggers attest to the use of land plants.

Faunal remains complete the picture of intensive exploitation of the varied habitats. Sea-lion bones, including some with stone points embedded in them, and bones of harbor seal, fur seal, and sea otter are represented in large numbers (Lippold, 1966). Fish bones correspond to the fish spears, as well as the fishhooks and sinkers. Large deposits of sea-urchin tests and spines document the continual collection of invertebrates from the reefs. Bird bones of many species—albatross, ducks, geese, swans, and especially large numbers of cormorants and some puffin (both cormorant and puffin were used for bird-skin parkas until recent time)—correspond to the long prongs of the three-pronged bird spear and the bolas, and of course may reflect several other methods of securing birds not indicated in surviving artifacts but known from ethnographic observations. The sinew snare with wood stake, used for trapping puffins in contemporary times, is one such method that leaves no artifactual record. Though Aleuts did not customarily heat their houses, lamps were used for light and may have been used for heating in earlier times. Burials in which artifacts have been placed with the skeleton complete the picture of the way of life of these Mongoloid people.

Many style changes occur throughout the site, and some expirations and new occurrences are also represented, along with basic artifacts that apparently have not changed in their recoverable portions. Changes occur in fishhook shanks, lamps, labrets, harpoon styles, insertion of chipped-stone end points (later styles have a basin rather than a slot), sinkers, harpoon sockets, and houses (early houses were possibly skin tents with low stone walls about their rounded or oval perimeters). Side prongs for the leister, a kind of fish spear, were a new kind of artifact added some 2,000 years ago, as were the ground-slate knife blades, added only a few hundred years ago.

In reading this record of artifacts, cultural features, faunal remains, and human skeletons, it is clear that at all levels there is a good correspondence between the archaeological materials and the exploitation of the natural resources. Two important points relevant to earlier people living on the land bridge can be derived from this evidence. First, though there have been many style changes, with some additions and deletions in the kinds of artifacts used, the overall way of life has not changed in 4,000 years. Whether halibut were caught with a compound fishhook whose shank was made from a seal rib (late style) or from an elbow-shaped piece of whalebone (earlier style) appears to have made no difference in the number of halibut caught. With W. G. Reeder, under whose direction the faunal analysis (Lippold, 1966) was made, I would emphasize that the evidence indicates that a variety of methods can be used for exploiting a single complex area, and that there is no tight or sensitive correspondence between artifacts and the particular objectives for which they were used. Any animal can be killed in a variety of ways, and any harpoon head can kill a variety of animals. Extrapolating this conclusion to a wider area in the Arctic, the toggle-head harpoon has no proved advantage over a simple detachable harpoon. Thus, the style changes do not necessarily represent improved or more efficient adaptation to the area. The second point consists of the observation by Reeder that the Aleuts lived well and in fact did not intensively exploit all possible resources (oral communication, October 16, 1965). For example, they generally did not bother with several species of fish common to the reef and shallow water, though small numbers may have been employed in times of scarcity.

Chaluka begins with an advanced culture, compared with that of Kodiak Island (Clark, 1966) or Cook Inlet (Laguna, 1934), which suggests a long preceding period of adaptation to this area. The level of culture displayed at Chaluka has even been used by one authority to claim an earlier date for a less evolved culture, thus: "However, if we accept a date of about 1,068 B.C. ± 280 years for a hearth well above the bottom of Laughlin's Chaluka site in the Aleutians, throwing in a few extra centuries to accommodate the first occupation, and if we are not afraid to argue by analogies and extrapolations, we may claim Kachemak Bay I as still older, since the Chaluka material gives every appearance of being more evolved" (Laguna, 1962, p. 166). Although the underlying assumption of the uniformity of rates of cultural evolution is dubious if not completely fallacious, the comparison of the Aleutian culture with the Eskimo culture of Kachemak Bay does illustrate the magnitude of difference separating Aleut from Eskimo culture.

Differences between Aleut and Eskimo in time depth, lithic industries, art styles, houses, physical characteristics, and language are too great to permit these two people to be lumped together for analytical purposes where both differences and similarities must be considered. Internal migrations are possible at many times in the past 4,000 years, but an immigration from other gene pools appears unlikely. Gene flow through intermarriage cannot be ruled out, but the amount of gene flow from the neighboring Koniag Eskimos was evidently minor.

The first real break in the archaeological record is the 3,000-to-4,000-year gap between the Anangula occupation and the lower levels of Chaluka. Four kinds of evidence suggest continuity between the populations of Chaluka and Anangula. These include artifactual similarities, a significant decline in the proportion of lamellar blade tools between the lower and upper levels of Chaluka, the proximity between the two sites, and the buffering or insulation against intrusions from the outside that is a consequence of their remote position.

The artifactual similarities between Anangula and early Chaluka consist of lamellar blades, retouched knives, pointed blades (gravers), and scrapers. In addition to these, there are the one stone vessel from Anangula, rubbing stones, red-ochre stains, and both scoria and pumice abraders. Though this list of shared artifacts is adequate for an interpretation of continuity between the two sites, there are absences from Chaluka that require explanation. No polyhedral cores, no platform tablets, no blades with transverse blows, and none of the larger lamellar blades have been found to date. Assuming a ratio of some 40 blades to one core, few cores would be expected in the Chaluka mound on the basis of the small numbers of lamellar blade tools recovered. Manufacture of the blades away from the excavated area might also be a factor. Absence of blades with transverse blows (angle burins), may be due either to small sample size or to the fact that their manufacture had been abandoned.

## EVIDENCE FOR CONTINUOUS OCCUPATION OF THE EASTERN ALEUTIANS SINCE LAND BRIDGE TIMES

The proposition that living Aleuts are descendants of people who lived on the Bering Land Bridge is a hypothesis sufficiently likely and important to require continued and intensive investigation. Many lines of evidence support this hypothesis. Working backward in time from the present inhabitants, we have established that modern Aleuts at Nikolski, some of whom still occupy a part of the Chaluka mound, are in fact living on top of their own ancestors. Some Unalaska Aleuts moved onto Umnak at least as early as the first half of the nineteenth century; but these and other relocations effected directly or indirectly by early Russian administrators were merely rearrangements within the Eastern Aleut gene pool. Ethnographic information, historical records, the archaeological evidence of continuity between living and dead, the contents of the mummy caves, comparisons of measurements of the living Aleuts with those of the skeletons, and comparison of a variety of discontinuous traits such as many found on the dentition, leave no doubt about the continuity, genetic and cultural, between living and dead Aleuts.

Uninterrupted continuity back through the preceding 4,000 years is indicated in the archaeological record of Chaluka. Though the early and late Aleut skulls differ in general proportions (Fig. 3), the magnitude and nature of the differences do not indicate a major genetic change; such differences could easily be induced by a single alteration in growth rate. Many more skeletons, especially from the earlier levels, are needed for finer analysis. Neither the early nor the late Aleutian skulls closely resemble those of early or late Koniag Eskimos.

Lamellar blade tools are found in Chaluka primarily in the lower levels; they constitute 77 per cent in the lower two-fifths of the deposits and only 6 per cent in the top fifth, for a series of 69 such tools (Laughlin and Marsh, 1956, p. 10). A similar decline is even more evident in more recent excavations at Chaluka. This decline in frequency indicates the possibility of a greater frequency during the period between Chaluka and Anangula, and therefore that the Anangula tradition for working stone diminished over the millennia.

The proximity of the two sites, occupying, as they do, opposite sides of a single body of water (Nikolski Bay and the channel now defined by Anangula and Umnak), in itself suggests a relationship. They are visible to each other and were separated by only a short walk at the time that Anangula was inhabited. The most reasonable interpretation is that the ancestral Aleuts have inhabited the south end of Umnak and Anangula continuously throughout the last 8,500 years and that other sites in this region will eventually fill in the intervening period, though some of the early sites in the sequence are undoubtedly submerged.

The Umnak region is now remote in terms of the multiple contacts ordinarily available to people who live on one section of a continental coast or who are surrounded by other people. Occupation at Port Moller, which is identified as Aleut on the basis of chipped-stone artifacts and two Paleo-Aleut skulls, begins only about 3,000 years ago (Laughlin, 1966b). The occupation of Kodiak Island extends back at least 5,500 years (radiocarbon dates communicated to Laughlin from R. J. Stuckenrath, 1965; also see Clark, 1966). Whether glacial ice prevented earlier population spread to the east from the land bridge, whether the land simply failed to develop sufficiently attractive habitats after becoming free of ice, or whether too few investigations have been made, the occupation of the Alaska Peninsula–Kodiak area appears to be substantially more recent than that of the Aleutians. The earliest artifacts found in the Koniag area clearly belong to other traditions than those in the Aleutians. Thus the Aleuts inhabiting the Umnak corner of the land bridge were rather isolated and well buffered against contact with other peoples. That their isolation ward consisted of a zoological garden rather than an impoverished marginal area contributes substantially to the explanation of their continuation in the area and their florescence.

## Significance of the Commander Islands for the Coastal Route

The idea of a migration into the western Aleutian Islands directly from Asia by way of the Commander Islands enjoyed popularity earlier in this century and still recurs occasionally in the literature. Conversely, a migration from the Aleutians to Kamchatka has also been postulated. In fact, a direct Asiatic origin has been employed to explain the dissimilarities between Aleuts and Eskimos. A map illustrating such a migration, from Kamchatka to the Commander Islands of the USSR to the Aleutian Islands, may be seen in *Anthropological Survey in Alaska* (Hrdlička, 1930, p. 360). Hrdlička revised his views following his researches in the Aleutian Islands and during his 1938 study of the Commander Islands (Hrdlička, 1945). He found no evidence of occupation prior to the stranding of Bering's party in 1741. Various Russian scholars who have studied the Commander Islands also failed to find evidence of pre-Russian occupation. Resident sea lions, seals, and sea otters, established rookeries of fur seals, Arctic fox, a large variety of birds and fish, as well as the sea cow, made these two islands among the richest in the world. Thus, the absence of any evidence of occupation suggests that no one had lived here prior to the arrival of the *St. Peter* in 1741.

There is some positive evidence that pre-Russian occupation of these two islands, Bering and Medni, is quite unlikely. Steller (Golder, 1922, 1925) observed that the animals of Bering Island (the westernmost of the Commander Islands) appeared never to have seen human beings before the arrival of Bering's party. These observations can be extended and given considerable time depth by the presence of two extremely vulnerable endemic animals when the Commander Islands were first discovered. The most remarkable was the great northern sea cow or manatee, *Hydrodamalis gigas*. This large, gluttonous, and edentulous seaweed eater was easily killed by hunters and became extinct in 1768 or soon after; its tasty meat had the additional advantage of keeping well, even in warm weather. Another form unique to the Commander Islands was the Spectacled Cormorant (*Phalocrocorax perspicillatus*), a practically flightless bird exterminated about 1850. It is safe to say that these two forms would not have long survived efficient marine hunters, such as the Aleuts or the Eskimos.

The riverine focus of the inhabitants of the Kamchatka Peninsula provides still another important argument against any intercontinental migrations by way of the Aleutian and Commander Islands. There is no tradition for marine hunting, and the archaeological record has now been extended to 10,675 ± 360 years at the Ushki site, inland on the Kamchatka River (Dikov, 1965, p. 13).

Finally, the Aleutian evidence in itself—linguistic, morphological, serological, archaeological, and ethnographic—indicates clearly that the flow of both genes and culture traits has been from the eastern Aleutian area of high population density to the western Aleutian area of low population density. The Aleutian Islands have never served as "stepping-stones" from Asia to America. The Commander Islands afford adequate evidence that the aboriginal inhabitants of Kamchatka at no time succeeded in spanning the 97 miles between Cape Kamchatka and Bering Island.

## Summary

The Bering Land Bridge provided two distinct kinds of routes for migration into North America—coastal and interior. It also provided permanent residence, especially for inhabitants of the southern coast. Conditions in the interior were severe, and likely only a few of its inhabitants found their way into North America; these wanderers probably became the ancestors of the American Indians. Coastal conditions were ecologically much more congenial to human occupation; coastal settlers became the ancestors of the Aleuts and the Eskimos. All migrations into the New World moved across either the coastal or the interior portions of the land bridge, or both; none entered by way of the Commander and Aleutian Islands.

The biological affinities of the Aleuts and Eskimos, in both serological and morphological traits, lie more with Asiatic Mongoloids than with American Indians. Coastal and interior adaptations on the Bering platform provided the geographical separation necessary for the evolutionary divergence of Mongoloids and American Indians. These two kinds of ecological adaptation precluded extensive contact between the two kinds of people. The entire coastal area of Bering Sea is now, and likely has been since its initial occupation, occupied only by Mongoloid peoples.

American Indians display unambiguous evidence of their Asiatic origin in the biology of both the living populations and the earliest skeletons. American Indians are "proto-Mongoloid" or less "Mongoloid" than are contemporary Asiatic Mongoloids, including Aleuts and Eskimos. Though important regional and temporal variations do exist, a provisional estimate of the amount of variation of Indians, coupled with their resemblance to Asiatic peoples, does not indicate any considerable antiquity. The variation within American Indians is compatible with an estimated time depth of 15,000 years. This degree of homogeneity, however, cannot be taken to exclude earlier occupation. Ecological separation, rather than separation in migration time, is adequate to explain the differences between Indians on the one hand, and Aleuts and Eskimos on the other.

Anangula and adjacent Umnak Island once formed the western terminus of the old Alaska Peninsula, which was then (ca. 11,000 years ago) the southern margin of the Bering Land Bridge. Samalga Pass and the Peninsula promontory were an "ecological magnet" for sea mammals, fish, invertebrates, and birds throughout the time when the Bering Land Bridge was emergent. Some of the sea mammals, such as the sea otter, sea lion, and fur seal, evolved in the North Pacific, and along with the walrus they have always been available to coastal peoples. The amount and direction of ocean currents and the extensive upwelling have been key elements enriching the ecosystem. The annually sustained abundance of intertidal foods and their accessibility to women, children, and aged persons, as well as to the more active hunters, have been of crucial importance in the maintenance of the coastal economy, and explain in large part the relatively high population densities of the Aleuts.

The Anangula Island unifacial core and blade industry of 8,000 years ago is in some respects more similar to pre-ceramic industries of Japan and Siberia, dated at 9,000 to 13,000 years ago, than to later Alaskan industries such as the Denbigh microblade and bifacial complex, an antecedent or early component of Eskimo culture.[e] The hypothesis that the Aleuts are descended from early occupants of the part of the Bering Land Bridge now forming the Aleutian Islands east of Samalga Pass is supported by indications of continuity between the village site of Chaluka, occupied for some 4,000 years, and the Anangula site, occupied 8,400 years ago. Owing to the ecological wealth of the area and to steep shore profiles, the ancestral Aleuts were able to remain in the eastern Aleutians as the western portion of the peninsula was converted into islands by rising water levels. People inhabiting lower and more northerly portions of the land bridge were forced to withdraw, and most of the old sites probably are now under water.

Occupants of the land bridge coast could retain their coastal-marine ecology and their Mongoloid racial identity by remaining at the coast as it moved inland. In fact, the southern perimeter of the Bering Land Bridge expanded greatly in absolute length as well as in proportion to the shrinking land mass. This increase in coastline favored the numerical expansion of the coastal-adapted ancestors of the Bering Sea Mongoloids—the Aleuts, Eskimos, Chukchi, and Koryak, and possibly some of the Kamchadals.

Occupants of the interior, the ancestors of the American Indians who may still have been living in the interior, were automatically forced to move into America or Siberia or in both directions at the time of the creation of Bering Strait. There is no evidence that any moved back into Siberia, and sparse skeletal evidence suggests that they had already moved into America.

Though Aleut and Eskimo are mutually unintelligible, the linguistic similarities between the two languages suggest that there may have been a common language at a remote period. The greatest linguistic diversity among Eskimos occurs within southern Alaska, and this is indicative of greater time depth of Eskimo occupation there than elsewhere in the Eskimo area. There is some evidence, tentative and not yet rigorously analyzed, that

---

[e] Thus, Müller-Beck (this volume, p. 403) characterizes the Anangula assemblage as a subgroup of his "Aurignacoid" industries. Bifacial tools and, especially, bifacial projectile points are lacking.

Chukchi and Koryak are related to Aleut and Eskimo. This is consistent with the anthropological and ethnological evidence and suggests the possibility that the ancestors of the Chukchi also may have lived on the coast of the Bering Land Bridge.

The hypothesis that the ancestors of the Aleuts and the Eskimos actually lived on the coast of the Bering Land Bridge and that the ancestral Eskimos were forced to withdraw as the water levels rose while the Aleuts remained in place is supported by a diverse body of evidence, including the early lithic remains, the biological differences between Aleuts and Eskimos, the linguistic distinctiveness, the ecological base, and the evidences of continuity in one area. Nevertheless, this is a hypothesis that needs considerably more evidence, such as actual skeletons of older inhabitants and earlier sites closer to or on the old coastlines, and it needs to be examined and tested in company with alternative hypotheses.

NOTE ADDED IN PROOF: After this essay was completed, Miss Jean Aigner and I had the good fortune to examine several artifacts and photographs of others from the oldest levels of the Onion Portage site on the Kobuk River in northwestern Alaska. "Band 8" at Onion Portage is about 8,000 years old, and the "Kobuk Terrace" component may be somewhat older. Both assemblages differ significantly from the Anangula materials and reinforce the conclusion that differently oriented regional cultures coexisted in Alaska at an early time level. The Onion Portage materials were provided through the kindness of Mrs. E. B. Giddings and Mr. Douglas Anderson.

## REFERENCES

Anderson, J. E. 1965. Human skeletons of Tehuacan: Science, v. 148, p. 496–497.

Birdsell, J. B. 1951. The problem of the early peopling of the Americas as viewed from Asia, p. 1–68a, *in* W. S. Laughlin, ed., The physical anthropology of the American Indian: Viking Fund, New York.

Black, R. F. 1966. Late Pleistocene to Recent history of Bering Sea–Alaska coast and man: Arctic Anthro., v. 3, p. 7–22.

Black, R. F., and W. S. Laughlin. 1964. Anangula: A geological interpretation of the oldest archeologic site in the Aleutians: Science, v. 143, p. 1321–1322.

Clark, D. W. 1966. Perspectives in the prehistory of Kodiak Island, Alaska: Am. Antiquity, v. 31, no. 3, pt. 1, p. 358–371.

Colinvaux, P. A. 1964. The environment of the Bering Land Bridge: Ecol. Monogr., v. 34, p. 297–329.

Collins, H. B. 1964. The Arctic and Subarctic, p. 84–114 *in* J. D. Jennings and E. Norbeck, eds., Prehistoric man in the New World: Univ. Chicago Press.

Dahlberg, A. A. 1963. Analysis of the American Indian dentition, p. 149–177 *in* D. R. Brothwell, ed., Dental anthropology: Macmillan, New York.

Dikov, N. N. 1965. The stone age of Kamchatka and the Chukchi Peninsula in the light of new archaeological data: Arctic Anthro., v. III, no. 1, p. 10–25.

Fewkes, J. W. 1912. The problems of the unity or plurality and the probable place of origin of the American aborigines: Am. Anthropologist, v. 14, n.s., p. 1–30.

Gessain, R. 1960. Contribution à l'anthropologie des Eskimo d'Angmagssalik: Med. om Grønland, v. 161, no. 4.

Giddings, J. L. 1964. The archaeology of Cape Denbigh: Brown Univ. Press, Providence, R.I., 331 p.

Golder, F. A. 1922, 1925. Bering's voyages: American Geographical Society, Research Series, New York, 2 v.

Haldane, J. B. S. 1956. The argument from animals to men: an examination of its validity for anthropology: Jour. Roy. Anthro. Inst., v. 86, pt. 1, p. 1–14.

Hopkins, D. M. 1959. Cenozoic history of the Bering Land Bridge: Science, v. 129, p. 1519–1528.

Hopkins, D. M., F. S. MacNeil, R. L. Merklin, and O. M. Petrov. 1965. Quaternary correlations across Bering Strait: Science, v. 147, p. 1107–1114.

Howells, W. W. 1959. Mankind in the making: the story of human evolution: Doubleday, New York, 382 p.

Hrdlička, A. 1930. Anthropological survey in Alaska: Forty-sixth Annual Report (1928–29): U.S. Bur. Am. Ethnology, p. 21–654.

———— 1945. The Aleutian and Commander Islands and their inhabitants: The Wistar Institute of Anatomy and Biology, Philadelphia, 630 p.

Irving, W. N. 1962. A provisional comparison of some Alaskan and Asian stone industries, p. 55–68 in John M. Campbell, ed., Prehistoric cultural relations between the arctic and temperate zones of North America: Arctic Inst. N. Am. Tech. Paper no. 11.

Jørgensen, J. B. 1953. The Eskimo skeleton: Med. om Grønland, v. 146, no. 2, 154 p.

Laguna, Frederica de. 1934. The archaeology of Cook Inlet, Alaska: University Museum, Univ. Penn. Press, Philadelphia.

———— 1962. Intemperate reflections on Arctic and Subarctic archaeology, p. 164–169 in J. M. Campbell, ed., Prehistoric cultural relations between the arctic and temperate zones of North America: Arctic Inst. N. Am. Tech. Paper no. 11.

Laughlin, W. S. 1950. Blood groups, morphology and population size of the Eskimos: Cold Spring Harbor Symposia on Quantitative Biology, v. XV, p. 165–173.

———— 1951. The Alaska gateway viewed from the Aleutian Islands, p. 98–126 in W. S. Laughlin, ed., The physical anthropology of the American Indian: Viking Fund, New York.

———— 1956. Human skeletal material from Kawumkan Springs midden: Trans. Am. Philos. Soc., v. 46, n.s., pt. 4, p. 475–480.

———— 1958. Neo-Aleut and Paleo-Aleut prehistory: Proc. 32d Internat. Cong. Americanists (Copenhagen), 1956, p. 516–530.

———— 1962. Generic problems and new evidence in the anthropology of the Eskimo-Aleut stock, in J. M. Campbell, ed., Prehistoric cultural relations between the arctic and temperate zones of North America: Arctic Inst. N. Am. Tech. Paper no. 11.

———— 1963. Eskimos and Aleuts: their origins and evolution: Science, v. 142, p. 633–645.

———— 1965a. Review of H. B. Collins, The Arctic and Subarctic, p. 84–114 in J. D. Jennings and E. Norbeck, eds., Prehistoric man in the New World (Univ. Chicago Press, 1964): Am. Antiquity, v. 30, p. 501–503.

———— 1965b. Review of J. L. Giddings, The archaeology of Cape Denbigh (Brown Univ. Press, Providence, R.I., 1964) : Arctic, v. 18, no. 1, p. 61–64.

———— 1966a. Genetical and anthropological characteristics of Arctic populations, p. 469–495 *in* J. Weiner and P. Baker, eds., The biology of human adaptability: Oxford Univ. Press (in press).

———— 1966b. Paleo-Aleut crania from Port Moller, Alaska Peninsula: Arctic Anthro., v. 3, p. 154.

Laughlin, W. S., and Jean S. Aigner. 1966. Preliminary analysis of the Anangula unifacial core and blade industry: Arctic Anthro., v. 3, p. 1–56.

Laughlin, W. S., and G. H. Marsh. 1954. The lamellar flake manufacturing site on Anangula Island in the Aleutians: Am. Antiquity, v. 20, no. 1, p. 27–39.

———— 1956. Trends in Aleutian chipped stone artifacts: Anthro. Papers Univ. Alaska, v. 5, no. 1, p. 5–21.

Laughlin, W. S., and W. G. Reeder. 1962. Revision of Aleutian prehistory, Science, v. 137, p. 856–857.

Levin, M. G. 1963. Ethnic origins of the peoples of northeastern Asia: page *in* H. N. Michael, ed., Anthropology of the North, Translations from Russian Sources, no. 3: Arctic Inst. N. Am., Univ. Toronto Press.

Lippold, Lois K. 1966. Chaluka: the economic base: Arctic Anthro., v. 3, p. 125–131.

Marsh, G. H., and W. S. Laughlin. 1956. Human anatomical knowledge among the Aleutian Islanders: Southwestern Jour. Anthro., v. 12, no. 1, p. 38–78.

Mayr, E. 1963. Animal species and evolution: Harvard Univ. Press, Cambridge, 797 p.

Merbs, C. F. 1963. The Sadlermiut Eskimo vertebral column: master's thesis (anthropology), University of Wisconsin, 1963.

Milan, F. A. 1963. An experimental study of thermoregulation in two Arctic races: Ph.D. thesis (anthropology), Univ. Wisconsin; University Microfilms, Inc., Ann Arbor, Mich.

Moorrees, C. F. A. 1957. The Aleut dentition: A correlative study of dental characteristics in an Eskimoid people: Harvard Univ. Press, Cambridge, 196 p.

Neel, J. V., and F. M. Salzano. 1964. A prospectus for genetic studies of the American Indian: Cold Spring Harbor Symposia on Quantitative Biology, v. XXIX, p. 85–98.

Newman, M. T. 1953. The application of ecological rules to the racial anthropology of the aboriginal New World: Am. Anthropologist, v. 55, no. 3, p. 311-327.

———— 1957. The physique of the Seneca Indians of Western New York State: Jour. Washington Acad. Sci., v. 47, no. 11, p. 357–362.

———— 1958. A trial formulation presenting evidence from physical anthropology for migrations from Mexico to South America: p. 33–46 *in* R. H. Thompson, ed., Migrations in New World culture history: Univ. Arizona Soc. Sci. Bull. no. 27.

———— 1962. Evolutionary changes in body size and head form in American Indians: Am. Anthropologist, v. 64, no. 2, p. 237–257.

Sapir, E. 1916. Time perspective in aboriginal American culture: a study method: Canada Department of Mines, Geological Survey, Memoir 90, Anthropological Series no. 13, Ottawa. 87 p.

Scheffer, V. B. 1958. Seals, sea lions, and walruses: a review of the Pinnepedia: Stanford Univ. Press, Stanford, Calif.

Spuhler, J. N. 1951. Some genetic variation in American Indians: p. 177–202 *in* W. S. Laughlin, ed., The physical anthropology of the American Indian: Viking Fund, New York.

Stewart, T. D. 1960. A physical anthropologist's view of the peopling of the New World: Southwestern Jour. Anthro. v. 16 no. 3, p. 259–273.

Stewart, T. D., and M. T. Newman. 1951. An historical resume of the concept of differences in Indian types: Am. Anthropologist, v. 53, no. 1, p. 19–36.

Suzuki, H. 1958. Changes in the skull features of the Japanese people from ancient to modern times: Selected papers of the Fifth International Congress of Anthropological and Ethnological Sciences, Philadelphia, 1956. Univ. Pennsylvania Press, Philadelphia.

Swadesh, M. 1962. Linguistic relations across Bering Strait: Am. Anthropologist, v. 64, no. 6, p. 1262–1291.

Turner, C. G. 1965. Aleut dental evolution: paper delivered at 34th Annual Meeting, American Association of Physical Anthropologists, Pennsylvania State University.

Udvardy, M. D. F. 1963. Zoogeographical study of the Pacific Alcidae, p. 85–111 in J. L. Gressitt, Pacific Basin Biogeography: Bishop Museum Press, Honolulu, 561 p.

Zenkevich, L. 1963. Biology of the seas of the U.S.S.R.: Interscience, New York.

Zolatoreva, I. M. 1965. Blood group distribution of the peoples of Northern Siberia: Arctic Anthro., v. III, no. 1, p. 26–33.

## 24. The Cenozoic History of Beringia—A Synthesis

DAVID M. HOPKINS
*U.S. Geological Survey, Menlo Park, California*

The many threads of evidence, inquiry, and hypothesis in the preceding pages pattern themselves into a tapestry of history—the history of the landscape of Beringia, at one time dominated by the sea, at another by a great plain; once covered with forest, then with treeless tundra; once populated by mammoth, horse, and bison, hunted by paleolithic man; now the home of seal, walrus, and polar bear, sought by the world's most skillful sea-mammal hunters. Beringian history is now known much more completely and in far greater detail than when I attempted an earlier historical synthesis in 1959; yet the new data, by the very process of solving old problems, present new questions. And paradoxes still exist, alerting us that much of our present data remains incomplete and in need of amplification. In this final, synthetic chapter, I shall try to set down the Cenozoic history of Beringia more or less chronologically, pointing out along the way the more painful paradoxes and the more exciting questions to be resolved by the next few years' research.

### Early Tertiary Land Bridges and Land Barriers

Compelling biogeographic evidence has long indicated the former existence in high northern latitudes of land connections between the Old and New Worlds. In 1959 it seemed to me that the only plausible site for a former land connection at *any* time within the Tertiary Period lay in Beringia. The evidence that has accumulated in the years since emphatically supports the existence of a Beringian land bridge from Asia to northwestern North America throughout the early and middle parts of the Tertiary Period. But, indeed, the case for an early Tertiary transatlantic or transarctic land connection between Europe and northeastern North America has also been greatly strengthened.

## POSSIBILITY OF TRANSATLANTIC OR TRANSARCTIC
## LAND CONNECTIONS

The virtual proof of ocean-floor spreading provided by recent geomagnetic studies (Vine, 1966) has heightened the plausibility of the conclusion, reached earlier by many biogeographers (Löve and Löve, 1963; MacNeil, 1965, p. 6–7; Kurtén, 1966) that Europe and northeastern North America were once either contiguous or joined by a land connection on the site of the present North Atlantic Ocean. The demonstration that the mid-Atlantic rift and its symmetrical systems of parallel magnetic anomalies extend into the Eurasian Basin of the Arctic Ocean (Heezen and Ewing, 1961; King, Zietz, and Alldredge, 1966) suggests that the Arctic Ocean itself may be a relatively recent feature of the earth's crust. But it is clear that there were no transatlantic or transarctic land connections between North America and Europe after early Tertiary time (Lindroth, 1957).

### EARLY TERTIARY HISTORY OF BERINGIA

Regardless of what was happening elsewhere in the north, Beringia evidently lay above sea level throughout most of early and middle Tertiary time. Land mammals were exchanged more or less freely between the Old and New Worlds (Simpson, 1947), and temporary inhibitions of these migrations seem to have been caused more by the presence of an epicontinental sea that bisected the Eurasian continent just east of the Urals, than by any extensive marine invasion of Beringia (Kurtén, 1966, Fig. 1).

Newly evolved species of land plants were exchanged across Beringia as readily as land-mammal populations, and during early and middle Tertiary time, when climatic zonation was not pronounced, Beringia was part of a series of floral provinces that extended far southward on both shores of the Pacific Ocean (Wolfe and Hopkins, 1967). As late as Seldovian (late Oligocene to middle Miocene) time, Beringia lay at the apex of a region of mixed mesophytic forests that encircled the North Pacific from Japan to present-day Oregon (Wolfe and Leopold, this volume).

The total lack of a seaway across Beringia during early and middle Tertiary time resulted in a partitioning of marine faunas that is dramatized by the occurrence at Camden Bay, on the Arctic coast of Alaska, of a small middle or late Miocene molluscan fauna of pronounced Atlantic affinities and by a complete lack of mollusks of Atlantic affinities in marine sediments of the same age along the Pacific coast of Alaska (MacNeil, 1957, 1965; Miller, 1957) (Fig. 1). Equally dramatic is the partitioning of marine mammalian faunas into a Pacific group that included desmostylians, sea lions, and ancestral walruses but no true seals of the family Phocidae, and an Atlantic group that included true seals but no desmostylians, sea lions, or walruses (Mitchell, 1966; C. A. Repenning, oral communication, March 10, 1967).

And what of the topography? Southern Alaska, the Aleutian Islands, and the Kamchatka Peninsula seem to have been restless, tectonically active regions throughout the Cenozoic Era, the sites of repeated and dramatic topographic changes brought on by intense crustal movements. These regions were no doubt largely mountainous throughout early and middle Tertiary time, just as they are today. But most of Beringia, to the north, seems to have been tectonically quiet during early and middle Tertiary time and to have undergone little topographic change except prolonged, slow subaerial erosion, a process that had reduced much of the region to a peneplain by late Miocene time.

The Beringian peneplain probably attained its highest perfection in the now-submerged areas of Bering and Chukchi Seas, where a monotonous, gently undulating erosional topography was relieved only by a few low monadnocks. These local hills, many of them formed of granite, later became the islands of Bering Strait and northern Bering Sea and the isolated, island-like highlands that rise from a low, wave-abraded surface on the eastern part of St. Lawrence Island. The peneplain extended far eastward and northeastward into the present area of central and northwestern Alaska and probably northwestward into Chukotka as well, a rolling terrain punctuated by broad local highlands underlain by granite, and by sharp ridges perhaps a hundred meters high, marking outcroppings of more resistant limestone. Remnants of this widespread erosional surface persist today in the unglaciated areas of Alaska as an undulating upland surface deeply scored by the modern stream valleys.

The tectonic stability of Beringia was not complete, even during early Tertiary times. Downwarping created local basins in which alluvium and peat began to accumulate; mild uplift created local highlands that became sources of coarse sediment. Basins of nonmarine sedimentation existed in northeastern Alaska and probably in the upper Yukon Valley during the Paleocene, along the middle course of the Yukon River during the Oligocene, and on western St. Lawrence Island, on Seward Peninsula, and along the middle course of the Yukon River during Seldovian (late Oligocene to middle Miocene) time (MacNeil et al., 1961; Wolfe et al., 1966; J. A. Wolfe, unpublished studies of Alaskan fossil floras). Secular subsidence of the southern margin of the Bering continental platform moved the shoreline northward at least as far as the Pribilof Islands, and part of the sediment resulting from the slow denudation of Beringia settled there to form a layered sequence, now gently deformed, that includes marine sandstone of Miocene age (Scholl *et al.*, 1966; D. E. Gershanovich, written communication, April 12, 1966). Tectonic activity in the area of the present Gulf of Anadyr also seems to have begun well before the end of the Miocene Epoch. The apparent lack of mollusks of Atlantic ancestry in the faunas of the Pestsov Suite (Petrov, this volume) suggests that a marine embayment extending far northward from the continental margin had developed there well before the first opening of Bering Strait in late Miocene time.

## Late Tertiary Land Bridges and Seaways

The sudden appearance in early Pleistocene beds in England of a flood of mollusks of Pacific ancestry led me to conclude in 1959 that the first opening of Bering Strait must have taken place just before the beginning of the Quaternary Period, at a time level that we would now place as 3.0 to 3.5 million years ago. But closer scrutiny of the paleontologic record has since made it clear that a temporary seaway must have come into existence much earlier. A very few marine organisms of Atlantic ancestry reached the North Pacific Ocean and a substantially larger number of Pacific organisms reached the North Atlantic during late Miocene time, in an exchange of marine faunas that took place about 10 to 12 million years ago, when endemism in the land-mammal faunas in the Old and New Worlds was becoming acute, evidently in response to the development of a geographic barrier in Beringia. Interestingly enough, both events seem to coincide with the time when the forests of northwestern North America and northeastern Asia first became isolated from one another.

The late Miocene seaway must have been short-lived, however, for there is a clear record of repeated dispersals of land mammals across Beringia through much of Pliocene time, during the interval from 10 to 4 million years ago. When the seaway finally opened again, about 3.5 million years ago, topographic changes seem to have taken place elsewhere on the shores of the Arctic Ocean, permitting a much more dramatic dispersal of marine faunas from the Pacific into the Atlantic Ocean.

### LATE MIOCENE PALEOGEOGRAPHY OF BERINGIA
### AND THE ARCTIC BASIN

Late Miocene land-mammal faunas in North America and Eurasia show a pronounced endemism, in contrast to the more cosmopolitan faunas shared by the two continents earlier in Miocene time and later during Pliocene time. Simpson (1947) showed that the Clarendonian land-mammal age of North

America, now estimated (by potassium-argon dating) to have extended from 12 to 9.9 million years ago (Turner, 1967), was an interval during which very few dispersals took place between the Old and New Worlds; he suggested that the avenue of land communication across Beringia was briefly disrupted at that time.[a] The paleobotanical record supports Simpson's conclusion: Wolfe and Leopold (this volume) show that the temperate mesophytic forests that previously had formed a rather homogeneous belt around the North Pacific were replaced during late Miocene time by rich boreal forests with rather different species compositions in northeastern Asia and northwestern North America.

The evidence that land biotas were partitioned by a late Miocene geographic barrier in Beringia accords nicely with the new evidence that marine organisms were dispersed across Beringia at that time. The late Miocene exchange of marine populations is attested mainly by study of fossil molluscan faunas (Durham and MacNeil, this volume), but the molluscan evidence is dramatically reinforced by the appearance of *Prorosmarus alleni*, an early walrus, on the Atlantic coast of North America in the Yorktown Formation of late Miocene age (Berry and Gregory, 1906).[b] A closely related fossil walrus has been found in beds of about the same age in Baja California (R. A. Tedford and C. A. Repenning, written communication, April 1, 1967). A dispersal of marine mammals through the Beringian seaway in the opposite direction seems to be recorded by the recent discovery near Santa Cruz, California, of the remains of a possible phocid seal in the Santa Margarita Formation[c] of late Miocene and early Pliocene age (C. A. Repenning, written communication, April 24, 1967).

Because no late Miocene or early Pliocene marine beds have yet been discovered within Beringia proper, the precise position of the late Miocene seaway is not yet known. However, the general lack of marine sediments older than late Pliocene along the present coasts of Chukotka and western Alaska suggests that the first seaway across Beringia was narrower than and generally restricted within the present areas of Bering and Chukchi Seas.

---

[a] Simpson followed the practice, current until recently, of assigning the Clarendonian land-mammal age to the early part of the Pliocene Epoch. A reevaluation of the position of the Miocene-Pliocene boundary in North America shows, however, that the Clarendonian land-mammal age is of late Miocene age in the European time scale (Repenning, 1967, and unpublished manuscript).

[b] Interestingly enough, modern large-necked clams of the genus *Mya*, a principal food of the walrus, also make their first Atlantic appearance in the Yorktown Formation (Durham and MacNeil, this volume). These beds have also recently yielded a horse, *Hipparion* cf. *eurystyle*, of Clarendonian or early Hemphillian age (U.S. Geological Survey, 1965, p. A71).

[c] A primitive *Hipparion* of Clarendonian age has been found lower in the same formation.

It seems rather likely that the trans-Beringian seaway was initiated by a
quickening of tectonic movements in late Miocene time, causing a preexisting marine embayment in the area of the present Gulf of Anadyr to expand
gradually northeastward toward the northwestern tip of St. Lawrence Island

Fig. 1. Speculative reconstruction of the paleogeography of Beringia and the Arctic Ocean during late Miocene time. ○ Middle Miocene marine beds in Arctic
area that lack Pacific immigrants. ● Middle Miocene marine beds in Pacific area
that lack Atlantic immigrants. △ Late Miocene and Pliocene marine beds in Atlantic and Arctic area containing Pacific immigrants. ▲ Late Miocene and Pliocene marine beds in Pacific area containing Atlantic immigrants. Cross-hatched
area marks the location of the nonmarine Beaufort formation of Miocene and
Pliocene age in the Queen Elizabeth Islands.

(Fig. 1). The long, low, curvilinear scarps extending north-northeastward from northwestern St. Lawrence Island to and through Bering Strait (Creager and McManus, this volume, Fig. 5) are strongly suggestive of fault scarps—delineating, perhaps, an area of tectonic subsidence along which the late Miocene seaway extended northward through Bering Strait. Alternatively, the Kolyuchinsk-Mechigmen depression, a lowland extending northwestward across northeastern Chukotka (Petrov, this volume, Fig. 1) may have provided a short-lived connection to the Arctic Ocean.

The Arctic coasts of Siberia and Canada extended much farther north during Miocene time than they do today. Until the late Miocene opening of the seaway across Beringia, the adjacent basin of the Arctic Ocean was a nearly enclosed bay of the Atlantic Ocean, a sort of cul-de-sac all but separated from Atlantic waters by North American and Siberian land areas extending as far north as latitudes of 81 to 83°. The northernmost point on the Siberian coast probably lay near the present north tip of the Severnaya Zemlya archipelago; the straits separating these islands from Mys Chelyuskin, the northernmost point on the present-day Siberian mainland coast (lat. 78°), did not come into existence until late Pliocene or early Pleistocene time (Strelkov, 1965, plates 1 and 2). And the Queen Elizabeth Islands, a vast archipelago extending northward to latitude 83°, were a part of the Canadian mainland throughout middle and late Tertiary time. The outermost islands to the northwest are mantled by a thick sheet of alluvial sand and gravel, the Beaufort Formation, which contains pebbles that can only have come from the inner islands of the archipelago and from the Canadian mainland far to the southeast (Tozer, 1956; Craig and Fyles, 1960; Thorsteinsson, 1961; Fyles, 1965). Pollen floras in the lower part of the Beaufort Formation are suggestive of a Seldovian (late Oligocene to middle Miocene) age, but the upper part contains pollen floras of late Pliocene or early Pleistocene age (Terasmae, 1956; Craig and Fyles, 1965). The channels separating the Queen Elizabeth Islands from the Canadian mainland coast developed no earlier than late Pliocene time.

## REESTABLISHMENT OF THE BERING LAND BRIDGE
## DURING THE PLIOCENE EPOCH

The numerous faunal dispersals between the Old and New Worlds during Hemphillian (Pannonian) time and during Blancan (Csarnotan and early Villafranchian) time clearly indicate that the Bering Land Bridge was reestablished throughout most of the Pliocene Epoch (Repenning, this volume). Potassium-argon age determinations on early Hemphillian and early Blancan faunas indicate that dispersals took place across Beringia as early as 10 million and as late as 4 million years ago (Turner, 1967; Repenning, this volume). Dispersals of such forest animals as beaver and flying squirrel indicate that woodlands existed on the land bridge during the Pliocene Epoch. Forests dominated by diverse conifers persisted at high latitudes in both North America and eastern Asia during this interval, but the forests of the two continents had almost no coniferous species in common. The land bridge itself was probably clothed in a mixture of muskeg vegetation and woodlands dominated by birch, aspen, alder, and willow.

One can only speculate on the nature of the geologic events that resulted in the destruction of the late Miocene seaway and the restablishment of land connections across Beringia during much of the Pliocene Epoch. Evidence has recently been uncovered suggesting that worldwide sea level was lowered during Hemphillian time (Webb and Tessman, 1967), presumably in response to some tectonic event that resulted in a change in the total volume of the ocean basins. Such a eustatic reduction in sea level might have sufficed to disrupt the continuity of a shallow trans-Beringian seaway. Alternatively, a narrow and sinuous late Miocene seaway may have been obstructed about 10 million years ago by the development of a delta at the mouth of one or another of the large streams that must have drained western Alaska and northeastern Siberia. Or the seaway may have been disrupted by continuing crustal movements, such as those that have repeatedly affected the Gulf of Anadyr during Tertiary and Quaternary time (Petrov, this volume).

REOPENING OF THE SEAWAY IN LATE PLIOCENE TIME

The reopening of Bering Strait near the end of the Pliocene Epoch is recorded both by local stratigraphic evidence and by biogeographic evidence in more remote areas on the shores of the Atlantic and Pacific Oceans. The presence of *Fortipecten hallae*, a diagnostic Pliocene mollusk, on Alaskan shores both north and south of Bering Strait assures us that the strait was open and that Bering and Chukchi Seas had assumed approximately their present forms by late Pliocene time (Hopkins, this volume). This evidence is supported by the occurrence of *Acila cobboldiae*, a diagnostic late Pliocene–early Pleistocene mollusk, on Sakhalin Island in the Pacific and in Belgium in the Atlantic, as well as by the appearance of many other new immigrants in late Pliocene beds in Belgium, Iceland, and Oregon (Krishtofovich, 1964; Durham and MacNeil, this volume). The new marine immigrants are associated in the North Sea region with mammals of Villafranchian age and in Oregon with mammals of Blancan age; it would seem, then, that Bering Strait must have reopened between 3.5 and 4.0 million years ago, shortly after the Blancan–Csarnotan–early Villafranchian dispersal of land mammals.

EFFECTS OF CHANGING PALEOGEOGRAPHY ON
MOLLUSCAN DISPERSALS

The late Miocene transarctic faunal dispersal was a subtle event, one that had only a minor effect upon either North Atlantic or North Pacific marine faunas. Only a few marine organisms were exchanged, and those few were boreal in affinities and not suggestive of especially warm waters along the migration route. But the North Atlantic molluscan fauna was transformed by the arrival of boreal mollusks from the west at the end of the Pliocene Epoch, even though, oddly enough, the molluscan fauna of the North Pacific again remained little affected. Molluscan dispersals across Beringia may have been inhibited during late Miocene time by a series of critical thresholds and ecological barriers—low-salinity segments, shallow segments, and segments with uniform substrate—in a narrow seaway of great length. But ecological barriers in Beringia can hardly be invoked to explain the gross imbalance in the ratio of Pacific-to-Atlantic and Atlantic-to-Pacific dispersals when the seaway opened again in late Pliocene time. The answer must be sought, instead, in the changing geography of the Arctic Basin.

At the time the late Miocene seaway opened across Beringia, we may assume that a rich Pacific-related fauna resided in the area of present-day Bering Sea; each ecological niche would have been occupied by mollusks closely adapted to its special environmental conditions. The western Arctic Ocean, on the other hand, probably contained only a small and impoverished fauna; only those Atlantic species capable of surviving as far north as latitudes 81 to 83° would have been able to disperse along the Siberian coast past Severnaya Zemlya or along the Canadian coast past the north tip of Ellesmere Island. The low degree of competition would have permitted individual taxa to occupy ecological niches to which they were relatively poorly adapted, and some ecological niches may have been entirely unoccupied. If these assumptions are correct, then dispersals through the late Miocene seaway across Beringia were probably limited not so much by the character of the seaway itself as by the ability of the species constituting each fauna to compete in the new environment on the other side. Only a few of the mollusk species resident in the adjacent reaches of the Arctic Ocean would have been able to establish themselves in the rich and diversified fauna of southwestern Bering Sea and the northern Pacific Ocean, whereas many of the mollusk species invading from the North Pacific would have readily established themselves in the Arctic Ocean because the ecological niches to which they were best adapted were either vacant or occupied by poorly adapted Atlantic species. However, only those few Pacific taxa capable of surviving the rigors of the Ellesmere or Severnaya Zemlya coasts would have been able to penetrate to the North Atlantic Ocean.

Thus, the North Pacific, Arctic, and North Atlantic faunas reached a new equilibrium—probably in a very short time—in which North Pacific and North Atlantic faunas remained relatively little affected by the opening of the late Miocene seaway across Beringia, but the fauna of the adjacent Arctic Ocean probably became essentially Pacific in character. This equilibrium would have persisted with little change during the lengthy interval, from 10 until 4 million years ago, during which the Bering Land Bridge was reestablished, isolating the Arctic Ocean once more from southwestern Bering Sea and the Pacific Ocean.

Because the fauna of the adjacent reaches of the Arctic Ocean was now largely Pacific in character, the reopening of Bering Strait near the end of the Pliocene Epoch would have had little effect upon either Pacific or Arctic faunas. The spectacular dispersal, just before the Pleistocene began, of many Pacific-related mollusks into the North Atlantic Ocean must have resulted not from the reopening of Bering Strait but from the development of the passages through the Queen Elizabeth Islands, providing continuous channels of marine communication at latitudes as low as 72° N. A corresponding westward dispersal of Atlantic mollusks would have been inhibited by the fact that those most suited to life in high latitudes were already represented by closely related forms that had reached the western Arctic much earlier in Tertiary time. Moreover, the imbalance of migrations would have been heightened by the prevalence of east-setting currents through the Queen Elizabeth Islands and by the length and rigor of the dispersal route along the north coast of Siberia past Severnaya Zemlya, where east-setting currents would otherwise favor Atlantic-to-Pacific migrations.

## Quaternary Paleogeography

### SHIFTING SEA LEVELS AND INTERMITTENT LAND CONNECTIONS

The Quaternary Period, which evidently began at least 3 million years ago, saw the beginning of a series of major worldwide climatic fluctuations that repeatedly produced large glaciers in land areas (Einarsson et al., this volume; Curry, 1966; MacDougall and Wensink, 1966; Opdyke et al., 1966; Savage and Curtis, 1967) and concomitant fluctuations in sea level. On at least six occasions, sea level stood high enough during interglacial or interstadial episodes to recreate the seaway across Beringia and through Bering Strait (Hopkins, this volume; McCulloch, this volume).

Sea level would have to fall only 46 meters below its present position to expose a narrow land connection between Chukotka and Alaska by way of St. Lawrence Island; a reduction to −50 meters would expose a second narrow connection north of Bering Strait; and a reduction to −100 meters would expose almost the entire area of the Bering-Chukchi continental platform (Creager and McManus, this volume, Fig. 5). Sea level did fall considerably below −100 meters during parts of the Wisconsin (or Würm) and Illinoian (or Riss) Glaciations, and a broad land bridge undoubtedly existed across Beringia at these times. But less is known of the extent of sea-level lowering during early and middle Quaternary glacial episodes. The possible effects of changes in the topography of northwestern Bering Sea must also be considered; regional uplift and the deposition of glacial moraines there have resulted in a slight reduction in the amount by which sea level would have to be lowered in order to establish intercontinental land connections during the Illinoian and Wisconsin Glaciations, as compared with earlier Pleistocene glacial episodes (Creager and McManus, this volume; Petrov, this volume; Sainsbury, this volume). Indeed, we cannot be certain that the land bridge was reestablished during each of the ten or more worldwide glaciations inferred by Einarsson *et al.* (this volume). However, the record of dispersals of land mammals between the Old and New Worlds assures us that the land bridge was reestablished on at least several occasions during early and middle Pleistocene time (Repenning, this volume; Vangengeim, this volume; Flerow, this volume).

Sea level was lowered to at least −135 meters and perhaps to −160 meters during the Illinoian Glaciation (Donn *et al.*, 1962), and a broad land bridge undoubtedly connected Siberia and Alaska at that time. The southwestern margin of the land bridge was deeply indented by an embayment in the present area of the Gulf of Anadyr and Kresta Bay (Petrov, this volume), and a large, shallow lake may have occupied a closed depression on the present site of Bering Strait, but the Chukchi shelf was probably entirely exposed and free of significant water barriers (Fig. 2). The land bridge was then disrupted again for a lengthy interval during the Sangamon Interglaciation.

Knowledge of sea-level history during the Wisconsin Glaciation is important for an understanding of the time and circumstances of man's first arrival in North America, but critical details of this history remain uncertain or unknown. Sea level is estimated to have fallen at least to −115 meters and possibly to −135 meters during early Wisconsin time (Donn *et al.*, 1962), and sea level fell to at least −120 meters at the height of the late Wisconsin Glaciation about 20,000 years ago (Curray, 1965). These two episodes of lowered sea level were separated by the Woronzofian transgression (Hopkins, this volume), a mid-Wisconsin high-sea-level episode that took place within the interval 25,000–35,000 years B.P., coinciding with at

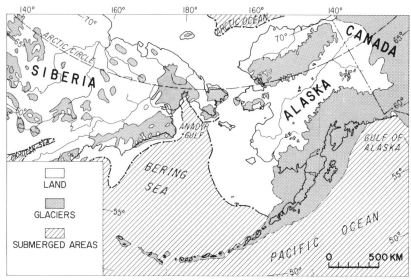

Fig. 2. Geography of Beringia during the height of the Illinoian or Riss Glaciation. Glacial boundaries in Canada from Owen Hughes and John Fyles, Geological Survey of Canada (unpublished preliminary compilation); glacial boundaries in Alaska from Coulter *et al.* (1965); glacial boundaries and shorelines in Siberia from Baranova and Biske (1964). Vegetation boundaries are omitted because of lack of adequate data.

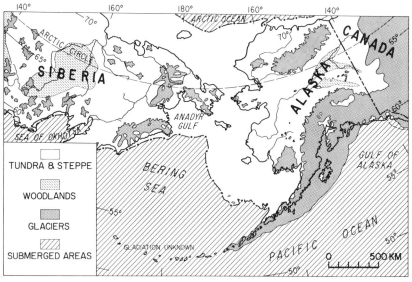

Fig. 3. Geography of Beringia during the height of the Wisconsin or Würm Glaciation. Glacial boundaries in Canada from Owen Hughes and John Fyles, Geological Survey of Canada (unpublished preliminary compilation); glacial boundaries in Alaska from Coulter *et al.* (1965); glacial boundaries and shorelines in Siberia from Baranova and Biske (1964); shoreline in Gulf of Anadyr after Jousé, this volume; woodland in Siberia from Giterman and Golubeva, this volume.

least part of the Karginsky Interstade of Siberia (Kind, this volume) and at least part of the Port Talbot Interstade of central North America (Dreimanis *et al.*, 1966). Sea level was certainly low enough to bring the land bridge into existence during both early and late Wisconsin time (Fig. 3), and was certainly high enough to drown the land bridge during the Woronzofian transgression. But the precise time of emergence of the late Wisconsin land bridge, following the Woronzofian transgression, has not yet been established.

Sea level began an oscillating rise about 18,000 to 20,000 years ago (Shephard, 1962; Curray, 1965), long before there is any evidence of substantial shrinkage of the outer margins of the continental glaciers and long before the paleobotanical record shows any evidence of warming climates[d] (Flint and Brandtner, 1961; various papers in Wright and Frey, 1965; Colinvaux, this volume; Giterman and Golubeva, this volume; Kind, this volume). By 4,500 years ago, sea level had risen to within a few meters of its present position. The most conservative interpretation of the history of the late Wisconsin–early Holocene rise in sea level would have the last Siberian-Alaskan land connection permanently drowned more than 11,000 years ago (Shephard, 1962). However, submerged shoreline features at altitudes of −38 meters, −20 to −25 meters, −15 meters, and −10 meters in Bering and Chukchi Seas suggest a more complex history and seem to record a series of successively younger stillstands during this late Wisconsin and early Holocene rise in sea level. Each of these submerged strandlines is at a level high enough to drown land connections between Alaska and Siberia, but sea level may have retreated briefly after each stillstand, possibly to levels that would have permitted the land connection to be renewed (Fig. 4).

The oldest of the submerged shorelines is that represented by the delta lying at a depth of −38 meters at the head of the Hope Seavalley in Chukchi Sea; radiocarbon analyses of specimens from this delta indicate an age of about 14,000 years (Creager and McManus, this volume).[e] Next younger is the −25-meter shoreline, recognized on continuous seismic-profile records obtained in Bering and Chukchi Seas by D. G. Moore (1964); specimens

[d] This striking paradox must indicate that continental glaciers had begun to thin substantially in their central areas long before noticeable retreat took place at the margins, and that the resulting moisture was somehow delivered back to the sea. Perhaps the continental ice masses of North America, Europe, and Siberia had generally negative water budgets after 18,000–20,000 years ago, owing to reduced atmospheric moisture circulation, and, perhaps, the positions of the glacier margins were nevertheless maintained by flow from the thinning central areas. This concept finds support in the evidence cited by Giterman and Golubeva (this volume) of cold, dry conditions during the middle parts of the Siberian glacial cycles.

[e] Creager and McManus, however, believe that the dated materials are contaminated by old carbon and that the delta is actually about 12,000 years old.

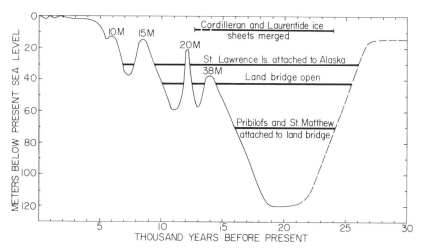

Fig. 4. Speculative reconstruction of the history of sea level during the last 30,000 years in the Bering Sea area.

bearing on the age of this shoreline have not yet been obtained, but I speculate that it may represent the position of sea level about 12,000 years ago, during the Two Creeks Interstade of the Wisconsin Glaciation. Submerged beach ridges representing the −15- and −10-meter shorelines can be recognized in several places in nearshore waters adjoining the coast of Seward Peninsula (for example, on U.S. Coast and Geodetic Survey Chart 9367, "Port Clarence and approaches," 1957). A maximum limit is placed on the possible age of these shorelines by peat dredged by Wendell Gayman of Ocean Sciences and Engineering, Inc., in water 16 to 18 meters deep near Nome. The peat is 9,700 ± 350 years old (Spec. W-1800, Meyer Rubin, written communication, 1966) and represents a bog soil that developed prior to the first Holocene marine inundation at that level; it is covered by marine mud less than one meter thick. The −15- and −10-meter submerged beaches are higher and were evidently formed less than 10,000 years ago. I speculate that the −15-meter beach was formed during the early Holocene warm period recognized in northwestern Alaska by McCulloch and Hopkins (1966), and that the −10-meter beach was formed during the postglacial hypsithermal interval.

A synthesis of this scanty data, along with that available from other parts of the world, suggests the following sequence of events (Fig. 4):

1. Sea level may have risen to −38 meters as early as 14,000 years ago, briefly severing land connections between Siberia and Alaska. St. Lawrence Island would have remained attached to the Alaskan mainland at that time.

2. Sparse data from other regions suggest that sea level may have lain below −50 meters about 13,000 years ago, permitting a renewal of land connections between Chukotka and Alaska.

3. Sea level rose to at least −38 meters and possibly to −25 meters about 12,000 years ago, once more drowning land connections between Alaska and Siberia and probably isolating St. Lawrence Island from the Alaskan mainland. This high-sea-level episode coincided approximately with a brief but pronounced climatic warming in North America, Siberia, and Europe.

4. Sparse data from other regions suggest that sea level may once again have fallen below −50 meters about 11,000 years ago, briefly reestablishing the Bering Land Bridge at a time coinciding approximately with a renewed advance of glaciers in the Old and New Worlds.

5. A rapid rise in sea level drowned the Bering Land Bridge for the last time about 10,000 years ago and isolated St. Lawrence Island from the mainland soon afterward.

6. Brief stillstands at −15 and −10 meters during the interval between 10,000 and 4,500 years ago had no significant effect upon the Bering Land Bridge, but a lowering of sea level between these two stillstands may have resulted in the temporary reestablishment of land connections between St. Lawrence Island and the Alaskan mainland. This lowering of sea level probably would have coincided with a brief early Holocene cold cycle about 8,000 years ago that resulted in the Cochrane readvance of the remnant of the continental glacier in Canada (Hughes, 1965) as well as pronounced readvances of valley glaciers in Alaska (Péwé *et al.*, 1965).

GLACIAL BARRIERS AND ICE-FREE CORRIDORS

The distribution of glacial ice has been a critical factor affecting dispersals of land biota to and beyond Beringia during late Quaternary time and possibly during early Quaternary time as well. Again the data are incomplete. Little is known of the frequency, extent, and age of pre-Illinoian glaciations. The maximum extent of glaciation during Illinoian time is reasonably well known (Fig. 2), and the extent of Wisconsin glacial advances is very well established (Fig. 3). But critical details remain to be established concerning the retreatal chronology of the Cordilleran and Laurentide ice caps in northwestern Canada during late Wisconsin time.

A long record of multiple pre-Illinoian glaciations is well documented in southern Alaska (Miller, 1953; Karlstrom, 1964), but this record cannot be matched within Beringia. The scattered evidence of pre-Illinoian ice advances there (Péwé *et al.*, 1965; Kind, this volume) may refer to one or to several different glacial events. It seems certain, however, that at least one glacial cycle is recorded that was comparable in extent to the Illinoian Glaciation in Alaska and to the so-called Maximum Glaciation of northeastern Siberia. The evidence at Nome seems to indicate that such an event took place prior to the Anvilian transgression (Hopkins, this volume). If the polymict sediments in the Pinakul' Suite of Chukotka are indeed of glacial origin, as Petrov suggests elsewhere in this volume, then a second preIllinoian glaciation is proved and, paradoxically, is shown to coincide with an important marine transgression.

The glacial record becomes clearer during the Illinoian cold cycle, which took place at some time between 175,000 and 100,000 years ago (Hopkins, this volume), and which was represented in the Old World by the Maximum Glaciation of Chukotka, the Taz or Sanchugov Glaciation of western Siberia, and the Riss Glaciation of Europe. This cold interval saw the growth over central Canada of a continental ice cap that ultimately expanded to merge with the Cordilleran glacier system; the merged glacier systems extended continuously from the Arctic to the Pacific Ocean, temporarily blocking all channels of land communication between Beringia and central North America (Fig. 2). Glaciation was less extensive within Beringia itself; small glacier systems developed in the higher mountains, but lowlands and many upland areas remained ice-free. However, an important barrier to land communication developed in Chukotka, where an ice cap or an extensive valley glacier system discharged directly into the Anadyr embayment.[f] Dispersals of land biota between northeastern Asia and Alaska could have taken place only across the present area of the continental shelf of the East Siberian Sea, north of the Arctic Circle, during the height of the Illinoian Glaciation.

Glaciers were less extensive throughout the world during the sequence of cold cycles known as Wisconsin in the New World and Würm in the Old, but during the most intense phases the Laurentide ice sheet of central Canada again merged with the Cordilleran glacier system to form a barrier to land communication between Alaska and central North America (Fig. 3). Central and northern Alaska and northeastern Siberia supported only relatively small valley glacier systems in major highland areas.[g] The relatively limited

[f] Soviet authors hold conflicting opinions concerning the extent of glaciers in Chukotka during the Maximum Glaciation, but their disagreement does not affect the conclusions reached here. Baranova and Biske (1964) indicate the presence of several isolated glacier systems there, whereas Petrov (this volume) argues that almost all of Chukotka was flooded by glacial ice.

[g] Although the limited extent of glaciation in Beringia during Wisconsin or Würm time is well established, uncertainties persist concerning the relative extent of glaciation at different moments within the Wisconsin Glaciation, and concerning the correlation of certain morainal systems from one part of Alaska to another and from Alaska to Siberia. Recent studies in Alaska, summarized by Péwé et al. (1965), indicate that the maxi-

extent of glacierization and the disappearance of the Anadyr marine embayment[h] resulted in an avenue of communication through northeastern Siberia and across the Bering-Chukchi platform so wide that Alaska must have become biogeographically a part of Siberia rather than a part of North America during much of Wisconsin time.

OPPORTUNITIES FOR INTERCONTINENTAL LAND PASSAGE
DURING LATE WISCONSIN TIME

An understanding of the relative timing, during the Wisconsin Glaciation, of openings and closings of the Bering Land Bridge and closings and openings of an ice-free corridor between Alaska and central North America is of critical importance for discussions of man's early history in the New World. Unfortunately, little attention has been given to the detailed glacial chronology of the remote regions in Yukon Territory, northeastern British Columbia, and northern Alberta, where a conjunction of the Laurentide ice sheet and the Cordilleran glacier system would have formed earliest and persisted longest. One can state only that an ice-free corridor must have existed there during the mid-Wisconsin episode of mild climate that took place between 35,000 and 25,000 years ago; that waxing glaciation probably closed the corridor again earlier than 20,000 years ago; and that the corridor must have remained closed until at least 14,000 years ago and possibly until

---

mum extent of glaciation was very similar there during early (prior to 35,000 years ago) and late (after 25,000 years ago) Wisconsin time—a pattern that has also been demonstrated throughout much of the conterminous United States (Wright and Frey, 1965). In western Siberia, however, the early Wisconsin Zyrianka Glaciation is thought to have been vastly more extensive than the late Wisconsin Sartan Glaciation (Kind, this volume). This paradox may be resolved by a recent restudy of the Karginsky terrace in the Lena River valley and in the Lower Yenisei region, which suggests that the Sartan Glaciation was much more extensive and much more nearly comparable in extent with the Zyrianka Glaciation than earlier work had indicated (S. A. Strelkov, Inst. Geol. and Geophysics, Akad. Nauk SSSR, written communication, May 10, 1966).

In the absence of radiocarbon dating, I must also question the ages assigned to the Vankarem and Yskaten' Glaciations of Chukotka (early and late Wisconsin, according to Petrov, this volume) and to the York and Mint River Glaciations of western Seward Peninsula (early and late Wisconsin, according to Sainsbury, this volume). Alaskan glaciated areas that have yielded a radiocarbon chronology commonly exhibit late Wisconsin moraines about 11,000 years old nested just within moraines of early Wisconsin age; much nearer to the ice sources, one commonly finds another set of sharply defined moraines about 8,000 years old (Péwé et al., 1965). Petrov's and Sainsbury's descriptions suggest to me that the deposits of the Vankarem and York Glaciations include moraines corresponding to those dated about 11,000 years B.P., and that the deposits of the Yskaten' and Mint River Glaciations include moraines corresponding to those dated about 8,000 years B.P. farther east in Alaska.

[h] Jousé shows (elsewhere in this volume) that the shoreline lay at the southern edge of the Gulf of Anadyr near lat. 62° N. during the Wisconsin Glaciation, and Saidova (this volume) shows that shorelines generally lay near the edges of the continental shelves in Bering Sea and the Sea of Okhotsk.

almost 10,000 years ago. As I have noted, knowledge of sea-level history during the interval 25,000 to 10,000 years ago is equally imprecise. But despite the uncertainties, one can construct a plausible speculative model for the timing relationships of land connections across Beringia and ice-free passages to central North America, as follows.

It seems probable that the waxing phases of a glaciation are tightly linked to reductions in sea level, since glaciers can increase in mass only at the expense of the reservoir of water in the sea. During waning phases, on the other hand, sea level evidently begins to rise long before appreciable shrinkage of glacier margins occurs (see footnote *d*, p. 463). Thus, one might expect sea level to have fallen sufficiently to expose a land connection between Chukotka and Alaska well before the Cordilleran and Laurentide ice sheets expanded to a merger point. But sea level probably rose sufficiently to drown the land connection well before the ice-free corridor reopened.

This model agrees with Müller-Beck's assumption (elsewhere in this volume) that there was a substantial interval during the waxing phase of the late Wisconsin glaciation, shortly after 25,000 years ago, when land communications were simultaneously possible between Siberia and Alaska and between Alaska and central North America. The ice-free corridor would then have closed for 10,000 years or more, and would not have reopened again until at or well after the moment when the land bridge had been closed again by rising sea level. However, if, as seems possible, the land bridge was briefly restored during slight glacial readvances about 13,000 and 11,000 years ago, then there may have been brief intervals in latest Wisconsin time when land communications were again uninterrupted from Siberia through Alaska to central North America.

## Quaternary Climate and Vegetation

Though the evidence is not yet conclusive, the Quaternary Period seems to have been characterized in Beringia by progressively increasing climatic stresses that reached a climax during the Illinoian Glaciation. The forests were evidently richer in numbers of tree species during early and middle Quaternary time, but after the Illinoian Glaciation they differed little from the modern taiga. And the mammals dispersing across Beringia during the Blancan (Pliocene and early Pleistocene) and Irvingtonian (middle Pleistocene) mammal ages included many species adapted to a forest environment, whereas the Rancholabrean (late Pleistocene) mammalian dispersal was dominated by steppe and tundra forms (Repenning, this volume). In fact, there is no clear fossil record for the existence of tundra vegetation prior to middle Pleistocene time, though it is certain that a severe climate prevailed to the southern edge of the land bridge as early as two million years ago. Well-developed tundra floras are known, however, in deposits of Illinoian and Wisconsin age. Some fossil assemblages of animals and plants that lived during the Illinoian and Wisconsin Glaciations represent mixtures of tundra and steppe elements that cannot be matched in any present-day landscape; they seem to record a wider distribution of xeric climates and more extensive grasslands during certain phases of these glacial cycles.

## CLIMATE AND VEGETATION PRIOR TO THE ILLINOIAN GLACIATION

Giterman and Golubeva (this volume) show that several tree species suggestive of relatively temperate climates, including hemlock, fir, linden, oak, and walnut,[i] persisted in northern Siberia through early Quaternary time. But by the middle Quaternary, the vegetation of northeastern Siberia differed from the modern vegetation only in details of the distribution of taiga, steppe, and tundra floras and in the existence, at times, of curious mixtures of steppe and tundra species.

The early and middle Pleistocene floras in Alaska are much less well known than those of Siberia. However, scattered observations suggest that certain tree species now regionally extinct may have persisted in Alaska until the middle Pleistocene interglacial episode represented by the Kotzebuan transgression. Wood of a cedar (*Chamaecyparis* sp.) has been found associated with poplar (*Populus*) wood and with cones of white spruce (*Picea glauca*) and black spruce (*Picea mariana*) in peaty lenses in gravel that is of either Pliocene age or early Pleistocene age on the Kugruk River in northern Seward Peninsula, and alone in a peat bed that may be of Kotzebuan age at Cape Deceit on the south shore of Kotzebue Sound. Douglas fir (*Pseūdotsuga* sp.) was identified many years ago by Knowlton in interglacial peat beneath drift of Illinoian age at Iron Creek on Seward Peninsula (Smith, 1910, p. 108–109; Hopkins and Benninghoff, 1953), and a single grain of *Pseudotsuga* pollen was found in the interior of a bison skull apparently derived from Kotzebuan sediments on Baldwin Peninsula at the

---

[i] The walnut, *Juglans cinerea*, has been found associated with remains of subarctic coniferous forest in early Pleistocene beds at the mouth of the Lena River (Vangengeim, 1961), suggesting that this species, presently confined to temperate regions of eastern North America, may have been represented by a now-extinct strain capable of surviving in extremely severe environments. Miki (1955) has shown that walnut remains found in upper Pliocene and early Pleistocene beds in Japan are indeed varietally distinct from living *J. cinerea*.

east shore of Kotzebue Sound (Péwé and Hopkins, this volume). A fossil cone of hemlock (*Tsuga sp.*) from the Arctic slope of Alaska may have been washed out from nearby beds of the Anvilian transgression (Hopkins, this volume), and hemlock pollen is tentatively identified in the fossil bison skull just mentioned. Pine pollen is present in Beringian beds (probably Pliocene) at Kivalina, in Anvilian beds at Nome, and in the bison skull; pine wood has also been identified in the Kotzebuan beds.[j] Preliminary studies by J. A. Wolfe of two other pollen floras of pre-Illinoian age from Seward Peninsula record vegetation that did not differ greatly from modern boreal forest assemblages in central and western Alaska. All of these occurrences are in areas that lie north or west of the modern boundaries of spruce forest, and thus all would seem to reflect preglacial or interglacial conditions.

The presence on St. George Island of numerous layers of frost-broken rubble intercalated in a sequence of lavas ranging in age from one to two million years indicates that a severe climate prevailed on the land bridge during some parts of early Quaternary time (Hopkins, unpublished data), but the first clear fossil evidence of the existence of tundra vegetation is that provided by pollen floras in the lower Pinakul' Suite of middle Quaternary age in Chukotka (Petrov, this volume). Frost cracks covered by beach gravel of the Einahnuhtan transgression on St. Paul Island indicate that permafrost was present as far south as the Pribilof Islands at about this same time (Hopkins, this volume).

Recent studies of fossil vertebrate and freshwater mollusk faunas in central and eastern United States indicate that early and middle Quaternary glacial and interglacial climates differed qualitatively from those of late Quaternary time. The succession of climates during glacial episodes became progressively more severe there, with notably more severe winters during the Wisconsin Glaciation than during previous glacial cycles, and interglacial climates during early and middle Quaternary time were generally less continental than were those of late Quaternary time (Auffenberg and Milstead, 1965; Hibbard *et al.*, 1965; Taylor, 1965). The paleoclimatic data from Beringia may record a similar progression in the intensity of climatic stresses, but, if so, the climax was reached during the Illinoian Glaciation rather than during the Wisconsin Glaciation. However, much additional study must be devoted to the fossil floras of Alaska before we can evaluate the validity and climatic significance of the evidence for the persistence into middle Quaternary time of tree species that are now regionally extinct there. A speculative model can be constructed according to which certain tree species would be eliminated by the cumulative effects of identical climatic

[j] Most of the wood was identified by R. A. Scott, most of the pollen by E. B. Leopold, and the cones by J. A. Wolfe.

stresses applied repeatedly during successive glacial intervals, rather than by progressive qualitative changes in the nature of glacial and interglacial climates. The following is such a model.

Presumably, the forest vegetation that covered much of Alaska and Siberia prior to the first Pleistocene glaciation was richer in tree species than the present taiga. During each Pleistocene glacial cycle the forests of Alaska were greatly reduced in extent, and were isolated by glacial ice and broad areas of steppe and tundra from the forests of central North America and eastern Siberia. Reduction in the extent of forest would have caused a reduction in the total population of each tree species. Many habitats and microclimatic environments would have been eliminated, and the biotypes adapted to those habitats would have perished. Thus, each successive glacial cycle would have reduced the total variability of each tree species, thereby limiting the capacity of that species to expand throughout Alaska when the climate ameliorated. A point may have been reached during one of these middle or late Pleistocene glacial cycles when a small, surviving population of a given tree species could finally be exterminated by parasites, fire, or some minor local climatic event, such as a 50-year sequence of summers too cold to permit seeds to mature.

The modern distribution pattern of pines seems to corroborate the model. Jackpine (*Pinus banksiana*) now grows vigorously in northern Canada, locally approaching timberline (Hustich, 1953), and lodgepole pine (*Pinus contorta*) thrives in the Yukon Territory and southeastern Alaska; very likely both could thrive in central Alaska today if seed stock were available. It is easy to imagine that pine might have persisted in Alaska through Kotzebuan time, becoming regionally extinct during the Illinoian Glaciation in accordance with our model. But Alaska cedar and two species of hemlock reach their northern limits considerably farther south, at latitude 62° in southern Alaska, and Douglas fir now grows no farther north than latitude 55° in northern British Columbia. If these genera did indeed persist in central and northern Alaska until middle Quaternary time, they may indicate that climatic stresses were less severe during early and middle Quaternary time. But alternatively, these taxa, like the walnut in Siberia, may have been represented by now-extinct strains that were better able to tolerate severe climatic conditions.

## CLIMATE AND VEGETATION DURING THE ILLINOIAN AND WISCONSIN GLACIATIONS

The Illinoian and Wisconsin Glaciations represented severe refrigerations throughout Beringia. Snowline was lowered 400 to 500 meters in Alaska during the Illinoian Glaciation and at least 300 meters during the Wisconsin

Glaciation (Péwé et al., 1965), and was probably lowered to a similar extent in Siberia. Ice-wedge casts and solifluction debris in sediments of Illinoian and Wisconsin age on the Pribilof Islands indicate that permafrost had again set in, and mean temperatures, therefore, were once again well below freezing as far south as the southern margin of the land bridge (Hopkins, this volume and unpublished data). In Siberia, the continuous belt of taiga that now partitions steppe and tundra landscapes was disrupted, and the two kinds of treeless vegetation merged along a broad front. Timberline in Alaska was lowered by at least 400 meters during the Wisconsin Glaciation and perhaps by still more during Illinoian time (Repenning et al., 1964; Péwé, 1965, p. 24; J. V. Matthews, Jr., written communication, 1966). Taiga must have been restricted in Alaska to small, isolated areas in the Yukon and Tanana River valleys, areas that would have been separated from the forests of North America by continental ice and from the forests of Siberia by a vast area of treeless vegetation. Moreover, fossil pollen floras of full-glacial age in western Alaska record herbaceous vegetation almost devoid of shrubs; like taiga, then, shrub tundra must also have been confined to the eastern area.

Present-day Siberia is a region of low precipitation, and the region seems to have been even dryer during parts of the Illinoian and Wisconsin cold cycles, for plants and animals adapted to steppe conditions became important elements in some tundra biotas (Vangengeim, this volume; Giterman and Golubeva, this volume; Petrov, this volume). The movement of steppe-adapted mammals northward into tundra areas was facilitated, no doubt, by the disruption of continuity of the taiga, but the abundance of xerophytes, such as *Artemisia, Ephedra,* and the chenopods, in fossil tundra floras from Chukotka to the Lena River valley must indicate that moisture-deficient sites were far more abundant than in any modern tundra.

Precipitation is also low throughout present-day central and northern Alaska, and paleogeographic considerations suggest that it would have been lower still during the Illinoian and Wisconsin Glaciations. Most of the precipitation falling in Alaska today is borne by air masses moving northeastward from the Pacific Ocean and Bering Sea; conspicuous rain shadows lie to the north and northeast of each of Alaska's mountain ranges. The pattern was similar during the Illinoian and Wisconsin Glaciations; the distribution of end moraines shows that glaciers were much more extensive in southern Alaska than in northern Alaska, and much more extensive on the south flanks than on the north flanks of individual mountain ranges (Coulter et al., 1965). The lowered moisture capacity of cooler air masses, the increase in the height of mountain barriers resulting from the presence of extensive ice fields and ice caps in southern Alaska, and the conversion of the northeastern half of Bering Sea to a land area are all factors that would have contributed to a reduction in the total amount of moisture carried annually into central and northern Alaska.

Compelling paleontologic evidence confirms the assumption that Alaska, like Siberia, had a dryer climate and supported xeric vegetation unlike any modern tundra, during parts of the Illinoian and Wisconsin Glaciations. Colinvaux (this volume) reports high abundances of *Artemisia* in many Alaskan pollen spectra of full-glacial age.[k] The dramatic occurrence in central Alaska of the steppe antelope, *Saiga*, and of the yak-like *Bos poephagus* also strongly suggests the presence of extensive steppelike environments throughout Beringia (Flerow, this volume). But even more impressive is the fact that three obligatory grazers—horse, bison, and mammoth—are by far the most common fossil land mammals collected in late Pleistocene beds in Alaska; in fact, these three animals compose 85 to 95 per cent of the total individuals in four large fossil mammal faunas collected near Fairbanks (Guthrie, manuscript). Feral horse and bison can survive in present-day Alaska only in a few restricted, highly specialized, and isolated environments.[l] Grasslands must have been far more widely distributed when large herds of bison, horse, and mammoth were present.

The presence in Alaska of extensive areas of shifting sand and of active deposition of loess (windblown dust) during the Illinoian and Wisconsin Glaciations would have augmented the effects of a dryer climate in favoring the development and perpetuation of local grasslands. Most of the major rivers in Alaska drained extensive glaciated areas, and most of them, therefore, were heavily charged with sand and finely ground rock flour. Winds swept sand and silt from the abundant shifting bars and islets of the braided outwash plains and deposited them on nearby uplands, creating extensive sand-dune areas and even more extensive areas that became mantled with

---

[k] In an earlier paper, Colinvaux (1964, Table 3) distinguished without comment a *Selaginella sibirica* zone that coincides approximately with his *Artemisia*-rich zone J in the Imuruk Lake core. Giterman and Golubeva (this volume) report that this club moss is characteristic of the xeric phases of full-glacial pollen spectra in Siberia.

[l] The chronicles of early Alaskan scientific-exploration parties that employed horses as pack animals make frequent reference to the fact that after the first frost the grasses and sedges lost their nutritional value. The horses lost weight, and often their condition became so poor that some had to be sacrificed before the party returned to the nearest town.

Bison have been introduced and pack horses have been left to fend for themselves through the winter in small areas in central and southern Alaska where wild grass is abundant on the braided outwash plains of modern glacial rivers and on adjoining areas of active sand movement or active deposition of windblown silt. The horses survive and the bison thrive, but neither animal has been able to expand its feral range beyond the highly specialized and isolated environment in which it has been placed.

windblown silt many meters thick (Péwé *et al.*, 1965). It is precisely in areas of this type that bison and feral horse survive in Alaska today.

Many lines of evidence, then, point to the existence of a more xeric climate and to a much wider distribution of grasslands in Beringia during the Illinoian and Wisconsin Glaciations. Grasslands were probably favored, because of the greater abundance of well-drained sites that would have resulted from dryer climates, by a greater frequency of tundra fires during the dry summers, by trampling and heavy grazing by the large herbivores themselves, and most of all by the existence of areas of active sand movement and active loess deposition. But even though grasslands must have been widely distributed, it does not follow that herbaceous tundra and marsh vegetation disappeared from the landscape. Johnson and Packer (this volume) show that trans-Beringian plant dispersals during the Pleistocene favored tundra plants adapted to lowland environments. The Bering Land Bridge was an area of low slopes and extremely low relief; it must have been cloaked largely in lowland marsh and meadow vegetation.

### CLIMATE AND VEGETATION DURING LATE QUATERNARY INTERGLACIAL, INTERSTADIAL, AND POSTGLACIAL INTERVALS

Little evidence is available concerning the nature of winter climates during the Sangamon Interglaciation and the equivalent Kazantsevo Interglaciation in western Siberia, but summer temperatures can be presumed to have been conspicuously warmer than at present, because forests extended well beyond their present limits in both Alaska and Siberia (Giterman and Golubeva, this volume; Colinvaux, this volume), and because Sangamon-aged deposits commonly show evidence of melting-out of ice wedges and lowering of the permafrost table far below its present position.

A complex of relatively warm episodes between 38,000 and 25,000 years ago within the Wisconsin Glaciation saw, once again, a marked northward expansion of forests in Siberia (Kind, this volume). Except for the marine deposits of the Woronzofian transgression, interstadial deposits of comparable age have not yet been clearly recognized in Alaska, but a brief mid-Wisconsin warming may be recorded by moderate increases in the abundance of birch pollen in zone J-2 of the Imuruk Lake core (Colinvaux, this volume). Brief late-glacial warmings are also well recorded and well dated 11,000 to 12,000 years B.P. in Siberia (Kind, this volume); these warmings evidently coincide with similar climatic oscillations in Europe and the United States. Colinvaux shows that dwarf birch had expanded into western Alaska at least as early as 13,000 years ago, but the detailed climatic chronology of late Wisconsin time has not yet been worked out there.

Details of climatic history during Holocene time are rather more clearly established. The Kotzebue Sound–Seward Peninsula area in western Alaska experienced a pronounced climatic amelioration between 10,000 and 8,300 years ago, during which the forest biota expanded far beyond its present limits (McCulloch, this volume; McCulloch and Hopkins, 1966). This warming is also recorded, though less conspicuously, at Point Barrow in northwestern Alaska, and it may be represented in southwestern Alaska by a peak in the abundance of spruce pollen, evidently blown from a distant source, at the 10,000-year-old level in a pollen profile from the Pribilof Islands (Colinvaux, 1967). But no evidence has yet been recognized for a comparable warming in Siberia during this interval. The so-called Hypsithermal Interval, a widely recognized warm episode that occurred about 6,000 to 4,000 years ago, was marked in Siberia by an expansion of the taiga far north of its present limits (Kind, this volume; Giterman and Golubeva, this volume). The Hypsithermal Interval in western Alaska was a relatively minor event compared with the early Holocene warming that took place prior to 8,300 years ago, but it is evidently recorded by the peak abundances of alder appearing midway in Holocene pollen profiles from northern Alaska, St. Lawrence Island, and the Alaska Peninsula (Colinvaux, this volume; Heusser, 1963).

These apparent regional differences in Holocene climatic and vegetational history may result in part from incomplete data, but McCulloch and Hopkins (1966) suggest that paleogeographic factors may also play a major role. The onset of a relatively warm postglacial climate at a time when sea level was still low and the shoreline relatively distant may have induced appreciably warmer summers in western Alaska during the early Holocene warm interval, 10,000 to 8,300 years ago, than during the Hypsithermal Interval, 6,000 to 4,000 years ago. Because the shoreline in southwestern Alaska reached its present position earlier, bringing with it summer fog and cooler summer temperatures, the early Holocene warm interval may have been terminated earlier there, as is suggested by the early termination of the spruce peak in Colinvaux's (1967) Pribilof pollen profile.

## Late Quaternary Mammalian Migrations and Extinctions

An imbalance has persisted throughout Cenozoic time in the direction of flow of land-mammal dispersals across the Bering Land Bridge. During late Tertiary time, the number of taxa that dispersed from the Old to the New World was roughly twice the number that dispersed from the New to the Old, and from Blancan or Villafranchian time onward, the imbalance became still more pronounced (Simpson, 1947; Repenning, this volume).

A climax was reached in the Late Quaternary Rancholabrean dispersal, during which more than 20 taxa migrated from the Old World to the New; with the possible exceptions of reindeer and musk-ox, none can be proved to have dispersed in the opposite direction (Flerow, this volume; Repenning, this volume). The giant supercontinent of Eurasia contains a much larger diversity of landscapes and climates than North America, and consequently a far larger total fauna. This difference in total size of the two reservoirs of species may be quite sufficient, as Repenning (this volume) suggests, to explain the relatively modest imbalance that prevailed throughout much of the Tertiary Period, but the progressive intensification of the imbalance during the Quaternary seems to require a more sophisticated explanation. The complex chronological sequence of openings and closings of a land bridge from Siberia to Alaska and of closings and openings of an ice-free corridor from Alaska to central North America has evidently operated increasingly as a set of one-way valves, allowing unrestricted flow of biota in one direction and extremely limited flow in the other.

The abrupt extinction of much of the large-mammal fauna about 10,000 years ago is as mysterious an event in Beringia as it is in other parts of the world. Most earlier waves of extinction have resulted from the displacement of older species by new species better adapted to their particular ecological niches, but the late Wisconsin event was a case of *extinction without replacement*. The total number of species in temperate and northern faunas was drastically reduced, and the eliminated species were *not* replaced by others occupying the same ecological niches. The great end-Pleistocene extinction coincided with a time of rapid climatic change, and the regional restriction or even disappearance of certain types of environment undoubtedly played an important role in the extinction of many large herbivores as well as their most specialized predators. But earlier episodes of rapid climatic change, such as the transition from the Illinoian Glaciation to the Sangamon Interglaciation, saw no comparable reduction in the diversity of temperate and arctic mammalian faunas. Some new factor must have been introduced at the end of Wisconsin time—probably the presence of human hunters.

## ONE-WAY TRAFFIC ACROSS BERINGIA

The pronounced imbalance in the direction of mammalian dispersals across Beringia during later Quaternary time must reflect differences in the rate of evolution of tundra-adapted animals in Eurasia and North America, but the effects of these differences have been intensified by the time phasing of land bridges and ice-free corridors between the two continents.

During interglacial times, tundra and diverse varieties of taiga have shared a common interface across the breadth of Eurasia. But during the Illinoian and Wisconsin Glaciations and perhaps during earlier cold cycles, as well, the continuity of the taiga belt was disrupted, and steppe and tundra merged along a broad front. Thus, during the later Quaternary, the mammals of both taiga and steppe have been provided abundant opportunities to evolve new forms adapted to life in the tundra, and these new forms have then dispersed eastward into Alaska at times when sea level was low. The xeric nature of the Siberian and Alaskan tundra during cold cycles probably facilitated the eastward dispersal of many herbivores and of the carnivores that preyed upon them. But until the climate warmed and sea level rose to drown the land bridge, Alaska's new residents were denied access to central North America by the Laurentide and Cordilleran ice sheets. Such characteristic Rancholabrean mammals as *Bison* reached Alaska long before they reached the conterminous United States (compare Péwé and Hopkins, this volume, and Hibbard *et al.*, 1965), and two mammals, *Bos* and *Saiga*, reached Alaska but became extinct there before the opportunity arose to expand their range into central North America. Dispersals from Siberia to Alaska during the Illinoian Glaciation were probably restricted to high arctic areas, because of the marine embayment in the Gulf of Anadyr and the extensive glaciation of Chukotka (Fig. 2); the probable limitation of *Saiga* and *Bos* in Alaska to deposits of Wisconsin age may indicate that opportunities for migration of steppe-adapted forms were better during the Wisconsin than during the Illinoian Glaciation.

The increasing floristic monotony and frequent isolation and areal restriction of the Alaskan taiga have curtailed the opportunities for taiga mammals to evolve new tundra-adapted forms there during later Quaternary time. Both taiga and prairie landscapes may have adjoined a narrow tundra belt in central North America during late Quaternary glacial cycles, but restrictions on the former extent of tundra areas limit the probability that new tundra-adapted mammals evolved there. Such new tundra mammals as may have appeared in central North America could only have reached Alaska about the time the land bridge was drowned by rising sea level, and in Alaska they would have encountered the rich and fully established tundra fauna that had dispersed from Siberia during the previous glacial cycle. Thus, there could have been little back-flow of tundra-adapted mammals from North America to Asia. Between the Rancholabrean dispersal and the Irvingtonian and Blancan dispersals, some offsetting of the chronological relationships between land bridges and ice-free passages must have taken place, because the imbalance in direction of migration was considerably less acute during the earlier dispersals.

EXTINCTION OF THE LARGE MAMMAL FAUNA AND
THE ARRIVAL OF MAN

Flerow (this volume) suggests that the rapid expansion of sphagnum bogs and muskegs caused the extinction of the late Pleistocene large-mammal fauna in northern regions. The climatic amelioration of late Wisconsin time indeed saw a rapid expansion of taiga, shrub tundra, cottongrass tundra, and sedge and sphagnum bogs, probably at the expense of grasslands, throughout Beringia (Colinvaux, this volume; Giterman and Golubeva, this volume).[m] These climatically induced changes in the vegetation undoubtedly placed great stress upon the large herbivores. But similar climatic and vegetational changes must have produced similar stresses during the Sangamon and earlier Pleistocene interglaciations, yet no comparable general extinctions took place. Martin (1958) shows that man, throughout the world, must have played a major role in the extinction of the late Pleistocene large-mammal fauna. In Beringia, it seems likely that man not only contributed substantially to the demise of mammoth, horse, and bison, but also, in so doing, hastened a drastic change in the vegetation pattern that resulted ultimately in a considerable net reduction in his total meat supply.

It is becoming increasingly certain that man was present in Beringia at least as early as 25,000 years ago,[n] and it is equally certain that his hunting techniques steadily increased in efficiency during the ensuing millennia. Guthrie (manuscript) shows that when the late Pleistocene large-mammal fauna was extant, the mammalian biomass must have been considerably greater in Alaska than it is today. The early human populations in Beringia probably increased steadily until some balance with the large-mammal population was reached. This balance would then have been sustained as long as the large-mammal count remained the same.

The rapid changes in climate and vegetation that took place at the end of the Wisconsin cold cycle must in themselves have resulted in a severe re-

[m] The paleobotanical evidence cited here is supported by observations of the nature of Wisconsin- and Holocene-aged deposits in many parts of western Alaska. Organic traces in thick layers of loess of Wisconsin age that I have examined near Fairbanks and in the Seward Peninsula–Kotzebue Sound area commonly consist of vertical root concretions, small erect twigs, and erect shreds of organic tissue suggestive of non-tussock-forming grasses or sedges. Compact lenses and layers of fibrous peat that might represent the remnants of buried turf layers first appear at levels that probably mark the onset of postglacial conditions. Thick masses of sedge and sphagnum peat, recording the development of bogs, are generally confined to interglacial and postglacial levels.
[n] Recent reports of evidence for human occupation in Saskatchewan about 25,000–35,000 years ago (Stalker, 1967), in Mexico about 24,000 years ago (José Lorenzo, paper read at Soc. Am. Archaeology, Ann Arbor, May, 1967), and in Idaho about 15,000 years ago (Ruth Gruhn, paper read at Soc. Am. Archaeology, Reno, May, 1966) lend weight to Müller-Beck's speculation (this volume) that man dispersed into Alaska and thence into central North America during or just after the mid-Wisconsin warm interval.

duction in the total area of suitable grazing land, and thus in a reduction in the total size of the large-mammal populations in Beringia—and the large-mammal population reduction took place at a time when efficient hunters were present in numbers that had been nicely adjusted to larger game resources. During earlier Pleistocene interglacial cycles, remnant populations of grazing animals probably survived in local arctic grassland areas, grasslands perpetuated in part by active sand movement and loess deposition near active glacial-outwash streams and in part by the interaction of the herbivores themselves with the treeless vegetation of the time. But during late Wisconsin time, hunting pressures upon the diminishing game resource probably reduced many species to population levels where their simple presence no longer tended to perpetuate grasslands. The last small herds, restricted to those very few, very small areas where grasslands still persist today, must have been as easy a quarry for stone-age men as the sea cows of the Commander Islands were for Bering's hungry sailors (Laughlin, this volume). This interaction of changing climate, changing vegetation, and relentless hunting by man resulted in the extinction of major elements of the Beringian mammalian fauna, and in a shift to a vegetation able to support only two herd animals—reindeer (or caribou) and musk-ox.

## Dispersal of Man into North America

The cultural, linguistic, and genetic differences between American Indians, on the one hand, and Eskimos and Aleuts, on the other, indicate clearly that Beringia has seen two waves of human migration (Laughlin, this volume). Müller-Beck shows (elsewhere in this volume) that the ancestors of modern American Indians must have been the earlier to arrive; most likely they were in Alaska as early as 25,000 years ago. Some bands no doubt dispersed into central North America along the temporarily ice-free corridor east of the Rocky Mountains, but others probably remained in Alaska through late Wisconsin time and through the period of rapidly changing climates, environments, and game resources that ushered in the Holocene. When the Laurentide and Cordilleran ice sheets merged again, the groups that had reached central North America would have been isolated from those that remained in Alaska, and their lithic technology would have evolved toward the anachronistic Mousteroid Llano industries of latest Wisconsin and earliest Holocene time, uninfluenced by the contemporary Aurignacoid industries that were developing in Eurasia.

Shortly after 25,000 years ago, the skilled bone-working techniques that distinguished Aurignacoid industries were diffused among the tundra- and steppe-hunting populations of Siberia, reaching the vicinity of Lake Baikal by about 14,000 or 15,000 years ago and northern Japan by about 12,000 years ago. Because game resources were *more* plentiful, not less so, during full-glacial time, I see no reason to believe that Chukotka, the Bering Land Bridge, and Alaska were vacated by man during the interval 20,000 to 10,000 years ago. But whatever men were there must have been influenced by the new Aurignacoid lithic and bone-working technology diffusing from the west. The paleo-Indian populations persisting in Alaska probably adopted Aurignacoid techniques. These men may have become the ancestors of the Athabaskan Indian groups of present-day Alaska and northwestern Canada, and they may have been the agents of change that introduced burins and microblades into Yukon Territory, British Columbia, and Texas when communications reopened in late Wisconsin and early Holocene time.

Though tundra and steppe hunters seem to have been the principal groups involved in the diffusion of Aurignacoid industries across Eurasia, their lithic techniques eventually reached groups exploiting the resources of river valleys in wooded areas. The appearance of Aurignacoid industries in Japan about 12,000 years ago was a sudden event that seems to mark the arrival of immigrants from the north,[o] but these groups would have come from nearby parts of easternmost Siberia that, like Hokkaido, were wooded even then (Giterman and Golubeva, this volume). The rapid environmental changes of latest Wisconsin and earliest Holocene time, and the consequent restriction and depletion of big-game resources, may have hastened the development of a new way of life among hunting populations living near the former boundaries between forest and tundra, and those who first learned to exploit the northern rivers and forests probably were able to exploit them most efficiently. Perhaps the modern Mongoloid populations, of which Eskimos and Aleuts are a part, evolved during late Wisconsin time in wooded valleys in Transbaikalia and Kamchatka, adapting the Aurignacoid flint- and bone-working techniques to the resources of a riverine environment there. Groups reaching the coast may have begun to exploit the special resources of estuaries and coastlines and eventually, those of the sea itself.

By early Holocene time, the riverine and shoreline-dwelling Mongoloids had dispersed into Alaska, reaching the southwestern coast at Anangula Island as early as 8,400 years ago, and reaching far up the Kobuk River by 8,000 years ago. The land bridge had long been drowned (Fig. 4), and until we find earlier evidence of their presence, we must be prepared to belive that these earliest Eskimos and Aleuts were descended from groups

[o] K. Hayashi, oral communication, April 20, 1967.

that had crossed from Chukotka to St. Lawrence Island and Seward Peninsula by boat.

Beringia, in its long history, has seen many striking geographic changes, some caused by the internal dynamics of the solid planet, some induced by the external dynamics of its atmosphere. But among the most dramatic landscape changes have been those caused by the interactions of the plants and animals that found in Beringia new pathways to new continents and new channels to new seas. And most striking of all as a living agent of change has been man, who dispersed across Beringia and found—quite unwittingly—a new world to conquer.

## REFERENCES

Auffenberg, Walter, and W. W. Milstead. 1965. Reptiles in the Quaternary of North America, *in* H. E. Wright and D. G. Frey, eds., The Quaternary of the United States: Princeton Univ. Press, p. 557–568.

Baranova, Yu. P., and S. F. Biske. 1964. Istoria Razvitiya Rel'efa Sibiri i Dal'nego Vostoka: Severo-Vostok SSSR (History of the development of the relief of Siberia and the Far East: Northeast Siberia): Akad. Nauk SSSR, Siberian Div., Inst. Geol. and Geophys., Izd-vo "Nauka," 288 p.

Berry, E. W., and W. K. Gregory. 1906. *Prororsmarus alleni*, a new genus and species of walrus from the upper Miocene of Yorktown, Virginia: Am. Jour. Sci., v. 21, 4th Ser., p. 444–450.

Colinvaux, P. A. 1964. The environment of the Bering Land Bridge: Ecol. Monogr., v. 34, p. 297–329.

——— 1967. Bering Land Bridge: Evidence of spruce in late Wisconsin times: Science, v. 156, p. 380–383.

Coulter, H. W., *et al.* 1965. Map showing the extent of glaciations in Alaska: U.S. Geol. Survey Misc. Geol. Inv. Map I-415.

Craig, B. G., and J. G. Fyles. 1960. Pleistocene geology of Arctic Canada: Geol. Survey, Canada, Paper 60–10, 21 p.

——— 1965. Chetvertichnye period v arkticheskikh oblastyakh Kanady (Quaternary of Arctic Canada), *in* Antropogenovye period v Arktike i subarktike (Anthropogene Period in the Arctic and Subarctic): Nauchno-Issled. Inst. Geol. Arktiki Trudy, v. 143, p. 5–33. English summary.

Curray, J. R. 1965. Late Quaternary history, continental shelves of the United States, *in* H. E. Wright and D. G. Frey, eds., The Quaternary of the United States: Princeton Univ. Press, p. 723–735.

Curry, R. R., 1966. Glaciation about 3,000,000 years ago in the Sierra Nevada: Science, v. 154, p. 770–771.

Davies, J. L. 1958. Pleistocene geography and the distribution of northern Pinnipeds: Ecology, v. 39, p. 97–113.

Donn, W. L., W. R. Farrand, and Maurice Ewing. 1962. Pleistocene ice volumes and sea-level lowering: Jour. Geology, v. 70, p. 206–214.

Dreimanis, A., J. Terasmae, and G. D. McKenzie. 1966. The Port Talbot Interstade of the Wisconsin Glaciation: Canadian Jour. Earth Sci., v. 3, p. 305–325.

Flint, R. F., and Friedrich Brandtner. 1961. Climatic changes since the last inter-glacial: Am. Jour. Sci., v. 259, p. 321–328.

Fyles, J. G. 1965. Surficial geology, western Queen Elizabeth Islands, *in* Report of activities: Field, 1964: Geol. Survey of Canada Paper 65-1, p. 3–5.

Guthrie, R. D. Manuscript. Paleo-ecology of the large mammal community in in-terior Alaska during the Wisconsin Glaciation: submitted for publication.

Heezen, B. C., and Maurice Ewing. 1961. The mid-oceanic ridge and its extension through the Arctic Basin, *in* G. O. Raasch, Geology of the Arctic, vol. 1: Univ. Toronto Press, 2 vols., p. 622–642.

Heusser, C. J. 1963. Postglacial palynology and archaeology in the Naknek River drainage area, Alaska: Am. Antiquity, v. 29, p. 74–81.

Hibbard, C. W., D. E. Ray, D. E. Savage, D. W. Taylor and J. E. Guilday. 1965. Quaternary mammals of North America, *in* H. E. Wright and D. G. Frey, eds., The Quaternary of the United States: Princeton Univ. Press, p. 509–525.

Hopkins, D. M. 1959. Cenozoic history of the Bering Land Bridge: Science, v. 129, p. 1519–1528.

Hopkins, D. M., and W. S. Benninghoff. 1953. Evidence of a very warm Pleisto-cene interglacial interval on Seward Peninsula, Alaska (abs.): Geol. Soc. America Bull., v. 64, p. 1435–1436.

Hughes, O. L. 1965. Surficial geology of part of the Cochrane District, Ontario, Canada: Geol. Soc. America Spec. Paper 84, p. 535–565.

Hustich, Ilmari. 1953. The boreal limits of conifers: Arctic, v. 6, p. 149–162.

Kalstrom, T. N. V. 1964. Quaternary geology of the Kenai Lowland and glacial history of the Cook Inlet region, Alaska: U.S. Geol. Survey Prof. Paper 443, 69 p.

King, E. R., Isadore Zietz, and L. R. Alldredge. 1966. Magnetic data on the struc-ture of the central Arctic region: Geol. Soc. America Bull., v. 77, p. 619–646.

Krishtofovich, L. V. 1964. Mollyuski Tretichnykh Otlozheniy Sakhalina (Mol-lusks in the Tertiary sediments of Sakhalin): Vsesoyuznogo Neftyanogo Nauchno-Issled. Geol.-Razvedochnogo Inst. (VNIGRI), Trudy, v. 232, 343 p.

Kurtén, Björn. 1966. Holarctic land connections in early Tertiary: Commenta-tiones Biol. Soc. Scient. Fennica, v. 29, p. 1–5.

Lindroth, C. H. 1957. The faunal connections between Europe and North Amer-ica: Wiley, 1957, 344 p.

Löve, Doris, and Askell Löve, eds. 1963. North Atlantic biota and their history: Pergamon, Oxford, 430 p.

McCulloch, D. S., and D. M. Hopkins. 1966. Evidence for an Early Recent warm interval in northwestern Alaska: Geol. Soc. America Bull., v. 77, p. 1089–1108.

McDougall, Ian, and H. Wensink. 1966. Paleomagnetism and geochronology of Pliocene-Pleistocene lavas in Iceland: Earth and Planetary Sci. Letters, v. 1, p. 232–236.

MacNeil, F. S. 1957. Cenozoic megafossils of northern Alaska: U.S. Geol. Survey Prof. Paper 294-C, p. 99–126.

——— 1965. Evolution and distribution of the Genus *Mya*, and Tertiary migra-tions of Mollusca: U.S. Geol. Survey Prof. Paper 483-G, 51 p.

MacNeil, F. S., J. A. Wolfe, D. J. Miller, and D. M. Hopkins. 1961. Correlation of Tertiary formations of Alaska: Bull. Am. Assoc. Petrol. Geol., v. 45, p. 1801–1809.

Martin, P. S. 1958. Pleistocene ecology and biogeography of North America, chap. 15, *in* C. L. Hubbs, ed., Zoogeography: Am. Assoc. Adv. Science Publ. 51, p. 375–420.

Miki, Shigiru. 1955. Nut remains of Juglandaceae in Japan: Jour. Inst. Polytechnics, Osaka City Univ., Ser. D, vol. 6, p. 131–144.

Miller, D. J. 1953. Late Cenozoic marine glacial sediments and marine terraces on Middleton Island, Alaska: Jour. Geology, v. 61, p. 17–40.

———— 1957. Geology of the southeastern part of the Robinson Mountains, Yakataga District, Alaska: U.S. Geol. Survey Oil and Gas Inv. Map OM-187.

Mitchell, Edward. 1966. Faunal succession of extinct north Pacific marine mammals: Norsk Hvalfangst-Tidende, No. 3, p. 47–60.

Moore, D. G. 1964. Acoustic-reflection reconnaissance of continental shelves; eastern Bering and Chukchi Seas, *in* R. L. Miller, ed., Papers in Marine Geology, Shephard Commemorative Volume: Macmillan, New York, p. 319–362.

Opdyke, N. D., B. Glass, J. D. Hays, and J. Foster. 1966. Paleomagnetic study of Antarctic cores: Science, v. 154, p. 349–357.

Péwé, T. L. 1965. Fairbanks Area, *in* Guidebook F, Central and Southern Alaska: Internat. Assoc. Quaternary Res. (INQUA), 7th Cong. (Boulder), 1965, p. 6–36.

Péwé, T. L., D. M. Hopkins, and J. L. Giddings. 1965. The Quaternary geology and archaeology of Alaska, *in* H. E. Wright and D. G. Frey, eds., The Quaternary of the United States: Princeton Univ. Press, p. 355–374.

Repenning, C. A. 1967. Miocene-Pliocene correlations based upon vertebrate fossils (abs.) : Geol. Soc. America, Cordilleran Section, Program, 63d Ann. Meeting, Santa Barbara, p. 58–59.

Repenning, C. A., D. M. Hopkins, and Meyer Rubin. 1964. Tundra rodents in a late Pleistocene fauna from the Tofty placer mining district, central Alaska: Arctic, v. 17, p. 177–197.

Savage, D. E., and G. H. Curtis. 1967. Villafranchian age and its radiometric dates (abs.) : Bull. Am. Assoc. Petr. Geol., v. 51, p. 479–480.

Scholl, D. W., E. C. Buffington, and D. M. Hopkins. 1966. Exposure of basement rock on the continental slope of the Bering Sea: Science, v. 153, p. 992–994.

Shephard, F. P. 1962. Thirty-five thousand years of sea level, *in* Essays in Marine Geology in honor of K. O. Emery: Univ. Southern Calif. Press, Los Angeles, p. 1–10.

Simpson, G. G. 1947. Holarctic mammalian faunas and continental relationships during the Cenozoic: Geol. Soc. America Bull., v. 58, p. 613–688.

Smith, P. S. 1910. Geology and mineral resources of the Solomon and Casadepaga Quadrangles, Seward Peninsula, Alaska: U.S. Geol. Survey Bull. 433, 234 p.

Stalker, A. M. 1967. Quaternary studies in the southwestern prairies, *in* Report of Activities, Part A: May to October, 1966: Geol. Survey of Canada, Paper 67-1, p. 113–114.

Strelkov, S. A. 1965. Istoriya razvitiya rel'efa Sibiri i Dal'nego Vostoka: Sever Sibiri (History of the development of the relief of Siberia and the Far East: Northern Siberia) : Akad. Nauk SSSR, Siberian Div., Inst. Geol. and Geophys., Izd-vo "Nauka," 334 p.

Taylor, D. W. 1965. The study of Pleistocene nonmarine mollusks in North America, *in* H. E. Wright and D. G. Frey, eds., The Quaternary of the United States: Princeton Univ. Press, p. 597–611.

Terasmae, J. 1956. Palynological study of Pleistocene deposits on Banks Island, Northwest Territories, Canada: Science, v. 123, p. 801–802.

Thorsteinsson, R. 1961. The history and geology of Meighan Island, Arctic Archipelago: Geol. Survey Canada Bull. 75, 19 p.

Tozer, E. T. 1956. Geological reconnaissance, Prince Patrick, Eglinton, and western Melville Islands, Arctic Archipelago, Northwest Territories: Geol. Survey Canada Paper 55-5, 32 p.

Turner, D. L. 1967. Review of radiometric dates pertaining to the Miocene-Pliocene boundary problem (abs.) : Geol. Soc. America, Cordilleran Section, Program, 63d Ann. Meeting, Santa Barbara, p. 65–67.

U.S. Geological Survey. 1965. Geological Survey Research in 1965, Chapter A: U.S. Geol. Survey Prof. Paper 525-A, 376 p.

Vangengeim, E. A. 1961. Paleontologicheskoe obosnovanie stratigrafii Antropogenovykh otlozhenii Severa Vostochnoy Sibiri (Paleontological basis of the stratigraphy of the Anthropogene deposits of Northeast Siberia) : Akad. Nauk SSSR, Trudy Geol. Inst. (GIN), v. 48, 181 p. Partial translation available from Am. Geol. Inst.

Vine, F. J. 1966. Spreading of the ocean floor—new evidence: Science, v. 154, p. 1405–1415.

Webb, S. D., and N. Tessman. 1967. Vertebrate evidence of a low sea level in the Middle Pliocene: Science, v. 156, p. 379.

Wolfe, J. A., and D. M. Hopkins. 1967. Climatic changes recorded by Tertiary land floras in northwestern North America, in Kotora Hatai, ed., 11th Pac. Sci. Congress, Tokyo, 1966, Symposium 25: Tertiary Correlations and climatic changes in the Pacific: Sasaki Printing and Publishing Co., Sendai, Japan, p. 67–76.

Wolfe, J. A., D. M. Hopkins, and E. B. Leopold. 1966. Tertiary stratigraphy and paleobotany of the Cook Inlet Region, Alaska: U.S. Geol. Survey Prof. Paper 398-A, 29 p.

Wright, H. E., and D. G. Frey, eds. 1965. The Quaternary of the United States: Princeton Univ. Press, 922 p.

# General Index

Acheuloid industries, 383
Acoustic-reflection survey, 10, 14–16, 23
Afontova Gora II, 182
Ainu of Japan, 415
Alacaluf Indians, 412
Alapah Mountain Glaciation, 114
Alaska: climatic history of, 469–79 *passim*; earthquake, 1964, 80; floral record of, 193–200, 205, 469–79 *passim*; glacial boundaries of, 462; Quaternary events in, 111–17
Alaskan Peninsula, 81–82, 420, 428–29, 431, 443, 445
Aldan River, 232, 283
Aleutian Arc, 33–40 *passim*, 364
Aleutian Basin, 33, 38, 41, 364
Aleutian Islands, 67, 71; as habitat of man, 409, 411, 420, 442–43, 445f, 453; marine terraces, 77, 82
Aleutian Trench, 33, 41
Aleuts; 409–21 *passim*; adaptation of, to marine life, 423–28; at Chaluka site, 483–43; migration summary of, 445–47, 479–80
Algonquin Indians, 419
Alleroed interval, 186, 378
Amchitka Island, 62, 65
Amersfoort interval, 377
Amguem beds, 81f, 159, 166f
Amurians, 415
Amur River valley, 198f
Anadyr: depression, 147; Gulf, 18, 39, 41, 146f, 371, 430, 454–58 *passim*, 462–67 *passim*, 477; Range, 144; River, 424; Strait, 13, 20f, 25–28 *passim*
Anangula site, 392, 395, 402f, 428, 431–43, 445–46, 480
Anchorage, 79–80, 82, 109

Andean Indians, 413f
Anderson Creek, 136
Anikovik River, 136
Anivik Lake Glaciation, 114
Antler Valley readvance, 113
Anvilian marine transgression, 48, 51, 61–67, 93, 97–101, 466, 470
Apache Indians, 413
Araya site, Honshu, 438
Archaeology: American, 393–97; Bering region, 398–403; European, 381–91; Siberian, 391–93. *See also archaeological sites and industries by name*
Arctic coastal plain of Alaska, 49, 61f, 70–77 *passim*
Arctic Ocean (Basin), 13, 19, 107, 151, 230, 322f, 452–60 *passim*
Arctic Small Tool Tradition, 438
Arikara Indians, 414
Astian Stage, 295
Astoria Formation, 340
Athabascan Indians, 419, 422, 480
Atkasuk, 63, 67, 81
Aurignacian industry, 385, 387
Aurignacoid industries, 373, 390–403 *passim*, 479f
Australian aborigines, 415

Baikal region, 232, 242; Lake Baikal, 241, 283, 480
Baldwin Peninsula, 109, 111, 116, 227, 268, 469f; cross section of, 101; as Kotzebuan locale, 73, 75
Baltic Sea, 44
Barents Sea, 44
Bathymetry, 10–16, 22, 24
Beaufort Formation, 456f
Belaya River, 183

# Index to Faunal and Floral Taxa